Property of:

Warez Devil

295-9984

ELECTRONIC FUNDAMENTALS

ELECTRONIC FUNDAMENTALS

by

Philip E. Wheeler

Edited by: Mike Sewell

Library of Congress Cataloging-in-Publication Data

Wheeler, Philip, 1961-
    Electronic fundamentals.

    Includes index.
    1. Electronics. I. Title.
TK7816.W44 1989   621.381   88-34713
ISBN 0-87119-188-1

Model EB-200

Published by

Heath Company
Benton Harbor, Michigan 49022

Copyright © 1989

Heath Company
Not Affiliated with D.C. Heath

All rights reserved. No part of this book may be reproduced in any form, electronic or mechanical, including photocopy, recording, or any information storage and retrieval system, without permission *in writing* from the publisher. The Heathkit Educational System logo is a registered trademark of Heath Company.

First printing—1989
Printed in the United States of America

595-4146-02   ISBN 0-87119-188-1

Preface

Heathkit began producing electronics instruction back in 1975, offering a series of Individual Learning courses in basic electronics. Heathkit discovered that many people found electronics technology to be vastly interesting, but at the same time mysterious and lofty. Heathkit's Individual Learning courses brought this subject down to its most basic levels, to the point where understanding the electronics industry was within the capabilities of a greater number of people. The wording was simple, the instructions and experiments moved in easy-to-understand sequences, and although the topics seemed very complex at the beginning, the conversation-like writing style put the reader at ease, providing a comfortable and relaxed learning environment.

This venture proved so successful that Heathkit adapted the Individual Learning courses for the classroom environment in 1978. Gradually, the variety of subjects that Heathkit offered increased, where Heathkit Educational Systems began offering courses on almost any subject, from an introduction to electricity, to microprocessors, artificial intelligence, and laser technology.

Throughout all of this expansion, however, the Basic Electronics Series remained the cornerstone of Heathkit's training reputation. Late in 1986, the need arose to teach basic electronics in a rather brief period of time. Schools wanted to instruct their students on basic electronic principles, and avoid much of the complex mathematics and design requirements too often emphasized in other electronics courses. From here, Electronic Fundamentals was born.

Electronic Fundamentals takes the material from our popular Basic Electronics Series, and presents it in the smooth, easy-to-read approach that is Heathkit's trademark. In addition, since the material is taught step-by-step, the transition from one subject to the next is logical and anticipated. This is Heathkit's first hardbound textbook, which succeeds in bringing the comfort of home-study right into the classroom.

The textbook consists of 30 chapters divided into four sections. "DC Electronics" introduces the student to the terms and concepts that surround all electronics, and teaches the principles behind the basic electrical circuit. "AC Electronics" explores the wide variety of applications that surround the sine wave; from its generation to its effect on different electrical components. "Semiconductors" discusses the operation of components such as diodes, transistors, and solar cells, and illustrates to the student the important role these components continue to play in today's high-tech advances. The final section—"Electronic Circuits"—combines the first three sections, and leads the student through the basic circuits found in almost every piece of electronics equipment, ranging from amplifiers and power supplies, to oscillators and multivibrators.

At the end of each chapter, there is a "Chapter Self-Test", where the student can determine how well he/she understands the material from that chapter. To assist the student, the answers to the odd-numbered questions are located in Appendix D. Also, Appendices A and B provide supplemental information on scientific notation and trigonometry; two portions of mathematics that directly aid in learning electronics. Finally, a glossary is located in Appendix E as a quick reference for the many electronic terms introduced in this textbook.

Keep in mind, however, that electronics is much more than terms, concepts, and formulas. Many of the individuals who enter the electronics field do so as a technician, who spends his or her time wiring circuits, building circuits, troubleshooting, soldering, and all those other tasks that comprise the "hands-on" aspect of electronics. For this reason, Electronic Fundamentals is supplemented with a Student Workbook, which contains 40 hands-on experiments that students can perform. Since the workbook is geared directly to the textbook, the student can read about a circuit or component, then build, test, and analyze the same circuit using sophisticated test equipment and most importantly, his or her own hands. This allows the student to discover the material aspect of electronics.

And to assist the instructor, an extensive Instructor's Guide provides the answers to all the questions within the textbook and workbook, as well as two separate final exams, a lesson plan, and a number of other features to ensure a smooth flow of material.

In closing, I would like to say that this book was written in a fashion that I would like to teach: simple, straight-forward, and without all the "gee-whiz" material which serves only to confuse students. Electronic Fundamentals is a wide, straight path through the forest of electronics, and I would like to thank the many people who helped me blaze the trail.

Philip E. Wheeler

Contents

SECTION I: DC ELECTRONICS

Chapter 1: The Physics of Electronics 1

    Composition of Matter..................................5
    Electrostatics..9

Chapter 2: Current and Voltage 19

    Current Flow...23
    Electrical Force.....................................29
    Producing EMF..33

Chapter 3: Properties of Electrical Circuits 44

    The Electrical Circuit...............................45
    Batteries..48
    Voltage Rises and Voltage Drops......................54
    Concept of Ground....................................57

Chapter 4: Resistance 63

    Electron Opposition..................................67
    Resistors..72
    Connecting Resistors.................................78

Chapter 5: Electrical Measurements 89

    Measuring Current....................................93
    Measuring Voltage....................................98
    Measuring Resistance................................101

Chapter 6: Ohm's Law 107

    Determining Current.................................111
    Finding Voltage.....................................116
    Finding Resistance..................................119
    Using Ohm's Law.....................................121

Chapter 7: Power 125

    Work and Power 129
    Power, Current, and Voltage 131
    Power Dissipation 134

Chapter 8: DC Circuits 139

    Simple DC Circuits 143
    Bridge Circuits 151
    Kirchhoff's Law 157

Chapter 9: Magnetism 163

    The Magnetic Field 167
    Electricity and Magnetism 171
    Induction 176

Chapter 10: Reactive Components 183

    Inductance 187
    Capacitance 195
    Capacitors 199

SECTION II: AC ELECTRONICS

Chapter 11: Introduction to AC 211

    The Importance of AC 215
    Generating AC 218
    The Sinusoidal Waveform 229
    AC Values 235
    Nonsinusoidal Waveforms 244

Chapter 12: Measuring AC 251

    Using Meters to Measure AC 255
    The Oscilloscope 259

Chapter 13: AC Resistance ... 271

    Basic Calculations ... 275
    Circuit Calculations ... 277
    Power in AC Circuits ... 281

Chapter 14: AC Capacitance ... 285

    Review ... 289
    Capacitors in AC Circuits ... 293
    RC Circuits ... 300
    RC Circuit Applications ... 315

Chapter 15: AC Inductance ... 327

    Review ... 331
    Inductors in AC Circuits ... 337
    RL Circuits ... 342
    RL Circuit Applications ... 353

Chapter 16: RLC Circuits ... 361

    Simple RLC Circuits ... 365
    Resonance ... 371
    Series Resonance ... 373
    Parallel Resonance ... 383
    LC Filters ... 392

Chapter 17: Transformers ... 401

    Transformer Action ... 405
    Transformer Theory ... 414
    Transformer Ratios ... 418
    Transformer Losses ... 425
    Transformer Applications ... 429

SECTION III: SEMICONDUCTORS

Chapter 18: The Physics of Semiconductors 437

    Semiconductor Materials 441
    Semiconductor Atoms and Crystals 444
    Conduction in Intrinsic Germanium and Silicon .447
    Conduction in Doped Germanium and Silicon 450

Chapter 19: Junction Diodes 457

    The PN Junction 461
    Diode Biasing. 464
    Diode Characteristics 470

Chapter 20: Zener Diodes 481

    Zener Diode Characteristics 485
    Current Limitations 489
    Zener Diode Impedance. 492

Chapter 21: Bipolar Transistors. 499

    Basic Transistor Action. 503
    Transistor Amplification. 510
    Transistor Amplifier Circuits 513

Chapter 22: Field Effect Transistors 531

    The Junction FET 535
    The Insulated Gate FET 543
    FET Amplifier Circuits 552

Chapter 23: Control Devices. 559

    Silicon Controlled Rectifiers 563
    Bi-Directional Triode Thyristors. 571
    Unijunction Transistors 577

Chapter 24: Light-Sensitive Devices 589

    Basic Principles of Light 593
    Light Sensing Devices 596
    Light-Emitting Devices 604
    Liquid Crystal Displays 609

SECTION IV: ELECTRONIC CIRCUITS

Chapter 25: Basic Amplifiers 615

    The Importance of Amplifiers 619
    Amplifier Biasing 623
    Amplifier Circuit Configurations 632
    Saturation and Cutoff 654
    Amplifier Coupling 660

Chapter 26: Amplifier Applications 673

    Audio Amplifiers 677
    Video Amplifiers 693
    Differential Amplifiers 704

Chapter 27: Operational Amplifiers 717

    Operational Amplifier Characteristics 721
    Closed-Loop Operation 726
    Applications 738

Chapter 28: Power Supplies 749

    Rectifier Circuits 753
    Power Supply Filters 763
    Voltage Regulation 769
    Series Voltage Regulation 772

Chapter 29: Oscillators 787

    Oscillator Fundamentals. 789
    Transformer Oscillators 793
    LC Oscillators 798
    Crystal Oscillators. 804
    RC Oscillators 812
    Nonsinusoidal Oscillators 818

Chapter 30: Waveshaping Circuits 827

    Waveshaping 831
    Rectangular Wave Generators 845
    Ramp Generators 857

SECTION V: APPENDICES

Appendix A: Powers of Ten. 867

Appendix B: Right Triangles and Trigonometry. 879

Appendix C: Table of Trigonometric Functions 887

Appendix D: Answers to Odd-Numbered Questions. 889

Appendix E: Glossary. 905

Index. 945

SECTION 1

- THE PHYSICS OF ELECTRONICS
- CURRENT AND VOLTAGE
- PROPERTIES OF ELECTRICAL CIRCUITS
- RESISTANCE
- ELECTRICAL MEASUREMENTS
- OHM'S LAW
- POWER
- DC CIRCUITS
- MAGNETISM
- REACTIVE COMPONENTS

The first section of this book lays the groundwork for a career in electronics technology. Before there were computers, integrated circuits, robots, and sophisticated electronic communications systems, there was the simple electrical circuit. And regardless of the make-up of modern-day electronics, the same basic principles that exist in the simple electrical circuit still apply. This section introduces you to the terms and concepts that surround electricity. You will learn what electricity is, how it is made, how it works, how it is measured, and how it can be applied. You'll learn the four fundamental terms that surround all electronics—voltage, current, power, and resistance—and how they inter-relate. You'll be introduced to basic electrical circuits and the components that are used in such circuits. You'll also learn how magnetism and electricity are related. The knowledge here will last an electronics technician or engineer throughout his or her career.

DC ELECTRONICS

CHAPTER 1

The Physics of Electronics

Contents

Introduction 3

Chapter Objectives 4

The Composition of Matter 5

Electrostatics 9

Chapter Self-Test 16

Summary 17

Introduction

Electronics is the branch of science that studies the behavior of electrons (if you'll notice, the word "electronics" is taken from the word "electron"). To understand electronics, you must first understand the basic behavior of the electron. In this chapter, you will see what the electron is, and how it behaves. You'll learn basic atomic structure, and how the electron plays such an important part in electronics.

Near the end of the chapter, there is a series of self-test questions for you to see how well you understand the information in this chapter.

DC ELECTRONICS

Chapter Objectives

When you have completed this chapter, you will be able to:

1. Define the following terms: matter, element, compound, molecule, atom, proton, neutron, electron, nucleus, ion, balanced state, and electrostatic field.

2. Identify the three basic atomic particles using a diagram of Bohr's atomic model.

3. Identify the difference between a balanced atom and an ion.

4. Describe the atomic structure of a positive and a negative ion.

5. State Coulomb's Law and the effects of the behavior of like and unlike charges.

6. State the three methods of transferring electrical charges from one object to another.

The Composition of Matter

Controlling the behavior of electrons is what electronics is all about. Therefore, an understanding of the electron is vitally important in order to understand electronic fundamentals. Electrons are tiny particles that carry the energy to light our homes, cook our food, and do much of our work. To understand what an electron is, we must investigate the make-up of matter.

Matter is generally described as anything that has weight and occupies space. Thus, the earth and everything on it is classified as matter. Matter exists in three different states—solid, liquid, and gas. Wood, water, and oxygen, are all forms of matter.

Elements and Compounds

Elements are the basic building materials from which all matter is constructed. Some examples of elements are iron, carbon, hydrogen, and gold. Over one hundred elements are presently known. Of these, 92 occur in nature. These are called natural elements. In addition, there are about a dozen elements that are man-made, such as plutonium and curium. These elements can only be created in scientific laboratories.

As you look around, it becomes obvious that there are many more types of matter than there are elements. For example, substances like salt, steel, water, and protein are not elements, but combinations of elements called compounds. A compound is composed of two or more elements. Just as the letters of the alphabet can be arranged in various combinations to form millions of different words, the elements can be arranged in various combinations to form millions of different compounds. For example, water is a compound that is made up of the elements hydrogen and oxygen; sugar is composed of hydrogen, carbon, and oxygen; and salt is composed of sodium and chlorine.

To better understand how compounds are related to their elements, let's investigate the structure of a compound with which you are familiar—water. Suppose you divide a drop of water into two parts. Then, you divide each part again and again. After a few dozen divisions, you have a drop so small that it can be seen only with a microscope. If you divide it even further into smaller and smaller particles, you will eventually get a particle so small that it can not be divided further and *still be water.* The smallest particle of water that still retains the characteristics of water is called a molecule. The water molecule can be broken into still smaller pieces, but the pieces will not be water. Thus, if you break up the water molecule, you will find that the pieces are the elements hydrogen and oxygen.

Atoms

The smallest particle to which an element can be reduced is called an atom. Molecules are made up of atoms that are bound together. The water molecule is shown in Figure 1-1 as three atoms. The two smaller atoms represent hydrogen, while the larger one represents oxygen. Therefore, a molecule of water consists of two atoms of hydrogen (H) and one atom of oxygen (O). This is why the chemical formula for water is H_2O.

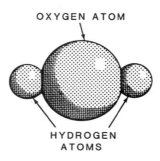

Figure 1-1 The water molecule.

Electrons, Protons, and Neutrons

As small as the atom is, it can be broken into even smaller particles. If you investigate the structure of the atom, you'll find that it contains three distinct types of particles: **electrons, protons,** and **neutrons**. These are the three basic building blocks that make up all atoms, and therefore, all matter. Electrons, protons, and neutrons are very different from each other. However, as far as we know, all electrons behave identically. The same can be said for all protons and neutrons.

Bohr Model of the Atom

Figure 1-2 shows how electrons, protons, and neutrons are combined to form an atom. This particular atom is a helium atom. The center of the atom, which is composed of protons and neutrons, is called the **nucleus**. Depending on the type of atom, the nucleus will contain from one to about 100 protons. Also, in all atoms except hydrogen, the nucleus contains neutrons. The neutrons and protons have approximately the same weight and size. Because they are much heavier than electrons, the overall weight of the atom is determined primarily by the number of protons and neutrons in the nucleus.

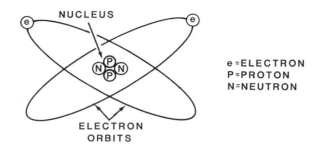

Figure 1-2 Bohr model of the helium atom.

Orbiting around the nucleus are the electrons. Notice that the helium atom has two electrons. The electrons are extremely light and they travel at fantastic speeds. The atom can be compared to the solar system with the nucleus representing the sun and the electrons representing the planets. The electrons orbit the nucleus in much the same way that the planets orbit the sun.

Any picture of an atom must be based on assumptions rather than actual observation. Figure 1-2 is a very simple model of the atom based on these assumptions. Today, much more complex models of the atom have been proposed. But all these models have several things in common, one of which is a basic structure of electrons orbiting a nucleus composed of protons and neutrons. Figure 1-2 will suit our purposes. This model of the atom is called the Bohr model after Danish physicist Niels Bohr, who proposed the model.

The Difference Between Elements

So why do different elements exist? If both iron and oxygen contain electrons, protons, and neutrons, why are they so different? The answer lies in the *number* of protons, neutrons, and electrons that each element contains. Every element is made up of atoms that contain a unique number of protons, electrons, and, with the exception of hydrogen, neutrons. Let's look at Figure 1-3 to see exactly what we mean.

DC ELECTRONICS

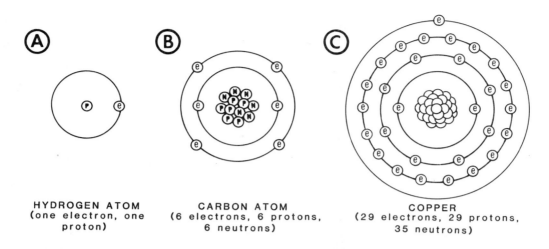

HYDROGEN ATOM
(one electron, one proton)

CARBON ATOM
(6 electrons, 6 protons, 6 neutrons)

COPPER
(29 electrons, 29 protons, 35 neutrons)

Figure 1-3   The difference between atoms is the number of electrons, protons and neutrons that they have.

The simplest of all elements, hydrogen, is shown in Figure 1-3A. It consists of a single electron orbiting a single proton. This is the only atom that contains no neutrons. Because it is made up of the fewest number of particles, hydrogen is the lightest element. Figure 1-3B represents the carbon atom. Notice, this atom is made up of 6 electrons that orbit a nucleus of 6 protons and 6 neutrons. The last element shown in Figure 1-3 is copper. It consists of 29 electrons, 29 protons, and 35 neutrons. However, the most complex natural element is uranium, which has 92 electrons, 92 protons and 146 neutrons.

The Balanced Atom

In the examples shown, you may have noticed that the number of electrons is always equal to the number of protons. This is normally true of any atom. When this is the case, the atom is said to be in its normal, balanced, or neutral state. As you will see later, this state can be upset by an external force. However, normally the atom is considered to contain equal numbers of electrons and protons.

THE PHYSICS OF ELECTRONICS

Electrostatics

The difference between electronics and electrostatics is very slight. Electrostatics is the branch of physics that deals with electrical charges at rest, or *static electricity*. Electronics is the study of electrical charges in motion. Before you can begin to understand charges in motion, you must first get an idea of how electrical charges behave at rest.

The Electrical Charge

You have examined the structure of the atom and learned about some of the characteristics of the electron, proton, and neutron. However, you have not yet learned the most important characteristic of these particles. This characteristic is their *electrical charge*. An electrical charge is a property associated with the electron and the proton. It is this electrical charge that makes the electron useful in electrical and electronic work.

The electrical charge is difficult to visualize because it is not an object, like a molecule or an atom. Rather, it is a property or characteristic that electrons and protons have that cause them to behave in certain predictable ways.

There are two distinct types of electrical charges. Because these two types of charges have opposite characteristics, they have been given the names *positive* and *negative*. The electrical charge associated with the *electron* has been arbitrarily given the name *negative*. As a result, the electrical charge associated with the *proton* is considered to be *positive*. The neutron has no electrical charge at all. It is electrically neutral and, therefore, plays no known role in electricity or electronics.

An electron revolving around the nucleus of an atom acts the same as a ball that is attached to the end of a string and twirled in a circle. If the string breaks, the ball flies off in a straight line. Thus, it is the restraining action of the string that holds the path of the ball to a circle. We also compared an electron's orbit to a planet's orbit around the sun. The gravitational force of the sun and the centrifugal force of the planet offset each other. Thus, the planet orbits the sun in essentially a circular manner.

Electrons orbit around the nucleus of the atom at a fantastic speed. What force keeps them from flying off into space? It is not gravity because the gravitational force exerted by the nucleus is much too weak. Instead, the force at work here results from the charge on the electron in orbit and the charge on the proton in the nucleus. The negative charge of the electron is *attracted* by the positive charge of the proton. We call this attraction *electrostatic force*.

DC ELECTRONICS

The attraction of the unlike charges (positive proton and negative electron) creates a force that tries to pull the protons and electrons together. The speed at which the electron is moving keeps the two particles from being pulled together. To explain this force, science has adopted the concept of an electrostatic field. Every charged particle is assumed to be surrounded by an electrostatic field that extends for a distance outside the particle itself. It is the interaction of the fields surrounding the charged particles that cause the electron and proton to attract each other.

Figure 1-4A shows a diagram of a proton. The plus sign in the center of the illustration represents the positive electrical charge. The arrows that extend outward represent the lines of force that make up the electrostatic field surrounding the proton. Notice that the lines are arbitrarily assumed to extend outward away from the positively charged particle. Compare this to the electron shown in Figure 1-4B. Here, the minus sign represents the negative charge while the arrows that point inward represent the lines of the electrostatic field.

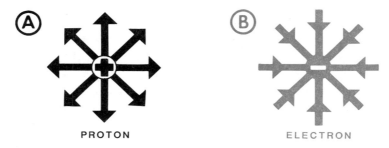

Figure 1-4 Fields associated with protons and electrons.

THE PHYSICS OF ELECTRONICS

Law of Electrical Charges

There is a basic law that describes the action of electrical charges. It is called Coulomb's Law, after Charles A. de Coulomb, who discovered this relationship.

Quite simply, Coulomb's Law states:

1. Like charges repel.

2. Unlike charges attract.

Because like charges repel, two electrons will repel each other, as will two protons. Figure 1-5A illustrates how the lines of force interact between two electrons. The directions of the lines of force are such that the two fields cannot interconnect. The net effect is that the electrons attempt to move apart, or repel each other. Figure 1-5B shows that the same is true of two protons. In Figure 1-5C, an electron and a proton are shown, and the two fields interconnect. As a result, the two charges attract and tend to move together.

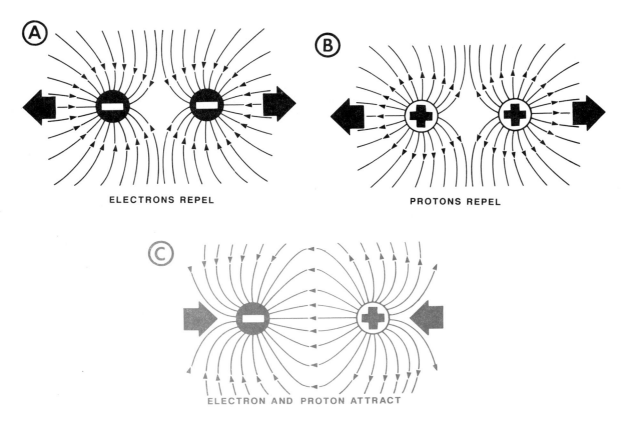

Figure 1-5 Action of like and unlike charges.

These examples show only individual charged particles. However, Coulomb's Law holds true for concentrations of charges as well. In fact, it holds true for any two charged bodies. An important part of Coulomb's Law is the equation that allows you to determine the force of attraction or repulsion between charged bodies. The equation states that:

$$F = \frac{q_1 \times q_2}{d^2}$$

Where:

F = the force of attraction between unlike charges or the force of repulsion between like charges.

q_1 = the charge on one body.

q_2 = the charge on the second body.

d^2 = the square of the distance between the two bodies.

If you look at the equation, you can see some interesting relationships. For example, if either charge doubles (all other factors being equal), the force between them doubles. If both charges double, the force is increased by a factor of four. On the other hand, increasing the distance between charges decreases the force by the square of the increase in distance. If the distance between charges is doubled, the force is reduced to one-fourth its former value.

The magnitude of the electron's negative charge is exactly equal to the magnitude of the proton's positive charge. That is, the negative charge of the electron is exactly offset by the positive charge of the proton. Thus, a balanced atom (an atom with the same number of protons and electrons) is electrically neutral.

Since a balanced atom is electrically neutral, it cannot attract or repel another balanced atom. Nor is a balanced atom attracted or repelled by charged particles such as electrons and protons. However, this balanced condition can be easily upset by external forces.

The Ion

Atoms are affected by many outside forces such as heat, light, electrostatic fields, chemical reactions, and magnetic fields. Quite often the balanced state of the atom is upset by one or more of these forces. As a result, an atom can lose or gain one or more electrons. When this happens, the number of negative charges is no longer exactly offset by the number of positive charges. An atom that is no longer in its neutral state is called an ion. The process of changing an atom to an ion is called ionization.

There are both negative and positive ions. Figure 1-6 compares a neutral atom of carbon with negative and positive ions of carbon. Figure 1-6A shows the balanced or neutral carbon atom. This balanced atom has six electrons and six protons. Since their numbers are the same, their charges offset, which gives the atom its neutral charge.

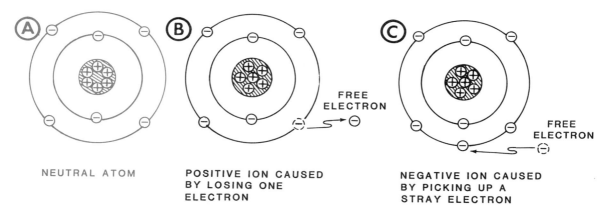

Figure 1-6 Carbon atom and ions.

Figure 1-6B shows the condition that exists when the carbon atom loses an electron. Notice that the carbon atom now has a greater number of protons than electrons. Thus, there is one positive charge that is not canceled by a corresponding negative charge. The atom now has a net positive charge and is called a positive ion.

Figure 1-6C shows the carbon atom with an added electron. Now the number of electrons is greater than the number of protons. Since an electron has a negative charge, we have an overall negative charge. Thus, this is an example of a negative ion.

It is important to note that the ion still has all the basic characteristics of carbon, because the nucleus of the atom has not been disturbed. Therefore, an atom can give off or pick up electrons without changing its basic characteristics.

DC ELECTRONICS

Changing atoms to ions is an easy thing to do and everything you see around you contains ions as well as atoms. The material around you also contains a large number of **free** or **stray electrons**. These are electrons that have escaped from atoms leaving behind positive ions. As you will see later, the electrical characteristics of different types of material are determined largely by the number of free electrons and ions within the material.

Action of Electrostatic Charges

At one time or another you have seen or felt the effects of an electrostatic charge. One spectacular effect is lightning. Less spectacular examples are often seen when you remove clothes from a dryer, comb your hair, or touch a metal object after scuffing your feet on a rug. In each of these cases, two different bodies receive opposite electrical charges. This occurs when one of the bodies gives up a large number of electrons to the other. The body that gives up the electrons becomes positively charged while the body receiving the electrons becomes negatively charged.

When you comb your hair vigorously with a hard rubber comb, your hair gives up electrons to the comb. This causes the comb to become negatively charged while your hair becomes positively charged. That is, the comb collects a large number of free electrons from your hair. This is an example of **charging by friction**.

There are other ways in which an object can become charged. For example, the charge on the comb can be partially transferred to another body simply by touching the comb to the uncharged body. When the charged comb comes in contact with the uncharged object, many of the excess electrons leave the comb and collect on the other object. If you now remove the comb, the object will have a charge of its own. This is called **charging by contact**.

You can also charge an object by **induction** or **electrostatic induction**. This method uses the electrostatic field which exists in the space surrounding a charged body. In this way, you can charge an object without actually touching it with a charged body. For example, if a negatively charged comb is placed close to an aluminum rod, the excess electrons in the comb repel the free electrons in the rod, and they gather at the end away from the charged comb. This causes that end of the rod to acquire a negative charge. The other end of the rod acquires a positive charge because of the deficiency of electrons. If you touch the negative end of the rod with a neutral body, some of the electrons leave the rod and enter the neutral body. This leaves the rod with a net positive charge. Thus, you have induced a positive charge into the rod without touching it with a charged body.

It is also possible to neutralize electrical charges. When a glass rod is rubbed with a silk cloth, the glass gives up electrons to the silk. As a result, the glass becomes positively charged while the silk becomes negatively charged. If the rod is then brought back into contact with the cloth, the negative electrons in the silk are attracted by the positive charge in the glass. The force of this attraction pulls the electrons back out of the silk so that the charge is neutralized. Thus, if two objects having equal but opposite charges are brought into contact, electrons flow from the negatively charged object into the positively charged object. The flow of electrons continues until both charges have been neutralized.

DC ELECTRONICS

Chapter Self-Test

At the conclusion of each chapter, there will be a number of questions concerning the information within that chapter. We recommend that you answer these questions on a separate piece of paper. In order to assist you, the answers to the odd numbered questions are located at the back of the book, under Appendix D.

1. Which particle of an atom has a positive charge?

2. Which two particles of an atom must have the same quantity for the atom to be balanced?

3. What is the name of the force that attracts protons and electrons?

4. State Coulomb's Law.

5. Which of the factors that determine the amount of force between two bodies will decrease the amount of force, if increased?

6. What is the term associated with an atom that no longer contains the same number of protons and electrons?

7. What is the term associated with an atom that has lost an electron?

8. These items wander freely through material, although they are not atoms. What are they?

9. When you comb your hair vigorously with a hard rubber comb, your hair gives up many electrons to the comb. What is the polarity of your hair?

10. Name the three methods of transferring an electrical charge.

Summary

Matter is anything which has weight and occupies space.

All matter is composed of one or more elements.

A compound is a substance composed of two or more elements. The smallest component of a compound is the molecule. A molecule consists of two or more atoms bound together.

The atom is the smallest particle into which an element can be divided. 92 different elements occur in nature, while another dozen are man-made.

The characteristics of an element is determined by the number of protons, neutrons, and electrons it contains.

Electricity is a property that protons and electrons have that cause them to behave in certain predictable ways. The electron is said to have a negative charge, while the proton has a positive charge.

An electrostatic field surrounds every charged particle.

Coulomb's Law describes the interaction between two charged particles. It states that like charges repel, and unlike charges attract.

An atom is said to be in its balanced, or neutral state when it contains the same number of protons and electrons.

When an atom gains or loses an electron, it is called an ion.

A positive ion is an atom with a deficiency of electrons.

A negative ion is an atom with an excess of electrons.

Electrons can be transferred from some objects to others by friction or contact. Other materials can transfer or induce a charge simply by coming close to another object.

DC ELECTRONICS

CHAPTER 2

Current and Voltage

Contents

Introduction 21

Chapter Objectives 22

Current Flow. 23

Electrical Force 29

Producing EMF 33

Chapter Self-Test 38

Summary. 39

CURRENT AND VOLTAGE

Introduction

In the previous chapter, you learned that electrons can be released from an atom to form positive ions, and excess electrons can gather in an atom to form negative ions.

This movement of electrons, called **current flow**, can be predicted and put to use within an electrical circuit. You will learn how some elements are well suited for current flow, while others are not, and that a force must be applied to these electrons to begin their movement.

The force that creates current flow is called **electromotive force,** or **voltage**. You will study how a potential can exist between two charges, how this potential can move electrons, and several different methods of creating this force.

DC ELECTRONICS

Chapter Objectives

When you have completed this chapter, you will be able to:

1. Define the following terms: current flow, orbital shells, valence shell, valence electrons, conductor, insulator, random drift, directed drift, electromotive force, potential difference, voltage, volt, magnetoelectricity, electrochemistry, triboelectricity, photoelectricity, piezoelectricity, and thermoelectricity.

2. Describe how electrons can be freed from the valence shell of an atom.

3. State the difference between a conductor, an insulator, and a semiconductor.

4. State the difference between random drift and directed drift.

5. State whether a potential difference exists between two charges, given their magnitude and polarity.

6. State the direction of electron movement between two charges, given their magnitude and polarity.

7. Match the method of producing EMF to its corresponding application(s).

CURRENT AND VOLTAGE

Current Flow

In electronics, **current** is defined as the flow of electrical charge from one point to another. You will remember that when a negatively charged body is touched to a positively charged body, electrons flow from the negatively charged object to the positively charged object. Since electrons carry a negative charge, this is an example of electrical charges flowing. Before an electron can flow from one point to another, it must first be freed from the atom. The following is a discussion of the mechanism by which electrons are removed from an atom.

Freeing Electrons

You have learned that electrons revolve around the atom's nucleus at very high speeds. Two forces hold the electron in a precarious balance. The centrifugal force of the electron that thrusts it away from the nucleus is exactly offset by the attraction of the protons in the nucleus. This balanced condition can be upset very easily so that the electron is dislodged from its orbit.

Not all electrons can be freed from the atom with the same ease. Some are dislodged more easily than others. To see why, you must study the concept of **orbital shells**. It has been proven that electrons orbit the atom's nucleus according to a certain pattern. For example, in all atoms that have two or more electrons, two of the electrons orbit relatively close to the nucleus. The area in which these electrons travel is called a **shell**. The shell closest to the nucleus contains two electrons. This area can support only two electrons and all other electrons must orbit in shells further from the nucleus.

A second shell somewhat further from the nucleus can hold up to eight electrons. There is a third shell that can contain up to 18 electrons and a fourth shell that can hold up to 32 electrons. The first four shells are illustrated in Figure 2-1. Although not shown, there are additional shells in the heavier atoms.

DC ELECTRONICS

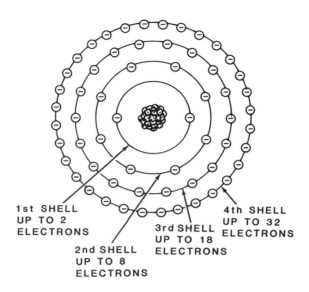

Figure 2-1 Arrangement of orbital shells in an atom.

Of particular importance in the study of electronics is the outer electron shell of the atom. Hydrogen has one electron in its outer shell, while helium has two. In this case, the outer shell is the first and only shell. For atoms that have three to ten electrons, the outer shell is the second shell. Regardless of which shell it happens to be, the outer shell is called the valence shell and the electrons in this shell are called valence electrons.

Electrons are arranged in such a way that the valence shell never has more than eight electrons. This may be confusing since the third shell can contain up to 18 electrons. The following example shows why both statements are true.

An atom of argon contains 18 electrons--2 in the first shell, 8 in the second shell, and 8 in the third shell, It might seem that the next heavier element, potassium, would have 9 electrons in its third shell. However, this would violate the valence rule stated above. Actually, what happens is that the extra electron is placed in a fourth shell. Thus the 19 electrons are distributed in this manner--2 in the first shell, 8 in the second shell, 8 in the third shell, and 1 in the fourth shell. Notice that the outer or valence shell becomes the fourth shell rather than the third. Once the fourth shell is established as the valence shell, the third shell can fill to its full capacity of 18 electrons.

The valence electrons are extremely important in electronics. These are the electrons that can be easily freed and used to perform work. To understand why the valence electrons are easy to free, consider the structure of an atom of copper. Figure 2-2A shows how the electrons are distributed in the various shells in the copper atom. Notice that the valence shell contains only one electron. This electron is further from the nucleus than any of the other electrons. From Coulomb's Law you know that the force of attraction between charged particles decreases dramatically as the distance between the particles increases. Therefore, the valence electrons experience less attraction from the nucleus. For this reason, these electrons can be easily separated from the atom.

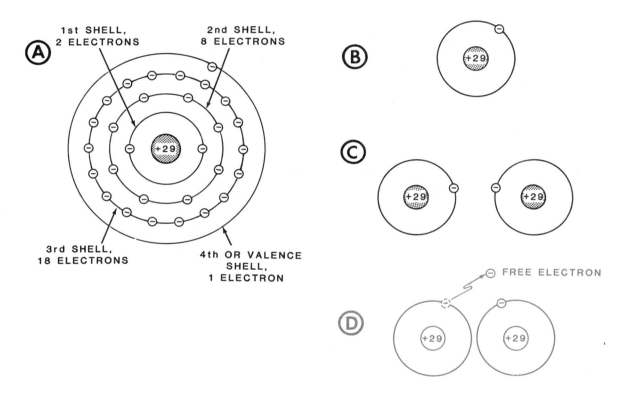

Figure 2-2 Freeing an electron from a copper atom.

Since you are only concerned with the valence electrons, the atom can be depicted in the simplified form shown in Figure 2-2B. Figures 2-2C and 2-2D use this simplified form to illustrate one way in which a valence electron can be freed. Here, two copper atoms are shown as they might appear in a copper wire. Each valence electron is held in orbit by the attraction of the nucleus. However, the force of attraction is quite weak because the orbits are so far from the nucleus. If these two atoms are close together, the electrons in the valence shells may be closer to each other than either electron is to its nucleus. At certain points in their orbits, the two electrons may come very close together. When this happens, the force of repulsion between the two electrons is stronger than the force of attraction exerted by the nucleus. Thus, one or both of the electrons may be forced out of orbit to wander as a free electron. Notice that when the electron leaves, the atom becomes a positive ion.

As the free electron wanders around the atomic structure, it may be eventually captured by another positive ion or it may come close enough to other valence electrons to force them from orbit. The point is that events like these occur frequently in many types of material. Thus, in a piece of copper wire containing billions and billions of atoms, there are also billions of free electrons.

Conductors and Insulators

The importance of the valence electrons cannot be emphasized too strongly. Both the electrical and chemical characteristics of the elements depend on the action of the valence electrons. An element's electrical and chemical stability are determined to a great extent by the number of electrons in the valence shell. You have seen that the valence shell can contain up to eight electrons. Those elements with five or more valence electrons tend to be stable. That is, they tend to maintain their atomic structure rather than give up or accept electrons.

For example, the elements neon, argon, krypton, xenon, and radon have 8 electrons in their valence shell. The valence shell is completely filled and, as a result, these elements are so stable that they resist any sort of chemical activity. They will not even combine with other elements to form compounds. Furthermore, atoms of these elements are very reluctant to give up electrons. Because they do not react with other elements, these elements are called inert gases.

Elements that have their valence shells almost filled also tend to be stable, although they are not as stable as those whose valence shells are completely filled. These elements strive to fill their valence shell by capturing free electrons. Consequently, elements of this type have very few free electrons in their atomic structure.

CURRENT AND VOLTAGE

Substances with five or more valence electrons are called **insulators**. In addition to pure elements, there are many compounds that exhibit the same characteristic. Thus, they act as insulators also. By opposing the production of free electrons, these substances resist certain electrical actions. Insulators are important in electrical and electronics work for this reason. The plastic material on electrical wires is an insulator that protects you from electrical shock.

Elements in which the valence shell is almost empty behave quite differently than insulators. Elements with three or less valence electrons tend to give them up very easily. For example, copper, silver, and gold each have one valence electron. In these elements, the valence electrons are easily removed. Consequently, a bar of any one of these elements has a greater number of free electrons.

Substances that have a large number of free electrons are called **conductors**. In addition to silver, copper, and gold, iron, nickel, and aluminum are also good conductors. Notice that all of these elements are metals. Most metals are good conductors. Conductors are important because they act as current paths and allow electrical current to move from one place to another.

Some elements, such as silicon and germanium, have four electrons in their valence shell. These elements are neither good conductors nor good insulators, and thus are called **semiconductors**. Semiconductors are used extensively in electronics, and will be covered in greater detail later in the text.

The Battery

In order to have current flow, you must first have free electrons. Free electrons can be formed when valence electrons are separated from atoms, with the atoms themselves becoming positive ions. This occurs when you comb your hair or rub a glass rod with a silk cloth. However, to perform a useful function, you must free tremendous numbers of electrons and concentrate them in one area. This requires more sophisticated techniques. One device for doing this is the ordinary battery. The two most common batteries are the dry cell, or flash-light battery, and the wet cell, or automobile battery.

Although batteries come in many shapes and sizes, they have several points in common. They all have two terminals or poles to which an electrical circuit can be connected. In addition, they employ a chemical reaction that produces an excess of electrons at the **negative** terminal and a deficiency of electrons at the **positive** terminal. This excess of electrons is called a **negative charge,** while a deficiency of electrons is called a **positive charge.** If a conductor is placed between these terminals, the large opposite charges of the terminals will have a specific effect on the movement of electrons. Batteries will be discussed in greater detail in Chapter 3.

DC ELECTRONICS

Random Drift and Directed Drift

The large number of free electrons within a conductor do not stand still. Instead, they drift about in random motion. This motion is referred to as **random drift,** and occurs in all conductors. But random drift has little practical use. To do work, the free electrons must be forced to drift in the same direction rather than at random.

You can influence the drift of electrons so that all or most electrons move in the same direction through the conductor. This is done by placing electrical charges at opposite ends of the conductor. If a negative charge is placed at one end of a conductor and a positive charge is placed at the other, the negative charge will repel the free electrons, while the positive charge attracts them. As a result, all of the free electrons within the conductor move or drift in the same general direction—from the negative charge to the positive charge.

Here, the application of the electrical charges at the ends of the conductor changes random drift to **directed drift.** The directed drift of free electrons is called **current flow** because electric current is flowing through the conductor. If the electrical charges are small in magnitude, the flow of electrons will quickly cancel both charges and only a momentary current will flow. However, if the two electrical charges are large in magnitude, current flow will continue longer. The symbol for current is **I**. Current is measured in **amperes,** or **amps**. The symbol for amps is **A**.

The difference in charge between the poles is the **force** that causes the electrons to flow. If a length of copper wire is connected between the terminals of a battery, a large amount of current will flow from the negative terminal of the battery to the positive terminal. An electron at the negative terminal will be repelled by the negative charge and attracted by the positive charge at the positive terminal. Thus, the electrons flow through the wire from the negative terminal to the positive terminal. When they enter the positive terminal of the battery, they are captured by positive ions. The chemical reaction of the battery is constantly releasing new free electrons and creating positive ions to make up for the ones lost by recombination.

It should be pointed out that in practice, you never connect a conductor directly across the terminals of a battery. The heavy current would quickly exhaust the battery and could cause the battery to rupture or explode. This is an example of a "short circuit" and is normally avoided at all cost. It was used in this context to illustrate the concept of current flow. In actual circuits, a physical resistance is placed in the current path. Resistance is opposition to current flow and always causes a reduction in the amount of current that can flow.

CURRENT AND VOLTAGE

Electrical Force

We previously stated that the terminals of a battery create a force which moves electrons from the negative terminal to the positive terminal. This electrical force is also known as electromotive force, and is abbreviated EMF.

You will recall that Coulomb's Law states that like charges repel while unlike charges attract. The battery, by chemical action, produces a negative charge at one terminal and a positive charge at the other. The negative charge is simply an excess of electrons while the positive charge is an excess of positive ions. If a circuit is connected across a battery as shown in Figure 2-3A, a path for electron flow exists between the battery terminals. Free electrons are repelled by the charge on the negative terminal and are attracted by the charge on the positive terminal. The two opposite charges exert a pressure that forces the electrons to move. That is, the force or pressure results from the attraction of the unlike charges. To summarize, EMF is the force that sets electrons in motion.

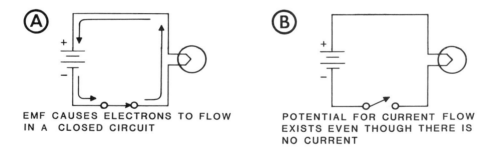

EMF CAUSES ELECTRONS TO FLOW IN A CLOSED CIRCUIT

POTENTIAL FOR CURRENT FLOW EXISTS EVEN THOUGH THERE IS NO CURRENT

Figure 2-3 EMF and potential difference.

Potential Difference

Another name for electrical force is potential difference. In Figure 2-3B, electrons cannot flow because the switch is open. Nevertheless, the battery still produces the same electrical pressure or force as before. Thus, the potential for producing current flow exists even though no current is presently flowing. Potential, used in this context, means the possibility of doing work. In physics, work is defined as FORCE through a DISTANCE. If the switch is closed, current flows, the lamp lights, and work is done. Therefore, whether a battery is connected to a circuit or not, it has the potential for doing work.

Actually, any charge has the potential for doing work. For example, a charge can move another charge either by attraction or repulsion. Even a single electron can repel another electron. If one electron moves as the result of the action of another electron, some small amount of work is done. In the battery, you are concerned with two different types of charges rather than a single charge. The electrons at the negative terminal are straining to rush to the positive terminal and cancel out the positive charge there. In the same way, the positive ions at the other terminal are straining to draw the electrons. We call this force **potential difference**. It is the **potential** for doing work that exists between two **different** charges.

Charges can differ in two ways. First, they can be of opposite **polarity**. This simply means that one is positive and the other is negative. Figure 2-4 shows a negative charge a small distance away from a positive charge. If a conductor is placed between the two charges, electrons will flow from the negative charge to the positive charge. The work that is done here is the transfer of electrons. Current will flow until the two charges cancel each other.

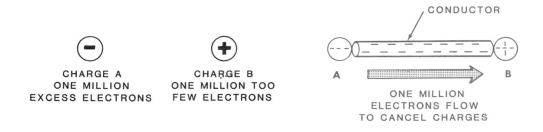

Figure 2-4 A potential difference exists between charges with opposite polarities.

A potential difference can also occur if the two charges have the same polarity, but different amounts of charge. This means they have a difference in **magnitude**. For example, Figure 2-5 shows two charges that have the same polarity, but different magnitudes: charge A is more negative than charge B because it has more excess electrons. If a conductor is connected between the two charges, electrons will flow from the greater negative charge to the lesser negative charge. The amount of electrons will be the exact number needed to equalize the two charges. In this example, charge A originally has three million excess electrons while charge B has an excess of only one million electrons. To equalize the two charges, one million electrons will flow from charge A to charge B. Electron flow ceases as soon as the two charges become equal. Notice that the direction of current flow is from the more negative charge to the less negative charge.

CURRENT AND VOLTAGE

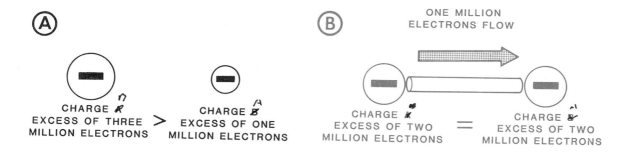

Figure 2-5 A potential exists between two charges with different magnitudes.

A potential difference can also exist between a terminal with a positive or negative charge and a terminal with no charge. This is shown in Figure 2-6. Terminal A has an excess of one million electrons, which gives it a negative charge. Terminal B has no charge: it contains the same number of free electrons and positive ions. When the two terminals are compared, terminal A is considered to be more negative than terminal B. If a conductor is placed between these two terminals, electrons will flow from terminal A, the more negative terminal, to terminal B.

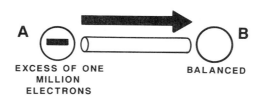

Figure 2-6 A potential exists between a terminal with a neg charge and a terminal with no charge.

31

Voltage

Another term that is often used interchangeably with EMF and potential difference is voltage. Voltage is the measure of EMF or potential difference. The unit of measurement for EMF, potential difference, or voltage, is the volt, and its abbreviation is V. For example, the battery in your car has a potential difference of 12 volts. The EMF supplied by wall outlets is 115 volts, while that required by most electric stoves is 220 volts. A large screen color TV receiver produces a voltage to light the picture tube that may be 25,000 volts or higher. High tension power lines often have a difference of potential as high as 500,000 volts.

At this point, it is difficult to visualize exactly how much EMF constitutes one volt. However, as you work with electronics, this point will become clear.

Producing EMF

EMF is produced when an electron is forced from its orbit around the atom. An electric pressure exists between the free electron and the resulting positive ion. Thus, any form of energy that can remove electrons from atoms can be used to produce EMF. In no case is energy actually created. It is simply changed to electrical energy from other forms. For example, a battery converts chemical energy to electrical energy, while a generator converts mechanical energy to electrical energy.

Magnetism

The most popular method of producing EMF is quite simple. When a conductor is moved through a magnetic field, an EMF is produced. This is called **magnetoelectricity**. The passing of a magnetic field through a conductor provides the energy necessary to free electrons in the conductor. If the conductor forms a closed loop, then the electrons will flow through the conductor. If the conductor does not form a closed loop, a potential difference is still present.

The basic requirements for producing an EMF using this method are a magnetic field, a conductor, and relative motion between the two. The requirement for relative motion means that either the conductor or the magnetic field may be moving, but the field must pass through the conductor. Producing EMF with magnetism will be covered in greater detail later in the text.

Chemical

The second method of generating EMF is by chemical means, sometimes called **electrochemistry**. Automobile and flashlight batteries are two examples of producing an EMF through chemical action. There are many chemical reactions that will transfer electrons to produce EMF. Some of these will be discussed in more detail later in this chapter.

Figure 2-7 illustrates how a basic battery or cell is made. A glass beaker is filled with a solution of sulfuric acid and water. This solution is called the *electrolyte*. In the electrolyte, the sulfuric acid breaks down into hydrogen and sulfate. Because of the chemical action involved, the hydrogen atoms give up electrons to the molecules of sulfate. Thus, the hydrogen atoms exist as positive ions, while the sulfate molecules act as negative ions. Even so, the solution has no net charge since there are the same number of negative and positive charges.

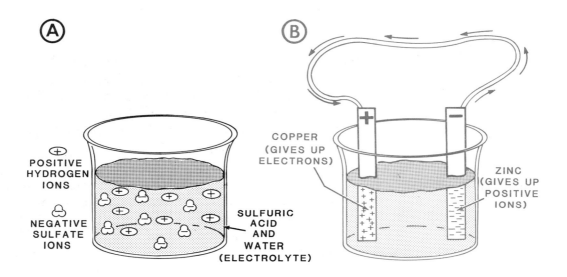

Figure 2-7 Electricity from chemicals.

Next, two bars called *electrodes* are placed in the solution. One bar is copper, while the other is zinc. The positive hydrogen ions attract the free electrons in the copper. This causes the copper bar to give up electrons to the electrolyte. Thus, the copper bar is left with a net positive charge.

The zinc reacts with the sulfate in much the same way. The sulfate molecules have a negative charge. Thus, positive zinc ions are pulled from the bar. This leaves the zinc bar with a surplus of electrons and a net negative charge.

If a conductor is connected between the zinc and copper bars, electrons will flow from the negative to the positive terminal. When current flow is always in the same direction, it is called *direct current* or *DC*. The EMF supplied by the battery is referred to as a DC voltage or volts DC.

CURRENT AND VOLTAGE

Friction

The oldest known method for producing electricity is by friction. Some examples of this were discussed earlier. Rubbing a glass rod with silk results in friction between the two materials. This produces a positive charge on the glass rod and a negative charge on the silk. Different materials can also produce a charge. If fur is rubbed on a hard rubber dowel, the fur becomes positively charged, while the dowel becomes negatively charged.

You have probably experienced this phenomenon yourself many times. When you scuff your feet across a nylon or wool rug, your shoes develop a charge that is transferred to your body. When you touch a neutral object, such as a metal door knob or another person, a discharge occurs. Frequently, there is a tiny arc between your finger and the neutral body. The arc is the visible movement of electrons between two objects.

In many cases, static electricity produced by friction is troublesome or annoying. However, a device called a Van de Graff generator is used in physics laboratories. The Van de Graff generator uses friction to develop voltages up to 10 million volts. Producing electricity from friction is called the triboelectric effect.

Light

Light energy can be converted to electrical energy in large enough quantities to provide limited amounts of power. A familiar example of this is the solar cells frequently used on spacecraft. Recent scientific advances have significantly reduced their cost. In fact, solar cells already have many applications, such as microwave towers in remote locations, and calculators.

To create a solar cell, also called a photocell, a photosensitive material is needed. A photosensitive material is one that develops a charge when it is subjected to light. Some substances that behave in this way are cesium, selenium, germanium, cadmium, and sodium. When the photosensitive material is struck by light, some of its atoms absorb the light energy and release electrons. This is known as the photoelectric effect.

A photocell works this way: photosensitive material is placed between a glass plate and an iron plate. The glass plate allows light to strike the photosensitive material. These plates also serve as the electrodes where the voltage will be developed. Light passes through the glass plate and strikes the photosensitive material underneath. The atoms within the photosensitive material give up electrons and a charge develops between the two plates.

When exposed to sunlight, a single cell can provide a fraction of a volt. When used as a power source, hundreds of the cells should be connected so that they produce large voltage levels.

Pressure

A small electrical charge develops in some materials when they are subjected to pressure. This is referred to as the piezoelectric effect. It is especially noticeable in substances such as quartz, tourmaline, and Rochelle salts, all of which have a crystalline structure. Figure 2-8 illustrates how the charge is produced. In the normal structure, negative and positive charges are distributed so that no overall charge can be measured. However, when the material is subjected to pressure, electrons leave one side of the material and accumulate on the other side. Thus, a charge develops. When the pressure is relieved, the electrons are again distributed so that there is no net charge.

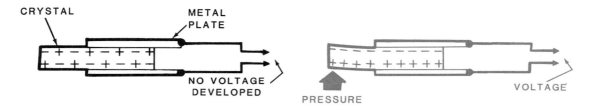

Figure 2-8 Electricity from pressure.

This effect is put to good use in crystal microphones, phonograph pickups, and precision oscillators. In all of these examples, the crystals are subjected to varying pressure. The pressure can be supplied by sound waves traveling through the air, as in the case with the microphone; or directly, as with a the phonograph pickup. At any rate, the voltage produced is very small and it must be amplified before it is used.

Heat

As with most other forms of energy, heat can be converted directly into electricity. The device used to do this is called a thermocouple. A thermocouple consists of two dissimilar metals joined together. Typical example of these metals are copper and zinc.

CURRENT AND VOLTAGE

You have seen that copper will readily give up electrons. This is especially true when the copper is heated. If zinc is then placed in contact with the heated copper, the free electrons from the copper are transferred to the zinc. Thus, the copper develops a positive charge, while the zinc develops a negative charge.

Since more heat will cause more electrons to transfer, the charge developed is directly proportional to the heat applied. This characteristic allows the thermocouple to be used as a thermometer in places that are too hot for conventional thermometers. A specific voltage across the thermocouple corresponds to a specific temperature. Therefore, the voltage can be measured and compared to a chart to find the corresponding temperature. The process by which heat is converted directly to electricity is called **thermoelectric effect**.

Effects of EMF

You have learned that an EMF can be produced by light, heat, magnetism, pressure, and chemical activity. It is interesting to note that the reverse is also true. That is, an EMF can be used to produce light, heat, magnetism, pressure, and chemical activity.

The electric light is an example of light produced by electricity. Current flowing through the light's filament causes it to glow. The light is a result of the EMF applied to the circuit. Here, the electrical energy is converted to light.

The toaster and electric stove are examples of electricity producing heat. In both of these, current flowing through heating elements produces a great amount of heat. In these instances, the energy applied to the components is given off as heat.

Whenever current flows through a wire, a magnetic field is produced. This field surrounds the wire and can be put to many practical uses. Some examples of these uses are motors, loud speakers, and solenoids.

Recall that a crystal produces a voltage when it is bent or twisted. The reverse is also true. When a voltage is applied to a crystal, the structure bends or twists. It can therefore be said that an EMF can produce pressure.

Finally, you already know that an EMF can be produced by chemical activity. It is also possible to produce chemical activity by applying an EMF to some materials. An example of this is the electrolysis of water. When an electric current flows through water, the water is broken down into its component parts of hydrogen and oxygen. Electroplating is another example of chemical activity caused by electricity. Here, ions of the plating material are removed from a solution and "plated" on the desired surface by the application of an electrical potential.

Chapter Self-Test

1. What are electrons in the outermost shell of an atom called?

2. For an element to be a good conductor, what is the *maximum* number of valence electrons it can contain?

3. When a battery is connected across a conductor, electrons will begin to flow from what terminal?

4. What is the term associated with free electrons flowing in a specific direction through a conductor?

5. What is the general name given to the force that moves electrons?

6. What are the two ways in which charges can differ?

7. Terminal A has one million excess electrons, and terminal B has three million excess electrons. When these terminals are connected, will there be electron movement? In which direction (A to B, or B to A)?

8. What are the three requirements for producing EMF by magnetism?

9. What is the name for producing EMF by bending a crystalline structure?

10. Name the method for producing EMF that causes each of the following applications to work.

    a. phonograph pickup
    b. battery
    c. solar cell
    d. thermocouple

CURRENT AND VOLTAGE

Summary

Electrons orbit around the nucleus of an atom in orbital shells.

The outermost orbital shell for any atom is referred to as the valence shell, and any electrons it contains are valence electrons.

The number of valence electrons that an element or compound contains determines its electrical characteristics.

An element with three or less valence electrons can easily lose or gain electrons, which causes it to be a conductor. Examples of conductors are copper, aluminum, gold, and silver.

An element or compound with five or more valence electrons is very stable, and tends not to conduct current. These elements and compounds are called insulators.

An element with four valence electrons is a semiconductor, because it is neither a good conductor, nor a good insulator.

In order to have electron movement from one point to another, there must be a source of free electrons. Conductors normally have billions of free electrons floating within the material.

A battery has two terminals: a negative terminal, which is a source of free electrons, and a positive terminal, which contains a vast supply of positive ions.

When opposite, or potentially different, charges are connected to two points within a conductor, electrons will always move from the more negative charge to the more positive charge.

The movement of electrons from one point to another is called current flow. Current flow can also be considered as a movement of charges.

In order to have current flow, you must first have a force which forces electrons to move, and a path for the electrons to travel.

EMF is the force that moves electrons. This force is a natural example of Coulomb's Law.

Potential difference is another name for EMF. It represents the potential for moving electrons.

EMF or potential difference exists between any two charges that are not exactly alike. A difference of potential exists between any uncharged body and any charged body. An EMF exists between two unequal positive charges, between unequal negative charges, and between any negative charge and any positive charge.

Voltage is another name for EMF or potential difference. The unit of measurement for electrical force is the volt.

EMF can be produced in several different ways. The most common method uses magnetism and mechanical motion. Other methods use chemical reactions, friction, light, pressure, or heat.

CHAPTER 3

Properties of Electrical Circuits

DC ELECTRONICS

Contents

Introduction 43

Chapter Objectives 44

The Electrical Circuit 45

Batteries 48

Voltage Rises and Voltage Drops 54

Concept of Ground 57

Chapter Self-Test 60

Summary 61

PROPERTIES OF ELECTRICAL CIRCUITS

Introduction

Now that you have studied what current and voltage are, we are ready to put them to work. To do this, an electrical circuit must be devised. In this chapter, you will learn what makes up an electrical circuit, and how current and voltage behave within a circuit. You will learn how schematic diagrams allow complex circuits to be represented on paper.

You will also study a common voltage source—the battery. We will tell you how different batteries are created, and how arranging them in different configurations allow us to provide different voltages. You will also learn about voltage rises and voltage drops, and the difference between the two. You will then be introduced to the concept of ground, and how it plays such an important part in electrical circuits.

DC ELECTRONICS

Chapter Objectives

When you have completed this chapter, you will be able to:

1. Define the following terms: load, schematic, secondary cell, primary cell, voltage source, voltage rise, voltage drop, earth ground, chassis ground, and frame ground.

2. Name the basic parts of an electrical circuit.

3. Identify the schematic symbols for a battery, lamp, resistor, conductor, and switch.

4. Identify the difference between the wet cell, dry cell, and the NICAD batteries.

5. State the relationship between a voltage rise and a voltage drop.

6. State the difference between earth ground, chassis ground, and circuit ground.

PROPERTIES OF ELECTRICAL CIRCUITS

The Electrical Circuit

In its simplest form, an electric circuit consists of a power source, a load, and conductors that connect the power source to the load. Often the power source is a battery.

The load is generally some kind of electrical device that performs a useful function. It might be a lamp that produces light, a motor that produces physical motion, a horn that produces sound, or a heating element that produces heat. Regardless of the type of load used, it performs a useful function only when electric current flows through it.

The third part of the circuit is the conductor that connects the power source to the load. It provides a path for current flow. The conductor may be a length of copper wire, a strip of aluminum, or the metal frame of an automobile.

Figure 3-1 shows an electric circuit consisting of a battery, a lamp, and connecting copper wires. The battery produces the force (voltage) necessary to cause the free electrons in the conductor to flow through the lamp in the direction shown. The free electrons are repelled by the negative charge and are attracted by the positive charge. Thus, the electrons flow from negative to positive. The negative and positive charges in the battery are constantly being replenished by the chemical action of the battery. Therefore, the battery maintains a current flow for a long period of time. As the electrons flow through the lamp, they heat up the wire within the lamp. As the wire becomes hotter, the lamp emits light. The lamp will glow as long as a fairly strong current is maintained. Thus, the battery is the power source, the lamp is the load, and the wires between the lamp and the battery are the conductors, or current path.

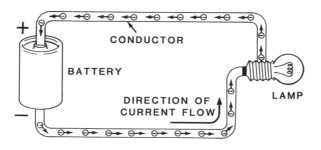

Figure 3-1 Simple electric circuit.

DC ELECTRONICS

You know from your experience with flashlights that a battery cannot maintain a constant current flow forever. As time passes, the chemical reaction within the battery slows down, and eventually, the battery produces no current at all. At this time, it is said to be dead, or run down. For this reason, it is necessary to add a component to your simple electrical circuit.

The circuit can be made much more practical by adding a control. In this case, the control is a switch that provides an easy way to start and stop current flow. Since some type of control is necessary, it can be said that a practical elementary circuit consists of four parts; the power source, the load, the conductor, and the control.

Figure 3-2 shows the circuit after the switch has been added. For simplicity, a "knife" switch is shown. It consists of two metal contacts to which conductors may be connected, a metal arm that can be opened and closed, and a base. Current does not flow through the base of the switch because it is made from an insulator material. Current can flow only through the arm and then only if the arm is closed.

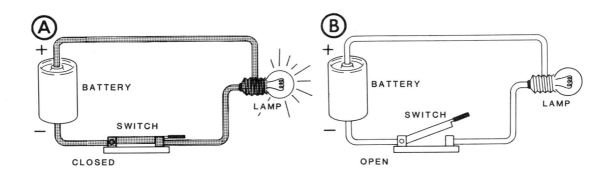

Figure 3-2 Circuit with switch.

Figure 3-2A shows the switch in the closed position. With the switch closed, the path for current flow between the terminals is complete. Current can flow in the circuit, and cause our lamp to glow. When the switch is opened, as shown in Figure 3-2B, the path for current flow is broken or open, and the lamp does not glow.

While simple circuits can be drawn in this manner, it is very difficult for drawing complex circuits. For this reason, the schematic diagram was developed. A schematic diagram is a drawing in which symbols are used to represent circuit components. The first step to understanding the schematic diagram is to learn the symbols used to represent the various components.

PROPERTIES OF ELECTRICAL CIRCUITS

Figure 3-3 compares the schematic symbol with the pictorial representation of the components used up to this point. Most of the symbols are self-explanatory, with the exception of the battery. Since the battery has positive and negative terminals, the symbol should distinguish between the two. The plus (+) and minus (−) symbols will not always be present on the battery. So, the long line is the positive terminal, and the short line is the negative terminal. The same symbol can be used regardless of the type of battery. These symbols are combined to create a schematic diagram for an electrical circuit.

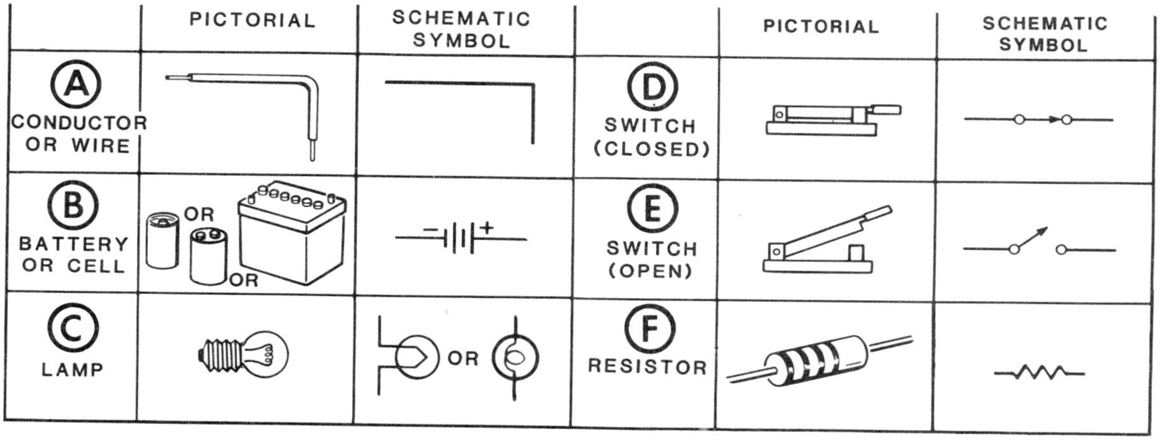

	PICTORIAL	SCHEMATIC SYMBOL		PICTORIAL	SCHEMATIC SYMBOL
Ⓐ CONDUCTOR OR WIRE	⌐	⌐	Ⓓ SWITCH (CLOSED)	▭─	—o→o—
Ⓑ BATTERY OR CELL	OR ▯▯ OR 🔋	—⊢⊢⊢⊢+	Ⓔ SWITCH (OPEN)	▭╱	—o ╱ o—
Ⓒ LAMP	💡	⊶ OR ⊕	Ⓕ RESISTOR	▭▭▭	—⋀⋀⋀—

Figure 3-3 Pictorial representations compared with the schematic symbols.

The schematic diagram for the closed circuit in Figure 3-2A is shown in Figure 3-4. This diagram could represent a simple flashlight, or the headlight system for an automobile. In reality, this diagram can represent any system with a battery, lamp, and switch.

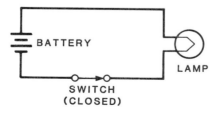

Figure 3-4 Schematic diagram of a simple circuit.

# Batteries

One type of battery, or cell, was discussed in the previous section. It consisted of zinc and copper electrodes inserted into an electrolyte of sulfuric acid and water. This section presents the construction and operation of several more common types of batteries. But first, you should know the difference between a battery and a cell.

A **cell** is a single unit that contains negative and positive electrodes immersed in an electrolyte. A **battery** is a combination of two or more electrochemical cells. Thus, what you call a flashlight battery is really a cell, since it contains only one unit for producing an EMF. In spite of this technical definition, battery is loosely used to describe a single cell.

Battery Types

There are two basic types of cells. One type can be *recharged* and is called a **secondary cell**. The other type, called a **primary cell,** *cannot* be recharged. No matter which type of cell you examine, they all store energy in a chemical form that can be released as electricity.

Although there are many different kinds of batteries in the field of electronics, the majority you will run across will be the dry cell, lead-acid, and the nickel-cadmium batteries.

Dry Cell

You've probably dealt extensively with the dry cell battery. Most flashlight, radio, and camera flash batteries are dry cells. Actually, the term dry cell is rather misleading, since the electrolyte within the dry cell battery is a moist paste. This paste is attached to the two terminals; zinc is used for the negative terminal, and steel is used for the positive terminal. This is shown in Figure 3-5. The entire battery, except for the terminals themselves, is then sealed with a wax film to insulate the user from the electrolyte.

PROPERTIES OF ELECTRICAL CIRCUITS

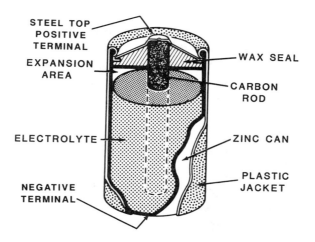

Figure 3-5 Construction of the dry cell.

Many different voltages are available with the dry cell battery, depending upon the type of material used for the terminals and the electrolyte. Thus, a small pen light cell can supply the same voltage as a large D cell battery. However, since there is more electrolyte in the D cell, it is capable of maintaining the voltage for a longer period of time.

As convenient as the dry cell battery is, it does have its disadvantages. The major disadvantage of the dry cell is that it is a primary cell, and thus, cannot be recharged. Secondly, as the electrolyte becomes exhausted, it is not able to supply as many free electrons or positive ions as when it was brand new. As a result, the voltage of the battery gradually declines.

Lead-Acid Battery

The common automobile battery is an example of a lead-acid battery. In this battery, the electrolyte is a solution of sulfuric acid and water; the negative terminal is composed of a material called spongy lead, and the positive terminal is composed of a material called lead dioxide. The reaction between these compounds creates both free electrons and positive ions, which creates the potential difference between the terminals.

The main advantage of the lead-acid battery is that it is comprised of secondary cells. The chemical process which creates the free electrons and positive ions can be reversed. That is, you can restore the exhausted chemical compounds by passing an electrical current through the battery.

DC ELECTRONICS

The Nickel-Cadmium (NICAD) Battery

Another type of secondary cell is the nickel-cadmium battery, or the NICAD battery. The plates in a NICAD battery are made of nickel powder heat-bonded to a metal screen. This results in an extremely porous plate that is then impregnated with electrically active materials. It is these active materials that actually produce EMF.

Certain characteristics of the NICAD battery make it very unique, which allows you to use it in a variety of ways. First, the NICAD battery is completely sealed. The fact that it is completely sealed makes it much safer than, for example, the lead-acid battery.

Second, unlike the dry cell and lead-acid batteries, the voltage created by the NICAD remains relatively constant throughout most of its discharge period. Only when the battery is nearly exhausted will its voltage begin to drop off significantly.

Despite all of its advantages, the expense of the NICAD battery is a limiting factor in some applications.

Connecting Batteries

Cells can be connected together to increase either the voltage or current rating. There are four different ways that cells or batteries can be connected: series aiding, series opposing, parallel, and series parallel. Most of these configurations have their own advantages and uses. The following is a discussion of these configurations.

Series Aiding Connection

In a 12-volt lead-acid battery (commonly found in automobiles) 6 cells are connected together so that the individual cell voltages are additive. In a 6-volt battery, three cells are connected in the same direction. This arrangement is called a **series aiding** connection. An example of this connection is shown in Figure 3-6A, and the schematic diagram is shown in Figure 3-6B.

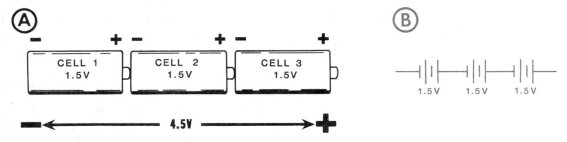

Figure 3-6 The series aiding connection.

PROPERTIES OF ELECTRICAL CIRCUITS

In Figure 3-6A the cells are connected so that the positive terminal of the first cell connects to the negative terminal of the second, and the positive terminal of the second cell connects to the negative terminal of the third. This type of connection forms a single current path through all three cells. This is called a series connection. It is an aiding connection because the EMF of each cell moves current in the same direction. Hence, the cells aid each other in current movement. Since the individual EMF of each cell is 1.5 volts, the overall EMF is 4.5 volts.

With the series aiding connection, the total voltage across the battery is equal to the sum of the individual values of each cell. However, the current capacity of the battery does not increase. Since the total circuit current flows through each cell, the current capacity is the same as for one cell.

Series Opposing Connections

The series aiding connection just described, is widely used. The **series opposing** connection of cells, on the other hand, has little practical use. We mention it here because an inexperienced person may inadvertently connect cells in this way.

If two cells are connected in series, with the *like* terminals of the cells connected together (as shown in Figure 3-7A), the cells provide EMF in opposite directions. If the two voltages are of equal value, they will cancel each other out, and the overall EMF is 0 volts. Because the voltages cancel, this arrangement cannot produce current flow. This connection is normally avoided. The schematic diagram for the series-opposing connection is shown in Figure 3-7B.

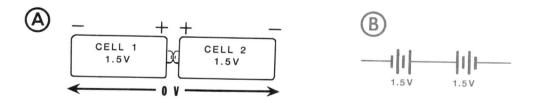

Figure 3-7 The series opposing connection.

Parallel Connection

You have learned that the series aiding connection of cells increases the output voltage, but not the current capabilities of the cells. There is a way to connect cells so that their current capabilities add together. This is called a parallel connection, and it is shown in Figure 3-8A. Here, like terminals are again connected, but in a manner different than the series-opposing connection. That is, all the positive terminals are connected together as are all of the negative terminals.

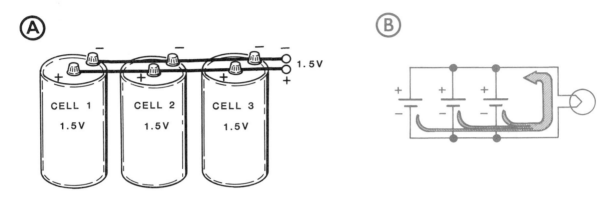

Figure 3-8 The parallel connection.

Figure 3-8B shows why the current capacities of the cells are added together. Notice that the total current through the lamp is the sum of the individual cell currents. Each cell provides one third of the total current. Thus, the total current capacity is three times that of any one cell. However, connecting cells in this way does not increase the voltage. That is, the total voltage is the same as that for any one cell. If 1.5-volt cells are used, then the total voltage is 1.5 volts.

Series-Parallel Connection

When both a higher voltage and increased current capacity are required, cells can be connected in series-parallel. For example, suppose you have four 1.5-volt cells, and you wish to connect them so that the EMF is 3 volts, and the current capacity is twice that of any one cell. Connect two of the cells in a series aiding configuration. This provides 3 volts, but it does not increase the current capacity. To double the current capacity, you then connect the other two cells in a second series aiding string. Finally, you connect the two series strings in parallel with each other, in parallel with the first. The result is a series-parallel arrangement, as shown in Figure 3-9A. The schematic diagram for this arrangement is shown in Figure 3-9B.

PROPERTIES OF ELECTRICAL CIRCUITS

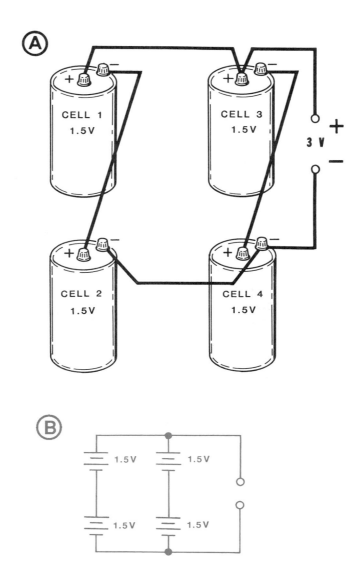

Figure 3-9 The series-parallel connection.

DC ELECTRONICS

Voltage Rises and Voltage Drops

In electronics and electrical work, there are two kinds of EMF. Both are expressed in volts but they have somewhat different characteristics. One type of EMF is called a voltage rise, while the other is called a voltage drop. The following sections discuss voltage rises and drops in detail.

Voltage Rise

You know that a battery provides a voltage by chemically producing an excess of electrons at the negative terminal and an excess of positive ions at the positive terminal. When a load is connected across the battery, electrons flow through the load. Each electron that leaves the negative terminal is replaced by another electron from within the battery. At the positive terminal, each electron arriving from the load cancels one positive ion. Each ion that is canceled is then replaced with another positive ion from within the battery. Thus, the voltage between the two terminals remains constant even though electrons are constantly flowing from the negative terminal and into the positive terminal.

Energy is required to move the electrons through the load. The battery gives each electron the energy required to make the trip. The energy comes from the chemical reaction within each terminal, which is soon converted to electrical energy. This energy has a capacity to do work, and the amount of work is determined by the potential difference between the two terminals, or the voltage of the battery. After all, it is the voltage of the battery that causes the electrons to flow in the first place. Since the battery is the source, or location, of the voltage, it can be referred to as a voltage source.

The battery, along with an AC generator (discussed later), are the two most common types of voltage sources. However, solar cells and thermocouples also produce an EMF, and may be considered as voltage sources. Any voltage introduced into a circuit by a voltage source is called a voltage rise. Thus, a 10 volt battery has a voltage rise of ten volts.

PROPERTIES OF ELECTRICAL CIRCUITS

Voltage Drop

Electrons that leave the negative terminal of a battery have been given energy by the battery. As the electrons flow through the load, they give up their energy to the load. Often, the energy is given up as heat. However, if the load is a light bulb, both heat and light are given off. In a motor, the energy is given off as heat and motion.

Since the energy introduced into the circuit is called a voltage rise, the energy removed from the circuit by the load is called a voltage drop. A voltage drop is expressed in volts like a voltage rise. In fact, if you attempt to measure both a voltage rise and a voltage drop, your test equipment (in this case a voltmeter) could not tell the difference.

There is, however, an important difference between a voltage rise and a voltage drop. If you hold a battery in your hands (a battery being a common voltage source), a potential difference *already exists* between the two terminals, even though the battery is not connected to a circuit. The voltage rise exists.

In order to have a voltage drop, electrons must flow through a load, and give their energy to the load. Therefore, you must have current flowing through a load to develop a voltage drop. In other words, current flow is required for a voltage drop, but not for a voltage rise.

Voltage Drops Equal Voltage Rises

By definition, a voltage drop across a load cannot exist unless there is a voltage source forcing electrons through the load. Thus, the question is raised: What is the relationship between the voltage rise and the voltage drop?

Consider the following example: A 10 volt battery is connected across a single lamp. The battery causes current to flow through the lamp, and the electrons give off their energy to the lamp in the form of light and heat. The lamp will consume the same amount of energy as the battery provides, so the voltage drop across the lamp must also be 10 volts. In conclusion, the voltage rise in an electrical circuit must equal the voltage drop.

If the same 10 volt battery is connected across two identical lamps, connected in series, the electrons must now give up their energy to each lamp. Keep in mind that these electrons have a finite amount of energy to pass. Since the two lamps are identical, they will consume (drop) the same amount of energy (voltage). Thus, the 10 volts supplied by the battery is divided equally in half; each lamp will drop 5 volts. When the two voltage drops are added together, they equal the voltage rise in the circuit.

If multiple voltage sources are used in a circuit, this rule still holds true. If we were to add a second 10 volt battery to our original circuit, in a series aiding configuration, the sum of the voltage rises would then be 20 volts. However, we have only one lamp which acts as the load. Since this one lamp is the only load, it must consume all the energy given to it; its voltage drop will equal 20 volts.

As a final example, consider the circuit in Figure 3-10. In this case, two 4.5 volt batteries are connected, in a series aiding configuration, with three identical lamps. The total voltage rise in the circuit is 9 volts. Since the lamps are identical, each drops one-third of the applied voltage or 3 volts. Once again, the sum of the voltage rises equals the sum of the voltage drop.

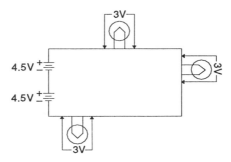

Figure 3-10 The voltage drops are equal to the voltage rises.

The term "connected in series" means there is only one path for the electrons to follow. All of the electrons leaving the negative battery terminal must flow through each component in the circuit before reaching the positive terminal of the battery. If an electron were to leave the negative terminal of the top battery, shown in Figure 3-10, it would have to pass through the bottom battery and the three lamps before it could reach the positive terminal of the original battery. There is no other path this electron could take. This describes components connected in series. When components are connected in series, an electron traveling through the circuit *must* pass through every component, one after the other. This concept will be discussed in greater detail later in the text.

PROPERTIES OF ELECTRICAL CIRCUITS

Concept of Ground

One of the most important points in the study of electricity and electronics is the concept of ground. Originally, ground was just what the name implies--the earth. In fact, in some countries, the name earth is used instead of ground. Earth is considered to have zero potential. Thus, ground, or earth, can be used as a reference point to which voltages, either positive or negative, are compared.

Figure 3-11 Ground reference.

Many electrical appliances in your home are grounded. This is often accomplished by connecting a heavy wire directly to a cold water pipe that is buried deep in the earth (ground). In other cases, a third prong on the power plug connects the metal frame through a third line in your house wiring to ground. The purpose of this is to protect you in case a short circuit develops in the appliance. It also places the grounded metal parts of different appliances at the same potential. This protects you from receiving an electrical shock from a difference in potential between two appliances. This type of ground is called earth ground. The schematic symbol for earth ground is shown in Figure 3-11A.

In some electronic equipment, the zero reference point, or ground point, is the metal frame, or chassis, on which the various circuits are constructed. In your automobile, the chassis, or metal body, of the automobile is considered as ground. All voltages are measured with respect to this chassis ground. In a schematic, chassis ground is depicted as shown in Figure 3-11B. In many instances, the schematic symbols shown in Figures 3-11A and 3-11B are interchangeable.

A ground may also exist without having any connection to earth. This type of ground is used more as a reference than for protection. This type of ground is called circuit ground. Circuit ground simply provides a zero reference point in electrical circuits. Usually, the circuit grounds in a piece of electronic equipment are interconnected. At any rate, all voltages measured within the circuit are measured with respect to circuit ground. The schematic symbol for circuit ground is shown in Figure 3-11C.

DC ELECTRONICS

The concept of ground is extremely important in electronics for two reasons. First, naming a specific point, or points, as ground provides you with a **reference**. This reference allows you to use, and measure, both negative and positive voltages. This is accomplished by measuring the potential at one point **with respect to** another point. Up to now, you have studied potential difference between two points, but not one point with respect to the other. For example, a 6 volt battery has an EMF between its two terminals of 6 volts. You have not yet learned to think of this as +6 volts or −6 volts, but simply 6 volts. By grounding one terminal of a battery, you can compare the potential of the other terminal, both in magnitude and polarity, to the grounded terminal.

Let's return to our 6 volt battery. If we attach the positive terminal to ground, or zero reference, then the negative terminal is 6 volts *more negative*. Thus, the voltage at this terminal **with respect to ground** is negative 6 volts, or −6 volts. If we ground the negative terminal, then the positive terminal is +6 volts. The overall EMF of the battery can be seen as −6 volts or +6 volts, depending on which terminal you reference to ground.

Let's consider Figure 3-12. Two batteries are connected in series, with the ground connection between them. This places the zero reference at point B. The top battery has an EMF of 10 volts. Since the negative terminal of this battery is connected to ground, the voltage at point C with respect to ground is +10 volts. The lower battery, on the other hand, has an EMF of 6 volts. Since the positive terminal for this battery is connected to ground, the EMF at point A, with respect to ground, is -6 volts. Remember, voltage is always the measure of the potential difference between two points. Therefore, whenever the voltage at a specific point in a circuit is measured, it is assumed that you are comparing the potential of that point to ground.

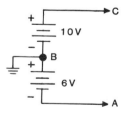

Figure 3-12 The polarity of the voltage depends upon the connection to ground.

PROPERTIES OF ELECTRICAL CIRCUITS

The second important application of ground is that it provides a means of **common return** in electrical equipment. Assume, for example, that you have a large piece of electrical equipment built surrounding a metal chassis. Each of the many circuits used in this equipment requires +6 volts to function. One possible method of fulfilling this need is to use one 6 volt battery for each circuit, but that wouldn't be practical. You then decide to take one 6 volt voltage source, and connect its terminals to each circuit.

Now, you have hundreds of wires running from the negative terminal of the battery to one side of each circuit, and hundreds of wires running from the other end of each circuit to the negative terminal. Again, this will work, but it's impractical.

The solution lies in the chassis. If you attach the negative terminal of the battery to chassis ground, you create +6 volts at the positive terminal with respect to ground. You still need to run wires from the positive terminal to each circuit, but you can attach the *negative* end of each circuit to the chassis. The voltage for each circuit, at the point where the positive terminal is connected, is now +6 volts. The path for current flow in each circuit begins at the negative terminal of the battery (current flows from negative to positive), through the chassis to the circuit, through the circuit itself, and then to the positive terminal of the battery. Since all the points within the chassis are common, there is no threat of electrical shock, provided you don't touch the chassis and the positive terminal of the battery.

DC ELECTRONICS

Chapter Self-Test

1. Describe the arrangement of a simple electrical circuit.

2. What is the general term associated with a device in a circuit that performs some useful function?

3. What does the short line in a battery symbol represent?

4. How does a secondary cell differ from a primary cell?

5. Four cells, each having an EMF of 2.1 volts, are connected in a series aiding configuration. What is the overall EMF produced?

6. What are the two types of EMF in an electrical circuit?

7. State the relationship between voltage drops and voltage rises in a single electrical circuit.

8. What are the two applications of ground?

9. What is another name for using the frame of a piece of equipment as ground?

10. What terminal must be grounded to provide a *positive* power source?

PROPERTIES OF ELECTRICAL CIRCUITS

Summary

A simple electrical circuit consists of a voltage source, a load, and conductors connecting the voltage source to the load.

A load is an electrical device that performs some useful function.

A schematic diagram is a drawing in which symbols are used to represent circuit components. Schematic diagrams allow complex electrical circuits to be drawn with relative ease.

A cell is an electrochemical device that contains negative and positive electrodes separated by an electrolyte.

A battery is a combination of two or more cells.

There are two basic types of cells: the primary cell and the secondary cell. The primary cell cannot be recharged. The secondary cell can be recharged.

The dry cell uses a paste-like electrolyte, and is a primary cell.

The most common type of wet cell is used in the lead-acid battery. This type of battery is used in almost all automobiles, and it is a secondary cell.

When two or more cells are connected in a series aiding configuration, a potential difference is produced, which is equal to the sum of the individual battery voltages. Thus, a 6 volt battery can be formed by connecting four 1.5 volt cells in the series-aiding arrangement.

When two cells are connected in a series opposing configuration, the voltage produced by one cell is subtracted from that produced by the other. This type of connection should be avoided.

When two or more cells are connected in parallel, the output voltage is the same as that for any one cell, but the current capacity increases.

Cells can be connected in a series-parallel arrangement so that both the voltage and the current capacity increase.

A voltage can exist at two different points in a circuit: where it is produced and again where it is used. A voltage that exists when EMF is being produced is called a voltage rise. A voltage that exists because energy is being used is called a voltage drop.

DC ELECTRONICS

A voltage rise can exist with or without current flow. A voltage drop exists across a component only when current flows through the component.

In a circuit in which current is flowing, the sum of the voltage drops is equal to the voltage rise, or the sum of the voltage rises.

Ground is the name given to the point in a circuit that is used as the zero reference. Often this is the metal chassis or frame of the electronic device.

Voltages may be negative or positive with respect to ground. A battery can be connected as a negative voltage rise or as a positive voltage rise, depending on how it is connected to ground.

CHAPTER 4

Resistance

DC ELECTRONICS

Contents

Introduction 65

Chapter Objectives 66

Electron Opposition 67

Resistors 72

Connecting Resistors 78

Chapter Self-Test 86

Summary 87

RESISTANCE

Introduction

Resistance is the name given to that property of any substance that causes it to oppose current flow. All materials have this property to some extent. Some materials, such as glass and rubber, offer a great deal of opposition to current flow. They allow almost no current flow through them. Thus, they are said to have a very high resistance. Other materials, such as silver, gold, and copper, offer very little opposition to current flow, and thus have a very low resistance.

This chapter discusses the characteristics of resistance. It describes what resistance is, how it is determined, and how it acts when connected in various configurations. This chapter also discusses the different types of resistors that are available, how they're constructed, and the differences between them.

Chapter Objectives

When you have completed this chapter, you will be able to:

1. Define the following terms: resistivity, resistance, conductance, ohm, thermistor, potentiometer, positive temperature coefficient, negative temperature coefficient, and rheostat.

2. Name four factors that determine the resistance of a substance and state the relationship that these factors have to the resistance of a substance.

3. Identify a resistor's value and tolerance using the color code.

4. Determine a resistor's power rating by examining its size.

5. Calculate the total resistance for resistors in series, parallel, and series-parallel.

6. State the relationship between resistance, voltage, and current.

7. State three types of fixed resistors.

RESISTANCE

Electron Opposition

You previously learned that the number of free electrons in a material will determine how well current flows within that material. In copper, there are many free electrons, and current flows easily. In rubber, there are few free electrons, and little or no current flows. However, all materials have some type of **opposition** to current flow. This opposition is called **resistance**. You may think that copper has no resistance, but this is not true. Current does flow easily through copper, but it does require some energy to cause this current. Rubber, obviously, has a high resistance, since it allows little or no current flow.

The letter **R** is used to represent resistance. The unit of measurement for resistance is the **ohm**. The symbol for the ohm is the Greek letter **Omega** (Ω). This unit is named for Georg Simon Ohm, a German physicist who also discovered the relationship between voltage, current, and resistance. This relationship is called Ohm's Law, and is discussed later in the text.

The most common way to define the ohm is in terms of voltage and current. If one volt causes one ampere of current to flow in a circuit, then the resistance of the circuit is one ohm. It may be helpful to think in terms of something with which you are more familiar. A length of normal copper wire that is 60 feet long has a resistance of about one ohm.

Resistivity

You cannot directly compare the resistances of two substances because their resistances will vary with their shape, size, and temperature. However, every substance has a property called **specific resistance** or **resistivity** that can be compared directly if all other characteristics are alike.

For wire, the resistivity is defined as the resistance of a one-foot length of wire. This one foot length must be exactly 0.001 inch (1 milli-inch, or one mil) in diameter and its temperature must be exactly 20 degrees centigrade. These requirements standardize the shape, size, and temperature of substances being compared so that any difference in resistance is caused solely by the material's atomic structure. A one foot length of wire with a 1 mil diameter is known as a mil-foot.

The resistivity of several substances is shown in Table 4-1. Notice that silver has the lowest resistivity, while copper is a close second. At the bottom of the list are glass and rubber. Silver and copper are the best conductors, while glass and rubber are two of the best insulators.

SUBSTANCES	RESISTIVITY IN OHMS/MIL FOOT
Silver	9.9
Copper	10.4
Gold	15.3
Aluminum	17.0
Tungsten	33.8
Iron	58.0
Steel	100.0
Constantan	295.0
Glass	10^{16}
Rubber	10^{20}

Table 4-1

As you have seen, both conductors and insulators are important in electronics. Conductors have many free electrons so they conduct current very easily. Thus, they are used to carry current from one place to another. Most metals such as silver, copper, gold, aluminum, tungsten, zinc, brass, platinum, iron, nickel, tin, and steel are good conductors.

Insulators, or nonconductors, are substances that have few free electrons. Therefore, they have very high values of resistivity. These substances are used to prevent electrical conduction. Most wires are coated with an insulator so that they do not accidentally touch another component. This could create an unwanted current path. Some examples of insulators are glass, rubber, plastic, mica, and dry air.

To summarize, it is the resistivity of a material that determines if the material is a conductor or an insulator. Resistivity is the resistance of a specific size and shape of a material at a specific temperature.

RESISTANCE

Conductance

Sometimes it is more convenient to think in terms of how well a material conducts current, rather than how well it opposes current. Because of this, a property called conductance is often used. Conductance is the opposite of resistance. It is defined as the ease with which a substance allows current to flow. Mathematically, conductance is the reciprocal of resistance. Stated as an equation:

$$\text{Conductance} = \frac{1}{\text{Resistance}}$$

The letter G is used to represent conductance. Therefore, the equation can be written:

$$G = \frac{1}{R}$$

The unit of conductance is the mho, pronounced "moe." Notice that this is ohm spelled backwards. The mho is the reciprocal of the ohm. Therefore,

$$\text{mho} = \frac{1}{\text{ohm}}$$

A substance with a resistance of 1 ohm has a conductance of 1 mho. However, a resistance of 2 ohms in a substance results in a conductance of 1/2, or 0.5, mhos. If the resistance is 1000 ohms, the conductance is 0.001 mhos, or 1 millimho.

In most cases, it is more convenient to think in terms of ohms (resistance) rather than in terms of mhos (conductance). Therefore, descriptions in this course will be related to resistance. However, it is important to remember the mho because it is a key characteristic of specific components we will discuss later.

Factors Determining Resistance

The most important factor in determining resistance is the resistivity of a material. However, three other factors also contribute to a substance's resistance. These are the length, cross-sectional area, and temperature of the material. This is why these three variables are carefully defined when determining resistivity. The following is a more detailed discussion of each of these factors.

Length

A 60-foot length of normal copper wire has a resistance of about 1 ohm. A 120-foot length of the same wire has a resistance of approximately 2 ohms. Thus, if you double the length of the wire, the resistance also doubles. In other words, the resistance of a conductor is **directly** proportional to its length. In fact, with any material, the greater the length, the higher the resistance. The reason for this is that the electrons must travel further through the resistant medium. Thus, if the length doubles, the resistance doubles; if the length triples, the resistance triples; and so forth.

Cross-Sectional Area

The cross-sectional area of a conductor is determined by its thickness or diameter. As you know, good conductors have a large number of free electrons. If the diameter of a conductor is increased, there will be more free electrons per unit of length. If there are more free electrons, current will flow easier; thus resistance is decreased. Therefore, large diameter conductors have less resistance than small diameter conductors.

All other things being equal, the resistance of a substance is inversely proportional to its cross-sectional area. If the cross-sectional area doubles, the resistance drops to one-half its former value. Also, if the area triples, the resistance drops to one-third of its previous value.

Temperature

Changes in length or cross-sectional area change the physical dimensions of the material, and these changes will affect the resistance of each material the same way. With a change in temperature, however, you change the actions of the atoms within the material. That's because each type of material has a different atomic structure, and a change in temperature will affect the resistance of each material differently.

In most materials, an increase in temperature causes an increase in resistance. As the temperature increases, the bonds between the nucleus of the atom and its valence electrons become stronger. The free electrons that roam through the material create new free electrons by colliding with valence electrons. With the increase in bonding strength between the nucleus and valence electrons within an atom, the free electrons will have a more difficult time breaking away these valence electrons. Thus, fewer free electrons exist within the material, and opposition to current flow increases. Materials that respond in this way are said to have a positive temperature coefficient. If a material has a **positive temperature coefficient**, its resistance increases as temperature increases and decreases as temperature decreases.

A few substances, such as carbon, have a **negative temperature coefficient**. This means that their resistance decreases as temperature increases. With these materials, the increase in temperature causes an increase in the activity of the free electrons. This increased activity makes it easier for the free electrons to collide with other valence electrons, thus making even more free electrons. With more free electrons available, the opposition to current decreases.

There are also some materials whose resistance does not change at all with changes in temperature. These materials are said to have a **zero**, or **constant temperature coefficient**.

In most circuits, the temperature coefficients of the components are not critical and are simply ignored. However, some circuits must consider temperature as a factor, and the coefficient of each component in the circuit should be evaluated.

A device called a **thermistor** uses the effect of temperature on resistance to great advantage. A thermistor is a special type of resistor that changes resistance when its temperature changes. Most thermistors have a negative temperature coefficient. In many thermistors, the resistance value can drop to one-half its former value for a temperature rise of 20 degrees centigrade. Thermistors are often used in temperature sensing circuits and as protective devices.

DC ELECTRONICS

Resistors

A resistor is an electronic component that has a certain specified opposition to current flow. Of course, other types of components also have resistance, but the resistor is designed specifically to introduce a desired amount of opposition into a circuit.

Wire-Wound Resistors

You have seen that copper has a resistivity of about 10 ohms per mil-foot. If you wrap a one foot length of 1 mil diameter copper wire on an insulated form and attach leads, as shown in Figure 4-1A, this gives you a 10 ohm wire-wound resistor. The process for producing practical wire-wound resistors is a little more involved, but the idea is the same.

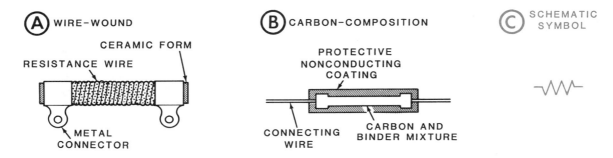

Figure 4-1   Types of resistors and their schematic symbols.

The resistance wire used is generally made from a nickel-chromium alloy, called nichrome, which is about 60 times as resistive as copper. The wire is often wound on a ceramic tube. After the leads are attached, the form and wire are covered with a hard protective coating to protect you from electrical shock.

This type of resistor is often used in high current circuits where relatively high amounts of power must be dissipated. The resistance range can vary from less than an ohm to several thousand ohms. The wire-wound technique is also used to produce precise value resistors. Such precision values are often required in meter circuits.

RESISTANCE

Carbon-Composition Resistors

Carbon, as you may remember, has only four valence electrons, which makes it a semiconductor. This makes carbon ideal as a material for resistors. By combining carbon granules with a powdered insulating material in various proportions, you can produce a wide range of resistors.

Figure 4-1B shows the construction of this type of resistor. Granules of carbon, insulating material, and a binder material are mixed together and shaped into a rod. Wire leads are inserted and the package is sealed with a nonconductive coating.

Carbon-composition resistors are inexpensive and, at one time, were the most common type used in electronics. Generally, they are used in low current circuits where they are not subjected to great amounts of heat. Values can vary from 10 ohms, or less, to 20 megohms, or more.

Deposited-Film Resistors

Presently, the most popular type of resistor is the film-type resistor. The construction of a film-type resistor is shown in Figure 4-2. In these devices, a resistance film is deposited on a nonconductive rod. Then the value of resistance is set by cutting a spiral grove through the film. This changes the appearance of the film to that of a long, flat ribbon spiraled around the rod. The groove sets the length and width of the ribbon so that the desired resistance value is achieved. Several metal-film types are available. One uses a nickel-chromium (nichrome[1]) film on an aluminum oxide rod, while another uses a tin-oxide film on a glass rod. Precision resistors can be created using carbon-film.

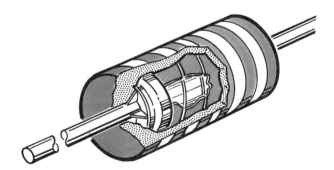

Figure 4-2 Deposited film resistor.

1-Trade name of Driver-Harris Co.

Wire-wound, carbon-composition, and deposited-film resistors are examples of *fixed* resistors. This means that their resistance value cannot be adjusted. Their value is determined when they are manufactured. The schematic symbol for a fixed resistor is shown in Figure 4-1C.

Resistor Ratings

Resistors have three very important ratings: resistance (in ohms), tolerance (in percent), and wattage (in watts). If you know what to look for, these ratings can usually be determined by examining the resistor. The following explains each of these ratings in more detail.

Resistance

You have already learned about resistance in some detail. You know that resistance is determined by the length, the cross-sectional area, and the resistivity of a material.

As an electronics technician, there will be times when you need to know the resistance of a particular resistor. There are three methods for determining this value. The first method is to look at the schematic diagram for the particular circuit you are examining. Most schematic diagrams will tell you the resistance of all the resistors in the circuit. The second method for determining resistance is to measure its value with a piece of test equipment called an **ohmmeter**. The next chapter covers this subject in greater detail.

The third method for determining the resistance, however, is to look at the resistor itself. With wire wound resistors, the value is normally written somewhere on the resistor. However, with carbon-composition and film resistors, the value is usually indicated by the color bands around the body of the resistor.

Figure 4-3 shows the color code normally used on resistors. As you can see, there are four bands on the resistor. The code is read starting with the band closest to the end of the resistor; band 1. The color of band 1 corresponds with a number from 0 to 9 as indicated in column 1 of Figure 4-3. This is the first digit of the resistor's nominal value.

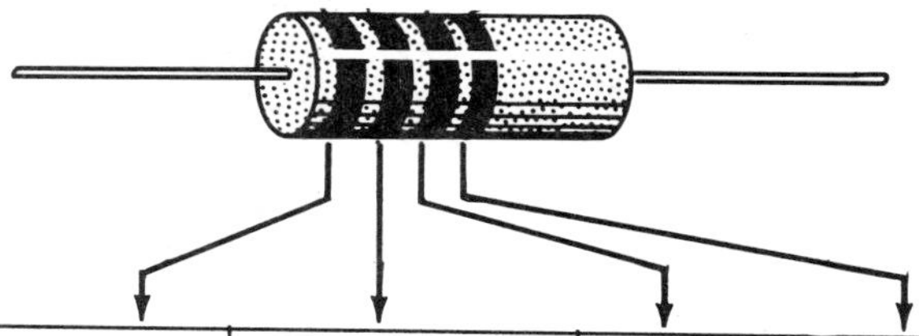

RESISTANCE

COLOR	BAND 1 FIRST NUMBER	BAND 2 SECOND NUMBER	BAND 3 MULTIPLIER	BAND 4 TOLERANCE
BLACK	0	0	1	
BROWN	1	1	10	
RED	2	2	100	
ORANGE	3	3	1,000	
YELLOW	4	4	10,000	
GREEN	5	5	100,000	
BLUE	6	6	1,000,000	
VIOLET	7	7	10,000,000	
GRAY	8	8	100,000,000	
WHITE	9	9	1,000,000,000	
GOLD			0.1	± 5%
SILVER			0.01	± 10%
NO BAND				± 20%

Figure 4-3 Color code for resistors.

Band 2, like Band 1, represents a number from 0 to 9. Band 2 represents the second digit in the resistor's nominal value. Thus, if the colors of the first two bands are green-blue, the first two numbers of the resistor's value are 56.

The third band on the resistor is the multiplier band. The number indicated in the first two bands is multiplied by the corresponding value given in column 3 of Figure 4-3. For example, a green-blue-orange color code is the number 56 multiplied by 1000, or 56,000. This means that the resistor has a nominal value of 56,000 ohms, or 56,000 Ω, or 56 k Ω.

The fourth band shown in Figure 4-3 is the tolerance band. We'll describe tolerance in a moment. Before that, we should point out that some resistors have five color bands. This coding system is often used for precision resistors, where an exact resistance must be known. With this coding system, the first three bands represent numbers and the fourth band is the multiplier band. The fifth band then becomes the tolerance band.

Tolerance

The actual resistance a resistor offers in a circuit is rarely the exact resistance indicated by the color code. The first three color bands indicate the initial value, and the fourth band represents the tolerance rating. This rating indicates the maximum and minimum values a resistor may be and still function properly. For example, a 1000 ohm resistor may have a tolerance of 10 percent. This means that the actual value of the resistor may be 10 percent above, or ten percent below 1000 ohms. Since ten percent of 1000 is 100, the actual value of the resistor can be anywhere from 900 ohms (1000 − 100) to 1100 ohms (1000 + 100).

Nowadays, most resistors have a tolerance of 5 percent, and 10 percent tolerance is a little less common. When a very precise resistance is required, precision resistors can be used, whose tolerance is 1 percent. This tolerance value is normally stamped on the body of the resistor.

Wattage

Wattage rating refers to the maximum amount of power or heat that the resistor can dissipate without burning up or significantly changing value. As you may recall, the load in an electrical circuit consumes the energy the electrons receive from the voltage source. If a resistor is the load, it dissipates this power in the form of heat. A larger resistor has more surface area than a smaller resistor, and thus can dissipate more heat. Thus, the larger the resistor, the higher the wattage rating. It is important that you know the wattage rating of a resistor, so the proper size can be used in a circuit with a particular current value.

Carbon composition resistors generally have fairly low wattage ratings. Ratings of 2 watts, 1 watt, 1/2 watt, and 1/4 watt are the most common. Wire-wound resistors can have much higher wattage ratings. A rating as high as 250 watts is not too uncommon for a wire-wound resistor.

Variable Resistors

The volume control on your TV receiver or radio is an example of a variable resistor. Variable resistors are resistors whose values can be changed in one way or another. As with fixed resistors, variable resistors can be constructed of many different materials. The schematic symbol for a variable resistor is shown in Figure 4-4A.

RESISTANCE

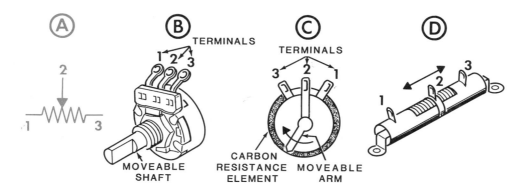

Figure 4-4 Variable resistors.

Figure 4-4B illustrates the construction of a variable carbon resistor. Figure 4-4C shows the rear-view of the inside of this device. A flat, circular strip of carbon is mounted between the two end terminals. A contact that moves along the resistance element connects to the center terminal. This contact is attached to a movable shaft. If the shaft is moved in a clockwise direction as shown by the arrow, the resistance between terminals 1 and 2 increases, while at the same time, the resistance between terminals 2 and 3 decreases. This is because the length of the resistive material between terminals 1 and 2 increases, while the length between terminals 2 and 3 decreases. This type of variable resistor is called a potentiometer, or simply a pot. A potentiometer has three terminals.

A variation of the potentiometer is called the rheostat. A rheostat is constructed in the same manner as a potentiometer, with the exception that only two terminals are used; one fixed and one adjustable. The resistance is adjusted the same way as with a pot, but without this third terminal, the possible uses of a rheostat are greatly reduced.

Wire-wound potentiometers are also common. Many have the same outward appearance as the potentiometer shown in Figure 4-4B. However, the internal construction is slightly different. Wire is wound around a circular-shaped insulating core, and a contact arm moves along the bare wire. Wire-wound potentiometers are often found in circuits where large amounts of current could damage a carbon-element pot.

Another type of variable resistor is shown in Figure 4-4D. This type is sometimes called a sliding contact resistor. It is used in high-power applications where the resistance value must be initially set or occasionally reset. The resistance value is changed by moving the sliding contact along the bare resistance wire.

Connecting Resistors

Rarely is only one resistor used as a load in an electrical circuit. Resistors are placed in different arrangements to perform different functions. These different arrangements—series, parallel, and series-parallel—determine the path for current flow in the circuit.

When more than one resistor is used, you have to perform one or more calculations to determine the total resistance for a circuit. Each arrangement has a unique method for calculating total resistance. The total amount of current flow in the circuit depends not only on the value of the voltage source, but also the amount of opposition to current, or resistance.

Series Connections

We mentioned earlier that a series circuit has only one path for current flow. An example of this is shown in Figure 4-5A. Notice that if an electron left the negative terminal of the battery, it would have to pass through R_3, R_2, and R_1, in sequence, before it could reach the positive terminal.

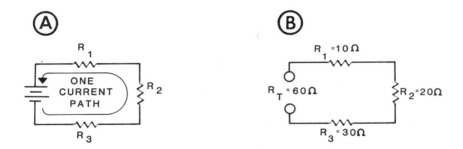

Figure 4-5 Resistors in series.

In Figure 4-5B, the values for these resistors are given. We can figure the total resistance our electron encountered. Keep in mind that this electron *must* pass through each resistor. When it passed through R_3, it felt 30 ohms of resistance. It then passed through R_2, which has a value of 20 ohms. Our electron then traveled through the 10 ohms of R_1.

The total resistance our electron felt, therefore, is the sum of all the individual resistor values. This is true for any series circuit. **The total resistance for a series circuit is the sum of the individual resistances.** This can be stated in the following equation:

$$R_T = R_1 + R_2 + R_3, \text{etc.}$$

In this equation, R_T stands for "total resistance."

Returning to Figure 4-5B, we can determine the total resistance by adding 30 ohms, 20 ohms, and 10 ohms, for a total of 60 ohms.

Parallel Connections

Parallel circuits behave quite differently from series circuits. An example of a parallel circuit is shown in Figure 4-6. A parallel circuit has more than one path for current flow. If an electron left the negative terminal, it could travel through R_1 or R_2.

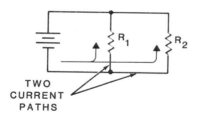

Figure 4-6 Resistors in parallel.

How does having more than one path for current affect total resistance? When we dealt with series circuits, total resistance could be found by adding the individual resistances. This was because current had to travel through every resistor in the circuit. With a parallel circuit, however, this is not true. An electron leaving the negative terminal of the battery could flow through R_1 without feeling the resistance offered by R_2. By the same token, an electron could pass through R_2 without passing through R_1.

Let's assume you have a battery and a lamp. When the lamp is connected to the battery, current flows through the lamp, and the lamp lights. Now let's add a second lamp to the existing circuit by placing it across the same terminals as the first battery. Now you have two paths for current flow. Current flows through both lamps, and both lamps light.

DC ELECTRONICS

When we added the second lamp, the total amount of current flow from the battery must have doubled. If the current from the battery doubles, the total opposition to current flow, or resistance, in the circuit must have decreased to half its former value. The total current in any parallel circuit is higher than the current for any individual path. Thus, we can draw an interesting conclusion about total resistance in parallel. **The total resistance in a parallel circuit must be less than any individual resistor in the circuit.**

Value Over Number

Because resistance in parallel is no longer the sum of the individual resistors, the methods for calculating total resistance in parallel are rather unique. The first method can only be used when all the resistors in the circuit are of equal value. This equation is known as "value over number," and is stated as follows:

$$R_T = \frac{\text{VALUE OF ONE RESISTOR}}{\text{TOTAL \# OF RESISTORS}}$$

Stated another way, the total resistance for a parallel circuit with equal value resistors is computed by dividing the value of one resistor by the number of resistors in the circuit.

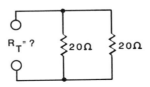

Figure 4-7 The value over number method.

As an example, consider the circuit shown in Figure 4-7. Since each resistor has a value of 20 ohms, the total resistance would be:

$$R_T = \frac{20}{2} = 10\ \Omega$$

Notice that the total resistance is less than the individual resistance values.

RESISTANCE

Product Over Sum

Obviously, there will be parallel circuits without equal value resistors. There are two methods for determining total resistance for these circuits.

The first method can be used if there are just two unequal resistors in a parallel circuit. This is called the "product over sum" method, and is stated as follows:

$$R_T = \frac{R_1 \times R_2}{R_1 + R_2}$$

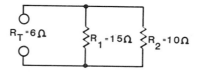

Figure 4-8 The product over sum method.

Figure 4-8 shows an example of a parallel circuit with two unequal resistors. Total resistance for this circuit would be calculated as follows:

$$R_T = \frac{15 \, \Omega \times 10 \, \Omega}{15 \, \Omega + 10 \, \Omega} = \frac{150}{25} = 6 \, \Omega$$

This means that the parallel circuit shown in Figure 4-8 has the same effect on current flow as a single 6 ohm resistor. This concept is very important when working with series-parallel circuits. Notice also that the total resistance for this circuit is less than any individual current path. Another term used to describe a current path in a parallel circuit is a **branch**.

Reciprocal Method

Obviously, there will be parallel circuits with more than two unequal resistors. Thus, we need an equation to determine total resistance for these circuits. This is known as the reciprocal method, and is stated as follows:

$$R_T = \frac{1}{\dfrac{1}{R_1} + \dfrac{1}{R_2} + \dfrac{1}{R_3}, \text{etc.}}$$

Actually, the value/number and product/sum equations are condensed versions of this equation.

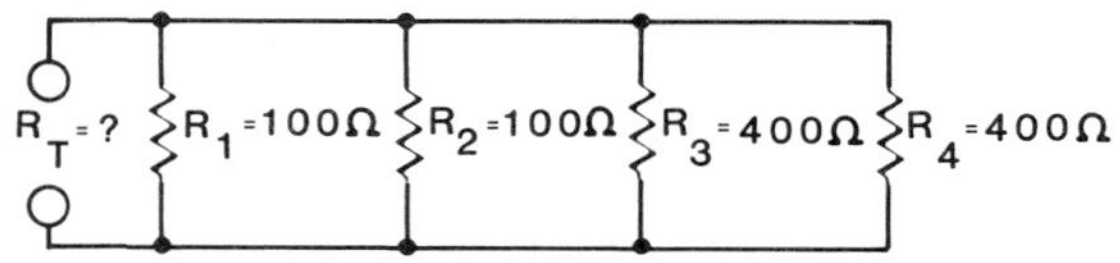

Figure 4-9 The reciprocal method.

Although this equation may look rather intimidating, it actually becomes quite easy when taken step by step. We'll use the circuit shown in Figure 4-9. First, insert the resistor values into the formula:

$$R_T = \frac{1}{\dfrac{1}{100\,\Omega} + \dfrac{1}{100\,\Omega} + \dfrac{1}{400\,\Omega} + \dfrac{1}{400\,\Omega}}$$

Next, find the decimal equivalent for each fraction located in the bottom of the formula. This is done by dividing each resistor value into 1:

$$R_T = \frac{1}{0.01 + 0.01 + 0.0025 + 0.0025}$$

From there, add the decimal values, and divide the sum into 1 again, for the final result:

$$R_T = \frac{1}{.025} = 40\,\Omega$$

Notice that, once again, total resistance is less than the value of any one resistor in the circuit.

This equation can be used for any parallel circuit, regardless of the amount or value of the resistors. The other equations, value/number and product/sum, are used in specific circumstances because they are easier.

Series-Parallel Connections

In many circuits, a parallel circuit is connected to series resistors in addition to the voltage source. The result is called a **series-parallel** circuit, because it contains both series and parallel resistors.

Because the series-parallel circuit contains both series and parallel components, finding total resistance for the circuit is done by combining both the series and the parallel methods. First, the resistance of the parallel portion is found. As stated earlier, the total resistance for a parallel circuit will have the same effect on current as a series resistor of the same value. Thus, we can then add the total resistance of the parallel section to any series resistance, allowing us to find the total resistance of the series-parallel circuit.

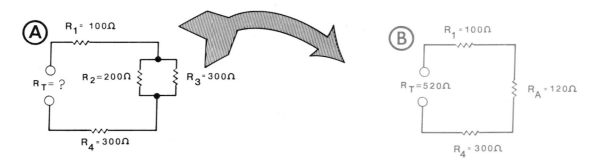

Figure 4-10 The series-parallel circuit.

We'll use the circuit in Figure 4-10 for an example. In this circuit, a parallel portion is connected in series with two series resistors. The first step in determining total resistance for this circuit is to find the equivalent resistance of the parallel portion. We'll call this equivalent resistance R_A. Since this parallel portion has two unequal resistors, we can use the product/sum method to find its equivalent resistance:

$$R_A = \frac{200\ \Omega \times 300\ \Omega}{200\ \Omega + 300\ \Omega} = \frac{60000}{500} = 120\ \Omega$$

Now we can add this equivalent resistance to the series resistors, and find the total resistance of the circuit:

$$R_T = 120\ \Omega + 300\ \Omega + 100\ \Omega = 520\ \Omega$$

This was an example of a simple series-parallel circuit. However, there are more complex series-parallel circuits in the field of electronics, and total resistance may have to be found for these.

A good example of a complex series-parallel circuit is shown in Figure 4-11. This is a variation of our original circuit. A fifth resistor, R_5, has been placed in series with R_2, in the left parallel branch.

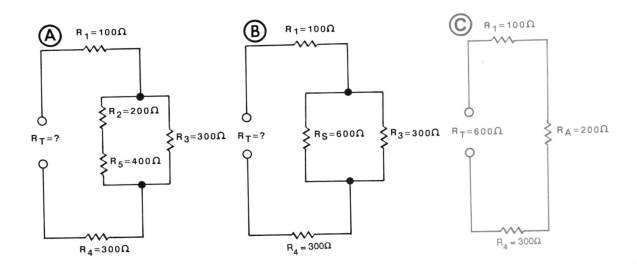

Figure 4-11 Complex series-parallel circuit.

The method for finding total resistance for this circuit is no different from the previous method. However, before we can find the equivalent resistance of the parallel portion, we must first find the resistance of the left branch. Since these two resistors are in series, their total resistance is found by adding them together. We'll call this resistance R_S:

$$R_S = 200\ \Omega + 400\ \Omega = 600\ \Omega$$

This is the total resistance of the left branch. Now, we can take this resistance, and combine it with the resistance of the right branch. This will give us the equivalent parallel resistance. Since there are two branches of unequal value, we can use the product/sum method:

$$R_A = \frac{600\ \Omega \times 300\ \Omega}{600\ \Omega + 300\ \Omega} = \frac{1800}{900} = 200\ \Omega$$

This equivalent resistance can then be added to the series resistors, finding the total resistance of the circuit:

$$R_T = 200\ \Omega + 300\ \Omega + 100\ \Omega = 600\ \Omega$$

Any calculations for total resistance in a series-parallel branch are accomplished using these methods.

DC ELECTRONICS

Chapter Self-Test

1. Define resistance.

2. What is the resistance of a material that has 1 amp of current with 1 volt of EMF placed across it?

3. What category is used for materials with low resistivity?

4. Define conductance.

5. Which of the factors that determine resistance are *directly* proportional to resistance?

6. Which factors are *inversely* proportional?

7. Describe the effect temperature has on the resistance of a material with a *positive* temperature coefficient.

8. What type of resistor is specifically designed to work with temperature?

9. Why do resistors have a tolerance rating?

10. Why does the physical size of a resistor assist in determining the wattage rating?

11. What is the formula for finding total resistance in a series circuit?

12. Name the three methods for finding total resistance in a parallel circuit.

13. What is the total resistance of a circuit containing a parallel portion, consisting of two 30 ohm resistors, in series with a 30 ohm resistor?

RESISTANCE

Summary

Resistance is the opposition to current flow. All materials offer some opposition to current flow although some materials are more resistive than others.

The unit of measurement of resistance is the ohm. The ohm can be expressed in terms of voltage and current. It is equal to the amount of resistance which will allow one ampere of current to flow when an EMF of 1 volt is applied. The Greek letter Omega (Ω) is the symbol used to represent the ohm.

Resistivity is defined as the resistance of a one mil-foot length of a material at 20 degrees centigrade. A mil-foot is a one foot length of wire which is one thousandth of an inch in diameter. The resistivity of a material determines if that material is an insulator or a conductor.

Most metals have a very low resistivity, and are therefore good conductors. Some examples are silver, copper, aluminum, nickel, iron, lead, and gold. The resistivity of other material is many times higher than those of metals. These materials are good insulators. Examples are glass, rubber, mica, and plastics.

The resistance of a material is determined not only by its resistivity but also by its size and shape. The resistance of a conductor is directly proportional to the length of the conductor and inversely proportional to its cross sectional area.

Temperature also affects resistance to some extent. If the resistance of a material increases with an increase in temperature, the substance is said to have a positive temperature coefficient. If the resistance decreases with an increase in temperature, the substance is said to have a negative temperature coefficient.

The thermistor is a special type of resistor whose resistance value changes with temperature.

There are three popular types of fixed resistors. These are: carbon-composition, wire-wound, and deposited-film. Wire-wound resistors generally have relatively low resistance values, but they can have high power ratings. Deposited-film resistors are the most common type of resistors in use today. They can be made more precise than composition resistors and they are cheaper than wire-wound resistors.

The resistance, tolerance, and wattage of most resistors can be determined by physical examination of the resistor. With many wire-wound resistors, this information is written on the body of the resistor. With the composition and film resistors, the resistance and tolerance are normally indicated by color bands. The wattage rating is indicated by the physical size of the resistor.

DC ELECTRONICS

Not all resistors have a fixed value; some are variable. A resistor whose value can be changed is called a potentiometer if it has three terminals, or a rheostat if it has only two terminals.

Resistors can be connected in series, in parallel, or in series-parallel combinations.

When connected in series, their resistance values add. Thus, if two 1000 ohm resistors are connected in series, the total resistance is 2000 ohms.

When resistors are connected in parallel, the total resistance decreases. This occurs because the resistors connected in parallel offer additional current paths in the circuit. Since more total current flows, the effect on the circuit is a decrease in total resistance. The total resistance of any parallel circuit is less than the value of any single resistor in that circuit.

When equal resistors are in parallel, their total resistance is found by dividing the value of one resistor by the total number of equal resistors in the circuit.

If two unequal resistors are in a parallel circuit, their resistance is found by the product/sum method. The equation for this method is as follows:

$$R_T = \frac{R_1 \times R_2}{R_1 + R_2}$$

For three or more unequal resistors, the reciprocal method must be used to find total resistance. The equation for the reciprocal method is:

$$R_T = \frac{1}{\frac{1}{R_1} + \frac{1}{R_2} + \frac{1}{R_3}, \text{etc.}}$$

To find total resistance in a series-parallel circuit, the parallel portion is combined to a single equivalent resistance using the applicable parallel equation, and this equivalent resistance is then added to any series resistor.

CHAPTER 5

Electrical Measurements

DC ELECTRONICS

Contents

Introduction 91

Chapter Objectives 92

Measuring Current. 93

Measuring Voltage 98

Measuring Resistance. 101

Chapter Self-Test 103

Summary. 104

ELECTRICAL MEASUREMENTS

Introduction

As you enter the field of electronics, you will discover that analyzing a circuit on paper occurs very rarely. More often you will have to measure the values for current, voltage, and resistance using test equipment. The values you measure allow you to accurately determine what is happening in the circuit.

In this chapter, you will be introduced to the ammeter, the voltmeter, and the ohmmeter. These meters, although they perform a similar function, are connected to an electrical circuit in different ways.

DC ELECTRONICS

Chapter Objectives

When you have completed this chapter, you will be able to:

1. Define the following terms: coulomb, scientific notation, ampere, ammeter, voltmeter, ohmmeter, short, and open.

2. State the reasons why scientific notation is used.

3. Describe the correct method for using an ammeter.

4. Describe the correct method for using a voltmeter.

5. Describe the correct method for using an ohmmeter.

6. Describe the effects that shorts and opens have on circuit operation.

ELECTRICAL MEASUREMENTS

Measuring Current

Current is the flow of electrons from a negative to a positive charge. To measure current flow, you must measure the number of electrons flowing past a point in a specific length of time. Before you learn how current is measured, you must first learn how the unit of electrical charge and the unit of current are defined.

The Coulomb

You have seen that the charge on an object is determined by the number of electrons that the object loses or gains. If the object loses electrons, the charge is positive. However, an object that gains electrons has a negative charge. In order to measure the magnitude of the charge, you must have a unit of measure.

The unit of measure for electrical charge is called the coulomb. The coulomb is equal to a charge of 6.25×10^{18} electrons. For those who are not familiar with expressing numbers in this way, the number is:

$$6,250,000,000,000,000,000.$$

An object that has gained 6.25×10^{18} electrons has a negative charge of one coulomb. On the other hand, an object that has given up 6.25×10^{18} electrons has a positive charge of one coulomb.

Powers of Ten and Scientific Notation

A word about powers of ten and scientific notation may be helpful at this point. The number 6,250,000,000,000,000,000 can be expressed as 6.25×10^{18}. This number is read "six point two five times ten to the eighteenth power." The expression "ten to the eighteenth power" means that the decimal place in 6.25 must be moved 18 places to the right in order to convert to a whole number. This number is expressed in powers of ten, however, because it is easier to write and remember 6.25×10^{18} than it is to write and remember 6,250,000,000,000,000,000.

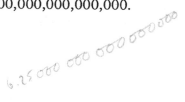

DC ELECTRONICS

This shorthand method of expressing numbers is known as **powers of ten**, or **scientific notation**. It is often used in electronics to express very large and very small numbers. Very small numbers are expressed by using negative powers of ten. For example, 3.2×10^{-8} is the scientific notation for 0.000000032. Here, "ten to the minus eight power" means "move the decimal place in 3.2 eight places to the left." More examples of scientific notation are shown below:

Positive Powers of Ten

$7.9 \times 10^4 = 79,000$
$9.1 \times 10^8 = 910,000,000$
$1.0 \times 10^{12} = 1,000,000,000,000$

Negative Powers of Ten

$7.9 \times 10^{-4} = 0.00079$
$9.1 \times 10^{-8} = 0.000\ 000\ 091$
$1.0 \times 10^{-12} = 0.000\ 000\ 000\ 001$

If the concept of scientific notation is still a little confusing, Appendix A, located at the end of this book, is devoted entirely to scientific notation and powers of ten.

The Ampere

The unit of measurement for current is the **ampere**. The ampere indicates the rate at which electrons move past a given point. As previously mentioned, 1 coulomb is equal to 6.25×10^{18} electrons. An ampere is equal to **1 coulomb per second**. That is, if 1 coulomb of electrons flows past a given point in one second, then the current past that point is 1 ampere. Coulombs indicate numbers of electrons; amperes indicate the number of electrons per second, or the rate of electron flow.

When 6.25×10^{18} electrons flow through a wire each second, the current flow is 1 ampere. If twice this number of electrons flows through a conductor each second, the current is 2 amperes. This relationship is expressed by the equation:

$$\text{amperes} = \frac{\text{coulombs}}{\text{seconds}}$$

If 10 coulombs flow past a point in two seconds, then the current flow is 5 amperes.

ELECTRICAL MEASUREMENTS

The term ampere is often shortened to **amp** and is normally abbreviated **A**. Many times the ampere is too large a unit to use to conveniently specify a rate of current flow. In these cases, metric prefixes are used to denote smaller units. The milliampere (mA) is one thousandth (0.001) of an ampere. The microampere (μA) is one millionth (0.000001) of an ampere. In other words, there are 1000 milliamperes or 1,000,000 microamperes in an ampere.

The Ammeter

The **ammeter** is a device for measuring current flow. "Ammeter" is actually "ampere meter" shortened. A typical ammeter is shown in Figure 5-1. It has a pointer that moves in front of a calibrated scale, and two terminals which are connected to the circuit being measured. The scale shown is calibrated in amperes, and with the pointer just past 6, the ammeter is measuring about 6.5 amperes. Many times, however, the scale is in milliamperes or microamperes. If the scale in Figure 5-1 were milliamperes, then the ammeter would be reading just over 6.5 milliamperes.

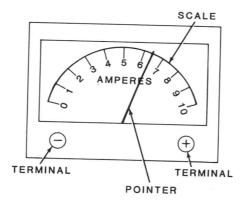

Figure 5-1 The ammeter.

Figure 5-2A shows a circuit in which an unknown amount of current is flowing. To measure the current in this circuit, the ammeter must be placed in the circuit. This is shown in Figure 5-2B. Notice that the schematic symbol for the ammeter is a circle with the letter A.

DC ELECTRONICS

Figure 5-2 Measuring current.

The ammeter measures current in an electrical circuit by having the current flow *through the meter*. The positive and negative terminals of the meter are connected so that circuit current enters the meter from the negative terminal, and exits the meter through the positive terminal. Notice that in order to accomplish this, the meter is placed in the circuit itself. The current in the circuit, in addition to traveling its normal route, simply flows through the meter as well. The ammeter is designed in such a way that when it is placed in a circuit, it has very little effect on the characteristics of the circuit. Another way to describe this method of measuring current is **placing the ammeter in series with the circuit**.

The ammeter is a very delicate instrument. Each meter is specifically designed to measure a maximum amount of current, and if that maximum is greatly exceeded, the meter may be permanently damaged. Ammeters that measure small amounts of current--anywhere in the milliampere or microampere range--are considered to be more sensitive than meters which measure amperes.

Because of the delicate nature of the ammeter, the method in which it is placed in a circuit must be very specific. This step-by-step method also protects you from possible electrical shock from the circuit. First, select a meter that can safely measure the approximate amount of current in the circuit. If you are working with an ammeter that can select different ranges, then select the range that will provide a half-scale reading. Sometimes, however, the current for a circuit is completely unknown. If this is the case, you would select the highest possible range (or meter), and work your way down until you have an accurate measurement.

Once you have selected the best meter or range, you are ready to make the actual measurement. Before you can place the ammeter in the circuit, however, you want to ensure there is no current flow. This is accomplished by removing the voltage source from the circuit. In battery powered circuits, this is done by removing the battery or by disconnecting one of the battery leads. The purpose of this step is to protect yourself from electrical shock as you connect the ammeter.

ELECTRICAL MEASUREMENTS

Once power has been removed, select the point in the circuit where the ammeter will be placed, and break the circuit at that point. The circuit must be broken in order to connect the ammeter in series with the rest of the circuit.

Fourth, the ammeter is connected to the circuit while observing polarity. The wire from the negative terminal of the battery must lead to the negative terminal of the ammeter. If the ammeter is connected backwards, the pointer will attempt to deflect backwards, which damages the meter.

Observing polarity simply means that the negative terminal of the ammeter is connected to the conductor that leads to the negative terminal of the battery. Naturally, the positive terminal of the ammeter is connected to the conductor that leads to the positive side of the battery.

Finally, power is reapplied to the circuit and a current reading is obtained from the ammeter scale. Figure 5-3 illustrates this step-by-step procedure.

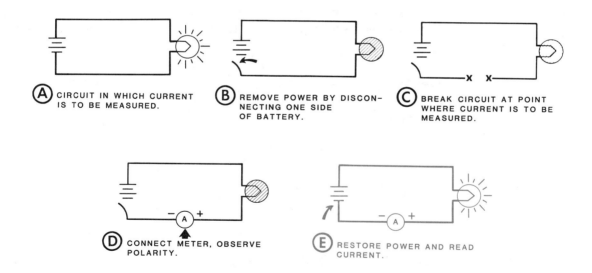

Figure 5-3 Procedure for measuring current.

Measuring Voltage

The device most often used for measuring voltage is the voltmeter. There are many different types of voltmeters in use today. Some use a mechanical meter movement like the ammeter previously described. Another type, called the digital voltmeter, is extremely popular, and uses a numeric display. This type of meter is easier to read, and is usually more accurate. When an analog meter is used, you must be careful to keep your eyes directly in front of the needle pointers. If you don't you may read a value that is off-center from the scale. This is known as operator interpolation. Many technicians feel that a numeric display is more accurate than a needle, because there is no operator interpolation.

To measure current, you had to connect the ammeter in series with the circuit being tested. This connection is not necessary with the voltmeter. The voltmeter is connected across the component. Another way of saying across is to say "parallel with". Although this will add a path for current that previously did not exist, it will not affect the circuit.

If you recall, voltage is the potential difference between two points. When you use a voltmeter, it is connected across the two points you wish to compare. You may also recall that to achieve a voltage drop, you must have current flow in the circuit. Therefore, the voltmeter is connected in parallel with the circuit, so as not to affect circuit current.

It is important to remember that when using either the ammeter or voltmeter *proper polarity must be observed.* You should also select a range that is capable of measuring more than the expected value.

REMEMBER

Ammeters are connected between (in series with) circuit components. Voltmeters are connected across (in parallel with) a circuit component.

Many digital meters have additional built-in safety features, such as auto-polarity and auto-ranging. The auto-polarity feature allows the meter to read either a positive or negative value and displays a "−" sign in front of the value if polarity is not properly observed. The auto-ranging meter automatically changes ranges or simply indicates an overflow. In either case, the meter is not damaged by reverse polarity or reasonable overload conditions.

ELECTRICAL MEASUREMENTS

CAUTION

Some voltmeters have their reference leads earth grounded. This type of voltmeter can <u>only</u> be used to measure voltages referenced to ground. This is because wherever you connect the reference lead you are inserting earth ground.

The schematic symbol for the voltmeter is shown in Figure 5-4A. Notice that one of the leads is marked negative, while the other is marked positive. As with the ammeter, polarity must be observed when using the voltmeter. This means that the negative lead must be attached to the more negative of the two points across which the voltage is to be measured.

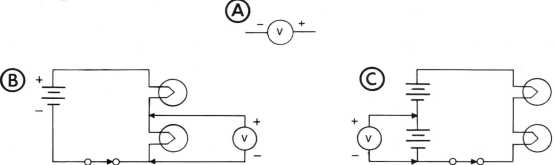

Figure 5-4 Connecting the voltmeter.

Fortunately, the voltmeter is much easier to use than the ammeter. When using the voltmeter, the circuit under test need not be broken nor disturbed in any way. To measure the voltage between two points, simply touch the two leads of the voltmeter to the two points. However, remember that you must observe polarity.

Figure 5-4B shows how the voltmeter is connected to measure the voltage drop across a lamp. Notice that the negative lead is connected to the more negative point. Also note that the meter is connected directly across the bulb. In this case, the meter is measuring a voltage drop. If the switch is opened so that current flow stops, the voltage drop disappears and the meter reading falls to zero.

Figure 5-4C shows a different circuit with the meter connected to measure the voltage rise of the bottom battery. Once again, polarity is observed and the meter is connected directly across the component. Here, the meter is used to measure a voltage rise. However, the voltmeter cannot distinguish between a voltage rise and a voltage drop. Also, since a voltage rise does not need current flow, this voltage could have been measured with the switch open.

As with the ammeter, you should always be certain the voltage you are going to measure does not exceed the range of the voltmeter. Again, this may be difficult if the voltage is unknown. If such is the case, you should always start at the highest range, and work your way down, until an accurate measurement is obtained. **Never attempt to measure a voltage higher than the voltmeter can handle, as this may cause permanent damage to the voltmeter.**

For your own safety there are other precautions that you must observe. For instance, when using test leads, hold them only by the insulated portions. Otherwise, you may receive an electrical shock.

As we stated earlier, voltages are very often measured with respect to ground. Therefore, when working on electronic equipment, it is a good idea to connect one lead to ground and leave it there. This way only one hand is required to make voltage measurements, and the other hand may be held away from the equipment. This greatly reduces your chance of receiving an electrical shock.

ELECTRICAL MEASUREMENTS

Measuring Resistance

The device used to measure resistance is called an ohmmeter. Ohmmeters come in both analog and digital styles. The analog ohmmeter uses the meter movement, and resistance is read from a scale on the meter's face; the digital ohmmeter uses a numeric display.

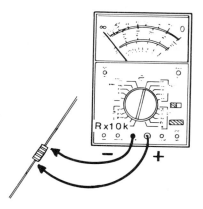

Figure 5-5 Using an ohmmeter.

Figure 5-5 shows how the ohmmeter is connected to a component. Notice that the leads are simply placed on either side of the component, similar to a voltmeter. However, the component being tested is removed from the circuit. Whenever possible, the component being tested should be removed. If this cannot be done, then *ensure that all power to the circuit is disconnected before connecting an ohmmeter*. An ohmmeter measures the resistance of a component or conductor by applying a small potential difference to either side of the component or conductor. This small voltage forces current through the component being tested, and that current is measured inside the ohmmeter to determine the resistance. If the component being tested is still connected to the circuit, inaccurate readings may result. First, the current from the ohmmeter may find a path other than through the component you wish to test; second, if the component is in the circuit and power to the circuit is applied, circuit current will be flowing through the component as well as the current from the meter. There is also the possibility of electrical shock from circuit current when you connect the ohmmeter, and if circuit current is very high, it may damage the ohmmeter.

When you are measuring an unknown resistance, you should always start at the *lowest range* on the ohmmeter. This is done for two reasons: First, most analog ohmmeters have full-scale deflection at 0 ohms. Thus, if you start at the lowest range, your initial deflection may be very slight. Second, the voltage applied from the ohmmeter increases as you increase ranges. Starting at a high range may apply a voltage to the component that could damage it.

It is possible to check components other than resistors with an ohmmeter. In fact, an ohmmeter can be used to test almost any component for resistance, as well as shorts or opens. Short and open are electrical terms for certain malfunctions. A short exists when a component which normally offers resistance malfunctions and offers no resistance. A short also describes a condition where a conductive path is created, sending current where it is not desired. This normally results from poor soldering techniques.

An open is completely opposite from a short. An open occurs when there is no longer a current path through either a component or conductor. When the resistance across an open is measured, it indicates infinite resistance, or total opposition to current flow.

If you attempt to measure the resistance of a conductor, it is very difficult to obtain a resistance reading above zero. Thus, if you get an extremely high reading when checking the resistance of a conductor, this usually indicates an open somewhere between the two points you measured. That is, at some point the conductor has separated and you are actually measuring the resistance of the air in the space between the conductor ends. For all practical purposes, the air presents an infinite opposition to current flow in this example.

Basically, the ohmmeter is a self-powered meter that is connected across (in parallel with) the component to be measured. Like the ammeter and voltmeter, it has different ranges, allowing you to measure anywhere from a few ohms to the hundred megohm range. Unlike the voltmeter and ammeter, ohmmeters are not usually considered polarity-sensitive, since they are self-powered and their polarity is already set by their internal power source.

ELECTRICAL MEASUREMENTS

Chapter Self-Test

Determine whether the following statements are true or false.

1. There is only one scale available for an ammeter.

2. When connecting an ammeter, you leave the power on, and place each test lead on either side of where you want to measure.

3. When first measuring an unknown current, always start at the lowest range, and work your way up.

4. When using an ammeter, it is important to observe polarity.

5. A voltmeter must be placed *in series* with the component or circuit you wish to test.

6. When using a voltmeter, connect the negative test lead to the more negative point.

7. Circuit current does not need to flow when you are measuring a voltage rise.

8. In order to properly measure the resistance of a component, the component must be in the circuit, and circuit current must be flowing through it.

9. You do not have to observe polarity when using an ohmmeter.

10. When measuring an unknown resistance, always start at the lowest range possible.

Summary

The coulomb is the unit of electrical charge. One coulomb of charge is equal to 6.25×10^{18} electrons.

Scientific notation is used to express extremely small or extremely large numbers.

Scientific notation is accomplished by multiplying a number by a power of ten. For example, 8.36×10^3 is equal to 8.36×1000, or 8,360.

Negative powers of ten are used to express small numbers. For example, 5.47×10^{-6} is equal to $5.47 \times .000001$, or .00000547.

The unit of measurement for current is the ampere, or amp. One ampere is equal to one coulomb of charge passing a given point in one second.

The relationship between amperes, current, and seconds are expressed in the equation:
$$\text{amperes} = \frac{1}{\text{coulombs}}$$

The device used to measure current is the ammeter.

The ammeter is an extremely delicate piece of equipment, and care should be taken not to damage the ammeter when measuring current.

Current is measured by placing the ammeter in series with the current you are testing.

To protect the ammeter from damage during use, you should first ensure the ammeter (or range) you select can withstand the amount of current being measured.

If the current is unknown, begin at the highest range of the meter.

After the proper range has been selected, remove power from the circuit, break the circuit at the point you want to insert the meter, insert the meter, and reapply power.

When inserting the ammeter, be sure to observe polarity. Connect the negative terminal to the more negative point, and the positive terminal to the more positive point.

ELECTRICAL MEASUREMENTS

The voltmeter is a piece of electrical measurement equipment used to measure voltage.

Like the ammeter, proper polarity must be observed when connecting a voltmeter to an electrical circuit.

The voltmeter is connected across, or in parallel with, the component being tested. The circuit being tested does not need to be changed or altered in any way.

The voltmeter cannot distinguish between a voltage rise and a voltage drop.

Current need not be flowing through the circuit for the voltmeter to measure a voltage rise.

For the voltmeter to measure a voltage drop, current must be flowing through the component being tested. If there is no current flow, the voltmeter will indicate 0 volts.

To measure the resistance of any component or conductor, an ohmmeter is used.

An ohmmeter measures the resistance of a component or conductor by applying a small voltage across the component or conductor, which forces a small amount of current through the component or conductor.

Because of the method an ohmmeter uses to measure resistance, accurate measurements can only be obtained if the component or conductor being tested is either 1) removed from the circuit completely, or 2) no power is applied through the circuit being tested.

Polarity need not be observed when using an ohmmeter.

An ohmmeter may be used to check a circuit for malfunctions. The two most common malfunctions are shorts and opens.

A short is an electronic malfunction that exists when a component that normally offers resistance in a circuit offers no resistance (0 ohms), or if there is a path for current which did not exist previously.

An open is an electrical malfunction that occurs when any path for current (through a component or conductor) no longer exists, and the resistance across the open is infinite.

CHAPTER 6

Ohm's Law

Contents

Introduction 109

Chapter Objectives 110

Determining Current 111

Finding Voltage 116

Finding Resistance 119

Using Ohm's Law 121

Chapter Self-Test 122

Summary 123

Introduction

Ohm's Law is the most important and basic law of electricity and electronics. It defines the relationship between the three fundamental electrical quantities: current, voltage, and resistance. Fortunately, the relationship between these three quantities is quite simple. Several implications of this relationship have already been discussed. Therefore, some of the information presented here will not be entirely new to you.

DC ELECTRONICS

Chapter Objectives

When you have completed this chapter, you will be able to:

1. State Ohm's Law.

2. Select the proper equation and calculate the current in any simple circuit in which voltage and resistance are known.

3. Select the proper equation and calculate the voltage in any simple circuit in which current and resistance are known.

4. Select the proper equation and calculate the resistance in any simple circuit in which voltage and current are known.

5. State the three equations for Ohm's Law.

Determining Current

Basic Calculations

Ohm's Law states that **current is directly proportional to voltage and inversely proportional to resistance.** As you know, voltage is the force that causes current to flow. Therefore, the higher the voltage applied to a circuit, the higher the current through the circuit, and a decrease in the applied voltage will result in a decrease in circuit current. This assumes that the circuit resistance, or opposition to current flow, remains constant.

However, the amount of current through a circuit is also determined by the resistance of the circuit. As you know, resistance is the opposition to current flow. Assuming that voltage is constant, an increase in resistance results in a decrease in current flow. On the other hand, lowering the resistance causes an increase in current.

These relationships can be expressed in the following equation:

$$\text{current} = \frac{\text{voltage}}{\text{resistance}}$$

Or, stated in terms of the units of measurement:

$$\text{amperes} = \frac{\text{volts}}{\text{ohms}}$$

When used in equations, letters of the alphabet are often used to represent current, voltage, and resistance. Resistance is represented by the letter **R**. Voltage may be represented either by the letter **V** (for voltage) or the letter **E** (for EMF). In this book, the letter E is used to represent EMF or voltage. Current is represented by the letter **I**. If you substitute the letters I, E, and R for current, voltage, and resistance, respectively, our equation is now stated as:

$$I = \frac{E}{R}$$

This equation may be used to find current in any circuit in which the voltage and resistance are known.

DC ELECTRONICS

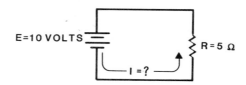

Figure 6-1 Current is determined by voltage and resistance.

Figure 6-1 shows a circuit in which the values of voltage and resistance are given. To determine the current, substitute the known values into the equation:

$$I = \frac{E}{R} = \frac{10V}{5\Omega} = 2A$$

Thus, a circuit with a voltage source of 10 volts and a total resistance of 5 ohms will have a total current of 2 amperes, or amps.

Figure 6-2A shows another circuit in which the quantities E and R are given. Solving for I, you find that the current is:

$$I = \frac{E}{R} = \frac{200V}{50\Omega} = 4A$$

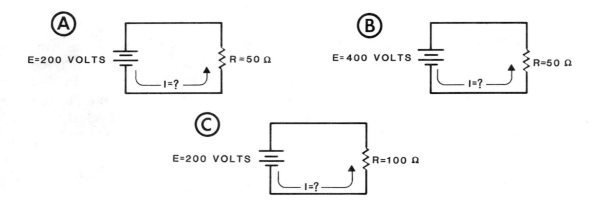

Figure 6-2 Current is directly proportional to voltage and inversely proportional to resistance.

OHM'S LAW

Notice what happens to the current if you double the applied voltage as shown in Figure 6-2B.

$$I = \frac{E}{R} = \frac{400V}{50\Omega} = 8A$$

Notice that when the voltage is doubled, the current also doubles. This is an example of the relationship between voltage and current in an electrical circuit. **Current and voltage are directly proportional.**

The relationship between current and resistance can be proven in the same manner. What would happen to the total current from our original circuit in Figure 6-2A, if we doubled the resistance in the circuit? The new circuit is shown in Figure 6-2C. Inserting our new values into the equation, we can find the value for current in this new circuit:

$$I = \frac{E}{R} = \frac{200V}{100\Omega} = 2A$$

Thus, when you double the resistance, the current is reduced to one-half its former value. If the resistance were decreased by half, the current would double. This shows the relationship between current and resistance. **Current and resistance are inversely proportional.**

Powers of Ten

Of the three electrical quantities—current, voltage, and resistance—current is the one quantity that is **dependent** upon the others. When electronic equipment is designed, the values for voltage and resistance can be selected and changed by the designer. The values he selects will determine the amount of current flow in the equipment. In order to have electronics perform work, the amount of current must be controlled. Although there are many circuits with high current, current is best controlled when it is a relatively small value. To achieve this small current (current in the milliamp or microamp range), high values of resistance, or small voltages, are required. It is quite common to see resistors in the kilohm and megohm range.

When high value resistors are used, the calculations for current must be slightly altered. The equation remains the same, but since we are working with kilo- and megohms, we need to find their equivalent for current.

For example, how much current flows through a 3 kilohm resistor when it is connected across a 9 volt battery? Initially, our only way to solve this is to convert 3 kilohms to 3000 ohms. Thus:

$$I = \frac{E}{R} = \frac{9V}{3000\,\Omega} = .003A$$

0.003A is equal to 3.00 x 0.001, and 0.001 is equal to 10^{-3}, if you recall our section on powers of ten. Combining, we can conclude that 0.003 amps is equal to 3.00 x 10^{-3} amps. Instead of writing 10^{-3}, however, we can use the metric prefix *milli*. So 0.003 amps is equal to 3.00 milliamps, or 3 milliamps. Anytime you divide kilohms into voltage, the answer is stated in milliamps.

For example, how much current flows when a lamp with a resistance of 2.4 kilohms is connected across a 120 volt line? The current in milliamperes is:

$$I = \frac{E}{R} = \frac{120V}{2.4K\Omega} = 50mA$$

You can check this by converting 2.4 kilohms to 2400 ohms and solving as you did earlier:

$$I = \frac{E}{R} = \frac{120V}{2400\,\Omega} = 0.05\ A$$

Remember that 0.05 amps is the same as 50 millamps.

The same idea can be applied to megohm resistors. What is the current when a 5 megohms resistor is connected across a 25 volt battery? You can convert 5 megohms to 5,000,000 ohms and solve for current in amperes:

$$I = \frac{E}{R} = \frac{25V}{5{,}000{,}000\,\Omega} = 0.000005\ A$$

0.000005 is equal to 5.0 x 0.000001, and 0.000001 is equal to 10^{-6}. The metric prefix for 10^{-6} is *micro*. We can conclude that 0.000005 amps is equal to 5 microamps. Thus, we have found our relationship for megohms: any time megohms is divided into volts, the answer will be in microamps.

Our original calculation could have been done this way:

$$I = \frac{E}{R} = \frac{25V}{5M\Omega} = 5\,\mu A$$

To summarize, the basic formula for current is:

$$I = \frac{E}{R}$$

If R is given in ohms, the equation becomes:

$$\text{amperes} = \frac{\text{volts}}{\text{ohms}}$$

However, if R is given in kilohms, the equation becomes:

$$\text{millamperes} = \frac{\text{volts}}{\text{kilohms}}$$

Finally, if R is given in megohms, we can think of the equation as:

$$\text{microamperes} = \frac{\text{volts}}{\text{megohms}}$$

Finding Voltage

Obviously, there will be times where the current is known, but voltage or resistance is unknown. We can alter our original current equation and come up with formulas we can use to solve for resistance and voltage. To alter our original formula, it must be **transposed**. Transposing simply means changing the equation from one form to another. We'll begin by transposing our formula to a formula for voltage. To do this, you must change the formula so that the value E is on one side of the equation by itself. This is easy to do if you remember a basic algebraic rule for transposing equations. The rule states that you can multiply or divide both sides of the equation by any quantity without changing the equality. The current equation is:

$$I = \frac{E}{R}$$

Multiplying both sides by R you have:

$$I \times R = \frac{E \times R}{R}$$

Notice that R appears in both the numerator and the denominator of the fraction on the right side of the equation. You will recall from basic mathematics that the two R's in the fraction can be canceled like this:

$$I \times R = \frac{E \times \cancel{R}}{\cancel{R}}$$

This leaves:

$$I \times R = E \quad \text{or} \quad E = I \times R$$

In other words, voltage is equal to current times resistance. Generally, the times sign (x) is omitted so that the formula is written:

$$E = IR$$

Figure 6-3A shows a circuit in which the resistance and current are known. To find the voltage:

$$E = IR = 0.5 \times 125\Omega = 62.5 \text{ V}$$

Notice that multiplying amperes times ohms results in an answer given in volts.

OHM'S LAW

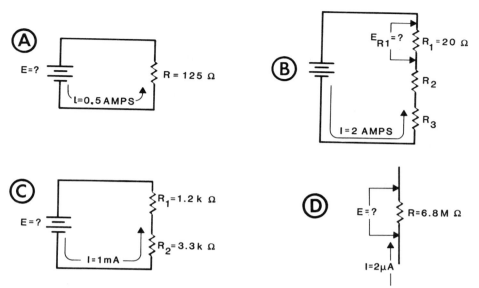

Figure 6-3 Finding voltage.

Figure 6-3B shows a slightly different problem. Here, you want to find the voltage drop across R_1. This voltage is represented as E_{R1}. You know from the information given in the figure that R_1 is 20 ohms, and the current through R_1 is 2 amps. Consequently, the voltage drop across R_1 is:

$$E_{R1} = I \times R_1 = 2 \text{ A} \times 20 \Omega = 40 \text{ volts}$$

In Figure 6-3C, the values of the resistors in the circuit and total current are given. To determine the battery, or source voltage, you must multiply the total resistance (R_T) times the current. In a series circuit, the current through an individual resistor is equal to circuit, or total current. However, once we enter parallel circuits, this will not be true. When making a calculation in Ohm's Law, you must use the known values that apply to the unknown you are seeking. If you are working with an individual component, you must use the known values for that component. If you are working with circuit totals, then you must use the known totals. The total resistance in a series circuit is the sum of the individual resistors. Therefore:

$$R_T = R_1 + R_2 = 1.2 \text{K}\Omega + 3.3 \text{K}\Omega = 4.5 \text{ k}\Omega$$

Once total resistance is known, we can use our equation for voltage:

$$E = I \times R_T = 1 \text{ mA} \times 4.5 \text{K}\Omega = 4.5 \text{ V}$$

Previously, when we multiplied ohms times amps, our result was volts. Notice, in this example, that multiplying kilohms times milliamps also gives volts. This is also true for megohms and microamps. To calculate the voltage drop for the resistor shown in Figure 6-3D, we simply plug the known values into the equation:

$$E = I \times R = 2\mu A \times 6.8 \text{ M}\Omega = 13.6 \text{ V}$$

Finding Resistance

The equation for voltage was found by transposing the original current equation. An equation for resistance can be found the same way, except instead of using our current equation, we'll use the voltage equation. The voltage equation is:

$$E = I \times R$$

To find a resistance equation, you must transpose so that R is on one side by itself. This can be done simply by dividing both sides by I:

$$\frac{E}{I} = \frac{IR}{I}$$

The I's in the fraction on the right cancel:

$$\frac{E}{I} = \frac{\cancel{I}R}{\cancel{I}}$$

So the equation becomes:

$$\frac{E}{I} = R \text{ or } R = \frac{E}{I}$$

This indicates that resistance is equal to voltage divided by current or:

$$\text{ohms} = \frac{\text{volts}}{\text{amperes}}$$

Using this formula, the resistance in a circuit can be found using known voltage and current values. A simple application for this is shown in Figure 6-4A. The resistance for the circuit is:

$$R = \frac{E}{I} = \frac{24V}{4.8\,A} = 5\Omega$$

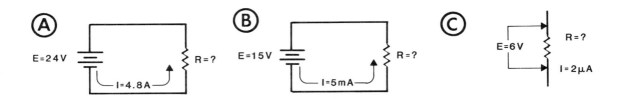

Figure 6-4 Finding resistance.

When you were calculating for current, you saw that if kilohms were divided into volts, the answer would be in milliamps. A similar relationship was described with megohms and microamps. When calculating for resistance, these relationships still exist. Thus, if milliamps are divided into volts, the answer is in kilohms. Likewise, if microamps are divided into volts, the answer is in megohms.

Figure 6-4B provides us with an example of the kilohm/milliamp relationship. The resistance of the circuit would be calculated as follows:

$$R = \frac{E}{I} = \frac{15V}{5mA} = 3K\Omega$$

Finally, Figure 6-4C demonstrates the megohm/microamp relationship. In this case, the resistance is not the resistance for the circuit, but the resistance for one particular resistor. Notice how the individual voltage drop for the resistor is used, instead of total voltage:

$$R = \frac{E}{I} = \frac{6V}{2\mu A} = 3M\Omega$$

Using Ohm's Law

Ohm's Law may be expressed by three different equation:

$$I = \frac{E}{R}$$

$$E = IR$$

$$R = \frac{E}{I}$$

In Figure 6-5A the three quantities are placed in a diagram in a way that may help you to remember these three equations. To use the diagram, place your index finger over the unknown value, and the diagram will indicate the equation for that value. Figures 6-5B, 6-5C, and 6-5D show how the unknown value is covered, with the rest of the diagram indicating the proper equation. For example, if you know the voltage drop and resistance of a resistor, but need to know how much current flows through the resistor, then you would cover the quantity "I", as demonstrated in Figure 6-5B.

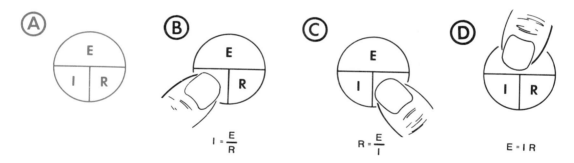

Figure 6-5 The three forms of Ohm's Law.

The remaining two symbols represent the equation used to find the unknown. Notice that the remaining quantities are:

$$\frac{E}{R}$$

Thus, current can be found using E/R.

DC ELECTRONICS

Chapter Self-Test

1. Ohm's Law defines the relationship between what three electrical quantities?

2. What is the Ohm's Law equation for current?

3. How much current flows through a 10 ohm resistor when 5 volts are applied?

4. If the EMF is given in volts and the resistance is given in kilohms, what quantity will current be?

5. What is the Ohm's Law equation for voltage?

6. Using Ohm's Law, calculate the voltage drop across a 120 ohm resistor when the current through the resistor is 0.25 amps.

7. When voltage is in volts, and current is in microamperes, what value must resistance be?

8. What is the Ohm's Law equation for resistance?

9. What is the ohmic value of a resistor which develops a voltage drop of 12 volts when a current of 0.75 amperes flows through it?

10. If resistance is given in megohms and the EMF is given in volts, what will the value for current be?

OHM'S LAW

Summary

Ohm's Law defines the relationship between the three fundamental electrical quantities. It describes how current, voltage, and resistance are related.

Equations are used to define the relationships between the three fundamental electrical quantities. In these equations, the letter I represents current; E represents voltage; and R represents resistance.

The current equation states that current is equal to voltage divided by resistance. Stated as an equation:

$$I = \frac{E}{R}$$

The unit of current in the Ohm's Law equations depends on the units of voltage and resistance. Assuming that EMF is expressed in volts, the current will be in amperes if the resistance is in ohms. However, the current will be in milliamperes if the resistance is in kilohms. Finally, the current will be in microamperes if the resistance is in megohms.

The voltage equation states that voltage is equal to the current multiplied by the resistance. That is:

$$E = I \times R$$

E is expressed in volts when I is in amperes and R is in ohms. E is also in volts when I is in milliamperes and R is in kilohms. Finally, E is in volts when I is in microamperes and R is in megohms.

The resistance equation states that resistance is equal to voltage divided by current. That is:

$$R = \frac{E}{I}$$

Assuming that E is given in volts, R will be in ohms when I is in amperes. However, R will be in kilohms when I is in milliamperes. Finally, R will be in megohms when I is in microamperes.

While it is handy to remember which of the metric prefixes go together when solving Ohm's Law problems, it is not absolutely essential. If you convert all quantities to volts, ohms, and amperes, you can solve any Ohm's Law problem without worrying about metric prefixes.

CHAPTER 7

Power

Contents

Introduction 127

Chapter Objectives 128

Work and Power 129

Power, Current, and Voltage 131

Power Dissipation 134

Chapter Self-Test 136

Summary 137

POWER

Introduction

In addition to the three basic electrical quantities (current, voltage, and resistance), there is a fourth quantity. This quantity is called power. In this chapter, you will learn what power is, how it is calculated, and its relationship with current, voltage, and resistance. Power becomes very important when analyzing electronic circuits, because if power ratings are ignored, damage to the equipment could occur.

DC ELECTRONICS

Chapter Objectives

When you have completed this chapter, you will be able to:

1. Define the following terms: joule, work, power, watt, dissipation.

2. Select the proper equation and calculate the power dissipated in a simple circuit in which two of the following values are known: voltage, current, and resistance.

POWER

Work and Power

To define work and power, we should review the definitions of other electronic terms.

Coulomb — The unit of measurement for electrical charge. One coulomb is equal to a charge of 6.25×10^{18} electrons.

Potential — A charge, either positive or negative, having the possibility of moving electrons and doing work.

Current — The movement of electrons from a more negatively charged point to a more positively charged point along a path.

Voltage — The measure of electromotive force, or EMF. One volt is the magnitude of EMF that forces one coulomb of charge to move from one point to another.

Voltage Rise — Any location in a circuit where energy, or voltage, is introduced to the circuit.

Voltage Drop — The energy (voltage) consumed (dropped) by the load as current passes through the load.

Work — The product of a force applied to an object, and the distance the object is moved. Electrically, when electrons are forced to move from a negative to a positive terminal, work is done.

When a voltage rise forces electrons through a load, work is done—electrons moving from one point to another. Notice that time does not enter into this definition. Thus, the same amount of work is done whether the electrons are moved in one second or one hour.

In electronic circuits, time is extremely important. Circuits are designed so they will work for very long periods of time, and the components used in the circuit should be able to perform their work without breaking down. **Power** is defined as the rate at which work is done. In other words, power refers to the amount of work done in a specific length of time. The unit of power is the **watt,** in honor of James Watt, who pioneered the development of the steam engine. When one volt forces one ampere of current to pass a point in one second, one watt of power has been used. The common term for power being used is **dissipated.**

Another term often associated with power is the joule, which is a unit of energy. We stated that one volt causes enough energy to move one coulomb of charge. The energy created is one joule. However, like work, joules do not relate to time. Power expresses joules per second.

The interrelation between volts, joules, amperes, ohms, and watts can get very complex. To summarize, let's re-describe a basic electrical circuit:

> If there is one volt of potential difference in a circuit, one joule of energy will cause one ampere of current to move through a resistance of one ohm, in one second, dissipating one watt of power.

Power, Current, and Voltage

You have learned that one watt is the amount of power used in a circuit when one volt forces one ampere of current through one ohm of resistance. If either voltage or current increases, the power for that circuit also increases. For example, if the resistance were decreased to .5 ohms, the current flow would increase to 2 amps (current and resistance, if you recall, are inversely proportional). The increase in current means that more electrons are passing through the circuit, therefore, more work is done per second. Thus, more power is dissipated.

If the resistance in our circuit increases to 2 ohms, the current would decrease to .5 amps. If we wanted the current to remain at 1 amp, we would have to increase the voltage to 2 volts. In other words, to keep a constant amount of current through an increased resistance, we must increase the force, or voltage. If the overall force in the circuit increases, we must be doing more work per second—more work, more power. Thus, an increase in voltage causes an increase in power.

We can conclude from the previous paragraphs that power is directly proportional to current and voltage. This can be expressed as:

   power = current x voltage

or, stated in terms of the units of the three quantities:

   watts = amperes x volts

Just as the letters "I" and "E" are used to represent current and voltage, respectively, the letter "P" is used to represent power. We can then express our formula as:

   $P = IE$

This formula is very easy to use, provided the current and voltage values are known. When you are calculating the power dissipated for an entire circuit, you should use total current and voltage values. For example, what is the power dissipated in a circuit with 50 volts applied and a total current of 3.2 amps?

   $P = IE = 3.2 \text{ A} \times 50 \text{ V} = 160 \text{ W}$

Notice how the total values are used to find the answer.

DC ELECTRONICS

If you are calculating the amount of power dissipated by an individual component, you must use the current and voltage values for that component. Another example can demonstrate this: how much power is dissipated by a resistor that drops 10 volts with 25 milliamps of current flow? The best way to solve this is to convert 25 milliamps to 0.025 amps:

$$P = IE = 10V \times 0.025A = 0.25 \text{ W} = 250 \text{ mW}$$

Notice how a small amount of current causes a small amount of power to be dissipated. We can conclude that when you multiply volts and milliamps, your answer will be in milliwatts.

When you worked with Ohm's Law, you were able to transpose the original formula into other formulas just as useful. Likewise, we can transpose the power equation and create other formulas. The first formula expresses voltage in terms of current and power. It is found by rearranging the equation:

$$P = IE$$

Dividing both sides by I:

$$\frac{P}{I} = \frac{IE}{I}$$

Canceling the two I's on the right side:

$$\frac{P}{I} = \frac{\cancel{I}E}{\cancel{I}}$$

We are then left with:

$$\frac{P}{I} = E \quad \text{or} \quad E = \frac{P}{I}$$

This states that voltage (in volts) is equal to power (in watts) divided by current (in amperes). This formula is just as easy to work with as our original power formula. For example, what is the voltage drop across a light bulb that dissipates 60 watts when the current through the bulb is 0.5 amps?

$$E = \frac{P}{I} = \frac{60W}{0.5A} = 120V$$

132

POWER

In addition, the original formula can be transposed the opposite way to give us a formula that expresses current in terms of power and voltage:

$$I = \frac{P}{E}$$

This formula is used when voltage and power values are known, but current is unknown. For example, how much current flows through a 75 watts light bulb which is connected across a 120 volt power line?

$$I = \frac{P}{E} = \frac{75W}{120V} = 0.625 \text{ A}$$

As you may have noticed, calculations involving power are just as simple as Ohm's Law calculations. However, later in this chapter we will be using other more complex formulas and calculations using power.

Power Dissipation

In resistors and most other electronic components, power is dissipated in the form of heat. In some cases, this heat is a desired result. For example, the purpose of the resistance element in a toaster, heating pad, and electric stove is to produce heat. However, in most electronic devices, the heat produced by resistors or other components represents wasted power. Remember, all power used in a circuit must be supplied by the power source. This includes the power lost in the heating of resistors or other components. Since power costs money, circuits are designed in such a way as to use the minimum amount of power to perform a task.

Since power is the result of forcing current through resistance, there must be some relationship between power and resistance. This relationship is actually second-hand: since power is directly proportional to current and voltage, and resistance is inversely proportional to current, resistance *must* be related to power in some way.

In order to express the relationship between resistance and power, we need a formula. Two separate formulas exist: one that expresses power in terms of resistance and voltage, and another that expresses power in terms of resistance and current. These formulas not only express the relationship between resistance and power, but are very helpful in calculating the power dissipation of a resistor when current or voltage is not known.

First, consider how P can be expressed in terms of E and R. The basic formula for power is:

$$P = IE$$

However, from Ohm's Law you know that:

$$I = \frac{E}{R}$$

Thus, you can substitute E/R for I in the basic power equation. When this is done, the equation becomes:

$$P = E \times \frac{E}{R} \quad \text{or} \quad P = \frac{E^2}{R}$$

or

$$P = \frac{E^2}{R}$$

This formula is used to calculate for power when voltage and resistance are known. For example, how much power is dissipated by a 22 ohm resistor if the voltage drop across the resistor is 5 volts?

$$P = \frac{E^2}{R} = \frac{(5V)^2}{22\Omega} = \frac{25V}{22\Omega} = 1.136 \text{ W}$$

In some cases, only the current and resistance are known. By combining the basic power formula with one of the Ohm's Law formulas, you can derive an equation in which P is expressed in terms of I and R. Again, we start with the basic power formula:

$$P = IE$$

From Ohm's Law you know that E = IR. Thus, you can substitute IR for E in the power formula. The equation becomes:

$$P = I \times IR$$

$$P = I^2 R$$

This is the formula that you use when you wish to find power and only current and resistance are known. For example, how much power is dissipated by a circuit that has a total of 40 ohms of resistance and 0.5 amps of current?

$$P = I^2 R = (0.5A)^2 \times 40\Omega = .25 \text{ A} \times 40 = 10W$$

These formulas are used if you want to make only one calculation. In the last example, you could have multiplied the current and the resistance to find the voltage for the circuit, then used P = IE to find the power. However, regardless of the method you use, all of these formulas are interchangeable, which makes double-checking your calculations very easy.

DC ELECTRONICS

Chapter Self-Test

1. What is the difference between power and work?

2. When one volt moves one coloumb of charge, how much energy is created?

3. What is the unit of measurement for power?

4. Name the formula that expresses the relationship between current, voltage, and power.

5. How much power is dissipated by a resistor which has a voltage drop of 3.2 volts with 0.75 amps of current flow through it?

6. What is the formula that expresses current in terms of voltage and power?

7. In a problem where current and resistance are known, what formula is used to find power?

8. The formula in Question #7 also expresses the relationship between resistance and power. What is the other formula that expresses this relationship?

9. How much power is dissipated by a circuit with a voltage source of 5 volts and a total resistance of 300 ohms?

10. How is power dissipated in most electronic circuits?

POWER

Summary

When a voltage rise forces current through resistance, work is done. However, this work does not relate to time.

Energy is expressed in terms of joules. One joule is equal to the energy created by one volt to move one coulomb of charge. Like work, joules do not relate to time.

Power is the measure of work over a length of time. The unit of measurement for power is the watt.

One watt of power is dissipated when one volt forces one ampere of current through one ohm of resistance. Watts can also be expressed as joules per second.

Power is directly proportional to both current and voltage. This is expressed in the formula:

$$P = I \times E$$

When calculating the power for an entire circuit, you must use total current and total voltage applied to the circuit. When calculating the power dissipated by a single component, you must use the current flow through that component, and the component's voltage drop.

When current is in milliamps, this indicates a very small amount of current. Thus, very little power will be dissipated. The resulting power value is expressed in milliwatts.

By transposing the original power formula, we can derive two formulas: one for voltage in terms of current and power, and one for current, in terms of voltage and power. These formulas are:

$$I = \frac{P}{E} \quad \text{and} \quad E = \frac{P}{I}$$

In most electronic circuits, power is dissipated in the form of heat. Occasionally, this heat is desired; but most heat in an electrical circuit is wasted power.

The relationship between power and resistance can be expressed in two formulas: one using voltage, and the other using current. These formulas are:

$$P = I^2 R \quad \text{and} \quad P = \frac{E^2}{R}$$

DC ELECTRONICS

CHAPTER 8

DC Circuits

Contents

Introduction 141

Chapter Objectives 142

Simple DC Circuits 143

Bridge Circuits 151

Kirchhoff's Law 157

Chapter Self-Test 160

Summary 161

DC CIRCUITS

Introduction

To this point, your knowledge of electrical circuits has been limited to very simple circuits. However, with Ohm's Law, and your knowledge of electrical characteristics, we can progress into more complex circuits. Although these circuits will be more complex than what you have already studied, the circuits covered in this chapter are the easiest to understand.

This chapter reviews the basic DC circuits: the series circuit, the parallel circuit, and the series-parallel circuit. You will learn that although each of these circuits follows Ohm's Law and all other known electrical characteristics, they each behave differently. You will also read about a voltage divider, one of the most common and useful circuits.

Following our study of these circuits, you will study the bridge circuit. The bridge circuit has many unique characteristics, and a wide variety of applications. You will then be shown a fundamental law used in analyzing circuits—Kirchhoff's Law.

All of the circuits studied in this chapter are known as DC circuits because the voltage source for all of these circuits is a battery, or some other type of direct current supply. This means the voltage and current values for these circuits are constant.

DC ELECTRONICS

Chapter Objectives

When you have completed this chapter, you will be able to:

1. Define the following terms: series circuit, parallel circuit, series-parallel circuit, voltage divider, bridge network, balanced bridge, and unbalanced bridge.

2. Given the resistor and voltage source values of a voltage divider, calculate the voltage felt at each point in the voltage divider.

3. Given the values of the resistors in a bridge network, determine whether the bridge is balanced.

4. Describe the operation of the wheatstone and temperature sensing bridge.

5. State Kirchhoff's Voltage and Current Laws.

DC CIRCUITS

Simple DC Circuits

Throughout your study of electronics, you will see certain circuits repeated over and over again. Some of the most commonly used circuits are the easiest to understand. You have already studied the characteristics of several types of simple circuits. These include the series circuit, the parallel circuit, and the series-parallel circuit. We will begin by reviewing the characteristics of these circuits. Later, you will learn about some more advanced concepts.

Series Circuits

In a series circuit, current flows from the negative terminal of the battery, then along a single path through each component towards the positive terminal. Total current flows through each component. Calculations for a series circuit are very simple. These calculations not only tell you the voltage, current, resistance, and power values for each component, but also allow you to analyze the circuit to ensure proper operation.

A typical series circuit is shown in Figure 8-1. As we begin our mathematical analysis of this circuit, it is best to start with total current. Since current is constant in a series circuit, knowing the total current for the circuit will tell us the current for each component. From there, you can calculate the voltage or power for all other components.

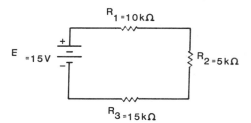

Figure 8-1 The series circuit.

Total current for this circuit is determined by dividing the total resistance into the applied voltage. The total resistance (R_T) in a series circuit is equal to the sum of the individual resistances. Thus, in Figure 8-1:

$$R_T = R_1 + R_2 + R_3 = 10k\Omega + 5k\Omega + 15k\Omega = 30k\Omega$$

Once the total resistance is known, total current (I_T) can be determined:

$$I_T = \frac{E}{R_T} = \frac{15V}{30K\Omega} = .5mA$$

Remember, anytime you divide kilohms into volts, the answer is in milliamps.

Since the current is the same through all resistors, total current and individual currents are equal. The voltage drop across any one resistor can be determined by multiplying resistance and current values of the resistor. The voltage drops for the three resistors are shown below:

$$E_{R1} = I \times R_1 = .5mA \times 10k\Omega = 5V$$

(Anytime milliamps and kilohms are multiplied together, the answer is in volts.)

$$E_{R2} = I \times R_2 = .5mA \times 5k\Omega = 2.5V$$

Notice that R_2, with half the resistance of R_1, drops half as much voltage as R_1, with the same current flow.

$$E_{R3} = I \times R_3 = .5mA \times 15k\Omega = 7.5V$$

Notice again that R_3, with three times as much resistance as R_2, drops three times as much voltage as R_2, with the same current flow.

With the voltage drops of the resistors known, another important characteristic of the series circuit can be shown. The applied voltage (known as E_A or E_T) is equal to the sum of the voltage drops. That is:

$$E_A = E_{R1} + E_{R2} + E_{R3} = 5V + 2.5V + 7.5V = 15V$$

Finally, the power dissipated by any resistor is equal to the current times the voltage drop across the resistor. Thus, the power dissipated by R_1 is:

$$P_{R1} = I \times E_{R1} = .5mA \times 5V = 2.5 \text{ mW}$$

We don't have to demonstrate the calculations for R_2 and R_3, since the procedure is the same as for R_1. Total power may be calculated by multiplying total current and voltage applied, or by adding the power of each component.

Parallel Circuits

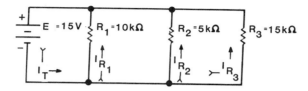

Figure 8-2 The parallel circuit.

A simple parallel circuit is shown in Figure 8-2. In a parallel circuit, the voltage applied to each branch is the same. That is, the voltage measured across any resistor in the parallel branch equals the voltage applied to the branch. Thus, each resistor in Figure 8-2 drops 15 volts. With the voltages known, we can progress straight to the current values. To find the current through any resistor in the parallel branch according to Ohm's Law, you must divide the voltage by the resistance value of that resistor. The current calculations are shown below:

$$I_{R1} = \frac{15V}{10k\Omega} = 1.5mA$$

$$I_{R2} = \frac{15V}{5k\Omega} = 3mA$$

$$I_{R3} = \frac{15V}{15k\Omega} = 1mA$$

In a parallel circuit, the only way to determine the current flow for each branch is mathematically. Although each branch has the same voltage drop, the resistance in each branch may be different, which would result in different currents through each branch.

The current flow for each branch, although they may be different values, must begin at the voltage source. Therefore, the total amount of current flow for the circuit is equal to the current flow from the voltage source. Since each branch current gets its current from the same voltage source, the total current is the sum of the branch currents:

$$I_T = I_{R1} + I_{R2} + I_{R3} = 1.5mA + 3mA + 1mA = 5.5mA$$

DC ELECTRONICS

The next unknown value is total resistance. You'll recall that there are a variety of methods used to calculate total resistance in a parallel circuit—the value/number method, the product/sum method, and the reciprocal method. Some of these methods, however, can only be used in specific circumstances. When we look at the individual resistor values, we can see that they are not the same value, which prevents us from using the value/number method. Secondly, since there are more than two resistors, we cannot use the product/sum method. The only alternative is to use the reciprocal method:

$$R_T = \frac{1}{\frac{1}{R_1} + \frac{1}{R_2} + \frac{1}{R_3}} = \frac{1}{\frac{1}{10,000} + \frac{1}{5,000} + \frac{1}{15,000}}$$

$$= \frac{1}{.0001\Omega + .0002\Omega + .00006667\Omega} = \frac{1}{00036667\Omega} = 2727\Omega$$

You've probably had enough practice with the reciprocal method to be able to make this calculation without too much difficulty. However, there is a much easier way to find total resistance in this circuit. Ohm's Law can be used to find total resistance, regardless of the number or values of the resistors. The only values you need are total current and the voltage applied:

$$R_T = \frac{15V}{5.5mA} = 2727\Omega$$

Since finding total resistance using Ohm's Law is the easiest method to use, you should use it whenever possible.

The only value left to determine is power. The power dissipated in any resistor can be found by multiplying the current through the resistor by the voltage dropped by the resistor. For example, the power dissipated by R_1 is:

$$P_{R1} = I_{R1} \times E_{R1} = 1.5mA \times 15V = 22.5mW$$

Again, total power can be found by multiplying total current and voltage applied, or by adding the power of each component. This does not change for parallel circuits.

DC CIRCUITS

Series-Parallel Circuits

In most electrical equipment, there are very few series and parallel circuits. The majority of circuits used are series-parallel, like the one shown in Figure 8-3A. These circuits have characteristics of both series and parallel circuits. Total current travels from the negative terminal through R_4, then splits. A portion of total current travels through R_3, and the remainder of total current travels through R_2. Current then rejoins, and total current runs through R_1. The applied voltage is divided in this manner: R_1 develops a certain portion, R_4 develops another, and the rest is developed by both R_2 and R_3. Because R_2 and R_3 are in parallel, they develop the same voltage.

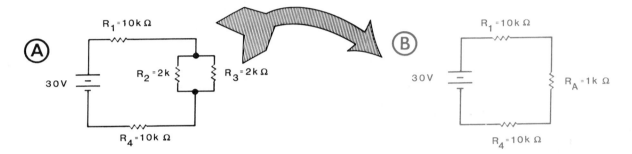

Figure 8-3 The series-parallel circuit.

Obviously, this creates some difficulties when calculations must be made. However, if the parallel portion is combined into an equivalent series resistance (R_A), these calculations become quite simple. Since R_2 and R_3 are the same value, we can use the value/number method:

$$R_A = \frac{2k\Omega}{2} = 1K\Omega$$

The resulting circuit shown in Figure 8-3B can now be handled like any other series circuit. The only difference is the method for determining the current flow through the parallel portion. Let's say, for example, that total current is 2 milliamps. Since R_2 and R_3 are of equal value, current will split evenly. Therefore, both R_2 and R_3 have 1 milliamp of current flow. However, if R_2 and R_3 were not equal, you would find their current by dividing their voltage drops by their resistance. Regardless of which method you use, you should remember that the sum of the two branch currents must equal total current.

DC ELECTRONICS

Voltage Dividers

One of the most commonly used circuits you will find is the series voltage divider. The series voltage divider, as the name implies, is a simple series circuit that takes its voltage source and divides it into different voltages.

The three schematics in Figure 8-4 all represent voltage dividers. Figure 8-4A shows the voltage source, in this case a 30 volt battery connected to the circuit consisting of three resistors of different values. Notice that the bottom of the circuit, which includes the negative battery terminal, is connected to ground. Figure 8-4B shows another way this arrangement can be shown. Both the negative terminal and the bottom of the circuit have a ground symbol. This implies that they have a path, along ground, which connects them to each other.

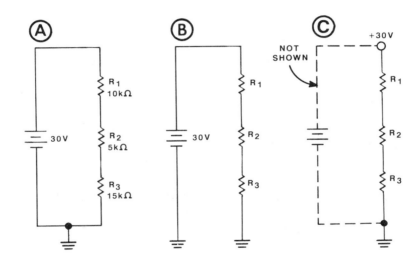

Figure 8-4 A grounded voltage divider.

Having one side of a voltage source connected to ground is a common condition in electronics. This allows each circuit that uses the voltage source to be connected to ground, instead of having separate wires traveling to that point on the voltage source (you may recall that providing a common return to the voltage source was an important function of ground). Figure 8-4C is a good example of how this voltage divider might actually be shown. Notice that the power source is indicated by a circle at the top of the schematic. "+30V" is written near the circle to indicate the polarity and magnitude of the voltage source. The positive polarity implies that the negative terminal is connected to ground. Our path for current remains the same, but the connection of the circuit to the negative terminal does not have to be drawn—it is implied by the positive power source.

DC CIRCUITS

Before we can analyze the operation of a voltage divider, we must determine the voltage drops of the resistors. Since this is a simple series circuit, the calculations will not be demonstrated. Total resistance for this circuit is 30 kilohms, divided into the 30 volt power source, gives us a total current of 1 milliamp. This 1 milliamp allows R_1 to develop 10 volts, R_2 to develop 5 volts, and R_3 to develop 15 volts. Notice that the sum of the voltage drops equals the voltage applied.

The difference between a simple series circuit and a voltage divider is not in the arrangement or placement of the components. The difference is in how voltages are *measured*. In a simple series circuit, the voltage drops are measured by placing the voltmeter across the component being measured. In a voltage divider, the voltages are measured **from a common point**. Usually, this common point is ground.

Figure 8-5 shows our original circuit with voltmeters connected to a voltage divider. Notice that the bottom of each voltmeter is connected to ground. When the voltmeter is connected to point A, it measures the voltage across R_3 only, which is 15 volts. When the voltmeter is connected to point B, it measures the voltage across R_3 *and* R_2. The two voltage drops are added together to give us a combined voltage of 20 volts.

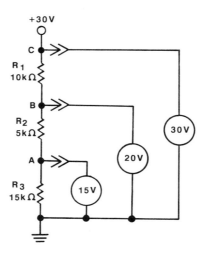

Figure 8-5 Measuring voltages.

Finally, the voltmeter is connected across all three resistors at point C. This also connects the voltmeter across the entire circuit. Thus, the voltage applied, 30 volts, is measured.

DC ELECTRONICS

The function of a voltage divider is to take one voltage source and obtain a number of different voltages. These multiple voltages can be used in other circuits. However, if other circuits are connected to the voltage divider, additional currents flow from the original source voltage. These additional currents affect the voltage divider, and the voltage drops of the resistors will change from the voltage drops when no other circuit is connected. Many variables must be taken into account when designing a voltage divider.

Bridge Circuits

We stated previously that the majority of circuits used in electrical equipment are series-parallel circuits. A good example of this is the bridge circuit. A simple bridge circuit consists of four resistors, as shown in Figure 8-6. The circuit has two input terminals and two output terminals. In DC applications, the input terminals are connected to a DC voltage source such as a battery. The output terminals usually have a voltmeter or ammeter connected between them.

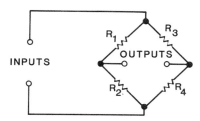

Figure 8-6 The bridge circuit.

Balanced Bridge

Before looking at some of the ways that bridge circuits can be used, you should first investigate the characteristics of a bridge circuit. A bridge circuit may be either balanced or unbalanced. A balanced bridge circuit is one in which the voltage measured between the two output terminals is 0 volts. Let's assume for a moment that all four resistors are the same value. Each parallel branch will have the same resistance, and therefore the same current flow. Thus, each resistor will have the same voltage drop. Since the voltage drops of R_2 and R_4 are equal, and the voltage drops of R_1 and R_3 are equal, there would be no potential difference between the output terminals, or bridge. If a voltmeter were connected across the bridge, it would read 0 volts. Likewise, if an ammeter were connected, it would indicate 0 amps—since there is no potential difference, no current can flow. Thus, the bridge is said to be balanced.

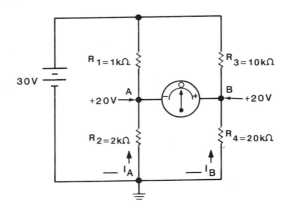

Figure 8-7 The balanced bridge.

Figure 8-7 shows that a balanced condition can exist even when all the resistors have different values. To prove this, we need to find the voltage drops of R_2 and R_4. The first step in finding these values is to find the current for each branch. For the left branch, current is the voltage divided by the resistance. Since the left branch is essentially one parallel branch, we know its voltage is equal to the voltage applied—30 volts. The resistance of the left branch is 3 kilohms, since the two resistors are in series. From there, the current for the left branch (I_L) can be calculated:

$$I_L = \frac{30V}{3k\Omega} = 10mA$$

Once we know the current through R_1 and R_2, we can calculate the voltage drop for R_2:

$$E_{R2} = I_L \times R_2 = 10mA \times 2k\Omega = 20V$$

The same procedure can be used to find the voltage drop of R_4. The voltage on the right branch is 30 volts (voltage is equal in parallel), and the resistance is 30 kilohms (resistance adds in series). Thus, I_R is equal to:

$$I_R = \frac{30V}{30k\Omega} = 1mA$$

Now that we know the current for R_3 and R_4, we can calculate the voltage drop of R_4:

$$E_{R4} = I_R \times R_4 = 1mA \times 20k\Omega = 20V$$

DC CIRCUITS

As you can see, the voltage drops of R_2 and R_4 are equal. When a voltmeter is connected across the bridge, there will be no potential difference. Thus, the bridge is balanced.

If you examine Figure 8-7 carefully, you will see that R_2 is twice the value of R_1, and R_4 is twice the value of R_3. R_1 and R_2 form a voltage divider that determines the voltage at point A. R_3 and R_4 form a divider for point B. If the ratio between R_1 and R_2 is equal to the ratio between R_3 and R_4, the bridge is balanced. Expressed as an equation, the bridge is balanced when:

$$\frac{R_1}{R_2} = \frac{R_3}{R_4}$$

Unbalanced Bridge

Figure 8-8A shows a variation of the basic bridge circuit. Here, R_4 is replaced with a potentiometer. If R_4 is set to the same value as the other three resistors, 200 ohms, the bridge is balanced. Both point A and point B are +15 volts with respect to ground.

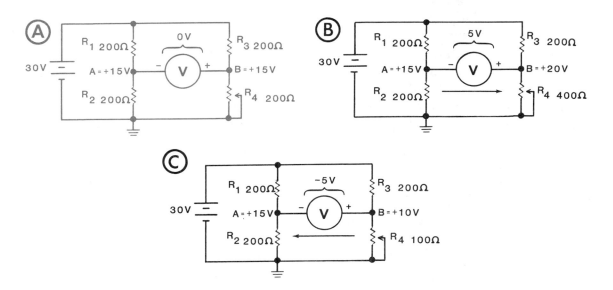

Figure 8-8 Unbalancing the bridge.

The balanced condition can be upset by changing R_4. If R_4 is adjusted to 400 ohms, for example, this will change the voltage at point B. Because R_4 is now twice as large as R_3, it drops twice as much voltage as R_3. Thus, R_4 now drops 20 volts while R_3 drops only 10 volts. Consequently, the voltage at point B is +20 volts with respect to ground. The bridge is no longer balanced, since a difference of potential exists between points A and B. Because point B is more positive than point A, current flows through the voltmeter from Point A to point B as shown. A voltmeter connected in this way indicates a 5 volts potential difference.

Figure 8-8C shows that the bridge can also be unbalanced by making the value of R_4 smaller than the value of R_3. Here, R_3 develops only 10 volts. Hence, the voltage at point B is only +10 volts. Thus, current will flow in the opposite direction through the meter.

Wheatstone Bridge

Now that you have an understanding of how the bridge circuit operates, it's time to examine some typical applications. The first practical application of the bridge circuit is the wheatstone bridge. The wheatstone bridge is a device to measure resistance.

A simple wheatstone bridge is shown in Figure 8-9. R_1 and R_2 are fixed matched precision resistors. R_S is a 1 kilohm potentiometer whose resistance is varied using a calibrated dial. At any setting, its value can be read directly from the dial. R_X is the unknown resistance that you want to measure. It is connected between the two test leads that form a part of the bridge. The meter is a very sensitive current meter called a galvanometer. Unlike the ammeter discussed earlier, the galvanometer can measure current flow in either direction. The center of the galvanometer scale is zero, and current flow in either direction is indicated by the deflection of the needle.

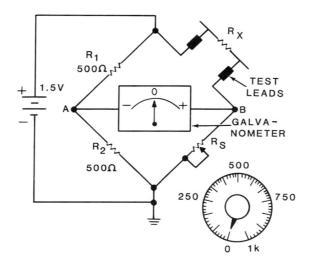

Figure 8-9 Wheatstone bridge.

To see how the bridge can be used to measure resistance, assume that a 210 ohm resistor is connected between the test leads. Since R_1 and R_2 are equal, the bridge will be balanced only if values R_X and R_S are also equal. Thus, R_S is set so that the galvanometer reads exactly zero. At this point the bridge is balanced, so R_S must equal R_X. The value of the unknown resistance can be read directly from the calibrated dial of R_S. This gives you a simple method for finding the value of the unknown resistor.

This simplified bridge can measure resistances up to 1 kilohm, the maximum setting of R_S. In actual bridges, R_S is replaced with a device called a decade resistor box. By setting rotary switches, any value of resistance from a fraction of an ohm to several megohms can be placed in the circuit. Thus, the bridge can be used to measure a wide range of resistances.

Temperature Sensing Bridge

If one of the resistors in a bridge circuit is replaced with a thermistor, the bridge can be used as a temperature sensor. Figure 8-10 shows a temperature sensing bridge.

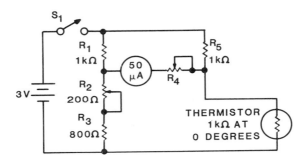

Figure 8-10 Temperature sensing bridge.

R_1 and R_5 are 1 kilohm resistors. At 0 degrees, the thermistor also has a resistance of 1 kilohm. If R_2 is set to 200 ohms, the fourth arm of the bridge also has a resistance of 1 kilohm. Consequently, the bridge is balanced and no current flows through the meter. The meter reads from 0 microamps to 50 microamps. Thus, the 0 microamp point on the meter scale can be labeled 0 degrees.

As temperature increases, the resistance of the thermistor decreases. This upsets the balanced condition and causes current to flow through the meter. As temperature increases, more current flows through the meter. Thus, current is an indication of temperature. The meter scale is calibrated in degrees rather than microamperes. Therefore, the temperature variations can be read directly from the meter.

R_4 is included to provide a means of calibrating the high end of the scale. For example, you may want the upper limit of your "thermometer" to be 100 degrees. In this case, the thermistor is exposed to a temperature of 100 degrees and R_4 is adjusted so that 50 microamps of current flows through the meter. This causes full scale deflection. Thus, the point of full-scale deflection can be labeled 100 degrees.

DC CIRCUITS

Kirchhoff's Law

Many series-parallel circuits can be analyzed using the techniques described earlier. However, more complex circuits cannot always be analyzed by such simple methods. Often a circuit will have several interconnected series-parallel branches and two or more voltage sources. Several techniques have been developed to help analyze circuits of this type. These techniques are generally grouped together under the name "network theorems."

A network is simply a circuit made up of several components such as resistors. Thus, the series-parallel circuits discussed earlier can be called **networks**. A network theorem is a logical method for analyzing a network. One of the most useful tools for analyzing a network is Kirchhoff's Law.

Kirchhoff's Voltage Law

One form of Kirchhoff's Law was discussed earlier. It states the relationship between the voltage rises and the voltage drops around the closed loop in a circuit. Recall that the sum of the voltage drops is equal to the sum of the voltage rises. This fact is referred to as Kirchhoff's Voltage Law.

Kirchhoff's Voltage Law can be stated in several different ways. Up to now, you've known it as: **around a closed loop, the sum of the voltage drops is equal to the sum of the voltage rises**. Another way of saying the same thing is: **around a closed loop, the algebraic sum of all the voltages is zero**. It becomes apparent that this statement is true when you trace the circuit in Figure 8-11.

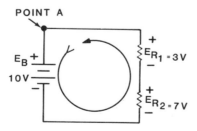

Figure 8-11 Kirchoff's Voltage Law.

DC ELECTRONICS

To keep the polarity of the voltages correct, it is helpful to establish a rule for adding the voltages. A convenient 3-part rule is:

1. Choose which direction you prefer to trace the circuit. Either clockwise or counterclockwise works equally well.

2. Trace around the circuit in the chosen direction. If the positive side of a voltage drop, or voltage rise, is encountered first, consider that voltage drop, or rise, to be positive.

3. If the negative side of a voltage drop, or voltage rise, is encountered first, consider that voltage drop, or rise, to be negative.

Starting at point A in Figure 8-11, trace counterclockwise as shown, recording each voltage encountered. The first voltage is E_B. Because the positive side of the battery is encountered first, the voltage is recorded as +10 V. Next, E_{R2} is encountered. This is recorded as -7 V because you arrive at the negative side of the resistor first. E_{R1} is recorded as -3 V for the same reason. Thus, the sum of the two negative voltages, and the positive voltage equal zero. This can be stated as:

$$E_B - E_{R2} - E_{R1} = 0V$$

In this example, the loop was traced in the counterclockwise direction. However, tracing in the clockwise direction, you'll find that the sum is still zero.

When you are working with a simple series circuit, applying Kirchhoff's Law is a rather simple process. In many cases, you will be able to see that the voltage drops equal the voltage rises without tracing a path, as previously described. However, when you deal with more complex circuits (such as a series-parallel circuit, or a circuit with multiple voltage sources), Kirchhoff's Voltage Law will allow you to analyze the circuit much easier than using Ohm's Law.

Kirchhoff's Current Law

Another form of Kirchhoff's Law involves current rather than voltage. Kirchhoff's Current Law can be stated in several different ways. One form states that in parallel circuits, the total current is equal to the sum of the branch currents. Stated another way, the current entering any point in a circuit is equal to the current leaving that same point.

Figure 8-12 illustrates these are simply two different ways of saying the same thing. Two branch currents of 1 A each are flowing in the circuit. Thus, the total current is 2 A. Now look at Point A. Notice that 2 A flows into this point. Consequently, 2 A must flow out of this point. One-half of the current flows through R_1 while the other half flows through R_2. Once again, an important law is simply common sense.

Nevertheless, this law can be used in much the same way as the voltage law to evaluate networks. However, Kirchhoff's Voltage Law is generally easier to use.

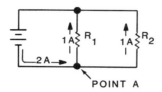

Figure 8-12 The current leaving point A is equal to the current entering point A.

Chapter Self-Test

1. What value is constant at all points in a series circuit?

2. Total resistance in a series circuit is found using what formula?

3. What value is constant at all branches in a parallel circuit?

4. What is the *easiest* method to find total resistance in a parallel circuit?

5. When is a bridge circuit balanced?

6. When the voltage across the output terminals is anything other than zero, what condition is the bridge?

7. Name a common application of the bridge circuit used to measure resistance.

8. What is the formula which states the preferred relationship between the resistors in a bridge circuit?

9. State Kirchhoff's Voltage Law.

10. State Kirchhoff's Current Law.

DC CIRCUITS

Summary

In a series circuit, the total resistance is found by adding the individual resistance values:

$$R_T = R_1 + R_2 + R_2 + R_3 + \text{etc.}$$

Also in a series circuit, the current is the same at all points in the circuit and the sum of the voltage drops is equal to the applied voltage.

In parallel circuits, the total resistance is less than the resistance of any individual branch.

The simplest method for finding total resistance in a parallel circuit is to use Ohm's Law.

Total resistance in parallel can also be found using either the value/number method, the product/sum method, or the reciprocal method.

In a parallel circuit, the voltage is the same across all branches, and the total current is the sum of the branch currents.

Simple series-parallel circuits can usually be reduced to simpler forms by combining resistors using both series and parallel formulas.

A voltage divider is used to produce two or more output voltages from a common higher voltage.

A bridge circuit generally consists of four resistances connected together so that there are two input terminals and two output terminals. Resistor values can be selected so that the bridge is balanced. In this condition, the voltage between the two output terminals is 0 V. The bridge can be used with a meter to measure resistance or temperature.

Kirchhoff's Law provides a way of analyzing circuits that cannot be analyzed using Ohm's Law alone. Kirchhoff's Voltage Law states that the sum of the voltages around a closed loop is zero. When a circuit has two or more loops, an equation can then be combined in such a way that only one unknown is left. The unknown quantity can then be determined.

Kirchhoff's Current Law states that the current entering a point is equal to the currents leaving that point. It too can be used to analyze complex circuits.

DC ELECTRONICS

CHAPTER 9

Magnetism

Contents

Introduction 165

Chapter Objectives 166

The Magnetic Field 167

Electricity and Magnetism 171

Induction 176

Chapter Self-Test 180

Summary 181

Introduction

You have probably seen magnetism at work while you were growing up. Perhaps you had a small horseshoe magnet, and found that it could pick up small nails and other metallic objects.

However, magnetism becomes very important in the study of electronics. Magnetism and electricity are very closely related. Current passing through a conductor produces a magnetic field, and when a conductor is passed through a magnetic field, current flows. This is a rather simple version of a very important idea.

In this chapter, you will be introduced to the terms and concepts that surround magnetism. You will learn about lines of flux, and how they can change, move, and affect different materials. You will also learn how magnetism is used to induce a voltage into a conductor—which is the foundation upon which most electricity is generated.

DC ELECTRONICS

Chapter Objectives

When you have completed this chapter, you will be able to:

1. Define the following terms: field, permanent magnet, temporary magnet, flux lines, permeability, reluctance, flux density, artificial magnet, natural magnet, magnetic induction, electromagnetic induction, residual magnetism, and retentivity.

2. List the four basic characteristics of flux lines.

3. State the left-hand rule for conductors, coils, and generators.

4. Describe the method of electromagnetic induction.

5. List the four factors that determine the amount of EMF induced into a conductor by electromagnetic induction.

6. State Faraday's Law of magnetic induction.

MAGNETISM

The Magnetic Field

In science, *action-at-a-distance* is explained in terms of fields. For example, you have seen that a charged particle can attract or repel another charged particle simply by coming close to it. This happens because a region of electrical influence extends outside each particle. This region of influence is called a field. An electrical field made up of lines of force is said to exist around every charged particle.

The field concept is also used to explain why certain metals can attract other metals. Everyone knows that a magnet can attract small pieces of iron or steel. A region of influence extends outside the magnet into the surrounding space. In this case, the region is called a magnetic field and is said to be made up of magnetic lines of force. Thus, a magnet is a piece of material that has a concentrated magnetic field surrounding it.

Magnets

Magnets may be classified in several different ways. First, they can be classified according to the method by which they obtain their magnetic field. The first known magnets were natural magnets, called magnetites or lodestones. These materials are surrounded by a magnetic field in their natural state.

Artificial magnets can be created from natural magnets. For example, if soft iron is rubbed repeatedly over a piece of lodestone, a magnetic field is transferred to the iron. Another type of artificial magnet is the electromagnet. Its magnetic field is produced by an electric current. The electromagnet will be discussed in much greater detail later. Magnets can also be classified by their shape. Thus, there are horseshoe magnets, bar magnets, and ring magnets.

Some materials readily retain their magnetic fields for long periods of time. These are called permanent magnets. Other materials quickly lose their magnetism and are called temporary magnets. Both of these types are widely used in electronics.

Finally, magnets are classified by the type of material used in their construction. Some examples of these are metallic magnets or ceramic magnets. Often this is carried even further and magnets are named according to the alloy from which they are made. Two popular classifications are Alnico (an alloy of aluminum, nickel, and cobalt) and Cunife (an alloy of copper, nickel, and iron).

The two ends of a magnet have different characteristics. One end is called a south (S) pole, while the other is called a north (N) pole. One reason for choosing these names is that a bar magnet will align itself in a north-south direction, if allowed freedom of movement. The magnet lines up in this way because the Earth itself is a huge magnet. The magnetic field that surrounds the Earth is shown in Figure 9-1. Note that the magnetic "north" and the geographic "north" are not at the same point, although it makes very little difference. The Earth's magnetic field is so powerful that it will influence any magnet on Earth. The north (N) pole of the magnet is defined as that end which points toward the north pole of the Earth. In reality, what we call the magnetic "north" is actually the south pole of the magnet, but it is called the "north" pole since it is close to geographic north.

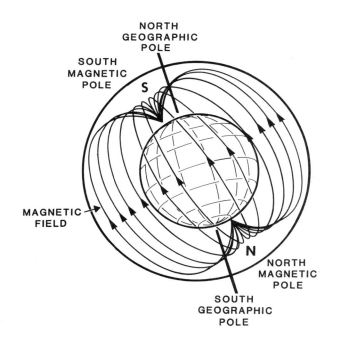

Figure 9-1 The Earth's magnetic field.

Magnets tend to align in a north-south direction because of a fundamental law of magnetism. This law states that like poles repel and unlike poles attract. Figure 9-2 illustrates this point. In this way, the north end of a magnet is attracted to the north magnetic pole, while the south end of the magnet points to the south magnetic pole. To see why magnets behave in this way, you must consider the nature of the magnetic lines of force.

Electricity and Magnetism

Electricity and magnetism are closely related. The electron has both an electrostatic field and a magnetic field. This may lead you to the conclusion that a charged object should have a magnetic field. However, this is not the case, since the magnetic field of about half the electrons will be opposite that of the other half. Nevertheless, the electron plays an important part in magnetism. It can be forced to produce a magnetic field in substances that are normally considered non-magnetic, such as copper and aluminum.

The key to creating a magnetic field electrically is motion. Motion is the catalyst that links electricity and magnetism. Anytime a charged particle moves, a magnetic field is produced. If a large number of charged particles are moved in a systematic way, a usable magnetic field is formed. You have learned that current flow is the systematic movement of large numbers of electrons. Thus, current flow, since it consists of a large number of charged particles moving in a systematic way, produces a magnetic field.

Current Flow and Magnetism

When current flows through a wire, a magnetic field is developed around the wire. The field exists as concentric flux lines as shown in Figure 9-6. While this field has no north or south pole, it does have direction. The direction of the field depends on the direction of current flow. The arrow heads on the flux lines indicate their direction. This does not mean that the flux lines are moving in this direction. It simply means that they are pointed in this direction.

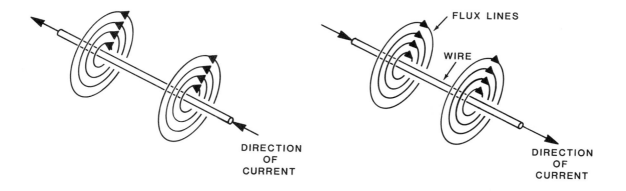

Figure 9-6 Flux lines exist as concentric circles around a current-carrying conductor.

The direction of the flux lines can be determined if the direction of current flow is known. The rule for determining this is called the **left-handed magnetic-field rule** or the **left-hand rule for conductors**. It is illustrated in Figure 9-7.

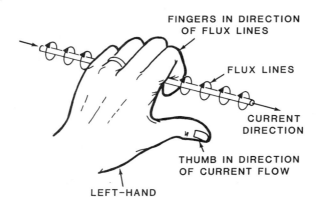

Figure 9-7 Left-hand rule for conductors.

Simply stated, if you grasp the conductor in your left hand with your thumb pointing in the direction of current flow through the conductor, your fingers now point in the direction of the flux lines.

In explaining some aspects of electromagnetism, it is helpful to show current flow in a third dimension. To do this, two new symbols are necessary. Figure 9-8A shows current flowing into the page. If the wire is viewed from the end, the tail of the arrow appears as a cross, as shown in Figure 9-8B. This cross is used to represent current flowing *into* the page. If the same wire is viewed form the other end, the head of the arrow appears as a round dot, as shown in Figure 9-8D. The dot is used to represent current flowing *out* of the page.

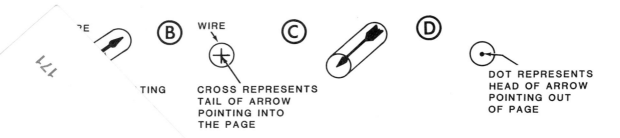

current flow in a third dimension.

In Figure 9-9, these new symbols are used to show how opposite and similar currents establish magnetic fields. In Figure 9-9A, opposite currents are shown. Using the left-hand rule, you can verify the direction of the two magnetic fields. Since the fields point in opposite directions, they tend to repel each other. Figure 9-9B shows that the opposite situation exists when the two currents flow in the same direction. This tends to draw the two fields together. Thus, they are free to connect and reinforce one another.

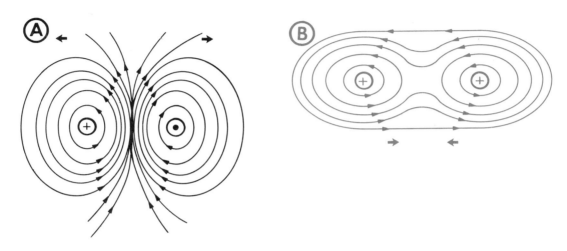

Figure 9-9 Opposite currents cause fields that repel; currents in the same direction cause fields that add and attract.

As long as the conductor is a straight piece of wire, the magnetic field produced is of little practical use. Although it has direction, it has no north or south pole. Also, unless the current is extremely high, the magnetic field has little strength. However, by changing the shape of a length of wire, you can greatly improve its magnetic characteristics.

Figure 9-10 shows two views of a short piece of wire twisted into a loop. Simply forming the loop helps the magnetic characteristics in three ways. First, it brings the flux lines closer together. Second, it concentrates the majority of the flux lines in the center or core of the loop. Third, it creates north and south poles. The north pole is the side where the flux lines exit the loop; the south pole, the side where they enter the loop. Thus, this loop of wire has the characteristics of a magnet. In fact, this is an example of a simple electromagnet.

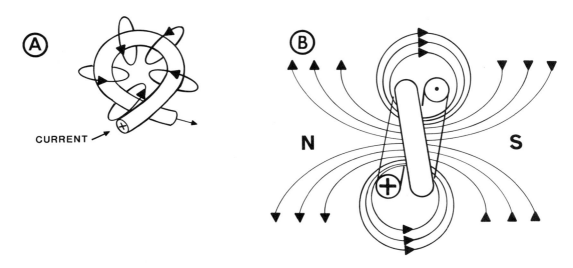

Figure 9-10 Flux lines around a loop of wire.

Magnetic Characteristics

Electricity, as you have learned, has many different characteristics which are described using such terms as voltage, current, resistance, conductance, and power. In much the same way, the study of magnetism requires that you learn several magnetic terms or characteristics. The most important of these characteristics are flux, flux density, permeability, and reluctance.

While these characteristics are very simplistic, the units of measure for these characteristics often become confusing. The reason for this is that three different systems of measurement have been used over the years. The first is the English system that uses the familiar inches, ounces, and pounds as units of measure. The other two systems are based on metric units. One is called the **cgs** system. Cgs stands for centimeter, gram, and second. The other system based on metric units is the **mks** system. Mks stands for meter, kilogram, and second. In the following discussion, the English units are used because you are probably most familiar with this system.

MAGNETISM

Flux
The complete magnetic field of a coil or a magnet is known as its flux. Thus, the flux is the total number of lines of magnetic force. In the English system, flux is measured in lines. A coil or magnet that produces 1000 lines of force has a flux of 1000 lines or 1 kiloline.

Flux Density
As the name implies, flux density refers to the number of flux lines per unit of area. In the English system, the unit of area is the square inch. Thus, flux density is expressed as the number of lines per square inch.

Permeability
Permeability is the ease with which a material can accept lines of force. It can also be thought of as the ability of a material to concentrate a large number of force lines in a small area. If you look back at Figure 9-5, the lines of flux passed through the iron bar easier than through air. Therefore, the iron bar has a higher permeability. Also, a 1-inch column of soft iron can hold hundreds of times more flux lines than a comparable column of aluminum. The Greek letter mu (μ) is used to represent permeability. Air is given an arbitrary permeability value of 1, and other materials are based on this standard.

Reluctance
The opposite or reciprocal of permeability is called reluctance and is represented by the letter R. Reluctance is generally defined as an opposition to flux. Thus, a material with high reluctance is reluctant to accept flux lines. Since reluctance is the reciprocal of permeability, it may be expressed by the equation:

$$R = \frac{1}{\mu}$$

For example, soft iron has a permeability of 2700. Thus, it has a reluctance of 1/2700. Since air has a permeability of 1, it has a reluctance of 1/1 or 1. Flux lines tend to follow the path of least reluctance.

Induction

Induction may be defined as the effect that the magnet has on the iron bar without any physical contact between them. Earlier, you saw that a charged body can induce a charge in another body simply by coming close to it. This is possible because an electrostatic field surrounds every charged body. Thus, the field of a charged body can affect another body without the two bodies actually touching. This is an example of electrostatic induction.

Magnetic Induction

Another type of induction is called magnetic induction. Everyone knows that a magnet can affect objects from a distance. A strong magnet can cause a compass needle to deflect even at a distance of several feet. A magnet can also induce a magnetic field in a previously unmagnetized object. For example, a magnet placed near a piece of iron can cause the iron to become magnetized.

Figure 9-11 shows a bar of soft iron close to a permanent magnet. Notice that part of the magnetic field passes through the iron bar. This happens because of the high permeability of the iron. Magnetic lines of force enter the left side of the iron and exit on the right. Thus, the piece of iron itself becomes a magnet.

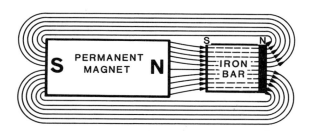

Figure 9-11 Magnetic induction.

In the iron bar, the south pole must be on the left since this is the end where the flux lines enter the iron. The north pole is on the right since the flux lines exit at this point. Notice that the north pole of the permanent magnet is closest to the induced south pole of the iron bar. Because these opposite poles attract, the iron bar is attracted to the magnet. Therefore, the attraction of a piece of iron by a permanent magnet is a natural result of magnetic induction.

When the piece of iron is removed from the magnetic field, the lines of flux no longer pass through the iron bar. However, the iron bar still retains a small amount of magnetism. Thus, the iron bar retains a weak magnetic field even after it is removed from the influence of the permanent magnet.

The magnetic field that remains in the iron bar is referred to as **residual magnetism**. The ability of material to retain a magnetic field even after the magnetizing force has been removed is called **retentivity**. Soft iron has a relatively low value of retentivity. Thus, it retains little residual magnetism. Steel has a somewhat higher value of retentivity. Therefore, its residual magnetism is also higher. Some materials, such as alnico, have a very high value of retentivity. In these materials, the residual magnetic field is almost as strong as the original magnetizing field.

Electromagnetic Induction

Electromagnetic induction is the action that causes electrons to flow in a conductor when the conductor moves across a magnetic field. Figure 9-12 illustrates this action. When the conductor moves up through the magnetic field, the free electrons are pushed to the right end of the conductor. This causes an excess of electrons at the right end of the conductor and a deficiency of electrons at the other end. The result is a potential difference that develops between the two ends of the conductor.

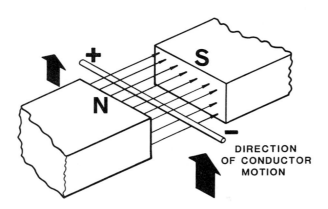

Figure 9-12 Electromagnetic induction.

DC ELECTRONICS

The potential difference developed in the conductor exists only while the conductor is moving through, or cutting, the flux lines of the magnet. When the conductor moves out of the magnetic field, the electrons return to their original positions and the potential difference disappears. The potential difference also disappears if the conductor stops in the magnetic field. Thus, there must be relative movement between the conductor and the magnetic lines of flux before a potential difference develops in the conductor.

Motion is essential to electromagnetic induction. Some outside force must be applied to the conductor in order to move it through the magnetic field. This mechanical force is converted to an electromotive force (EMF) by electromagnetic induction. In other words, an EMF is induced into the conductor. The potential difference across the conductor is called an induced EMF or an induced voltage.

The amount of EMF induced in a conductor is determined by four factors:

1. **The strength of the magnetic field.** The stronger the magnetic field, the greater the number of lines of flux in the field per unit area. If the magnetic field has a greater number of lines of force, then the moving conductor can cut a greater number of lines. The greater the number of lines cut for a given time period, the greater the induced EMF.

2. **The speed of the conductor with respect to the field.** As a conductor moves through a magnetic field, it cuts the magnetic lines of force. Increasing the conductor's speed through the field results in a greater number of lines of flux being cut for a given time period. This results in an increase in the EMF induced in the conductor.

3. **The angle at which the conductor cuts the field.** This might be a bit harder to visualize than the previous two statements. If one conductor moves perpendicular, or at right angles, to the magnetic lines of force, it will cut a maximum number lines of force per second and produce a maximum EMF. This movement is shown in Figure 9-12. An identical conductor moving at the same speed, but at an angle other than perpendicular to the field, will cut fewer lines of force in the same time, and therefore induce less EMF.

4. **The length of the conductor in the field.** A longer conductor must be coiled in order for it to fit into the magnetic field. Each loop in the coil cuts the field. Thus, the lines of flux are cut a greater number of times per second resulting in a greater induced voltage.

All four of these factors are a natural consequence of a basic law of electromagnetic induction. This law is called **Faraday's Law** and it states:

> The voltage induced in the conductor is directly proportional to the rate at which the conductor cuts the magnetic lines of force.

In other words, the more flux lines cut per second, the higher the induced EMF.

The polarity of the induced EMF can be determined by another of the left-hand rules. This one is called the **left-hand rule for generators** and is illustrated in Figure 9-13. It involves the thumb and the first two fingers of the left hand. The thumb is pointed in the direction the conductor is moving. The index or forefinger is pointed in the direction of the magnetic field: from north to south. Now, the middle finger is pointed straight out from the palm at a right angle to the index finger. The middle finger now points to the negative end of the conductor. This is the direction that current flows if an external circuit is connected across the two ends of the conductor.

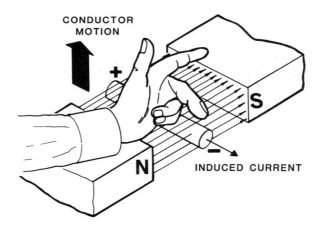

Figure 9-13 Left-hand rule for generators.

Chapter Self-Test

1. What is the name for the individual lines of force in a magnetic field?

2. State the interrelation between the two poles of a magnet.

3. State the four basic rules and characteristics of magnetic lines of force.

4. Using the left-hand rule for conductors, what does the thumb represent?

5. Why do the magnetic characteristics of a wire increase when the wire is formed into a coil?

6. Define permeability and reluctance, and state the relationship between the two.

7. When an iron bar is placed near a permanent magnet, the lines of flux pass through the iron bar, magnetizing it. When the iron bar is removed from the permanent magnet, a weak magnetic field remains within the iron bar. What is this remaining magnetic field called?

8. What term describes the ability of a material to retain a magnetic field after the magnetizing force has been removed?

9. What are the three requirements for electromagnetic induction?

10. State the four factors that determine the amount of EMF induced into a conductor by electromagnetic induction.

Summary

Action-at-a-distance is explained in terms of a field. Because a magnet can affect objects at a distance, it is said to be surrounded by a magnetic field. The field is assumed to be made up of lines of force called flux lines.

Magnets have north and south poles. Arbitrarily, the flux lines have been assigned a direction so that they leave the magnet at the north pole and enter the magnet at the south pole forming complete loops. The flux lines cannot cross each other and tend to form the smallest possible loops.

Like magnetic poles repel; unlike poles attract.

Magnetism and electricity are closely related. Current flow produces a magnetic field and a moving magnetic field can produce current flow.

The direction of the magnetic field caused by current flow can be determined by the left-hand rule for conductors. This rule states: Grasp the conductor in your left hand with the thumb pointing in the direction of current flow. Your fingers now point in the direction of the flux lines.

The magnetic field around the conductor can be strengthened and concentrated by winding the conductor as a coil. The result is called an electromagnet.

The north pole of the electromagnet can be determined by the left-hand rule for coils. It states: Grasp the coil in your left hand with your fingers wrapped around it in the direction of current flow. Your thumb now points toward the north pole of the coil.

Several magnetic quantities are important. Permeability is the ease with which a substance accepts lines of force. The reciprocal of permeability is called reluctance. Flux is the total lines of force around a magnet. Flux density refers to the number of flux lines per unit of area.

When one body has a magnetic field, it can induce a change in another body without actually touching the other body. This is called induction.

A magnet can induce a magnetic field into a body without touching it. This is called magnetic induction. When the magnet is taken away, a magnetic field will remain in the ferromagnetic body. This is called residual magnetism. The ability of a substance to retain a magnetic field after the magnetizing force has been removed is called retentivity.

When a conductor moves across a magnetic field, an EMF is induced into the conductor. This is called electromagnetic induction.

The magnitude of the induced EMF is proportional to the rate at which the conductor cuts the magnetic lines of force. The more lines per second that are cut, the higher the induced EMF.

The polarity of the induced EMF can be determined by the left-hand rule for generators. Using this rule, the thumb is pointed in the direction that the conductor is moving. The index finger is pointed in the direction of the flux lines (from north to south). If the middle finger is now placed at right angles to the thumb and index finger, it points in the direction in which current will flow through the conductor.

CHAPTER 10

Reactive Components

DC ELECTRONICS

Contents

Introduction 185

Chapter Objectives 186

Inductance. 187

Capacitance. 195

Capacitors 199

Chapter Self-Test 208

Summary. 209

REACTIVE COMPONENTS

Introduction

A resistor is known as a **resistive component**. As such, it offers a steady amount of opposition to current flow over a period of time—its opposition never changes. A **reactive component,** on the other hand, changes its characteristics. When a steady voltage is applied to a reactive component over a period of time, its opposition to current changes. Thus, the component **reacts** to the voltage.

The most common reactive components are capacitors and inductors. When you enter into AC circuitry, you will see the wide variety of applications these components have. However, before we begin with AC circuitry, there are certain applications in DC circuits that you should know.

This chapter introduces you to the inductor and the capacitor. You will read how these components operate, and learn the various ways these components can be arranged.

DC ELECTRONICS

Chapter Objectives

When you have completed this chapter, you will be able to:

1. Define the following terms: self-induction, steady-state, transient state, induction, inductance, CEMF, time constant, henry, dielectric, farad, capacitance, and polarized.

2. State the three factors that determine capacitance.

3. Identify the schematic symbol for the inductor and the capacitor.

4. Given the values of resistance and inductance, calculate the time constant for an inductive circuit.

5. Given the values of resistance and capacitance, calculate the time constant for an capacitive circuit.

6. Calculate the total inductance for series, parallel, and series-parallel circuits.

7. Calculate the total capacitance for series, parallel, and series-parallel circuits.

Inductance

Earlier you learned two rules that are quite important to the study of inductance. First, when current flows through a conductor, a magnetic field builds up around the conductor. Second, when a conductor is subjected to a moving magnetic field, a voltage is induced into the conductor. These two rules form the basis for a phenomenon called **self-induction**.

Self-Induction

Before we begin with self-induction, you must first learn the two conditions that can exist in any DC circuit. The first is called the **steady-state** condition while the other is called the **transient-state** condition. Up to this point, you have worked primarily with the steady-state condition.

Most DC circuits reach the steady-state condition within a fraction of a second after power is applied. In this condition, the current in the circuit has reached the value computed by Ohm's Law. That is, the current in the circuit equals the voltage applied to the circuit divided by the resistance of the circuit. However, because of other characteristics of the circuit, the current does not reach the steady-state value instantaneously. There is a brief period called the **transient time** in which the current builds up to its steady-state value. Thus, the transient condition only exists for an instant after power is applied to a circuit.

In circuits containing only resistors, the transient condition exists for such a short period of time that it can only be detected with sensitive instruments. However, if inductors or capacitors are used in the circuit, the transient condition may be extended so that it is readily apparent.

During the transient time, when the current is changing from zero to some finite value, **self-induction** occurs. Consider what happens during the transient time in the circuit shown in Figure 10-1A. When S_1 is closed, current begins to flow and a magnetic field builds up around the conductor as shown. However, the magnetic field does not just suddenly appear; it expands as current flow increases, starting at the center of the wire. If you look at a cross section of the wire (shown in Figures 10-1B and 10-1C), you see the gradual expansion of the magnetic field.

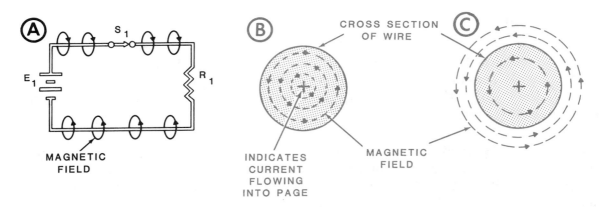

Figure 10-1 Self-induction.

As the magnetic field expands from the center of the wire, its lines of force cut across the wire. This fulfills the requirements for inducing a voltage into the wire; the relative motion between a magnetic field and a conductor. Thus, the sequence of events is as follows:

1. The switch is closed.

2. Current begins to flow through the wire.

3. A magnetic field begins to build up around the wire.

4. The moving magnetic field expanding from within the wire induces a voltage into the wire.

REACTIVE COMPONENTS

In Figure 10-2A, the tail of the arrow (indicated by the "+" sign) shows that current is flowing into the page. From the left-hand rule for conductors you know that the direction of the magnetic field is counterclockwise around the wire. Consequently, on the right side of the wire, the general direction of the field is as shown in Figure 10-2A. Also, as the magnetic field expands outward on the right, the relative motion is the same as if the conductor had moved to the left. Apply the left-hand rule for generators by pointing your thumb and forefinger as shown. Notice that your middle finger, which indicates the direction of the induced current, points out of the page. Thus, the induced current flows in the opposite direction to the original current.

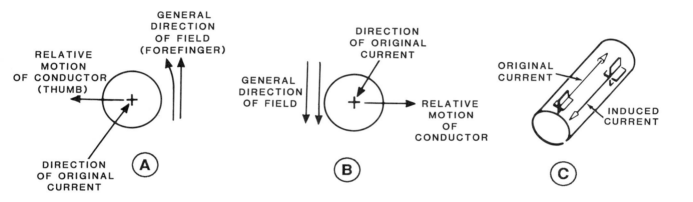

Figure 10-2 Determining the direction of the induced current.

Figure 10-2B shows that the same result is achieved if the left-hand rule is applied to the left side of the conductor. While it is true that the general direction of the field is reversed at this point, the relative motion of the conductor is also reversed so that the induced current still flows out of the page.

Figure 10-2C shows the relationship of the original current and the induced current. The original current induces a lower reverse current. These currents flow in opposite directions and the net result is that the original current is initially less than can be accounted for by Ohm's Law. That is, at least when power is first applied, there is more opposition to current flow than just the resistance of the circuit.

The induced current is caused by an induced EMF. The induced EMF attempts to force current counter to the original current. For this reason, the induced EMF is often called a **counter EMF**, or **CEMF**.

It should be pointed out that CEMF can only exist when the magnetic field is moving. This movement only occurs when the current through the conductor is changing in value—once the current steadies, the magnetic field is motionless. Therefore, CEMF can only exist during the transient time in a DC circuit.

However, the transient time of a circuit also occurs when current decreases. If the switch is opened, current will attempt to stop. This decrease in current **collapses** the magnetic field, and again, CEMF is induced. Obviously, once the switch is opened, current cannot flow. Nevertheless, CEMF attempts to keep current flowing.

The process by which CEMF is produced is called **self-induction.** The effect of self-induction is to oppose changes in current flow. If the original current attempts to increase, self-induction opposes the increase. If the original current attempts to decrease, self-induction opposes the decrease. Self-induction may be defined as **the action of inducing CEMF into a conductor when there is a change of current in the conductor.**

Inductance

Inductance is the ability of a device or circuit to oppose a change in current flow. Inductance may also be defined as the ability to induce CEMF with a change in current flow. **Induction** and **inductance** are easily confused, therefore it is necessary to discuss the difference between them.

Induction is the **action** of inducing CEMF with a change in current. Obviously then, induction exists only when a change in current occurs. Inductance, on the other hand, is the **ability** to induce CEMF with a change in current. If a circuit, or device, has this ability, it has it with or without current flow. Thus, inductance is a physical property. Like resistance, inductance exists whether current is flowing or not.

The unit of measurement for inductance is the **henry,** abbreviated with "H" or "h". It is named in honor of Joseph Henry; a nineteenth century physicist who did important research in this area of science. A henry is the amount of inductance that causes an EMF of 1 volt to be induced into a conductor when the current through the conductor changes at the rate of 1 ampere per second. In most electronics applications, the henry is an inconveniently large quantity. For this reason, **millihenry (mH)** and **microhenry (µH)** are more commonly used units of measurement.

The symbol for inductance is **L.** Thus, the statement "the inductance is 10 millihenries" can be written as the equation:

$$L = 10 \text{ mH}$$

Inductors

Since a magnetic field forms around any conductor when current flows through it, every conductor has a certain value of inductance. However, with short lengths of wire, the inductance value is so small that it can only be measured with sensitive instruments. In electronics, a specific amount of inductance is required in a circuit. A device that is designed for a specific value of inductance is called an *inductor*.

Inductors come in a variety of values from microhenries to several henries. The construction of the inductor is extremely simple. It consists of a length of wire coiled around some type of core. For this reason, the inductor is often called a *coil*.

Different inductance values can be obtained using different lengths of wire. Actually, the number of *coils* determines the amount of inductance. This is illustrated in Figure 10-3. In Figure 10-3A, a single loop is shown. As the magnetic field expands or contracts, it cuts this single turn of wire, and a small CEMF is induced. If two coils are used, as shown in Figure 10-3B, the field is twice as strong and both turns are cut by the entire field. Since both the field strength and the number of turns are doubled, CEMF increases by a factor of 4. Thus, the inductance is four times as great. Figure 10-3C shows three turns that produce a field strength three times as high as before. Now, three times the number of flux lines cut three times the number of turns. Thus, the inductance and the induced EMF increases by a factor of 3×3 or 9.

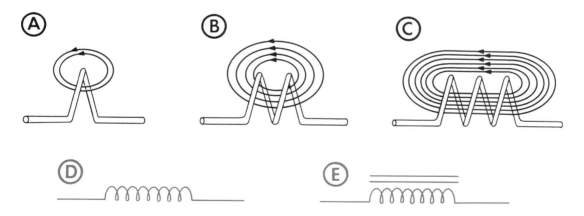

Figure 10-3 The inductor.

Another way to dramatically increase the inductance is to wind the coil around a core material that has a high permeability. For example, a coil wound on a soft iron core will have many times the inductance of an air-core coil. Figure 10-3D shows the schematic symbol for the air-core inductor, and Figure 10-3E shows the symbol for the iron-core inductor.

Time Constant of an Inductor

So far, we have determined that an inductor will oppose a change in current flow. In DC circuits, current only changes from power on to the point where it reaches its steady value. This was referred to as transient time. If a circuit has a small amount of inductance, the amount of CEMF induced is very small and the transient time is very short. If a circuit has a high amount of inductance, however, it will induce a large amount of CEMF, and the transient time will be longer.

But CEMF does not remain at a steady value, either. This is due to the complex relationship between inductance and current. As you know, changing current through an inductor produces CEMF. CEMF will then affect the current. Thus, the CEMF induced by an inductor is constantly changing. This relationship will not allow us to compute transient time. However, if the transient time is divided into 5 intervals, we can calculate the amount of time for one interval. This interval is known as the time constant. After the first time constant, current will have reached 63% of its maximum value. The time constant for a circuit is dependent upon the inductance of the circuit.

The time constant for a circuit is also affected by resistance. If there is a large amount of resistance, the total current value will be small. Thus, it takes less time to reach maximum. With less time to reach total current, the time constant will be smaller. If a circuit has a small amount of resistance, total current will be higher. Consequently, it takes longer for current to reach that value. Thus, there will be a longer time constant.

We can conclude that resistance and inductance affect time constants differently. Inductance is said to be directly proportional to the time constant—more inductance, the more time. Resistance, on the other hand, is inversely proportional to the time constant. If there is more resistance, there will be a smaller time constant. The relationship between the time constant (T), inductance (L), and resistance (R) can be expressed mathematically as:

$$T = \frac{L}{R}$$

For example, a circuit that consists of a 5 millihenry inductor and a 1 kilohm resistor will have a time constant of:

$$T = \frac{.005H}{1000\Omega}$$

T = .000005 seconds or 5 µs

REACTIVE COMPONENTS

For this circuit, 5 microseconds is the length of time it takes for the current traveling through the inductor to reach 63.2% of its maximum value. During each succeeding time constant, the current increases another 63.2% of the remaining difference between the present value and its maximum value.

If the maximum current through the circuit is 100 A, the current after one time constant is:

100 A x .632 = 63.2 A

Therefore, the current must increase another 36.8 amperes before it reaches maximum value. During the second time constant, the current increases by 63.2% of the remaining 36.8 amperes.

36.8 A x .632 = 23.25 A

This resulting 23.25 amperes is the amount of increase in current that occurs during the second time constant. To determine the current through the circuit after two time constants, you must add the second current increase to the first current increase:

63.2 A + 23.25 A = 86.45 A

This is the current through the inductor after two time constants.

During each succeeding time constant, the current continues to approach its maximum value. Theoretically, the current through the circuit never reaches its maximum value. However, for all practical purposes after five time constants the current through the inductor is considered to be at its maximum value.

Later, you will be introduced to time constant curves as a method of determining the percent of current or voltage after a given time constant.

Inductors in Combination

Inductors, like other components, can be connected in series, parallel, or series-parallel combinations. No matter what the combination, there are times when it is necessary to determine the total inductance of a circuit. The next two sections explain how this is done.

DC ELECTRONICS

Inductors in Series

If two inductors, each with 100 windings, are connected in series, they have the same overall effect on the circuit as one inductor with 200 windings. As the number of windings increases, the amount of inductance in the circuit increases. The inductance of the inductors is additive. That is, the inductance of any number of inductors in series can be determined by using the formula:

$$L_T = L_1 + L_2 + L_3 .....$$

As you can see, inductors in series are treated mathematically the same as resistors in series. The total inductance is equal to the sum of the individual inductances.

Inductors in Parallel

A single inductor in a circuit will have a given amount of inductance or opposition to a change in current flow. If another inductor of the same size is placed in parallel with the first inductor, the amount of inductance in the circuit decreases. Each inductor still opposes a change in current through the circuit. However, because there are now two current paths, the parallel inductive reactance decreases the total opposition to current flow.

The equations used to determine inductance in parallel are the same equations used to determine resistance in parallel. If all the inductors in a parallel circuit have an equal value, the value/number method can be used. If you have only two inductors, you can use the product/sum method. For any other circuit, you must use the reciprocal method. However, although we stated earlier that Ohm's Law was the easiest method to compute total resistance in parallel, you cannot use Ohm's Law to compute total inductance. The reason for this will be covered when you reach inductors in AC circuitry.

Inductors in Series-Parallel

If a number of inductors are connected in a series-parallel circuit, the total inductance is calculated in the same manner as the total resistance in a resistive circuit. First, the inductance of any parallel inductors is determined. This value is added to the inductance of any portion of the circuit in series with parallel inductance.

REACTIVE COMPONENTS

Capacitance

Capacitance is the property of a circuit or device which enables it to store electrical energy by means of an electrostatic field. A device especially designed to have a certain value of capacitance is called a capacitor. The capacitor has the ability to store electrons and release them at a later time. The number of electrons that it can store for a given applied voltage is determined by the capacitor's capacitance. Figure 10-4 shows the principle parts of a capacitor: two metal plates, separated by a non-conducting material called a dielectric. Often metal foil is used for the "plates", while the dielectric may be paper, glass, ceramic, mica, or some other type of good insulator.

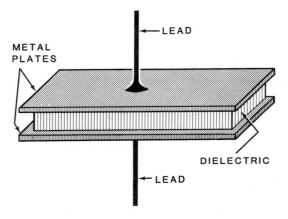

Figure 10-4 The capacitor.

Charging the Capacitor

One characteristic of the capacitor is its ability to store an electrical charge. In Figure 10-5A the capacitor is not charged. This means that there are the same number of free electrons in both plates. Thus, there is no difference of potential between the two plates and a voltmeter connected across the plates will indicate 0 volts. No current is flowing in the circuit because switch S_1 is open.

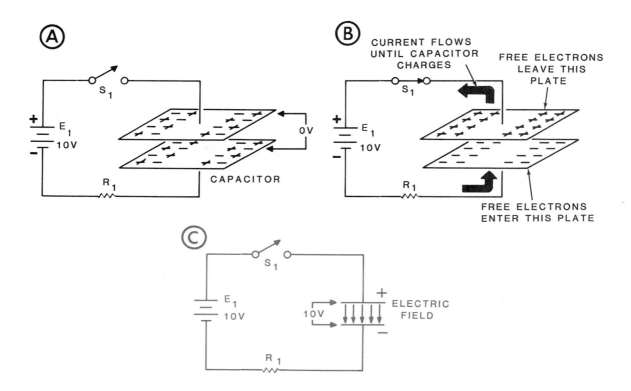

Figure 10-5 Charging the capacitor.

Figure 10-5B shows what happens when S_1 is closed. The positive terminal of the battery, connected to the upper plate of the capacitor, attracts the free electrons in the upper plate. These electrons flow out of the upper plate to the positive terminal of the battery.

At the same instant, the positive upper plate of the capacitor attracts the free electrons in the negative plate. However, because the two plates are separated by an insulator, no electrons can flow to the upper plate from the lower plate. Nevertheless, the attraction of the positive charge on the upper plate pulls free electrons into the lower plate. Thus, for every electron that leaves the upper plate and flows to the positive terminal of the battery, another electron leaves the negative terminal and flows into the bottom plate.

As the capacitor charges, a difference of potential begins to build up across the two plates. Also, an electrostatic field is established in the dielectric material between the plates. The capacitor continues to charge until the difference of potential between the two plates is the same as the voltage across the battery. In Figure 10-5B, current flows until the charge on the capacitor builds up to 10 volts. Once the capacitor has the same EMF as the battery, no additional current can flow because there is no longer a difference of potential between the battery and the capacitor. The capacitor is said to be **charged,** and will remain so until it is **discharged.** Our fully charged capacitor is shown in Figure 10-5C.

It should be emphasized again that although current flows in the circuit while the capacitor is charging, **current does not flow through the capacitor.** Electrons flow out of the positive plate and into the negative plate. However, electrons cannot flow through the capacitor because of the insulating dielectric. Moreover, if electrons did flow through the dielectric, the capacitor would not develop a charge in the first place. It would simply produce a voltage drop in the same way as a resistor.

Discharging the Capacitor

Theoretically, all of the energy stored in a capacitor can be recovered. Thus, a perfect capacitor will dissipate no power. It simply stores energy, and releases that energy at a later time. While a perfect capacitor cannot be built, it is possible to approach this condition. We previously stated that when a capacitor reaches the applied voltage, it is charged. The capacitor will remain in this condition for some time, unless the potential is released. Releasing the charge is known as **discharging** the capacitor.

Figure 10-6 illustrates the charge and discharge cycle. In Figure 10-6A, S_1 connects the battery and the capacitor. Current flows, and the capacitor charges to the battery voltage—10 volts. Notice the schematic symbol for the capacitor.

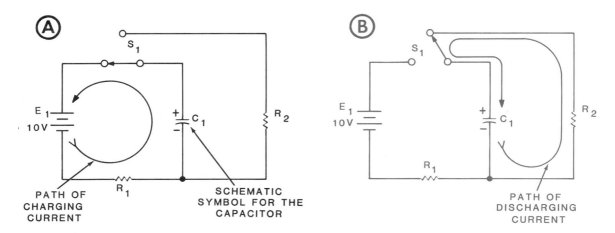

Figure 10-6 Discharging the capacitor.

When S_1 disconnects the battery, and connects the capacitor to R_1 (as shown in Figure 10-6B), the free electrons on the negative plate rush through R_2 to the positively charged plate. The flow of electrons continues until the two plates are once again at the same potential. At this time, the capacitor is said to be **discharged,** and current flow stops.

As the capacitor discharges, the voltage across it decreases. When completely discharged, the voltage across the capacitor is once again 0 volts. At this time all the energy which was initially stored has been released. The power consumed by R_2 is provided by the battery with C_1 acting as a temporary storage medium.

REACTIVE COMPONENTS

Capacitors

Unit of Capacitance

Capacitance is a measure of the amount of charge that a capacitor can store for a given applied voltage. The unit of capacitance is the farad, abbreviated F. This unit of measure gets its name from Michael Faraday; a scientist who did a great amount of research with capacitance.

One farad is the amount of capacitance that will store a charge of one coulomb of electrons when an EMF of one volt is applied. The farad is an extremely large value of capacitance. For this reason, the unit microfarad, µF, meaning one millionth of a farad, is more often used. Even the microfarad is frequently too large. In these cases the unit micro-microfarad, or picofarad (pF), is used.

There is an equation that expresses capacitance in terms of charge and voltage. The equation is:

$$C = \frac{Q}{E}$$

C is the capacitance in farads; Q is the charge in coulombs; and E is the EMF in volts.

Factors Determining Capacitance

A capacitor can be designed in many different ways in order to achieve a specific amount of capacitance. There are certain factors that must be taken into account, because they will directly determine the amount of capacitance.

The first of these factors is plate area. The amount of plate material in a capacitor will determine the maximum number of electrons the plate can hold. Obviously, a capacitor with a larger plate area (all other factors being equal) will be able to hold more electrons than a capacitor with a smaller plate area. If a capacitor can hold more electrons, it can store a higher potential difference, or voltage.

The second factor to be determined is plate distance. As one plate builds a positive potential, this potential attracts electrons to the other plate. This can only occur if the two plates are very close. As the plates get further and further apart, their ability to interact decreases. Thus, the wider the plate distance (all other factors being equal), the lower the ability to store a charge.

Capacitance can also be changed by changing the type of dielectric material. The dielectric material stores the electrostatic lines of force, and different insulators will perform this task differently. The ease with which an insulator supports electrostatic lines of force is indicated by its **dielectric constant.** Air is used as a reference, because it is a very poor dielectric, and most insulators will support the electrostatic field better than air. Air is arbitrarily given a dielectric constant of 1. Most insulators have a higher dielectric constant. For example, a sheet of waxed paper has a dielectric constant of about 3. This means that a 1 microfarad capacitor with an air dielectric would increase to 3 microfarads if waxed paper were substituted for air. Some typical dielectric constants for common types of insulators are given in Table 10-1. We can conclude that capacitance is directly proportional to the dielectric constant—the higher the dielectric constant, the higher the capacitance.

Material	Dielectric Constant (K)
Air	1
Vacuum	1
Waxed Paper	3 - 4
Mica	5 - 7
Glass	4 - 10
Rubber	2 - 3
Ceramics	10 - 5000

Table 10-1

Types of Capacitors

Capacitors are available in many different shapes and sizes. However, all capacitors can be placed in one of two categories: variable and fixed. An example of a variable capacitor is shown in Figure 10-7. The capacitance value of this type of capacitor can be changed by rotating the shaft. The capacitor plates are attached to the shaft. As the shaft is turned, the rotating plates change position in relation to the stationary plates. The rotating and stationary plates are electrically connected and each forms one plate of the capacitor. The rotating plates and stationary plates mesh together but do not touch. By moving the shaft, the area of the plates across from each other can be changed from maximum (when fully meshed) to minimum (when fully open). As you have seen, this changes the capacitance of the device.

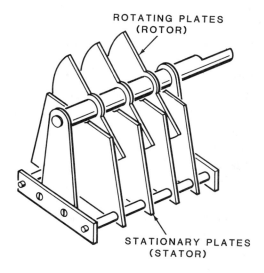

Figure 10-7 The variable capacitor.

Fixed capacitors are constructed by attaching leads to metal foils, and winding these foils with the dielectric between them. Figure 10-8A shows the construction of a common fixed capacitor, known as the **electrolytic** capacitor. Fixed capacitors are often named for the material used as their dielectric. Thus, in addition to electrolytic capacitors, there are paper, ceramic, and mica capacitors. Sometimes capacitors are classified according to their shape. Thus, there are disc capacitors and tubular capacitors.

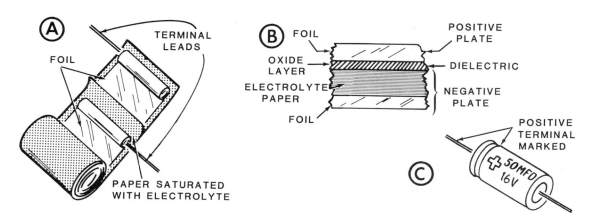

Figure 10-8 The electrolytic capacitor.

The construction of the electrolytic capacitor is different from the construction of an ordinary fixed capacitor. Sheets of metal foil are separated by a sheet of paper or gauze that is saturated with a chemical paste called an **electrolyte**. The electrolyte is a good conductor, and therefore, the paper is not the dielectric. Actually, the dielectric is formed during the manufacturing process. A DC voltage is applied across the foil plates; as current flows, a thin layer of aluminum oxide builds up on the plate that is connected to the positive side of the DC voltage source. As shown in Figure 10-8B, the oxide layer is extremely thin. Because the oxide is a fairly good insulator, it acts as a dielectric. The upper foil becomes the positive plate; the oxide becomes the dielectric; and the electrolyte becomes the negative plate. Notice that the bottom layer of foil simply provides a connection to the electrolyte.

Recall that capacitance is inversely proportional to the spacing between plates. Because the oxide layer is extremely thin, very high values of capacitance are possible with this technique. While most other capacitors have capacitances below 1 microfarad, the electrolytic capacitor may have capacitance values ranging anywhere from 1 microfarads to thousands of microfarads.

Because of its construction, the electrolytic capacitor is **polarized**. This means that the capacitor has a negative and a positive lead. When connected in a circuit, the positive lead must be connected to the more positive point. As shown in Figure 10-8C, the positive lead is marked on the electrolytic capacitor. At other times, the marking on the capacitor may indicate the negative lead.

CAUTION must be exercised when working with **polarized capacitors**.

When proper polarity is **NOT** observed, a polarized capacitor may explode. When a capacitor explodes, it can cause extensive damage to your circuit, but more importantly it is very dangerous to you and anyone who is nearby. An exploding capacitor sounds like a gun shot and has the same devastating results if anyone is in the line of fire.

Always check capacitors for polarity markings and proper installation, prior to applying power to a newly constructed circuit.

An important characteristic of the electrolytic or any other capacitor is its voltage rating. The voltage rating indicates the maximum voltage that the capacitor can withstand without the dielectric breaking down or arcing over. With electrolytic capacitors this value is generally printed on the capacitor, along with its capacitance value.

REACTIVE COMPONENTS

Since capacitors are made of two plates separated by an insulator (dielectric), they block DC current flow and will measure a very high resistance value once fully charged. Initially, when an ohmmeter is connected across a capacitor's plates, you may see the measured resistance begin at a low value, and increase until it stabilizes at a high resistance reading.

RC Time Constants

When a capacitor is connected across a DC voltage source, it charges to the applied voltage. If the charged capacitor is then connected across a load, it will discharge through the load. The length of time required for a capacitor to charge or discharge can be computed if certain circuit values are known.

During the discussion with inductors, we concluded that we could not compute the actual time, but an interval of time. The same is true for capacitors. Just as inductance and resistance determined the time constant for an inductive circuit, capacitance and resistance determines the time constant for a capacitive circuit. The time that it takes for the capacitor to charge or discharge is *directly* proportional to both resistance and capacitance.

It takes five time constants for a capacitor to completely charge. When charging first begins, the capacitor will charge to 63.2% of the applied voltage after the first time constant. During the second time constant, the capacitor will charge an additional 63% of the *remaining* voltage. The third time constant then charges to 63% of the remaining voltage, and so on. Actually, after the fifth time constant, the capacitor is not fully charged, but it is so close to fully charged that we can consider the capacitor to be fully charged.

The time constant for a capacitive circuit can be calculated if the capacitance and the resistance of the circuit are known. The equation is:

$$T = R \times C$$

In this formula, T is the time constant expressed in seconds; R is the resistance in ohms; and C is the capacitance in farads. As mentioned earlier, the farad is too large a volume to be practical and capacitance is most often expressed in microfarads. In the previous equation, if C is in microfarads and R is in ohms, then T will be in microseconds. If C is in microfarads and R is in kilohms, then T will be in milliseconds, Finally, if C is in microfarads and R is in megohms, then T will be in seconds.

To see how a capacitor charges, consider the following example. Figure 10-9 shows a 1 microfarad capacitor connected in series with a 1 megohm resistor. Using the previous formula, the time constant is equal to one second. Initially, the capacitor is completely discharged and the voltage across it equals 0 volts. When the arm of the switch is moved up so that the 100 volt power supply is connected to the RC network, the capacitor attempts to charge to the level of the applied voltage. However, the capacitor does not charge instantaneously. It takes a specific amount of time that depends on the circuit capacitance and resistance. In fact, it takes 5 time constants before the capacitor is considered to be fully charged. With a time constant of one second, we can conclude that it takes 5 seconds for the capacitor to become fully charged. Thus, after the first time constant, the capacitor will have charged to 63.2 percent of the applied voltage or to 63.2 volts.

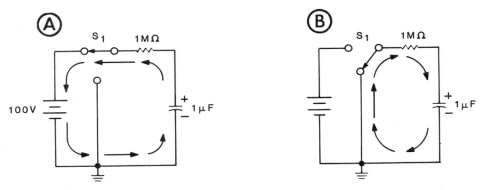

Figure 10-9 Working with time constants.

Figure 10-10 shows two time constant curves that can be helpful when working with time constants. Curve A shows how a capacitor charges. Initially, the capacitor charges rapidly, charging to 63.2 percent of the applied voltage during the first time constant. However, as time passes, the capacitor begins to charge more slowly.

REACTIVE COMPONENTS

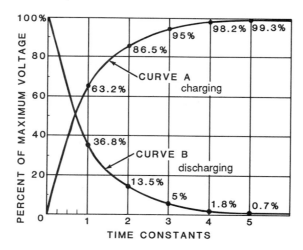

Figure 10-10 Time constant curves.

During the second time constant, the capacitor charges to 63.2 percent of the remaining voltage. In the example, the remaining voltage after 1 time constant is 100 V - 63.2 V = 36.8 volts. Now, 63.2 percent of 36.8 volts is about 23.3 volts. Thus, at the end of the second time constant, the voltage on the capacitor has risen to:

$$63.2 \text{ V} + 23.3 \text{ V} = 86.5 \text{ V}.$$

This is 86.5 percent of the applied voltage.

During the third time constant, the capacitor once again charges to 63.2 percent of the remaining voltage. Therefore, after 3 time constants, the capacitor has charged to 95 percent of the applied voltage. Once again, during the 4th time constant, the capacitor charges to 63.2 percent of the remaining voltage. Thus, after 4 time constants, it has charged to 98.2 percent; and finally, after 5 time constants, to more than 99 percent of the source voltage.

For most purposes, the capacitor is considered fully charged after five time constants. You may recognize this mathematical progression from the previous discussion of inductors. The percentages on the curves shown in Figure 10-10 are the same with inductors. Therefore, this figure can be applied to current calculations with inductors.

Because of its shape, curve A is called an exponential curve. The capacitor does not charge at the same rate each second, but at a rate governed by a power of a given factor. Therefore, the capacitor is said to charge exponentially.

DC ELECTRONICS

Curve B of Figure 10-10 shows the rate at which the capacitor discharges. At the first instant, the capacitor is fully charged. During the first time constant, the voltage drops by 63.2 percent to 36.8 percent of its original value. During the second time constant, the voltage drops an additional 63.2 percent or to only 13.5 percent of its original value. The voltage drops to 5 percent at the third time constant and to 1.8 percent after the fourth. After five time constants, the charge is less than 1 percent of its original value. For all practical purposes, the capacitor is considered fully discharged after five time constants.

Back in Figure 10-9A, the capacitor is charged to approximately 100 volts in 5 seconds. In Figure 10-9B, the capacitor is being discharged. According to curve B, the charge on the capacitor will be:

100 volts initially:

36.8 volts after 1 time constant (1 second);

13.5 volts after 2 time constants (2 seconds);

5 volts after 3 time constants (3 seconds);

1.8 volts after 4 time constants (4 seconds);

0.7 volts after 5 time constants (5 seconds).

As you can see, 0.7 volts is such a small fraction of the original 100 volts that, after five time constants, the capacitor can be considered as discharged.

Capacitors in Combination

Like other electronic components, capacitors can be connected in a variety of configurations. Thus, you should know how capacitors behave when connected together in different ways. Although the quantities differ, you will recognize the equations used to calculate capacitance values.

Capacitors in Parallel

When two capacitors are placed in parallel, the total capacitance increases. This is because when you connect both capacitors across the voltage source, you essentially increase the total plate area between the two terminals of the voltage source. In other words, when you connect capacitors in parallel, you are creating a single capacitor with a plate area of both capacitors combined.

Since total capacitance increases when capacitors are connected in parallel, we can determine the total amount of capacitance merely by adding the individual capacitance values. The formula for this is:

$$C_T = C_1 + C_2 + C_3, \text{etc.}$$

Capacitors connected in parallel will all charge to the same voltage. Remember, the voltage is the same across every section of a parallel network.

Capacitors in Series

Figure 10-11 shows what happens when capacitors are connected in series. A 225 picofarad capacitor with an air dielectric of 0.001 inch is shown in Figure 10-11A. When a second identical capacitor is placed in series with the first, the width of the dielectric is doubled. Thus, the overall distance is equivalent to a single capacitor with a width of 0.002 inches. Since plate distance and capacitance are inversely proportional, increasing the overall distance decreases total capacitance.

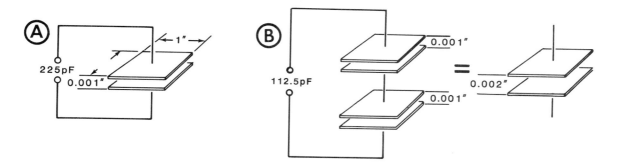

Figure 10-11 Connecting capacitors in series decreases capacitance.

The total capacitance of a group of series capacitors is calculated in the same way as the total resistance of parallel resistors. Or stated more simply, capacitors in series combine in the same way as resistors in parallel. The total capacitance of two capacitors in series can be calculated by using any of the three methods for total resistance in parallel—value/number, product/sum, or the reciprocal method.

DC ELECTRONICS

Chapter Self-Test

1. Define self-induction.

2. What is the difference between induction and inductance?

3. What is the unit of measurement for inductance?

4. Define CEMF.

5. What is the equation for finding total inductance in a series circuit?

6. How is electrical energy stored in a capacitor?

7. What happens to the current in a circuit while a capacitor is charging?

8. Name three factors that determine the amount of capacitance for a capacitor.

9. When a capacitor is charging or discharging, the amount of time is divided into five time constants. After the first time constant, what is the percentage that capacitor is charged or discharged?

10. What is the total capacitance for a series circuit with five 25 microfarad capacitors?

REACTIVE COMPONENTS

Summary

When current flows through a conductor a magnetic field builds up around the conductor. As the magnetic field builds up, a voltage is induced into the conductor. The induced voltage opposes the applied voltage, and is called counter EMF, or CEMF. The process by which the counter EMF is produced is called self-induction.

CEMF is always a polarity that opposes changes in current. It opposes the increase in current that occurs when power is applied to a circuit. It also opposes the decrease in current that occurs when power is removed. The ability of a device to oppose a change of current is called inductance. The unit of inductance is the henry.

A device designed to have a specific inductance is called an inductor. The inductor consists of turns of wire wrapped around a core. The greater the number of turns and the higher the permeability of the core, the greater will be the value of the inductance.

The capacitor consists of two metal plates separated by an insulator called a dielectric. It has the ability to store an electrical charge. This ability to store a charge is called capacitance. When connected to a voltage source, the capacitor charges to the value of the applied voltage. If the charged capacitor is then connected across a load, it will discharge through the load.

The unit of capacitance is the farad; although the microfarad and the picofarad are more commonly used. Three factors determine the value of a capacitor. They are: the area of the plates, the spacing between the plates, and the dielectric constant.

There are many different types of capacitors. They are generally classified by their dielectric. The most popular types are air, paper, mica, ceramic, and electrolytic.

The length of time required for a capacitor to charge is determined by the capacitance and the resistance in the circuit. A time constant is the length of time required for a capacitor to charge to 63.2 percent of its applied voltage. The equation for finding the time constant is:

$$T = R \times C$$

The time constant chart is used when working with time constants. It shows the manner in which capacitors charge and discharge. It plots the number of time constants against the percent of the applied voltage.

Capacitors may be connected in series or in parallel. When connected in parallel, the total capacitance is equal to the sum of the individual capacitor values. The equation is:

$$C_T = C_1 + C_2 + C_3, \text{etc.}$$

When capacitors are connected in series, the total capacitance decreases, much like resistors in parallel. Thus, to compute the total capacitance in a circuit, you may use the value/number, product/sum, or the reciprocal methods.

SECTION 11

INTRODUCTION TO AC
MEASURING AC
AC RESISTANCE
AC CAPACITANCE
AC INDUCTANCE
RLC CIRCUITS
TRANSFORMERS

This next section covers alternating current, or AC. While DC fulfills many necessary requirements in electronics, there are a large number of additional needs in today's technology where direct current is ill-suited, and alternating current is preferred. For instance, alternating current is much easier to produce and transmit for the electricity in homes and businesses. Also, a steady, unchanging voltage cannot be used for transmitting audio and video signals because the voltage level never changes. The constantly changing voltage and current levels of an AC signal make it perfect for just such applications.

This section introduces you to alternating current and alternating voltage. We will describe how AC differs from DC, and how it operates within a basic resistive circuit. We'll also discuss the various methods of measuring an AC signal, and introduce you to the oscilloscope; perhaps the single most important piece of test equipment for electronics. We'll discuss in detail how capacitors and inductors act when an AC voltage is applied, and introduce you to the transformer; a component which only performs in AC circuits. Because of the wide variety of uses, AC is an extremely important aspect to electronics which must be thoroughly understood.

AC ELECTRONICS

CHAPTER 11

Introduction to AC

AC ELECTRONICS

Contents

Introduction 213

Chapter Objectives 214

The Importance of AC 215

Generating AC 218

The Sinusoidal Waveform 229

AC Values 235

Nonsinusoidal Waveforms 244

Chapter Self-Test 247

Summary 248

Introduction

You have spent the entire first section of this book working with direct current, or DC, because it was much easier to define other electronic characteristics with DC. However, you may have noticed that DC, to this point, is only available with batteries. Actually, you *can* create a DC signal, but you need an AC signal to do so.

Alternating current is everywhere—it provides the power in your home, and makes possible an almost endless variety of applications. Electronics today could not exist without the AC signal.

This chapter will introduce you to the terminology associated with AC. It will show you how AC is produced, and the characteristics of the basic AC signal. You will also be introduced to other signals which, although they are not sine waves, still carry the basic characteristics of an alternating signal.

AC ELECTRONICS

Chapter Objectives

When you have completed this chapter, you will be able to:

1. Define the following terms: alternating current (AC), sine wave, cycle, alternation, instantaneous value, peak value, peak-to-peak value, effective value, average value, period, frequency, and hertz.

2. State the difference between alternating current and direct current.

3. Describe the operation of the basic AC generator.

4. Determine the peak value of a sine wave.

5. Determine the peak-to-peak value of a sine wave.

6. Determine the effective value of a sine wave.

7. Calculate the frequency of an AC signal.

8. Calculate the average value of a sine wave.

9. Determine the period of a sine wave.

10. List three types of AC waveforms, other than the sine wave.

INTRODUCTION TO AC

The Importance of AC

Because of its characteristics, **alternating current,** or **AC,** as it is popularly known, is suitable for a variety of commercial, industrial, and military applications. This makes AC far more useful than DC. The following discussion will briefly describe *what* AC is, and its many advantages over DC. This will explain to you why AC is used more frequently than DC.

What is AC?

Unlike direct current (DC), which flows only in one direction, alternating current (AC) periodically changes its direction of flow. In other words, alternating current flows first in one direction and then in the opposite direction.

This difference between DC and AC is illustrated in Figure 11-1. Figure 11-1A shows a resistor connected to a DC voltage source. The voltage source provides a fixed value of EMF to the circuit. Current flow, therefore, reaches its maximum value almost immediately after power has been applied. Current will then remain at that value until power is removed. Current will flow from negative to positive, and continue to flow in that direction, based on the polarities of the voltage source.

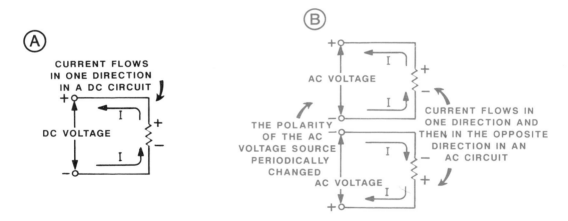

Figure 11-1 Current flow in AC and DC circuits.

Figure 11-1B shows how alternating current flows through a resistor. In the top figure, current flows in a counterclockwise direction while, in the bottom figure, current flows clockwise. Notice also that the polarity of the source voltage, as well as the polarity of the voltage drop across the resistor, changes in these illustrations. The change in the polarity of the voltage applied to the circuit results in a change in the direction of current flow, and a change in the polarity of the voltage drop. The voltage source shown here is generally referred to as an **AC voltage.** The resistor and the voltage source form a basic AC circuit.

AC ELECTRONICS

It is also important to note that direct current usually has a steady or constant value. This is because it is usually produced by a DC voltage that has a fixed value. However, momentary changes may result if the DC voltage is adjusted to a higher or lower value, or when the circuit resistance changes in value. However, in most DC circuits we are concerned with a steady current that always flows in one direction. By comparison, an alternating current changes in both value and direction. In other words, the current in an AC circuit will increase from zero to some maximum value, and then drop back to zero as it flows in one direction. The polarity of the applied voltage then changes, and current flows in the same manner, but the opposite direction. The exact manner in which the current increases and decreases in each direction can be controlled. This makes it possible to produce various types of AC signals. In fact, a variety of AC voltage sources are used to generate different types of AC signals that are suitable for a variety of applications. A variety of these signals will be examined in this chapter.

Why is AC Used?

Alternating current is widely used because of its versatility. Since AC changes in both value and direction, it has characteristics which can be utilized in a wider range of applications than is possible with DC currents. For example, when a large amount of electrical energy is required for a particular application, it is much easier to generate and transmit alternating current instead of direct current. In applications where large amounts of power are required, devices such as batteries (which produce DC voltages suitable for low power applications) cannot be used. In these applications, electromechanical devices known as generators are used to generate the high voltages and currents required.

A **generator** is defined as a device that converts mechanical motion into a voltage. Generators will be discussed in more detail later in this chapter. Generators can be used to produce both DC and AC electricity. AC generators (when compared to DC generators) are less complex, can be constructed in larger sizes, and are often more economical to operate. Therefore, AC electrical power is simpler, easier, and cheaper to produce than DC electrical power.

You can easily change an AC voltage to a higher or lower voltage by passing it through a device known as a **transformer**. The transformer is an AC component and it will be explained in its own chapter.

When needed, the alternating current can be easily converted into direct current by a process called **rectification**. The basic rectification process is accomplished using a solid-state component called a diode. Diodes are semiconductors, and will be covered in greater detail in the next section.

INTRODUCTION TO AC

While it is true that DC power can be converted into AC power, the process is much more complex, more expensive, and less efficient than converting AC into DC. Therefore, when AC is used as the primary source of electrical power, DC can be obtained.

The characteristics and features just described may seem to indicate that AC is useful only because it can serve as a source of electrical power to operate electronic equipment. However, this is not the case. Alternating current is also used extensively to transmit information from one location to another. This information carrying capability results because the characteristics of an alternating current or voltage can be made to vary in a desired manner. In other words, the magnitude or amplitude (maximum value in each direction) can be varied to represent intelligence or information. Even the rate at which the alternating current changes direction can be varied to represent intelligence. In this way, information can be inserted within an alternating current or voltage, thus making it possible for the AC to carry information. When alternating currents or voltages are used to carry information they are often referred to as **AC signals**. AC signals are widely used in electronics to carry information from one point to another within an electronic circuit. These signals can also be retransmitted over long distances by using long wires or transmission lines.

Alternating current may also be converted into electromagnetic waves (also called radio waves) which can radiate or travel through space. This is possible because a conductor which carries alternating current is surrounded by a magnetic field. The field expands and collapses as the intensity and direction of the current changes. If the current changes at a sufficiently high rate of speed, the magnetic field will actually radiate outwards and the radiated energy will vary in accordance with the alternating current. This means that AC signals (which contain information) can be transmitted from one location to another, without the use of wires or transmission lines. This cannot be accomplished with direct current.

The points just discussed illustrate a few of the reasons why AC is used. Although there are many additional factors to consider, we can summarize by saying that AC is primarily used to either provide electrical power, or to provide a means of transmitting information or intelligence from one point to another.

Generating AC

Although alternating current may be generated in a number of ways, the most basic means of obtaining AC is by using an electromechanical device known as an **AC generator** or **alternator**. An AC generator operates on the principle of **electromagnetic induction,** which was discussed in Chapter 9.

Electromagnetic Induction

In Chapter 9, you were introduced to the concept of electromagnetic induction, and the left-hand rule for generators. These concepts are extremely important in the AC generator, so we will examine them more closely.

Figure 11-2 illustrates both of these concepts. As we describe the action of the conductor and the magnetic field, compare their actions to the diagram of the left-hand generator rule.

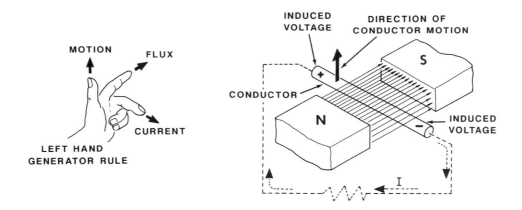

Figure 11-2 Voltage induced in a conductor by movement in a magnetic field.

The north and south poles of the magnet are separated. A fundamental law of magnetism states that unlike poles attract, and lines of flux exist between the two poles traveling from north to south. As the conductor passes upward through the lines of flux, the free electrons within the conductor are forced to move. However, this movement will only occur if there is a complete path for current *outside* the conductor, shown by the dotted line. If this path exists, current will travel from the negative end of the conductor to the positive end.

Note the similarity between the actions we described and the left-hand generator rule. If lines of flux are moving in the direction of the index finger, and a conductor cuts these lines in the direction of the thumb, the induced voltage will force current in the direction of the middle finger.

In order for electromagnetic induction to occur, you need a conductor, a magnetic field, and **relative motion** between the two. For Figure 11-2, the motion was carried out by the conductor. However, the motion can also occur by a moving magnetic field. In Chapter 9, we discussed **self-induction,** where a changing current moved the magnetic field through the conductor, inducing a voltage. However, for the AC generator, the conductor will be moving through a stationary magnetic field.

For an AC generator, both the polarity and the magnitude of the induced voltage are important. The polarity of the voltage induced within the moving conductor is determined by the direction that the conductor is moving, and the direction of the magnetic field. Notice the direction of conductor motion and the direction (north to south) of the magnetic flux in Figure 11-2. Note the polarity of the induced voltage and the direction that current will travel when a complete current path is provided. If either the direction of conductor motion or the direction of the magnetic field is reversed, the polarity of the induced voltage will be reversed, and current will flow in the opposite direction.

The magnitude of the induced voltage is determined by four factors. These factors, as we originally mentioned in Chapter 9, were:

1. The strength of the magnetic field.

2. The speed of conductor movement.

3. The length of the conductor in the field.

4. The angle at which the conductor cuts the field.

Each of these factors have one thing in common—they determine the number of flux lines the conductor cuts per second. If you will recall Faraday's Law from Chapter 9, the factors combined into a simple rule:

> The voltage induced in a conductor is directly proportional to the rate which the conductor cuts the magnetic lines of force.

AC ELECTRONICS

The word "rate" in this rule is used to indicate the number of lines of force cut per second. The rule indicates that the induced voltage is proportional to the number of lines of force cut per second. When more lines of force are cut per second, the induced voltage increases. When fewer lines are cut per second, the voltage decreases. The rate can be changed by the strength of the field, the speed of conductor motion, the length of the conductor, or the cutting angle. Let's discuss each of these factors in greater detail.

First, the induced voltage is affected by the strength of the magnetic field. A stronger magnetic field results in more lines of force per unit area. This means that there are more lines to cut and the induced voltage is increased. When the field strength is reduced, fewer lines of force exist, and the induced voltage decreases.

Induced voltage also depends on the speed of conductor movement. This particular factor should be rather obvious. The faster the conductor moves, the greater the induced voltage. This is because the faster moving conductor cuts more lines of force per second, increasing the rate. When the speed of the conductor is reduced, fewer lines of force are cut per unit of time and induced voltage decreases.

The length of the conductor within the magnetic field also affects the induced voltage. The longer the conductor, the greater the induced voltage. The longer conductor cuts more lines of force as it moves through the magnetic field. A shorter conductor cuts fewer lines of force, and induced voltage is decreased.

The angle at which the conductor cuts the magnetic field also affects the induced voltage. When a conductor moves at a right angle (90 degrees) with respect to the magnetic field, as shown in Figure 11-2, maximum voltage is induced. When the angle between the field and the direction of conductor motion decreases, induced voltage decreases.

The angle of movement is perhaps the single most important factor when working with an AC generator. When we discuss the generator later in this chapter, our other three factors will remain constant—the strength of the field, the conductor length, and the speed of the conductor. The angle at which the conductor cuts the field, however, constantly changes.

The differences in the cutting angles are shown in Figure 11-3. If the conductor (viewed edgewise) moves straight up from the starting position (direction A), it moves at a right angle (90 degrees) with respect to the magnetic field. The conductor is cutting the maximum number of lines per unit of time, and the induced voltage is maximum. Maximum voltage will also be induced if the conductor moves in direction E. However, since the conductor is moving in the *opposite* direction, the induced voltage is opposite. For direction A, current flows out of the page, while current flows into the page for direction E.

INTRODUCTION TO AC

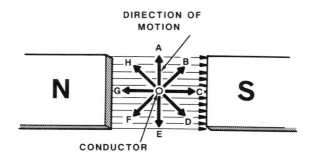

Figure 11-3 Cutting magnetic lines of force at different angles.

When the conductor moves in direction B, the cutting angle decreases below 90 degrees. This means that the conductor must travel further between lines of force. This results in fewer lines being cut per second (all other factors being equal). The induced voltage would be less than it is at points A or E. This is also true when the conductor moves in direction H.

In either direction, B or H, the cutting angle will be the same and the induced voltage will be the same. In each case, the induced current flows out of the page. The same cutting angle is obtained when the conductor moves in the direction of D or F. The amount of induced voltage is the same in both examples. However, in direction D or F, the induced current flows in the opposite direction.

When the conductor moves in the direction of C or G, the cutting angle is effectively zero. At this time, the conductor is parallel with the lines of force, and no lines are cut. Under these conditions, no voltage is induced in the conductor and no current flows.

A Simple AC Generator

A simple AC generator may be formed by bending a wire or conductor in the form of a loop, and then mounting the loop so that it can rotate within a magnetic field. When the wire loop rotates, an AC voltage is induced into the loop. The only other considerations are to provide a convenient means of extracting the AC voltage generated within the rotating loop, and applying this voltage to the load.

AC ELECTRONICS

Generator Construction

An AC generator is shown in Figure 11-4. It consists of a wire loop called an **armature**, which is mounted so that it rotates within a magnetic field. The magnet used for this purpose is commonly referred to as a **field magnet**. The field magnet is constructed so that it produces a strong, concentrated magnetic field between its poles. It can be either a permanent magnet or an electromagnet. The electromagnetic is preferred in applications where a high field strength is required to produce substantial output power.

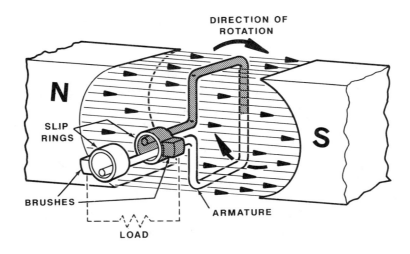

Figure 11-4 A basic AC generator.

The AC voltage induced in the rotating armature must be extracted at the ends of the wire loops, which form the armature. However, the armature constantly turns, thus making it impossible to permanently attach any wires, or leads, directly to the armature. For this reason, it is necessary to use some type of sliding contact at each end of the wire loop. As you can see in Figure 11-4, two cylindrical metal rings are attached to the opposite ends of the loop. These metal rings are called **slip rings**. An external circuit, or load, is connected to these slip rings through contacts which are held against the rings.

The contacts are made from a soft but highly conductive material (usually carbon) and are called **brushes**. The brushes slide against the slip rings as the armature turns. The brushes serve as two stationary contacts to which an external load can be connected. The brushes are the output terminals of the generator. Simply stated, the AC output voltage is applied from the armature to the load, through the brushes.

INTRODUCTION TO AC

Generator Operation

In order to function properly, an AC generator must operate so that its armature rotates at a constant speed. As the armature rotates in the magnetic field, one side moves down through the magnetic field while the other side moves up. It is important to note that during a revolution of the armature, each side must move down and then up through the field. Furthermore, each side of the armature always remains in contact with its respective brush, through a slip ring. Keeping these considerations in mind, let's examine the basic action that takes place during one complete revolution of the armature.

An armature is shown in four specific positions in Figure 11-5. These are intermediate positions which occur during one complete revolution of the armature. For the sake of explanation, one side of the armature and its associated slip ring and brush are shaded, and the other side is white. The shading is used to help you keep track of each side of the armature. Also, a resistive load is connected to the brushes so that a complete circuit is formed. The complete circuit allows current to flow through the armature to the load. The output voltage is monitored by a voltmeter. Notice that the voltmeter is connected across the load.

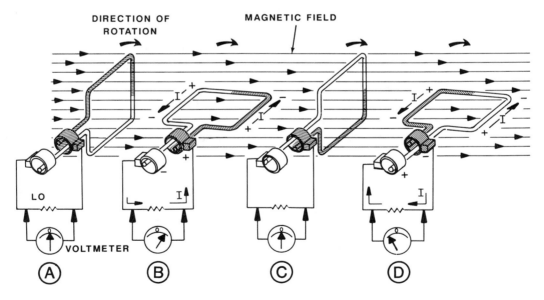

Figure 11-5 Generating an AC voltage.

AC ELECTRONICS

Assume that the armature starts rotating in a clockwise direction, from the initial position shown in Figure 11-5A. Notice that initially the shaded side of the armature is on top and the white side is on the bottom. As the armature moves from this starting position, the shaded side moves from left to right, and the white side moves from right to left. However, both sides are moving parallel to the lines of force. No lines are cut, and the voltmeter indicates zero.

As the armature rotates to the position shown in Figure 11-5B, the shaded side moves down through the field while the white side moves up. The two sides of the armature cuts the magnetic lines of force in opposite directions. The polarity of the voltage induced in the shaded side is opposite the polarity of the voltage induced in the white side. However, the voltages induced in each side are series-aiding, and the two sides of the armature form a complete loop. These induced voltages are equal in value. Therefore, the voltage which appears at the brushes is equal to the sum of the voltages induced in each side.

The polarity of these voltages are shown in Figure 11-5B, along with the resulting currents. Notice that the series-aiding voltages produce a current that circulates through the armature and the load.

As you examine Figure 11-5B, notice that the armature is horizontal. In other words, the two sides of the armature are cutting the magnetic lines of force at right angles, resulting in maximum induced voltage. At this time, the output voltage applied to the load is at its maximum value, as indicated by the voltmeter. It is important to note that the output voltage does not suddenly jump from zero to maximum. The output voltage increases at a specific rate. As the armature rotates from the position shown in Figure 11-5A to the position shown in Figure 11-5B, it cuts the magnetic lines of force, at an ever increasing angle, until maximum voltage is obtained. This causes the output voltage to increase smoothly from zero to its maximum value.

When the armature rotates from the position shown in Figure 11-5B to the position shown in Figure 11-5C, it cuts the magnetic lines of force at a continually slower rate. When the armature reaches the position shown in Figure 11-5C, the opposite sides of the armature are moving parallel to the lines of force and no flux lines are cut. This means that the output voltage decreases from maximum to zero as the armature moves to the position shown in Figure 11-5C. The corresponding load current also decreases to zero as the output voltage decreases.

INTRODUCTION TO AC

At this point in the discussion, the armature has completed one-half of a revolution, and has produced an output voltage that has increased from zero to a maximum value, and decreased back to zero. The corresponding load current also varied in the same manner. It is important to note that up to this point the output voltage has changed in magnitude, but *not* in polarity. This is defined as an **alternation**. To be more specific, from zero to the positive maximum and back to zero is described as the **positive alternation** of an AC waveform. Later, this will be described as being one-half of a sine wave.

As the armature continues its rotation from the position shown in Figure 11-5C to the position shown in Figure 11-5D, the opposite sides of the armature move across the magnetic lines of force, in opposite directions. However, the black side of the armature now moves up and the white side moves down. This is exactly opposite to the situation which occurred during the first one-half revolution. Therefore, the voltages induced into each side of the armature have polarities which are opposite to those induced earlier. These induced voltages are series-aiding to produce an output voltage.

The output voltage now has a polarity exactly opposite to the polarity that was produced earlier. As the armature moves to the position shown in Figure 11-5D, the new output voltage (of opposite polarity) increases to maximum because the armature cuts the lines of force at the fastest rate. This maximum voltage is indicated on the voltmeter. The load current is also maximum at this time and it flows in the direction shown. The amplitude or value is exactly the same as before, but the polarity of the voltage is opposite its initial direction.

The armature completes a full revolution by returning to its initial position as shown in Figure 11-5A. As the armature continues rotating toward its initial position, the armature cuts the lines of force at a decreasing rate until no lines are cut. This causes the induced voltage within the armature and the resultant output voltage to decrease to zero. The corresponding load current is also zero at this time.

As you can see, one complete revolution of the armature produces a voltage (and corresponding load current) that changes in both magnitude and polarity. Therefore, the output voltage and current are AC values. This discussion should not be considered a detailed description of exactly how these AC quantities vary. We have only shown what an AC output is, and how an AC output voltage is produced by generator action.

AC ELECTRONICS

Let's look again at the generator action that was just described and consider exactly how an AC output voltage is generated. We will again rotate the armature one complete revolution (360 degrees) through the magnetic field. This time, instead of examining only the maximum and minimum points, we will consider the action that takes place at a number of intermediate armature positions. This example is shown in Figure 11-6. Since the voltage induced in each side of the armature is series-aiding, we can observe one side of the armature. The one-half armature loop will rotate through 16 different positions in a clockwise direction.

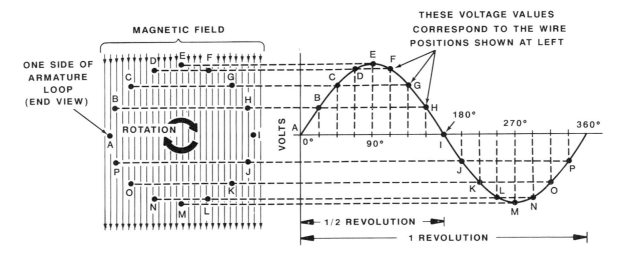

Figure 11-6 Plotting an AC output.

At position A, the wire moves parallel with the lines of force and no voltage is induced. However, as it moves through Position B, C, D, to E, it cuts lines of force at an increasing rate. At position E, the wire moves at a right angle (90 degrees), with respect to the field. The voltage induced in the wire is proportional to the rate at which the lines of force are cut. Since the cutting rate increases in a nonlinear manner, the voltage increases in the same way.

INTRODUCTION TO AC

This increasing nonlinear voltage is shown graphically on the right side of Figure 11-6. The voltage graph is plotted by marking off the angular rotation of the wire (in degrees) along a horizontal line. This horizontal line is calibrated for one complete revolution of the wire (360 degrees). The value or amplitude of the voltage that is induced within the wire is plotted vertically as shown. Notice that five voltage values have been plotted that correspond to wire positions A through E. Each successive value is higher and corresponds to the rate at which the magnetic lines of force are cut. These values are then connected to form a continuous line. The resulting curve shows how the induced voltage increases when the armature rotates one-quarter of a revolution which is 90 degrees. The voltage value increases rapidly from zero (at zero degree rotation) and tapers off to a maximum value (at 90 degrees of rotation).

As the wire continues its rotation past position E and moves through positions F, G, H, and I, the number of lines cut per second decreases because the armature is closer to moving in parallel with the flux lines. This causes the value of the induced voltage to decrease from a maximum value (at 90 degrees) to a minimum value (at 180 degrees). At position I, the wire is moving parallel to the magnetic flux lines. At this time, no lines are cut, and the induced voltage is again zero.

The induced voltage values which correspond to wire positions F through I are plotted on the voltage graph. These values (F through I) are successively lower, with the value at I being equal to zero. When these voltage values are placed in a continuous line, you can see that the voltage decreases slowly from its maximum value, at an increasing rate, until it reaches zero. The voltage curve between values E and I indicates how the voltage decreases when the armature rotates an additional 90 degrees, which is the second quarter of one revolution.

To this point, the armature has rotated one-half of one revolution (180 degrees) and has produced a voltage that varies from zero to a maximum value and back to zero. When the armature moves past position I, it starts cutting flux lines. This time the armature loop is moving in the opposite direction through the magnetic field. Therefore, the polarity of the voltage induced in the armature during this alternation is *opposite* the polarity of the voltage produced during the first one-half revolution (positive alternation).

As the armature moves from position I through positions J, K, L, and M, it cuts the lines of force at an increasing rate. The amplitude of the induced voltage is the same as it was when the armature moved from position A to position E. The only difference is that the induced voltage now has the opposite polarity. To show that the polarity is opposite, the voltage values which occur at positions J, K, L, and M are plotted below the horizontal line as shown. This portion of the curve shows that the voltage varies, from zero to a maximum value, in the opposite direction as the armature wire rotates from the 180 degree position to the 270 degree position (position M).

AC ELECTRONICS

When the armature moves to positions N, O, and P, it cuts the lines of force at a decreasing rate. When it reaches position A (its initial starting position), it has completed one complete revolution (360 degrees). At this point, the armature is again moving parallel to the lines of force and no flux lines are cut. The voltage drops from its maximum value back to zero, as shown. The voltage produced during the second half of rotation (between 180 degrees and 360 degrees), increases from zero to a maximum negative value and back to zero. Since the induced voltage between 180 and 360 degrees is opposite in polarity to the positive alternation (induced between 0 and 180 degrees), it is known as the **negative alternation**.

One complete revolution of the armature produces an AC voltage that varies in both amplitude and direction, as indicated on the right side of Figure 11-6. During one-half of the revolution (the positive alternation), the voltage increases from zero to maximum and back to zero. During the next half of a revolution (the negative alternation), the voltage increases from zero to a negative maximum and back to zero.

When this AC voltage is applied to a load, the resulting current through the load varies in the same manner. In other words, the current increases and decreases in one direction, then increases and decreases in the opposite direction.

One complete revolution (360 degrees) is made up of a positive (0 to 180 degrees) and a negative (180 to 360 degrees) alternation. These two alternations comprise what is called a **cycle** of AC. A cycle of AC is the time that it takes to generate the positive and negative alternations. This is also defined as the **period** of an AC waveform. Since the armature of an AC generator rotates at a constant speed, the AC output voltage produced by the device continually changes in magnitude and polarity.

The AC generator just described is the simplest device that can be used to generate AC voltages. The AC generators used to produce electrical power for commercial applications are more complex in construction. However, all AC generators operate on the principles just described. Practical AC generators utilize many loops of wire within their armatures to increase the induced voltage to a much higher value. These generators may also contain more than one pair of north-south magnetic poles. When more than one pair of poles are used, one revolution of the armature can produce more than one AC voltage variation.

INTRODUCTION TO AC

The Sinusoidal Waveform

In the previous portion of this chapter, you learned how an AC voltage could be produced by a simple AC generator. This AC voltage was graphically plotted in Figure 11-6 so that you could see exactly how it varies throughout one complete revolution of the generator's armature. When voltage (or current) values are plotted to form a continuous curve, they form a picture or pattern which is referred to as a waveform. The waveform shows exactly how the voltage varies over a period of time as the armature rotates. You may examine the waveform to determine the exact value and polarity of the AC voltage at any particular instant, when the armature is at a specific point.

The waveform shown in Figure 11-6 varies in a unique manner and is given a special name. It is called a sinusoidal waveform or simply a sine wave. The sine wave is the most basic and widely used AC waveform. It can be produced by an AC generator as previously shown or it can be produced by various types of electronic circuits.

The AC voltage that you use in your home to provide heat, light, and operate appliances, varies in a sinusoidal manner. The radio and television signals which carry sound and picture information are basically sine waves whose characteristics have been modified. A variety of waveforms that are more complex than the basic sine wave can be proven to be mathematically equivalent to combinations of sine waves.

AC ELECTRONICS

The Basic Sine Wave

The AC waveform is referred to as a sine wave because it changes in value according to the trigonometric function known as the sine. The sine is a trigonometric function that describes the relationship between the sides of a right triangle. An angle has a sine value that is equal to the ratio of the length of the opposite side (the side opposite the working angle) to the length of the hypotenuse (the side opposite the right angle). This relationship is shown in Figure 11-7. Notice that the sine of angle A is equal to the opposite side (side Y) divided by the hypotenuse (side Z). Angle B would have a sine value that is equal to its opposite side (side X) divided by the hypotenuse (side Z). Notice also that the hypotenuse is always larger than either of the other sides. Therefore, the value of the sine varies from 0 to 1, for angles between zero and 90 degrees.

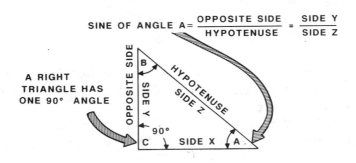

Figure 11-7 The sine (SIN) function.

The sine of angle A (or angle B) will vary as the angle varies because the length of the opposite side and the length of the hypotenuse both change as the angle changes. Angle A or angle B can be any value between 0 and 90 degrees. These are the extreme limits which can occur within the right triangle. Remember, the sum of angles A, B, and C must always equal 180 degrees.

As angle A (or B) varies from a minimum of 0 degrees to a maximum of 90 degrees, its sine value will vary from zero (when side Y is infinitely small) to 1 (when side Y and side Z are equal). The various sine values for all of the angles between 0 degrees and 90 degrees (in increments of 1 degree) are located in Appendix C at the end of the book. Figure 11-8 is an annotated table of sine functions only. Notice that for 0 degrees, sine is also 0. Then, as the angle increases, sine also increases. When the angle reaches 90 degrees, the sine function has reached its maximum of 1.

INTRODUCTION TO AC

Angle	Sine	Angle	Sine	Angle	Sine
0°	0.000	31°	.515	61°	.875
1°	.018	32°	.530	62°	.883
2°	.035	33°	.545	63°	.891
3°	.052	34°	.559	64°	.899
4°	.070	35°	.574	65°	.906
5°	.087				
6°	.105	36°	.558	66°	.914
7°	.122	37°	.602	67°	.921
8°	.139	38°	.616	68°	.927
9°	.156	39°	.629	69°	.934
10°	.174	40°	.643	70°	.940
11°	.191	41°	.656	71°	.946
12°	.208	42°	.669	72°	.951
13°	.225	43°	.682	73°	.956
14°	.242	44°	.695	74°	.961
15°	.259	45°	.707	75°	.966
16°	.276	46°	.719	76°	.970
17°	.292	47°	.731	77°	.974
18°	.309	48°	.743	78°	.978
19°	.326	49°	.755	79°	.982
20°	.342	50°	.766	80°	.985
21°	.358	51°	.777	81°	.988
22°	.375	52°	.788	82°	.990
23°	.391	53°	.799	83°	.993
24°	.407	54°	.809	84°	.995
25°	.423	55°	.819	85°	.996
26°	.438	56°	.829	86°	.998
27°	.454	57°	.830	87°	.999
28°	.470	58°	.848	88°	.999
29°	.485	59°	.857	89°	1.000
30°	.500	60°	.866	90°	1.000

Figure 11-8 Table of sine values.

AC ELECTRONICS

Let's see how this applies to the voltage output from an AC generator. As you know, the output voltage from an AC generator varies in a sinusoidal manner. Figure 11-9A is a representation of the output of an AC generator 30 degrees through the armature's rotation. In this illustration, the maximum output voltage for the generator is shown as the hypotenuse of the right triangle—100 volts. The actual output voltage at any instant is represented by the length of the side opposite the angle of rotation.

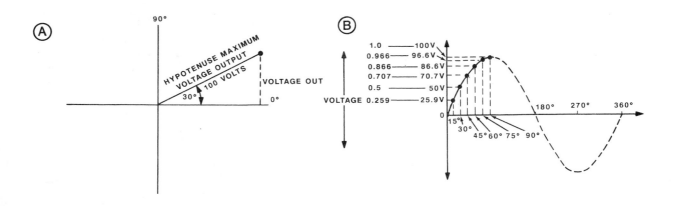

Figure 11-9 Amplitude plotted as a function of rotation.

Since the sine of any angle is the relationship of the opposite side divided by the hypotenuse, you can determine the output voltage of the generator 30 degrees through the armature's rotation. This is done by multiplying the length of the hypotenuse (100 volts) by the sine of 30 degrees (.5). In this case, the output voltage of the generator 30 degrees through the armature's rotation is 50 volts.

Figure 11-9B illustrates values for a number of armature positions. Notice that specific voltage values are shown at 15 degree intervals between 0 degrees and 90 degrees. Keep in mind that the horizontal line indicates the number of degrees of armature rotation.

The voltage is zero at the beginning of the waveform when the armature is at 0 degrees (parallel to the field). When the armature rotates 15 degrees, the voltage increases to a value equal to the sine of 15 degrees (.259) times the generator's maximum output voltage. Therefore, the output voltage after 15 degrees of rotation is 25.9 volts.

INTRODUCTION TO AC

The output voltage continues to rise for the first 90 degrees of rotation. The sine of the angles, along with the voltages present at the output, are shown in Figure B. At 30 degrees the output is 50 volts; at 45 degrees the output is 70.7 volts, and so on. At 90 degrees, the sine of the angle is 1 and the voltage output at this time is the maximum voltage output for the generator.

When you compare the angular position of the armature with the output voltage, you will find that the voltage increases according to the sine of the angle of rotation.

Between the 90 and 180 degree points on the waveform, the voltage drops from its maximum value to zero. During this portion of the waveform, the voltage decreases in the exact opposite manner that it increased. In other words, the sine is maximum at 90 degrees and decreases to zero at 180 degrees. Between the 180 and 360 degree points, the voltage varies in the same manner as it did between 0 degrees and 180 degrees. The only difference is that the polarity of the voltage is opposite. This is why the curve extends below the horizontal line.

The Cycle

Each time the armature of an AC generator rotates through one complete revolution, it generates an output voltage that increases and decreases in value in first one direction and then in the opposite direction. When the armature rotates one complete revolution, it completes one cycle of events. In other words, it produces one complete change in voltage values. If the armature rotates another 360 degrees, it will simply repeat the cycle or sequence of output voltages. The output voltage produced during one complete revolution of the armature is, therefore, referred to as one cycle. The armature will also produce one cycle of output current when there is a complete circuit through which current can flow.

The sine wave is made up of an infinite number of voltage values which have been plotted and joined together. At any particular point on the waveform, the voltage is generally referred to as an *instantaneous* value—a value that occurs at a specific instant of time.

The amount of time it takes for the armature to complete one revolution can also be determined with the sine wave. The horizontal line on which the waveform is plotted can be thought of as a time line, or a time base. The horizontal line can be calibrated in units of time, as well as in degrees of armature rotation.

AC ELECTRONICS

The horizontal line also serves as a zero reference line. Any voltage value that is plotted on this line will have a value of zero. Above the line, all values are assumed to be positive. Below the line, all values are assumed to be negative. Therefore, all of the instantaneous voltage values which make up the positive alternation have positive values; all of the instantaneous values which form the negative alternation have negative values.

The terms described in this discussion are extremely important. If you want to understand AC electronics, you must become familiar with them. These terms describe the various characteristics of one cycle of a sine wave. You must understand the exact meaning of these terms, since they will be used throughout your study of electronics.

It is also important to note that the sine wave can represent one complete cycle of output current, as well as a cycle of output voltage. When current is represented, the positive and negative alternations represent current that is increasing and decreasing in one direction, then increasing and decreasing in the opposite direction.

INTRODUCTION TO AC

AC Values

Since the value of a sine wave of voltage or current constantly changes, you must be specific when describing the value of the waveform. In other words, you cannot simply state that a voltage sine wave has a value of 100 volts. You must specify that this is the maximum value of the voltage waveform, or some value at a specific point between zero and maximum. There are several ways of expressing the value of a sine wave and you must be familiar with each of them.

Peak Value

Each alternation of a sine wave is made up of an infinite number of instantaneous values. These values are plotted at various points above and below the horizontal line to form a continuous waveform. It is the height (amplitude) of these instantaneous values above and below the line that represent the sine wave's actual value. The greater the amplitude, the higher the value. The sine wave's maximum instantaneous values, both above and below the horizontal line, are defined as peak values.

There are two peak values in one cycle of a sine wave. This is shown in Figure 11-10. One peak occurs during the positive alternation when the waveform reaches its maximum height. This point is appropriately identified as the positive peak. The positive peak is the maximum positive value that occurs during one cycle. The second peak value occurs during the negative alternation when the waveform again reaches its maximum height, but this time the peak is below the line. This point is referred to as the negative peak.

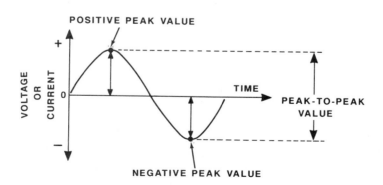

Figure 11-10 Peak and peak-to-peak values.

235

AC ELECTRONICS

The positive and negative peak values of a sine wave are equal in value. These values were obtained when the armature was cutting the magnetic field at it's maximum angle, 90 degrees. The only difference was the direction the armature was moving, which allows for the two polarities. For example, a voltage sine wave which has a positive peak value of +5 volts AC (VAC) would have a negative peak value of −5 VAC, and the widths of the alternations would be the same. Only the polarity of the voltage is different and is indicated by the + and − signs. The same relationship is also true for a sine wave of current. The peak value of the current during the positive alternation is equal to the peak value during the negative alternation. The only difference is that current does not have a negative polarity. In the case of current, the minus means that the current reverses, or flows in the opposite direction.

Peak values are sometimes described by using other terms which have the same meaning. For example, the peak value is sometimes called the peak amplitude or the maximum amplitude. Any of these terms used to describe maximum value or amplitude can be applied to any type of waveform.

Peak-To-Peak Value

Sometimes it is necessary to know the total height or value of a sine wave. The overall value of the sine wave (from one peak to the other) is called its peak-to-peak value and can be written as $E_{p\text{-}p}$. The peak-to-peak value of a sine wave is indicated in Figure 11-10.

The peak-to-peak value can be determined by adding the positive and negative peak values. Since two peaks are equal in value, it is necessary to know only one peak value. The peak-to-peak value can be determined by simply multiplying the peak value by 2. This relationship is shown mathematically as follows:

peak-to-peak value = 2 × peak value.

It is necessary to understand the meaning and relationship of the peak and peak-to-peak values of a sine wave. You will need these values when analyzing the characteristics of waveforms, and when measuring waveform values. Various test instruments (such as oscilloscopes and certain types of AC meters) are used to directly measure the peak-to-peak value of a waveform. The operation and use of the oscilloscope and the AC meter will be covered in the next chapter.

INTRODUCTION TO AC

Average Value

When we examine one alternation of a voltage or current sine wave, we find that it increases from zero to a peak value, and decreases back to zero, as shown in Figure 11-11. The voltage or current remains at its peak value for only an instant. Except for this one instant of time, the voltage or current is always lower than its peak value. Since the voltage or current remains at the peak value for only an instant, the average voltage for the entire alternation must be less than the peak value. This average voltage can be determined by taking a large number of the instantaneous values which occur during one alternation, and computing their average value. The average can also be calculated using integral calculus. Integral calculus is the branch of mathematics used to find volumes, areas, and equations of curves. In either case, you will find that the average value of one alternation is approximately equal to 0.636 of its maximum or peak value. This relationship between peak and average is shown in Figure 11-11. It can also be expressed by the following equation:

$$\text{average value} = 0.636 \times \text{peak value}.$$

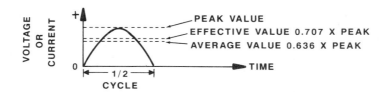

Figure 11-11 Average and effective (RMS) values.

For example, a voltage sine wave which has a peak value of 100 volts will have an average value of 0.636 x 100 V or 63.6 volts. This equation is used to determine the average value of either a voltage or current sine wave. For example, a current sine wave with a peak value of 10 amperes would have an average value of 0.636 x 10 A or 6.36 amperes.

This equation may also be transposed so that the peak value can be determined when the average value is known. In this case, the equation becomes:

$$\text{peak value} = \frac{\text{average value}}{0.636}$$

Suppose the average value of a voltage sine wave is 50 volts. Its peak value will be equal to 50 V/0.636 or 78.6 volts. A current sine wave, with an average value of 1 ampere, would have a peak value of 1 A/0.636 or 1.57 amperes.

AC ELECTRONICS

It is important to note that we have considered the average value of only one alternation or one-half cycle. We must add the average value of one alternation to the average value of the other alternation. Since each alternation has the same average value (0.636 x peak value) and because one value is positive when the other is negative, you can see that the two averages cancel. Therefore, the average value of a sine wave, as opposed to an alternation, is zero. When the average value of a sine wave is discussed, it is common practice to assume that the discussion concerns only one of the alternations.

Average values are not used extensively when dealing with voltage and current sine waves, but they do have certain special applications. As you proceed with your study of electronics, you will find it beneficial to have a knowledge of how these average values are determined.

Effective Value

When direct current (DC) flows through a resistor, a certain amount of power is dissipated by the resistor in the form of heat. A certain amount of heat is also produced if an alternating current is allowed to flow through the same resistance. However, the heat produced by an alternating current with a peak value of 1 ampere will not be as great as the heat produced by a DC current of 1 ampere. The alternating current produces less heat because it stays at its peak value of 1 ampere for only a short period of time. Therefore, an alternating current, with a peak value higher than 1 ampere, must be used to produce the same amount of heat that is produced by 1 ampere of direct current.

The AC current that will produce the same amount of heat in a specified resistance as a DC current that has a value of 1 ampere is considered to have an **effective value** of 1 ampere. In other words, a DC current of 1 ampere is equivalent to an AC current which has an effective value of 1 ampere, as far as their ability to produce heat is concerned. For this reason, the AC current must have a peak value higher than 1 ampere in order to be equivalent to the DC current.

The effective value of a sine wave of current can be determined by a mathematical process known as the **root-mean-square** or **rms**. Thus, the effective value of an AC sine wave is sometimes referred to as an **rms value**. Root-mean-square calculations are very lengthy, and need not be demonstrated. Once all the calculations are completed, the effective value of a sine wave is equal to 0.707 times its peak value. The relationship between peak and effective values of a sine wave is shown in Figure 11-11. This relationship is also shown in the following formula:

effective value = 0.707 × peak value.

For example, a current sine wave with a peak value of 10 amperes would have an effective value of 0.707 x 10 A or 7.07 amperes. This AC current, which has a peak value of 10 amperes, will produce the same heating effect as a direct current of 7.07 amperes.

Since an AC current is produced by an AC voltage, you may express the AC voltage in terms of its effective value. The effective value of a voltage sine wave is determined by using the equation (0.707 x peak value). For example, a voltage sine wave with a peak value of 40 volts would have an effective value of 0.707 x 40 V or 28.28 volts.

The peak value of a sine wave can be determined when its effective value is known by using the equation that follows:

peak value = 1.414 × effective value.

The number 1.414 is derived by taking the square root of 2 or $\sqrt{2}$. The rms value (0.707) is derived by taking the reciprocal of 1.414. Taking the reciprocal of a number means to divide the number into 1 (1/1.414 = 0.707). The terms effective value and rms value, are used interchangeably.

This equation may be used to determine the peak value of either a current sine wave or a voltage sine wave. For example, a current sine wave with an effective value of 7 amperes would have a peak value of 1.414 x 7 A, or approximately 10 amperes. A voltage sine wave with an effective value of 30 volts would have a peak value of 1.414 x 30 V or 42.4 volts.

Effective (or rms) values are used extensively when working with AC sine waves. It is important that you understand the relationship between effective and peak values of a sine wave. In most cases, when an AC voltage or current is specified, the effective value is used. In fact, effective values are used so extensively that they are often not specifically identified, but implied. For example, it is common practice to express an effective voltage value of 100 volts or simply 100 volts AC or an effective current value of 10 amperes as simply 10 amperes AC. The 120 volt AC electrical power that is used in your home has an effective value of 120 volts. Its peak value is approximately 170 volts. Most AC voltmeters and ammeters are calibrated to read effective (rms) values.

Period

When you are analyzing an AC sine wave, it is often necessary to know exactly how much time is required to generate one complete cycle. The time required to produce one complete cycle is called the **period** of the waveform. The period of a sine wave is shown in Figure 11-12. The period is usually measured in seconds, although other units of time can be used. The period is often represented by the letter T.

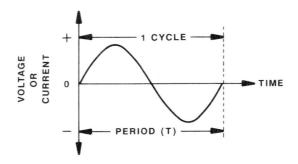

Figure 11-12 The period of a sine wave.

If a generator produces 1 cycle of output voltage in 1 second, the output sine wave has a period of 1 second. However, if 4 cycles are produced in 1 second, the output sine wave will have a period of 1/4 of a second (T = 0.25 seconds). It is important to remember that the period is the time of *one cycle*, and not the total time required to generate a given number of cycles.

The time of a cycle is also expressed in angular notation. Remember that an armature rotates 360 degrees to produce a complete sine wave; starting at zero, and increasing to maximum at 90 degrees. This portion of one revolution is equal to 1/4 of one revolution. From maximum back to zero completes one alternation, which is one-half of one cycle. The other half-cycle is the other alternation.

Frequency

It is often necessary to know how rapidly an AC waveform is changing in value. In other words, it may be important to know how many cycles of the waveform occur in a given period of time. The number of cycles that occur in a specified period of time is called the **frequency** of the waveform.

INTRODUCTION TO AC

Each time the armature of the simple AC generator completes one revolution, one cycle is produced. This means that the frequency of an AC waveform is determined by the speed at which the armature rotates. As the speed of rotation increases, more cycles are generated for each unit of time. Therefore, the frequency is increased.

The frequency of an AC sine wave is usually expressed in terms of the number of cycles generated per second. For example, an armature, rotating 1 complete revolution each second, produces 1 cycle of AC each second. The AC voltage would have a frequency of 1 cycle per second.

Although the frequency is the number of cycles produced each second, it is expressed in **hertz,** abbreviated **Hz.** A generator which produces an AC voltage that completes 1 cycle per second, is said to be operating at a frequency of 1 hertz. The term hertz is simply used in place of cycles per second, and can be used for both singular and plural values. If the AC generator produces 30 cycles of AC output voltage each second, it is operating at a frequency of 30 hertz, or 30 Hz. Likewise, an output of 60 cycles each second is expressed as 60 hertz, or 60 Hz.

There is a definite relationship between the frequency and the period of a sine wave. When the period of a sine wave is 1 second, the frequency is 1 hertz, as shown in Figure 11-13 If the period decreases to 0.5 seconds, or one-half of its original value, the frequency increases to 2 hertz. This is because exactly twice as many cycles would occur each second. Likewise, if the period doubled, the frequency would be cut in half.

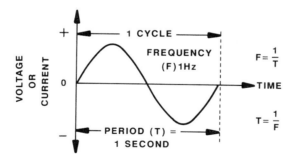

Figure 11-13 The frequency of a sine wave.

AC ELECTRONICS

Frequency is the reciprocal of time. This relationship is shown in the following equation:

$$F = \frac{1}{T}$$

This equation states that frequency, represented by the letter F, is equal to 1 divided by the period, T. Furthermore, if the period is expressed in seconds, the frequency will be in hertz. For example, when the period of a sine wave is equal to 0.05 seconds, the frequency of the waveform will be equal to:

$$F = \frac{1}{0.050 \text{ s}} \text{ or 20 hertz}$$

A period of 0.05 seconds would therefore correspond to a frequency of 20 hertz. However, if the period was cut in half or reduced to 0.025 seconds, the frequency would be:

$$F = \frac{1}{0.025 \text{ s}} \text{ or 40 hertz}$$

In other words, the frequency would increase to 40 hertz, or double. The equation shows that F and T are inversely proportional. When one increases, the other decreases proportionally.

Our equation can be transposed so that T can be determined when F is known. This transposed equation becomes:

$$T = \frac{1}{F}$$

The equation states that T is equal to 1 divided by F. If F is expressed in hertz, the value of T will be in seconds. For example, when F is equal to 100 hertz, T can be determined as follows:

$$T = \frac{1}{100 \text{ Hz}} \text{ or 0.01 seconds}$$

The period (T) is equal to 0.01 seconds or 10 ms.

INTRODUCTION TO AC

In the electronics industry, frequencies of all ranges are used, from just a few hertz to many millions of hertz. For example, the 120 volt AC electrical power used in your home has a frequency of 60 hertz. This 60 hertz power is used to operate your lights and appliances, and it may even provide heat. In many electronic applications, much higher frequencies are required. This is because high frequencies are needed to carry information or intelligence. The higher frequencies are easier to convert into electromagnetic (radio) waves, which can be transmitted more easily over long distances.

These higher frequencies cannot be produced by mechanical AC generators because they cannot rotate at the high speeds required to produce frequencies such as 10,000 Hz. To produce a frequency equal to 10,000 Hz requires a waveform with a period of (1 divided by 10,000). The generator would have to turn at the rate of 600,000 revolutions per minute (RPM). Therefore, electronic generators are used in place of mechanical generators to produce the frequencies that are required. Electronic circuits do not have moving parts, and are easily capable of producing frequencies many times greater that 10,000 Hz.

When working with frequencies that extend up to many millions of hertz, you must work with very large numbers. However, these large numbers can be reduced to a manageable size by using various metric prefixes and scientific notation. A frequency of 10,000 Hz can be expressed as 10 kHz. Likewise, a frequency of 4,000,000 Hz can be referred to as 4 MHz, and for frequencies at or larger than 1,000,000,000 Hz, the metric prefix *giga* (with a symbol of G) is used. Giga has a value of 10^9, or 1,000,000,000; so if a sine wave has a frequency of 4,000,000,000 Hz, it can be expressed as 4 GHz.

AC ELECTRONICS

Nonsinusoidal Waveforms

Although the sine wave is the most basic and widely used AC waveform, it is not the only type of waveform that is used in electronics. In fact, many different types of AC waveforms are used and these waveforms may have very simple or extremely complex shapes.

We will now briefly examine some of the other common AC waveforms that are used. These nonsinusoidal AC waveforms cannot be produced by the mechanically operated AC generators previously discussed. Nonsinusoidal waveforms are generated by electronic circuits. In this section, we are concerned with the actual types of waveforms, and how they relate to the sine wave.

The Square Wave

Figure 11-14 shows four different types of nonsinusoidal waveforms which represent either current or voltage. In each case, only one cycle of the waveform is shown. The waveform shown in Figure 11-14A is commonly referred to as a square wave. The square shape of each alternation indicates that voltage or current immediately increases to its maximum or peak value at one polarity, and remains there throughout that alternation. Then the voltage waveform immediately changes its polarity, or the current waveform reverses its direction. Notice that the waveform changes to a peak value almost instantly, and remains there for the duration of the alternation. When a continuous train of these square waves is produced, the voltage or current simply continues to fluctuate back and forth between its peak values.

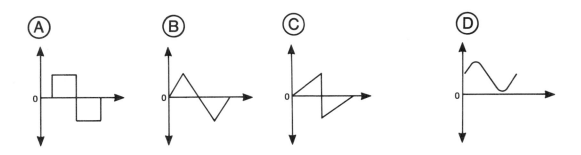

Figure 11-14 Nonsinusodial waveforms.

Not all square waves have alternations of equal length like the square wave in Figure 11-14A. In some cases, the positive half may be wider or narrower (longer or shorter time duration) than the negative half. Also, some square waves may have a positive peak value that is higher or lower (amplitude variations) than its negative peak value.

Although electrical power can be generated as square waves, the square wave is more useful as an electronic signal. The square wave is used to represent electronic data because its characteristics can be easily varied.

The Triangular Wave

The waveform shown in Figure 11-14B is called a triangular wave because its positive and negative alternations are triangular in shape. Notice that during the positive alternation the waveform rises at a linear rate from zero to a peak value and then decreases linearly back to zero. Then, on the negative alternation, its polarity reverses.

Triangular waves may have peak values that are higher or lower than those shown in Figure 11-14B. In other words, the positive and negative alternations may not always form a perfect triangle which has three equal sides. Triangular waves are used as electronic signals and are seldom used to provide electrical power.

The Sawtooth Wave

The waveform shown in Figure 11-14C is called a sawtooth wave. The sawtooth wave is similar to the triangular wave, but there are important differences. The sawtooth wave is formed when a voltage or current increases from zero to its positive peak value at a linear rate, and rapidly changes to its negative peak value. The waveform then decreases back to zero at a linear rate. This sequence of events represents one complete cycle of the waveform. When a number of cycles are graphically plotted, the waveform has a sawtooth appearance. The sawtooth waveform has a shape that includes characteristics of both the square wave and the triangular waveform. One common use of the sawtooth waveform is to provide a horizontal time base for an oscilloscope, which will be explained in the next chapter.

The sawtooth waveform may vary slightly from the shape shown in Figure 11-14C. For example, the change from the positive peak value to the negative peak value may not occur almost instantaneously, as shown. The change may require a small but discernible amount of time. Thus, the waveform appears to more closely resemble a triangular waveform than a square wave.

AC ELECTRONICS

Fluctuating DC Waves

There are many instances when an electronic signal does not change direction. In other words, the voltage level may constantly change, yet remain positive the whole time. These waveforms cannot be truly defined as AC waveforms, but they still behave as if they were signals. For example, the waveform in Figure 11-14D actually represents a DC voltage or current which fluctuates in value in the same manner as the sine wave. The DC voltage fluctuates or rides above the horizontal line which serves as a time base and a zero reference line. This fluctuating DC waveform can produce the same effect as an AC sine wave in certain applications. Such a waveform might appear at an intermediate point within an electronic circuit, and converted into a true AC signal before it reaches the output of the circuit.

The waveforms in Figures 11-14A, B, and C although they are non-sinusoidal, can still be considered AC signals because they have equal positive and negative portions. The fluctuating DC signal in Figure 11-14D, however, *cannot* be considered an AC signal since all of its instantaneous values are positive. We separated Figure 11-14D away from the others because it is somewhat different.

Chapter Self-Test

1. Of the four factors that determine the amount of EMF produced by an AC generator, which do *NOT* change?

2. What is the name of the AC generator's rotating loop?

3. The time of one complete revolution is defined as what?

4. The sine wave is maximum at which degrees?

5. Each time the armature makes one complete revolution (360 degrees), what is completed?

6. What is another name for the cycle of a waveform?

7. What is the term for those points on an AC waveform that are furthest from the reference line?

8. What is the formula for determining the peak-to-peak value of a sine wave, if the peak value is known?

9. A current sine wave with a peak value of 3 amperes would have an effective value of what?

10. What is the unit of measurement for frequency?

11. What is the equation that defines the relationship between the frequency and time of a waveform?

12. If the frequency of a signal is 5 kHz, what is the period?

13. What waveform's positive and negative alternations change at a linear rate?

AC ELECTRONICS

Summary

Unlike direct current (DC), which flows in only one direction and has a constant value, alternating current (AC) flows in one direction, then the other. The value or magnitude of an AC waveform varies as it flows in each direction.

Alternating current may be used as an electronic signal to carry information from one point to another. The AC signal may be carried by wires or transmission lines. It may be converted to electromagnetic waves which can be transmitted through space without the aid of wires and transmission lines to convey information or intelligence.

An AC generator is able to produce an alternating current because it makes use of a process known as electromagnetic induction. Electromagnetic induction is the process of inducing a voltage in a conductor. The induced voltage appears when the conductor moves through a magnetic field.

The voltage induced within a conductor is affected by the strength of the magnetic field, the speed of conductor movement, the length of the conductor in the field, and the angle at which the conductor cuts the field. When all of these factors are considered, a rule is formed that states that the voltage induced into a conductor is directly proportional to the rate at which the conductor cuts the magnetic field. Maximum voltage is induced when the conductor is moving perpendicular to the field (90 and 270 degree positions).

The polarity of the induced voltage is determined by the direction of conductor motion and the direction of the magnetic field.

A simple AC generator is formed by bending a wire into a loop (called an armature) and rotating the armature within a magnetic field. Slip rings and brushes are used to extract the AC voltage that is induced within the armature.

The AC output voltage produced by an AC generator varies from zero to maximum and back to zero again as the armature completes one-half of a revolution. Then during the next one-half revolution the voltage value is the same, but its polarity is reversed. One complete revolution of the armature produces an AC voltage that changes in both amplitude and polarity. If a load resistance is connected to the generator, the AC output voltage will cause a corresponding AC current to flow through the load.

When the output voltage values are graphically plotted with respect to armature position or time, a curve is formed which shows how the AC voltage varies. Such a curve is called a waveform.

INTRODUCTION TO AC

The AC voltage produced by the simple generator varies from zero to its maximum value according to the sine function, and decreases to zero in the exact opposite manner. The AC voltage changes in this manner in each direction. For this reason, the AC voltage is said to vary in a sinusoidal manner. When this AC waveform is graphically plotted, it is called a sinusoidal waveform, or simply a sine wave.

One complete revolution of the armature produces one cycle of output voltage. One cycle is further divided into two alternations. One alternation is called the positive alternation, and corresponds to the first one-half revolution of the armature. The other alternation is called the negative alternation, and corresponds to the second one-half revolution of the armature.

The maximum value reached during each alternation is called a peak value. The total value of a waveform, measured from one peak to the other, is called a peak-to-peak value.

The average value of one alternation is equal to 0.636 times the peak value. The actual average of a sine wave is zero.

The effective value of a sine wave is equal to 0.707 times its peak value. A current sine wave with an effective value of 1 ampere will produce the same amount of heat in a given resistance as a direct current of 1 ampere. The effective value is also called the root-mean-square or rms value.

The period of a sine wave is the time required to produce one complete cycle. The number of cycles that occur in a specified period of time is called the frequency. The period is usually measured in seconds and the frequency is measured in hertz (Hz), which stands for cycles per second. Time and frequency have a reciprocal relationship.

A variety of nonsinusoidal waveforms are also used in electronics. These waveforms are usually named for their shape. Such waveforms include the square wave, triangular wave, and sawtooth wave.

AC ELECTRONICS

CHAPTER 12

AC Measurements

AC ELECTRONICS

Contents

Introduction 253

Chapter Objectives 254

Using Meters to Measure AC 255

The Oscilloscope 259

Chapter Self-Test 269

Summary 270

AC MEASUREMENTS

Introduction

In Chapter 5, you were introduced to the various pieces of test equipment you will encounter as an electronics technician—the ammeter, the voltmeter, and the ohmmeter. We discussed how each of these measurements were made in a DC circuit.

Due to the characteristic differences between DC and AC, AC values are measured in a different manner than DC values. This chapter shows you how the ammeter and voltmeter are used to measure AC values.

However, in many cases, you may discover that a voltmeter cannot provide all the information you need. (After all, a voltmeter can only display a single voltage.) An AC signal is constantly changing, which quite often makes the voltmeter useless. This chapter will also introduce you to the oscilloscope; a piece of test equipment that allows you to see a changing voltage over a period of time.

AC ELECTRONICS

Chapter Objectives

When you have completed this chapter, you will be able to:

1. Define; Lissajous pattern, CRT, graticule, and in-phase.

2. Describe the methods for measuring AC current with an ammeter, and the differences between measuring AC current and DC current.

3. Describe the methods for measuring AC voltage with a voltmeter, and the differences between measuring AC voltage and DC voltage.

4. Explain the advantages of an oscilloscope when compared to an AC voltmeter.

5. Explain how the oscilloscope is used to measure a waveform's pulse width, period, amplitude, rate of change, and phase relationships.

AC MEASUREMENTS

Using Meters to Measure AC

You have previously seen how the ammeter and voltmeter are used to measure their specific values in a DC circuit. When working with AC, however, there are important differences in measuring these values. These differences are very slight, but their importance cannot be overemphasized. As an electronics technician, you must know *what* you are measuring, in order to apply the information these meters provide.

Measuring Current

When we discussed measuring DC current, we stated that it is necessary to have circuit current flowing through the meter itself. We suggested a five-step process for proper meter use. Let's briefly review this process:

1. **Select the proper ammeter.** You must ensure that the ammeter being used is capable of measuring the amount of current being measured. With a multi-range ammeter, select the range which will provide a half-scale reading. If the amount of current is unknown, select an ammeter (or range) that will provide for the highest amount of current. Then, if necessary, move to a lower scale.

2. **Remove power from the circuit under test.** This step is vital in protecting the meter from damage during the steps to follow. Also, failure to remove power at this point may cause electrical shock.

3. **Break the current path.** In order to have the desired current travel through the meter, the path that the current travels must be broken.

4. **Insert the meter.** The meter is placed within the broken path. This ensures that current will travel from one broken point, through the meter, and back into the circuit at the second broken point.

5. **Reapply power to the circuit.** Once the meter has been properly installed, power can be reapplied, and the current measured.

When this process was followed to measure DC current, step 4 required particular attention. The meter was installed with **proper polarity** being observed. Since every ammeter has a positive and a negative lead, you had to ensure that current entering the meter entered at the *negative* lead, and left the meter at the *positive* lead. Failure to do this could cause electrical shock to you, and very likely damage the ammeter.

AC ELECTRONICS

It is also in step 4 where the difference between measuring DC and AC currents occur. When an AC voltage is applied to a circuit, current travels in both directions, as the applied voltage switches between the positive and negative alternations. Therefore, the method for measuring AC current is actually simpler than measuring DC current. Each step outlined previously is still followed, with the exception of step 4. An AC ammeter can be connected *without polarity being observed*.

The difference between DC and AC ammeters is internal. Within the circuits for the AC ammeter, rectification takes place. If you'll recall from Chapter 11, rectification is the process by which one polarity of AC is removed. This protects the AC ammeter from current traveling in the wrong direction. The details behind this process will be explained when power supplies are discussed in Chapter 28.

However, this does not mean that you can connect an AC ammeter any way you desire. There are many AC ammeters which have one lead referenced to chassis ground. Similarly, many AC circuits are also referenced to chassis ground. If you are working with a chassis-grounded AC ammeter and chassis-grounded circuits, you should always connect the test lead of the ammeter that is referenced to chassis ground to that point in the circuit that is referenced to chassis ground. If you accidentally connect this ground-referenced lead to a point in the circuit that is not grounded, severe equipment damage may result.

DO NOT connect a ground-referenced test lead to any point that is NOT grounded.

Figure 12-1 illustrates how an ammeter is connected to an AC circuit. Notice that current (traveling in both directions) leaves the circuit, passes through the ammeter, and back into the circuit.

AC MEASUREMENTS

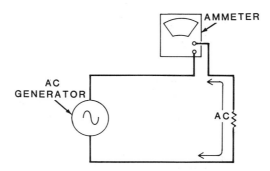

Figure 12-1 Measuring current with an ammeter.

The final point to measuring AC involves the type of current you will read. Back in Chapter 11, we mentioned that with an AC signal, there are peak values and effective values. An AC ammeter cannot measure the peak value of a signal because the signal remains at the peak value for only an instant. If an ammeter read peak values, the reading on the ammeter would be constantly changing. Therefore, any readings you obtain with an AC ammeter will be an *effective* value. If your reading in an AC circuit was 10 amps AC, it would be 10 amps rms.

Measuring Voltage

In Chapter 5, the method for measuring DC voltage was discussed. This method was very simple: the test leads of the voltmeter are connected in parallel with the voltage being measured. Power does not have to be removed, and the circuit does not have to be broken or altered in any way. When measuring voltage, you are simply measuring a potential difference between two points; for example, across a component.

However, there are precautions that must be observed before voltage can be measured. First, like the ammeter, be sure you select a voltmeter (or range, in the case of a multiple-range voltmeter) that is able to read the voltage without possible damage to the meter. If the voltage is unknown, select a voltmeter (or range) that will allow you to read the highest possible voltage. Then, if necessary, work your way down to a lower range for a more accurate reading.

Also, many voltmeters have a particular lead referenced to chassis ground. If you are using such a voltmeter in a circuit that *also* has a chassis ground, you must be certain to connect the ground-referenced test lead to the ground-reference point in the circuit. The same precaution was mentioned with the ammeter, and it cannot be stressed enough.

AC ELECTRONICS

DO NOT connect a ground-referenced test lead to any point that is NOT grounded.

In order to measure an AC voltage, the same precautions and methods are used as those for measuring DC voltage, but polarity need not be observed. AC voltage is constantly changing from a positive value to a negative value, so determining polarity is impossible. The test leads for the voltmeter are simply placed any way you desire, with the exception of a ground-referenced lead, as was mentioned.

Figure 12-2 provides us with two examples of measuring AC voltage. In Figure 12-2A, the voltmeter is connected across the voltage source. In this case, the source is still connected to the circuit. This is not always necessary, since you are dealing with a voltage rise. The test leads are placed on either side of the voltage source, and the reading is taken. Placing a voltmeter across two points, whether they are across a voltage rise or a voltage drop, will not affect the circuit.

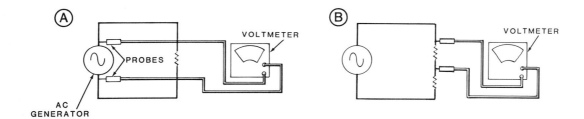

Figure 12-2 Measuring voltage with a voltmeter.

In Figure 12-2B, the voltage drop of a resistor is being measured. Power to the circuit remains applied to allow current to flow through the resistor. (If you'll recall, a resistor cannot develop a voltage drop unless current is flowing through it.) Again, the placement of the voltmeter's test leads will not affect circuit current.

The voltage measured with an AC voltmeter is not positive or negative, as was previously explained. But the voltage measured is not a peak value of AC, either. Remember, the AC signal remains at the peak value for only an instant. AC voltage is constantly changing in both polarity and magnitude. Therefore, the voltage measured with an AC voltmeter is an effective value.

AC MEASUREMENTS

The Oscilloscope

AC meters can provide reasonably accurate measurements of current and voltage, but these instruments do not allow you to see what the AC quantities actually look like. In most cases, they are calibrated to indicate the effective or rms value of a sine wave. When used to measure nonsinusoidal waveforms, they do not provide true rms readings.

When you are troubleshooting or analyzing electronic equipment, it is often helpful or even necessary to know exactly what an AC waveform looks like. In many cases, it is necessary to know the peak and peak-to-peak values, instantaneous values, and the waveform's period. These measurements, and many others, can be performed using a device known as an **oscilloscope.**

The oscilloscope may be used to analyze any type of AC waveform and measure its most important electrical characteristics. The oscilloscope, or **o'scope** as it is commonly called, is one of the most important test instruments used to measure AC quantities. In addition, an oscilloscope can also be used to measure DC voltages.

Oscilloscope Operation

Oscilloscopes do not contain moving parts like the AC meters previously described. They are electronic test instruments which utilize various types of electronic circuits. The early oscilloscopes used vacuum tubes as the principle controlling elements in their electronic circuits. The newer oscilloscopes use semiconductor components such as transistors and solid state diodes in place of the larger tubes. The tube versions were much larger and required much more power to operate than the modern day solid state oscilloscope.

An oscilloscope is capable of measuring AC or DC voltages and displaying the voltage on a graph. The AC or DC voltage appears as a picture on a screen similar to the type of screen used in a television set. The main difference in the oscilloscope's screen is that it is divided into equally spaced 1 centimeter squares. The oscilloscope contains a number of controls which are used to adjust the size and the number of complete waveforms that are displayed. Most oscilloscopes are calibrated so that the waveform can be visually analyzed, and its most important characteristics easily determined.

AC ELECTRONICS

A simplified block diagram of an oscilloscope is shown in Figure 12-3. The device has two input terminals which are used to measure AC or DC voltages. These terminals are connected in parallel with the voltage source that is to be measured. They are generally referred to as the **vertical input terminals.** The AC voltage at these terminals is applied to an amplifier circuit. The amplifier increases the amplitude (magnitude) of the voltage before it is applied to the **cathode ray tube** or **CRT.** The CRT is the device which graphically displays the waveform being measured.

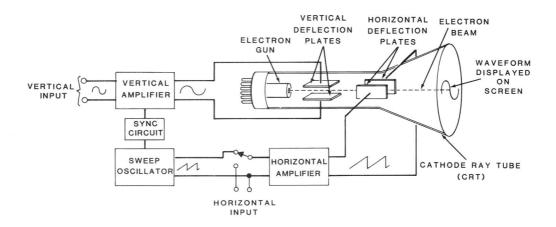

Figure 12-3 Basic oscilloscope.

As shown in Figure 12-3, the CRT contains an electron gun and two sets of deflection plates. These components are mounted inside a large glass tube which fans out at one end to form a screen. The air is pumped out of the tube and the end is sealed, so the components within the CRT operate within a vacuum. In this respect, the device is similar to an ordinary vacuum tube. The electron gun produces a stream of electrons which are focused into a narrow beam and aimed at the CRT screen. When the beam strikes the screen, it illuminates a phosphorus coating on the inside of the screen and a spot of light is produced. The electron beam is controlled by two sets of deflection plates. The beam passes between these plates before it reaches the CRT.

AC MEASUREMENTS

The AC voltage from the vertical amplifier is applied to the **vertical deflection plates**. This alternating voltage causes the plates to become positively and negatively charged. The polarity of these charges is continually reversed. The electrons in the beam are negatively charged and tend to deflect toward the positive plate and away from the negative plate. Like charges repel and unlike charges attract. This constant change causes the electron beam to bend. Since the charges on the vertical plates continually change direction, the electron beam is deflected (sweeping motion) up and down. The beam movement causes a vertical trace to appear on the CRT screen. The height of the vertical trace is the amplitude of the applied AC voltage being measured and the amount of amplification provided by the oscilloscope's vertical amplifier circuit.

When the electron beam is moved up and down, only a vertical line or trace appears on the screen. Such a display could indicate the peak-to-peak amplitude of a waveform but still not the exact shape of the waveform. In order to show the waveform's variations, the electron beam must also be deflected horizontally across the screen.

This is accomplished by a circuit known as a **sweep oscillator**. This circuit generates an AC sawtooth waveform. This waveform is then amplified by a **horizontal amplifier** and applied to the **horizontal deflection plates**. The sawtooth voltage increases at a linear rate from a negative peak value to a positive peak value and then almost instantly changes back to a negative value again. The positive and negative charges on the horizontal deflection plates varies at the sawtooth's linear rate. This causes the electron beam to move from left to right (be swept) across the screen at a linear rate. Once the beam reaches the right-hand side of the screen, it jumps immediately back to the left-hand side, and begins another sweep.

When the sawtooth waveform is applied to the horizontal plates with no voltage on the vertical plates, only the horizontal trace appears on the screen. This trace is the **horizontal time base** that is used to measure the time of a waveform. The electron beam moves from left to right in a specific period of time. The horizontal trace is continually repeated. A waveform that is generated and occurs continuously is the definition of the output of an **oscillator** circuit.

When the vertical signal voltage and the horizontal sawtooth voltage are both applied to the CRT, the signal waveform is displayed on the screen of the CRT. As the beam moves from left to right (at a linear rate), the vertical AC voltage causes the beam to move up and down following the AC variations. When the time required for the beam to move across the screen from left to right is equal to the time of one cycle of the AC input voltage, one cycle of the AC waveform will be displayed on the screen.

The relationship between the input AC sine wave, the sawtooth wave, and the displayed waveform is further illustrated in Figure 12-4. Notice that one completed cycle of the sine wave (Figure 12-4A) occurs in the time required to generate one complete cycle of the sawtooth wave (Figure 12-4B). When these conditions are met, one complete AC sine wave is displayed. This is shown in Figure 12-4C. In other words, the time (1/F) of the input AC waveform must be equal to the time (1/F) of the sawtooth waveform.

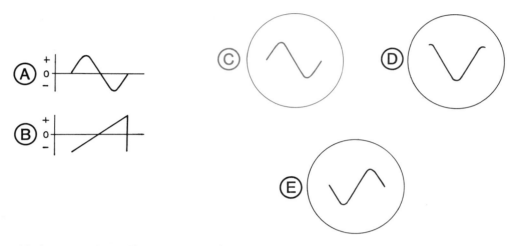

Figure 12-4 Typical oscilloscope waveforms.

Another important condition is that the sine wave and sawtooth wave must begin their cycles at the same time. If the two waveforms are not properly synchronized (start at the same time), the displayed waveform might appear as shown in Figure 12-4D, or Figure 12-4E. Although these waveforms are complete AC cycles, they are not properly oriented sine waves. Figure 12-4D is shifted right by 90 degrees, while Figure E is shifted by 180 degrees. Note that 180 degrees is the opposite polarity. This is easily seen when Figures 12-4C and 12-4E are compared.

To ensure that the input AC waveform and the sawtooth waveform are properly synchronized, a synchronization or **sync** circuit is included in the oscilloscope circuit. This circuit samples the incoming AC signal and produces a control signal that is applied to the sawtooth oscillator to make sure that the sawtooth generator begins its cycle at the proper time.

The input AC waveform and sawtooth waveform must occur repeatedly to produce a display on the screen. In other words, the electron beam follows the pattern of the waveform again and again. This results in a sine wave being displayed on the screen. The phosphor on the screen produces light for only an instant after the electron beam strikes it. Therefore, constant repetition (sweeping of the electron beam) is required to produce a pattern that is constantly illuminated.

AC MEASUREMENTS

As we stated when this discussion began, a thorough knowledge of oscilloscope operation is not necessary. However, a basic knowledge of how the oscilloscope works will enable you to operate it more efficiently.

Using the Oscilloscope

As you work more and more with electronics, you will find the oscilloscope to be an extremely useful piece of test equipment. The oscilloscope's ability to view signals allows a technician to perform accurate troubleshooting and precise circuit analysis. However, if you don't understand how to *use* the oscilloscope, troubleshooting and circuit analysis can become very difficult. The following section explains how to use the oscilloscope to measure different values.

Measuring Voltage

Since the oscilloscope displays an entire AC waveform, it can be used to determine instantaneous values as well as peak and peak-to-peak values.

A typical oscilloscope display is shown in Figure 12-5. Notice that the screen of the oscilloscope is marked with vertical and horizontal lines which form squares. This grid pattern is commonly referred to as a **graticule**. The squares are usually 1 centimeter high and 1 centimeter wide and are similar to a sheet of graph paper.

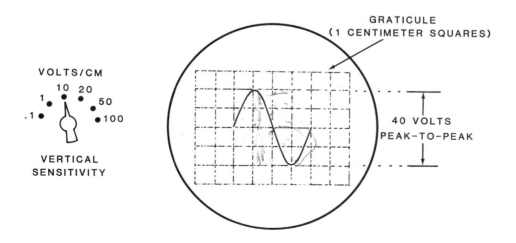

Figure 12-5 Measuring voltage with an oscilloscope.

AC ELECTRONICS

When observing an AC waveform, the vertical control must be used. **Vertical control** means that the vertical height (amplitude) of the waveform can be adjusted. The vertical controls are usually calibrated so that a given input voltage will produce a specific amount of vertical deflection on the screen. Vertical amplification is usually adjusted by a **vertical sensitivity** control or **vertical attenuator** control. For example, suppose the vertical sensitivity control was set to the 10 volts per centimeter position as shown in Figure 12-5. Each centimeter of vertical height or deflection would represent 10 volts at the vertical input terminals. The waveform being observed in Figure 12-5 is 4 centimeters high. Therefore, it has a peak-to-peak amplitude of 4 times 10 or 40 volts. The peak value of the waveform is equal to one-half of 40 volts, or 20 volts.

The value at any point on the waveform can be determined by referring to the squares on the graticule. The vertical sensitivity control can be reset to other positions to increase or decrease the vertical gain of the oscilloscope. When this is done, the squares represent values of voltage which are either lower or higher than 10 volts per square.

Measuring The Period

One of the most important uses of the oscilloscope is to measure the period of an AC waveform. The period, or time for 1 cycle, is determined by observing the horizontal width of the waveform. The oscilloscope's sawtooth oscillator can usually be adjusted so that the electron beam will move from left to right across the screen at a specific speed. The time required for the beam to move horizontally across the screen is referred to as the **sweep time**. The sweep time can be adjusted by a control that is mounted on the oscilloscope. This control sets the amount of time, in seconds, milliseconds, or microseconds, required for the trace to move horizontally a distance of 1 centimeter.

Let's assume that the oscilloscope's sweep time control (TIME/CM) is set to the 5 milliseconds per centimeter position, as shown in Figure 12-6. Each centimeter of horizontal deflection represents a time interval of 5 milliseconds. The waveform being displayed in Figure 12-6 is 4 centimeters wide. In other words, one complete cycle occupies 4 centimeters of the trace. The period of the waveform is equal to 4 times 5 milliseconds or 20 milliseconds.

AC MEASUREMENTS

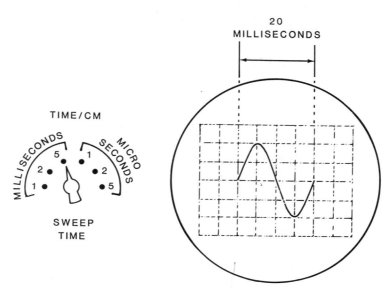

Figure 12-6 Measuring time with an oscilloscope.

The horizontal sweep control can be set to a variety of positions. This allows you to use the oscilloscope to measure waveforms that have very long or very short cycle times. In many cases, it is desirable to display only one cycle of the waveform as shown. However, the oscilloscope is capable of displaying many complete AC cycles at the same time. If the time for an AC signal is 4 milliseconds, the sweep time control can be set to 2 milliseconds per centimeter. Thus, one cycle will occur every two centimeters. With ten centimeters across the oscilloscope screen, five cycles of the waveform will be displayed.

The oscilloscope can be used to display a number of AC input cycles. When a number of cycles are displayed on the screen, it is only necessary to determine the time of one cycle in order to determine the period of the waveform.

Measuring Frequency

In order to determine the frequency of an AC waveform, you must know the period of the waveform. If you'll recall from Chapter 11, the frequency of an AC waveform is equal to 1 divided by the period of the waveform. In other words, time and frequency are reciprocal functions. Time is measured in seconds and frequency is measured in cycles per second. The unit of measure for frequency is the Hz and is expressed mathematically as:

$$F = \frac{1}{T}$$

AC ELECTRONICS

For example, the waveform in Figure 12-6 has a period (T) of 20 milliseconds (0.02 seconds). This waveform has a frequency (F) of:

$$F = \frac{1}{0.02s} \text{ or } 50 \text{ Hz}$$

Since the oscilloscope is used to measure time, you must convert time to frequency mathematically. Another piece of test equipment called the **frequency counter** is used to measure frequency directly. The frequency counter may have a period mode and a frequency mode. In the period mode, the counter displays the time of 1 cycle. In the frequency mode, the period is internally converted and the frequency is displayed.

Measuring Phase Relationships

In some cases, it is necessary to compare two AC waveforms of the same frequency to determine if the two waveforms coincide (occur at the same time). In many cases, two AC waveforms within the same circuit will be displaced in time or by a given number of degrees. For example, an AC signal may be shifted by approximately 90 degrees across a capacitor.

If two AC waveforms are in coincidence, they are said to be **in-phase** with each other. When the two waveforms are displaced (do not occur at the same time), they are said to be **out-of-phase**. The amount of phase displacement is usually measured in **degrees**.

Figure 12-7 shows a basic sine wave in a number of different phases. The two waveforms in Figures 12-7A and 12-7B are in-phase. Notice that the positive and negative peaks occur at precisely the same time.

AC MEASUREMENTS

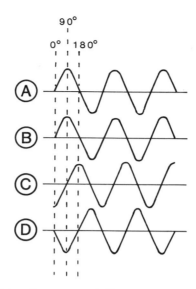

Figure 12-7 Phase relationships between AC sine waves.

The waveform in Figure 12-7C, on the other hand, is 90 degrees out-of-phase with the waveform in Figure 12-7A. When the positive peak for Figure A occurs (which means the armature has completed one-quarter of a revolution, or 90 degrees), the waveform in Figure 12-7C is just beginning. The waveform in Figure 12-7C will constantly be 90 degrees behind the waveform in Figure 12-7A.

The last waveform, Figure 12-7D, is precisely 180 degrees out-of-phase with the waveform in Figure 12-7A. Here, the peak values occur at the same time, but the positive peak for Figure 12-7A occurs when the *negative* peak for 12-7D occurs.

When two AC waveforms which are out-of-phase are compared, they share a phase relationship. This phase relationship is expressed in degrees. In order to compare two AC sine waves to determine their phase relationship, it is necessary to apply one sine wave to the oscilloscope's vertical deflection plates and the other sine wave to the scope's horizontal deflection plates. This is done by applying one waveform to the vertical input terminals and the other waveform to a set of horizontal input terminals. A switch is usually provided (shown previously in Figure 12-3) which disconnects the sweep oscillator and connects the horizontal input terminals directly to the horizontal amplifier.

When both AC sine waves are properly applied to the scope, and the vertical and horizontal controls are properly adjusted, the electron beam will deflect both AC waveforms. The resulting patterns displayed on the screen are referred to as Lissajous patterns. The phase relationship between the two waveforms can be determined by properly interpreting these unique patterns. A lissajous pattern is defined as a shape that is undefined.

267

AC ELECTRONICS

Several common Lissajous patterns are shown in Figure 12-8. These patterns occur at phase difference intervals of 45 degrees, starting at 0 degrees and extending to 360 degrees. The pattern shown in Figure 12-8A occurs when both sine waves are in-phase. This pattern is a diagonal line that extends from the lower left-hand portion of the screen to the upper right hand portion. The pattern in Figure 12-8B occurs when two waveforms are 45 degrees out-of-phase.

Figure 12-8 Typical Lissajous patterns.

At a phase difference of 90 degrees, a perfect circle is formed as shown in Figure 12-8C. Then at 135 degrees and 180 degrees, an ellipse and diagonal line are again formed, as shown in Figure 12-8D and Figure 12-8E, respectively. However, these last two patterns are slanted in the opposite direction.

Phase differences other than the values shown in Figure 12-8 will produce elliptical patterns that have shapes midway between the various shapes shown. It is important to realize that the Lissajous pattern changes from a diagonal line, to an ellipse, to a circle, then back to an ellipse, and back again to a diagonal line as the phase difference changes from 0 degrees to 180 degrees and 360 degrees. The sequence is repeated but in the opposite direction as shown in Figures 12-8D and 12-8E. These phase difference displays were produced using two sine waves that are equal in amplitude and frequency, but separated in-phase.

By using Lissajous patterns, the oscilloscope becomes a reasonably accurate phase measuring device. However, the accuracy of the measurements will depend largely on the skill of the individual using the scope. Also, the waveforms must be sinusoidal in shape to produce the patterns shown in Figure 12-8. Nonsinusoidal waveforms will produce irregular shaped patterns which are extremely difficult to analyze.

Many oscilloscopes have two vertical input channels. The two channels can be used to evaluate or compare two signals at the same time. This is especially helpful when monitoring the input to a circuit on one of the channels and observing the various outputs of the different circuits that the input is driving. The output waveform can be instantly compared to the input to determine possible amplification, polarity inversion, and phase shift all at the same time. As you progress through this course, you will have the opportunity to use many of the oscilloscope's capabilities.

AC MEASUREMENTS

Chapter Self-Test

1. What is the primary difference between measuring AC current (or voltage) and DC current (or voltage)?

2. Which component of the oscilloscope displays the signal?

3. The face of the CRT on an oscilloscope is divided by a graph of 1 centimeter squares. What is this graph called?

4. Where is the sawtooth wave within the oscilloscope applied?

5. Which axis of the oscilloscope measures voltage (amplitude)?

6. A scope is adjusted to have a vertical sensitivity of 20 volts per centimeter. The waveform on its screen has a total height of 6 squares. What is the waveform's peak-to-peak value?

7. A signal takes up 4 squares horizontally. The TIME/CM control is set to 2 ms. What is the frequency of the waveform?

8. The scope is adjusted so that the beam requires 2 ms to move 1 centimeter. Four cycles have a combined width of 8 centimeters. What is the frequency of *one* cycle?

9. It is also possible to determine phase relationships between sine waves using an oscilloscope. When doing so, what type of patterns does the oscilloscope produce?

10. If an oscilloscope displays a phase relationship pattern that is a circle, what is the phase relationship between the two signals?

Summary

A meter that is used to measure current is called an ammeter. Normally an ammeter must be connected in series with the current to be measured. The circuit under test must be turned off and the ammeter wired into the circuit. An AC ammeter can be connected without polarity being observed.

An AC voltmeter is connected to a circuit in the same manner as the DC voltmeter, with the exception that polarity need not be observed.

Both the AC ammeter and AC voltmeter measure effective values.

The oscilloscope is perhaps the most versatile of all test instruments. It is used to measure peak, peak-to-peak, and other instantaneous waveform values. This instrument is also used to measure the period of a waveform or the time of an alternation. It is used to compare AC signals, to determine their phase relationships, frequencies, and amplitudes.

Frequency counters are used to measure frequency. Many frequency counters have a period mode that is used to measure cycle time. The frequency counter measures frequency directly. Oscilloscopes measure time. Time is then converted into frequency by dividing the time of one cycle into 1. This is called the reciprocal relationship between time and frequency. Time is measured in seconds and frequency is measured in hertz (Hz).

An oscilloscope can measure the difference in time between two signals, known as the phase shift. The signals are applied to the horizontal and vertical inputs, and the oscilloscope displays a lissajous pattern. Different patterns correspond to different degrees of phase shift.

CHAPTER 13

AC Resistance

AC ELECTRONICS

Contents

Introduction 273

Chapter Objectives 274

Basic Calculations 275

Circuit Calculations 277

Power in AC Circuits 281

Chapter Self-Test 283

Summary. 284

AC RESISTANCE

Introduction

You have examined some of the basic test instruments that are used to measure AC values. Now it is time to analyze some fundamental AC circuits, and the rules which apply to these circuits. For now, we will confine our discussion to AC circuits which contain only resistance. We will discuss the relationship between current, resistance, and voltage in these circuits and compare them with similar DC circuits. Chapters 14 and 15 will examine more complex AC circuits involving inductance and capacitance. Before you reach those chapters, however, you must have a complete understanding of simple resistive circuits. While many of the relationships and concepts that you encountered in DC circuits will remain the same, there are vital differences when power is considered in an AC circuit.

AC ELECTRONICS

Chapter Objectives

When you have completed this chapter, you will be able to:

1. Describe the phase relationship between voltage and current in an AC resistive circuit.

2. Calculate the unknown values in an AC series resistive circuit.

3. Calculate the unknown values in an AC parallel resistive circuit.

4. Describe the difference between the voltage and current curves in an AC circuit, and the power curve in an AC circuit.

5. Define average power, and calculate for average power in an AC circuit.

AC RESISTANCE

Basic Calculations

A basic AC circuit can be formed by connecting a load resistance across an AC voltage source. This is shown in Figure 13-1A. Although a resistor is used in this circuit, any resistive component (a lamp or heating element, for example) will have the same effect. Such devices are considered purely resistive and have a negligible amount of inductance or capacitance.

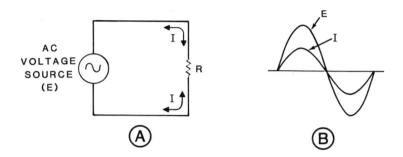

Figure 13-1 Current and voltage in an AC resistive circuit.

The voltage source could be an AC generator or an electronic circuit that produces an AC voltage. When this voltage is applied across the resistor, a corresponding AC current flows through the resistor. This current varies in amplitude and direction and is in-phase with the AC voltage applied. In other words, the current is zero when the voltage is zero and it reaches its maximum when the voltage is maximum. When the voltage changes its polarity, the current changes its direction. The voltage and current in a purely resistive AC circuit are in-phase.

This phase relationship between the voltage and current is shown graphically in Figure 13-1B. The illustration shows that the voltage (E) waveform and the current (I) waveform pass through zero and maximum at the same time. Both E and I change direction at the same time. The two waveforms do not have exactly the same peak amplitude because they represent different quantities. The quantities are also measured in different units—E in volts, and I in amperes. They are drawn together only to show that they occur simultaneously.

AC ELECTRONICS

The value of current flowing through the resistor in Figure 13-1A, at any instant, depends on the voltage at that instant, and the circuit resistance. In this example, the resistance is a physical resistor. The current at any instant can be determined by using Ohm's Law. When working with AC circuits, instantaneous values of voltage and current are seldom used in AC calculations. In most cases, **effective values** are used. The effective value of an AC voltage or current sine wave is equal to 0.707 times its peak or maximum value. An AC current sine wave with an effective value of 1 ampere effectively produces the same amount of heat, in a given resistance, as a DC current of 1 ampere. When we use effective values we are expressing the AC quantity in terms of its DC equivalent value.

Ohm's Law can be used with effective values just as easily as with instantaneous values. You must ensure, however, that all the values used have the same reference. In other words, to calculate the effective value for voltage, you must use the effective value for current, and vice versa. This also holds true for instantaneous values.

Let's consider a typical circuit that has an AC voltage source with an effective value of 100 volts and a resistance of 100 ohms. According to Ohm's Law, the effective value of current (I) is equal to:

$$I = \frac{E}{R} = \frac{100 \text{ V}}{100 \text{ }\Omega} = 1 \text{ ampere}$$

Therefore, an effective voltage (force) of 100 volts will cause an effective current of 1 ampere to flow through a resistance of 100 ohms.

Let's consider a circuit in which resistance and current are known and the voltage must be determined. For example, if the current in an AC circuit has an effective value of 3 amperes and a resistance equal to 50 ohms, the applied voltage must have an effective value of:

$$E = I \times R = (3 \text{ A})(50 \text{ }\Omega) = 150 \text{ volts}.$$

As a final example, if an AC circuit has a total effective current of 50 milliamperes when an effective voltage of 200 volts is applied, the total resistance of the circuit is:

$$R = \frac{E}{I} = \frac{200 \text{ V}}{50 \text{mA}} = 4\text{k}\Omega$$

As you can see, Ohm's Law continues to define the relationship between voltage, current, and resistance, regardless of whether you are working with a DC or AC circuitry.

AC RESISTANCE

Circuit Calculations

You will find very little difference between working with AC or DC resistive circuits. As we pointed out earlier, you have the choice of working with effective or instantaneous values with an AC circuit, provided you remain consistent throughout your calculations. It is common practice to use effective values when analyzing an AC circuit. The following section describes how calculations are made for AC series and parallel circuits.

Series Calculations

In addition to Ohm's Law remaining the same, other basic series circuit characteristics also remain the same. For example, the total resistance in the circuit is still the sum of all individual resistances. Also, the applied voltage (either effective or instantaneous value) will be the sum of the voltage drops (with the same reference) of each component. And finally, once total current is calculated, that current must flow through each component.

A typical series AC circuit is shown in Figure 13-2. Both resistor values are given, which allows us to determine that total resistance is 150 ohms. With an applied voltage of 150 volts rms, we can compute total current using Ohm's Law:

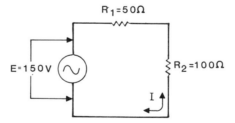

Figure 13-2 Calculating current and voltage in a series circuit.

$$I_T = \frac{E}{R} = \frac{150 \text{ V}}{150 \text{ }\Omega} = 1\text{A}$$

This current value of 1 ampere represents the effective value of circuit current. The current flowing through R_1 produces a voltage drop which, according to Ohm's Law, must be equal to the circuit current times the resistance of R_1:

$$E = I \times R_1 = (1 \text{ A})(50 \text{ ohms}) = 50 \text{ volts rms}$$

Since current is constant throughout a series circuit, it also flows through R_2. This produces a voltage across R_2 that is equal to:

$$E = I \times R_2 = (1\text{ A})(100\text{ ohms}) = 100 \text{ volts rms.}$$

Notice that these voltage drops are proportional to the resistors' values. In other words, the larger resistor drops the most voltage in a series circuit. The sum of these two effective voltages is equal to the effective value of the applied voltage. This is expressed mathematically as:

$$E_A = E_{R1} + E_{R2} = 50\text{ V} + 100\text{ V} = 150 \text{ volts rms.}$$

As you can see, AC effective values are used in the same way that DC values are used. The same rules apply to a series DC or AC circuit that contains only resistors (purely resistive circuit).

As we stated earlier, the current in a purely resistive circuit is always in-phase with the applied voltage. Since the current flows through both R_1 and R_2, the voltage across each resistor, at any given instant, is equal to the product of the circuit current at that instant and the resistor's value. The two voltages, are in-phase with the applied voltage because they are developed at the same time and vary in the same direction. Therefore, all of the voltage and current values are in-phase with each other. This is shown in Figure 13-3.

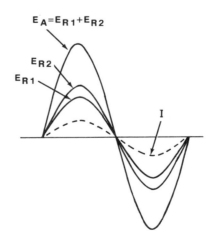

Figure 13-3 The phase relationship between values in a series circuit.

AC RESISTANCE

Parallel AC Circuit Calculations

Like a series AC circuit, there is very little difference between an AC parallel resistive circuit and a DC parallel resistive circuit. Ohm's Law still applies, as well as the basic characteristics of a parallel circuit. The voltage applied is felt across each parallel branch, and total current is equal to the sum of the branch currents.

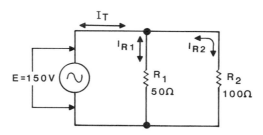

Figure 13-4 Calculating current and voltage in a parallel circuit.

Two resistors are connected in parallel with an AC voltage in Figure 13-4. The applied voltage is equal to 150 volts rms, and R_1 and R_2 have values of 50 ohms and 100 ohms respectively. You can determine the value of branch current I_{R1} by dividing the applied voltage by the branch resistance (R_1). The value is equal to:

$$I_{R1} = \frac{150 \text{ V}}{50 \text{ }\Omega} = 3\text{A}$$

The current through R_2 is equal to the same applied voltage divided by R_2.

$$I_{R2} = \frac{150 \text{ V}}{100 \text{ }\Omega} = 1.5 \text{ A}$$

Notice the proportions in the current values. R_2, with twice the resistance, only allows half the current of R_1, with the same voltage.

Once the two branch currents are found, total current can be calculated simply by adding the two branch currents. Therefore, total current (I_T) is equal to:

$$I_T = I_{R1} + I_{R2} = 3 \text{ A} + 1.5 \text{ A} = 4.5 \text{ A}$$

Now that total current is known, we can find total resistance using Ohm's Law:

$$R_T = \frac{150 \text{ V}}{4.5 \text{ A}} = 33.33\overline{3} \text{ }\Omega$$

The circuit functions as if it contained one resistor that has a value of 33.3 ohms. If such a resistor were connected across the voltage source, the same total current would result. In a parallel circuit, R_T is the equivalent resistance that results from combining all of the individual resistors into a single value.

There are other methods of calculating total resistance in parallel. With this circuit, we could have used either the product/sum method or the reciprocal method. But, as we stated in Chapter 8, Ohm's Law is the easiest method for finding total resistance in parallel.

Were the values we calculated effective or instantaneous? If you'll recall in the beginning of this section, we stated that the voltage applied was 150 volts rms. Since our initial value was an effective value, every value we calculated from that point must also be an effective value.

The phase relationship between parallel circuit values is illustrated in Figure 13-5. Total current and the voltage applied remain in-phase. The two branch currents, since they vary in direction and magnitude with total current, must also be in-phase with the voltage applied. At any given time, the sum of the branch currents will equal total current.

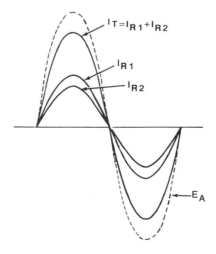

Figure 13-5 The phase relationship between values in a parallel circuit.

AC RESISTANCE

Power in AC Circuits

In an AC resistive circuit, power is consumed by the resistive component in the form of heat, just like in a DC circuit. When power in a DC circuit was discussed, you learned the equation for calculating power not only for specific components, but for entire circuits as well. The same relationship applies to AC circuits, but only those circuits that are *purely resistive*. In other words, the current at a specific instant can be multiplied by the voltage of that instant to determine the instantaneous power. When all of the instantaneous values are multiplied for a complete cycle of voltage and current, we find that the power curve follows the voltage and current changes. In other words, AC power is proportional to the product of voltage and current at any point on their curves.

A simple AC circuit is shown in Figure 13-6A. The power consumed by the resistor in this circuit varies with the product of the current through the resistor and the voltage dropped by the resistor. The relationship between power, current, and voltage is shown in Figure 13-6B. Notice that the power curve, or waveform, does not extend below the zero axis (horizontal line). This is because the power is effectively dissipated in the form of heat, no matter which direction current flows, which ensures power will be a positive value. Notice that power reaches its peak value when both E and I are maximum. Likewise, power drops to zero when both E and I equal zero.

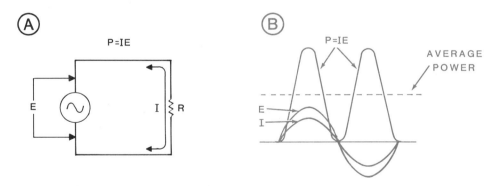

Figure 13-6 Power in an AC circuit.

Since the power fluctuates between a peak value and zero, the average power used by the circuit is midway between these two extremes. In other words, you draw a line midway between the peak and zero values, the line indicates the average power used. It is average power that is most important in AC circuits. Average power is the power that is actually used in AC circuits.

AC ELECTRONICS

The average power dissipated in an AC circuit can be calculated by multiplying the effective value of current times the effective value of voltage. Therefore, the power equation (P = IE) can be used in AC calculations as long as effective values of I and E are used. For example, suppose that the resistor in Figure 13-6A has a value of 100 ohms, and the applied voltage has an effective value of 100 volts. The current in the circuit would be equal to:

$$I = \frac{100 \text{ V}}{100 \text{ }\Omega} = 1 \text{ A}$$

This effective current value of 1 ampere is multiplied by the effective value of the applied voltage as follows:

$$P = IE = (1A)(100V) = 100 \text{ W}.$$

The resistor would consume (dissipate in the form of heat) 100 watts of power.

In Chapter 7, we found that power could be found using other formulas. These formulas, which also express the relationship between power and resistance, are:

$$P = \frac{E^2}{R} \quad \text{or} \quad P = I^2 \times R$$

Since the voltage and resistance values in Figure 13-6A were known, we could have used the first alternate equation to find power:

$$P = \frac{E^2}{R} = \frac{(100 \text{ V})^2}{100} = \frac{10,000}{100} = 100 \text{ W}.$$

If this formula had been used, it would not have been necessary to find total current. However, using two different formulas to find the same value is a good method of double-checking your calculations.

AC RESISTANCE

Chapter Self-Test

1. What is the phase relationship between current and voltage in an AC resistive circuit?

2. Which AC value is most commonly used when analyzing an AC circuit?

3. AC current, if drawn graphically, goes above and below the zero reference line. Does this indicate that AC current can be a negative value? Explain your answer.

4. An AC series resistive circuit has a total resistance of 10 kilohms, and 20 milliamps of current. What is the voltage applied? Is this effective or instantaneous voltage?

5. A series circuit contains a 10 ohm resistor and a 20 ohm resistor. When an effective voltage of 90 volts is applied across the resistors, what is total current?

6. When the two resistors in question 5 are connected in parallel, and 90 volts is applied across them, what is total current?

7. Why does power in an AC circuit never go below zero?

8. If an AC circuit has 40 volts peak-to-peak applied, and a total resistance of 7.07 kilohms, what is the average power?

Summary

AC circuits that contain only resistance (purely resistive circuits) are analyzed in much the same way as DC circuits which contain only resistance. When analyzing AC circuits it is common practice to use effective values of the currents and voltages.

In any AC circuit, Ohm's Law will still allow you to calculate for unknown values. When using Ohm's Law with effective values, the resulting value is also effective. If instantaneous values are used, the results will also be an instantaneous value.

When analyzing an AC series circuit, all of the characteristics of a series circuit remain the same. Current is constant throughout the circuit, and the sum of the voltage drops is equal to the voltage applied.

In an AC parallel circuit, total current can still be found by adding the individual branch currents, and the voltage applied is still felt across each parallel branch.

Power is dissipated in an AC circuit in the form of heat, just like a DC circuit. When the voltage and current curves dip below zero, power remains a positive value. The average power for an AC circuit is the value halfway between zero and the peak value for power. The average power can also be calculated by multiplying effective voltage and effective current.

CHAPTER 14

AC Capacitance

AC ELECTRONICS

Contents

Introduction 287

Chapter Objectives 288

Review 289

Capacitors in AC Circuits 293

RC Circuits 300

RC Circuit Applications 315

Chapter Self-Test 323

Summary 324

AC CAPACITANCE

Introduction

In Chapter 10, we introduced you to the capacitor. You were shown how a capacitor is constructed, and how it operates. You were introduced to many new terms, including **dielectric, capacitance, charging** and **discharging,** and the **farad.**

In a DC circuit, a capacitor simply charged to the value of the applied voltage, and at that point, current did not flow. When a capacitor is connected to an AC circuit, current constantly flows. This makes the capacitor an extremely useful component in AC circuits. The operating characteristics of a capacitor in an AC circuit are much more extensive than the capacitor's operation in a DC circuit.

In this chapter, we will first review the concepts you learned in Chapter 10. Then we will show you how a capacitor operates in an AC circuit, and the effects it has on voltage, current, and power. You will read about capacitive reactance; perhaps the most important characteristic a capacitor has in an AC circuit. We will also discuss many of the applications a capacitor has in AC circuitry.

AC ELECTRONICS

Chapter Objectives

When you have completed this chapter, you will be able to:

1. Define the following terms: orbital distortion, capacitive reactance, phase angle, vector, vector addition, impedance, reactive power, true power, apparent power, VARS, volt-amps, attenuation, filter, complementary angles.

2. Describe the phase relationship between voltage and current in an RC circuit.

3. List the factors that determine capacitive reactance.

4. Calculate the capacitive reactance for a capacitor, using capacitance and input frequency.

5. Calculate the total impedance for a series RC circuit using vector diagrams and Pythagorean's Theorem.

6. Calculate the applied voltage for a series RC circuit using vector diagrams and Pythagorean's Theorem.

7. Calculate total current in a parallel RC circuit using vector diagrams and Pythagorean's Theorem.

8. Calculate the apparent power for series and parallel RC circuits.

9. Calculate the phase angle for series and parallel RC circuits using trigonometric functions.

10. Calculate the power factor for series and parallel RC circuits.

11. Describe the operational characteristics of a capacitive voltage divider.

12. Describe the operational characteristics for both RC high-pass and RC low-pass filters.

13. Describe the operational characteristics for leading and lagging output RC phase shifting networks.

AC CAPACITANCE

Review

Before you learn how capacitors are used in alternating current circuits, you need to review the basic operation and characteristics of a capacitor.

This review will not cover all of the information covered in Chapter 10, but will simply review the terms and concepts surrounding capacitors.

Capacitor Operation

A capacitor is a component that stores electrical energy in the form of an electric field. In its simplest form, a capacitor consists of two conduction plates separated by an insulator called the **dielectric**.

When a DC voltage is applied to the plates of a capacitor, the capacitor becomes **charged**. Figure 14-1 shows a battery (DC source) connected to the capacitor. The positive terminal of the battery attracts electrons from the left-hand plate of the capacitor, which leaves the left-hand plate with a positive potential. Electrons flow from the negative terminal of the battery to the right-hand plate, causing it to be at a negative potential. While these electrons flow, the capacitor is **charging**.

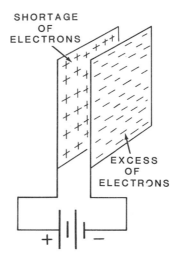

Figure 14-1 A capacitor connected to a DC voltage source.

Since the positive and negative electrical potentials on the plates attract one another, a force field is set up in the dielectric. However, there is no current flowing *through* the capacitor. The only time current flows is when the DC potential is initially connected. Current flow, or the movement of electrons, takes place only during the time it takes to charge the capacitor. Once the capacitor is charged to a value approximately equal to the DC source voltage, current ceases to flow, and the capacitor is completely charged.

When the battery in Figure 14-1 is removed from the capacitor, the electrical charges on the plates remain. The attraction of the positive and negative charges on the two plates across the dielectric holds the charges in place and the capacitor remains charged. As long as the two plates are insulated from one another, the capacitor will remain charged.

The capacitor can be **discharged** by shorting the plates together. This is accomplished by connecting a current path between the plates. When a capacitor is discharged, electrons flow from the negative plate to the positively charged plate. The excess of electrons on the right-hand plate neutralizes the positive charge on the left-hand plate. This discharge action gives the capacitor a neutral or zero charge. When a capacitor discharges, current flow is maximum when discharging begins, and decreases as the charge (difference of potential between the plates) decreases. Once the plates are neutralized, current ceases to flow.

As you can see, a capacitor stores electrical energy in the form of a charge. The capacitor is charged by an external voltage source, then retains that charge because of the attraction of the two plates across the dielectric. The capacitor can be discharged by connecting a path between the two plates. Current *does not* flow through the capacitor itself. Current does flow in a capacitive circuit during the time the capacitor is being charged or discharged. Figure 14-2 shows the electronic symbols used to represent capacitors in schematic diagrams.

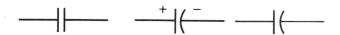

Figure 14-2 Schematic symbols for capacitors.

AC CAPACITANCE

Capacitance

The ability of an electronic component to store an electrical charge is referred to as capacitance. The unit of measurement for capacitance is the farad. One farad is the ability to store one coulomb of charge when one volt is applied. The farad is a very large unit of measure. A one farad capacitor would be physically very large. A capacitor of this size is much larger than is required in most electronic applications. Most capacitors used in electronic circuits have a capacitance in the microfarad or picofarad range.

Factors Affecting Capacitance

There are three factors that affect the amount of charge a capacitor can store. The first of these factors is the plate area. As a capacitor charges, electrons move from the negative terminal of the battery to one plate of the capacitor. If the plate has a larger area, it can accept and release more electrons than a plate with a smaller area, thereby allowing it to store a larger voltage. Therefore, plate area is directly proportional to capacitance.

The second factor that affects capacitance is plate distance. If the two plates of a capacitor are closer together, the repulsion between the opposite charges is greater, thereby moving more electrons. If the plates are moved further apart, the repulsion will not be as intense. Therefore, plate distance is inversely proportional to capacitance.

The third factor is the type of dielectric material. Different materials support electrostatic lines of force differently. Air, for example, is a very poor dielectric. Air is arbitrarily given a dielectric rating of 1, and materials with a greater ability to support lines of force are given higher ratings. The higher the rating, the higher the capacitance.

Time Constants

The amount of time a capacitor takes to charge and discharge depends upon the resistance within the circuit and the capacitance of the capacitor. The following formula defines this relationship:

$$T = R \times C$$

Once this value is calculated, however, it does not define the *total* amount of time for charging and discharging, but instead defines an interval of time. In order for a capacitor to completely charge or discharge, it takes five time constants.

AC CAPACITANCE

Capacitors in AC Circuits

Now that we've reviewed how a capacitor operates when connected to a DC voltage, we can now place a capacitor across an AC voltage, and see how it operates.

Capacitor Operation

When an AC voltage is applied to a capacitor, alternating current flows in the circuit. Figure 14-3 shows a sine wave generator connected to a capacitor. As the AC voltage varies, the current in the circuit follows a sinusoidal path. As the applied AC voltage rises and then falls, the capacitor charges and then discharges. The charge and discharge rate of the capacitor is determined by the frequency of the input voltage and the circuit's time constant. The current arrows indicate that current flows in both directions; a rise and fall in one direction, and a rise and fall in the opposite direction.

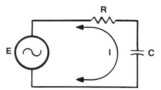

Figure 14-3 An AC voltage source applied to a capacitor.

Figure 14-4 illustrates the action of an electron in the dielectric of a capacitor when an AC voltage is applied. When the capacitor is charged, as shown in Figure 14-4A, the electrons orbiting in the dielectric are repelled by the negative plate (top) and attracted by the positive plate (bottom). This is referred to as **orbital distortion**. The amount of oribital distortion depends on the voltage applied to the capacitor. Figure 14-4B shows the orbital distortion when the capacitor is charged in the opposite direction.

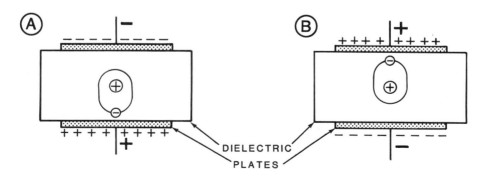

Figure 14-4 Electron distortion in the dielectric of a charged capacitor.

AC ELECTRONICS

When an AC voltage is applied to the capacitor, the polarity of the applied voltage alternates. This causes the electrons in the dielectric to change directions. While the amount of electron shift is small, it nevertheless constitutes a movement of electrons within the dielectric. While its true that none of the electrons actually break loose from their orbits and flow in the external circuit, we can say that the movement of electrons constitutes current flow. As the capacitor is charged and discharged by the AC voltage, the external circuit feels current flow as electrons move off one plate and onto the other. When the applied voltage is a sine wave, current flow in the circuit is also sinusoidal.

Current-Voltage Relationships

When an AC voltage is applied in a purely resistive circuit, the maximum and minimum values for current and voltage occur at the same time. In this case, current and voltage are in-phase. In other words, the positive and negative half cycles of voltage and current in a resistive AC circuit are in step with one another.

In a capacitive AC circuit, the capacitor constantly charges and discharges at the rate of change of the applied voltage. In other words, as the voltage applied increases in one direction, the capacitor will charge to the same magnitude. As the voltage decreases, the capacitor will discharge. Thus, the voltage drop of the capacitor is in-phase with the voltage applied.

Current flow, on the other hand, is a result of the amount of charge or discharge within the capacitor. As the voltage applied increases, the amount of the charge within the capacitor also increases, which *decreases* the amount of current flow in the circuit. When the polarity of the applied voltage switches, the capacitor discharges. As it discharges, its voltage decreases, while current flow within the circuit *increases*. Once the capacitor is completely discharged, it will charge in the opposite direction. When this occurs, current in the circuit will again decrease, while the voltage across the capacitor increases.

As you may have assumed by this point, the maximum and minimum values for voltage and current in a capacitive circuit will not occur at the same time. When voltage is maximum, current is minimum, and vice versa. If we were to graph the values for current and voltage, you would see a phase shift between the voltage and current in the circuit. The phase shift in a purely capacitive circuit is 90 degrees. We can say, therefore, that current leads voltage by 90 degrees.

AC CAPACITANCE

The exact relationship between the current and voltage in a capacitive circuit when a sine wave is applied is shown in Figure 14-5. Note that when the current is maximum, the voltage across the the capacitor is zero. As you can see, there is a phase shift between the current and voltage in the circuit. This phase shift is expressed in terms of degrees. Remember that one complete cycle of a sine wave contains 360 degrees. The amount of phase shift in the capacitive circuit is one-fourth or 90 degrees. We say that the current and voltage in a purely capacitive circuit are 90 degrees out of phase with one another. Another important fact to note is that the change in current **leads** the change in voltage. This is shown by current being at maximum at zero time, while voltage is minimum. 90 degrees later, voltage is maximum and current is minimum.

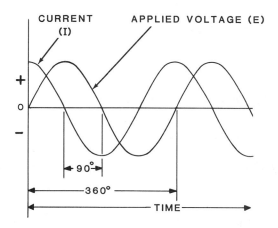

Figure 14-5 Current and voltage relationship in a purely capacitive AC circuit.

The following memory aid may help you remember the current-voltage relationships in capacitive and inductive circuits:

ELI the **ICE** man.

This memory aid states that in an inductive circuit (L), voltage (E) leads current (I). In a capacitive circuit (C), current (I) leads voltage (E).

Looking at Figure 14-5, you can see that the capacitor voltage change follows the current change in time. We say that current leads the voltage in a capacitive circuit. In other words, there is a 90 degree leading phase shift in a purely capacitive circuit.

AC ELECTRONICS

Capacitive Reactance

It is the basic nature of a capacitor to oppose changes in voltage. In a DC circuit, a capacitor will charge to the applied voltage. If the applied voltage is increased, the capacitor will then charge to the new higher voltage. If the applied voltage is decreased, the capacitor will discharge until the voltage across it equals the new lower applied voltage. Keep in mind that it takes a finite period of time for a capacitor to charge or discharge to a new voltage level.

In an AC circuit, the capacitor is constantly charging and discharging. The voltage across the capacitor is in constant opposition to the applied voltage. This constant opposition to changes in the applied voltage creates an opposition to current flow in the circuit. This opposition to the flow of AC current by a capacitor is called **capacitive reactance**. Capacitive reactance is represented by the symbol X_C and, like resistance, it is measured in **ohms**.

Factors Affecting Capacitive Reactance

Just as there are certain physical characteristics that determine the amount of capacitance for a capacitor, there are factors that will determine the amount of opposition, or capacitive reactance, that a capacitor offers in an AC circuit.

The first of these factors, and perhaps the most obvious, is the total amount of capacitance in the circuit. The amount of capacitance (in farads) determines the number of electrons flowing in the circuit. For example, if the capacitance value doubles, the amount of electrons (current flow) in the circuit must also double because the capacitor has a greater ability to store and release electrons. Since X_C is opposition to current flow in an AC circuit, and the current flow increased with the increase in capacitance, X_C must decrease. This shows the inverse relationship between capacitance and capacitive reactance. If capacitance increases, capacitive reactance decreases. If capacitance decreases, capacitive reactance increases. Capacitance and capacitive reactance are **inversely proportional**.

The frequency of the AC voltage also determines the amount of opposition, or X_C, in an AC circuit. This is because the frequency determines the rate of charge and discharge for the capacitor. If the input frequency increases tenfold, the capacitor must charge and discharge 10 times as often within the same period of time. Thus, current flow in the circuit increases. Since current has increased, X_C must have decreased with the increase in frequency. As with capacitance, frequency is **inversely proportional** to capacitive reactance.

AC CAPACITANCE

Capacitive reactance is calculated by using the equation:

$$X_C = \frac{1}{2\pi FC}$$

In this expression, X_C is the capacitive reactance in ohms, F is the frequency in Hz, C is the capacitance in farads and pi is a constant that is equal to approximately 3.14. Pi is used in the equation because you are dealing with an AC sine wave, which is the result of the revolution of an AC generator, and one revolution is equal to 360 degrees or 2π radians. Since $1/2\pi = 1/6.28$, which is approximately .1592, you can simplify the expression as shown below:

$$X_C = \frac{.159}{FC}$$

Since the farad is an unrealistically high unit of capacitance, this formula is difficult to use. If we assume that the capacitance in the formula is expressed in microfarads, the expression for capacitive reactance can be changed. Begin by substituting the units of measurement for F and C (assuming capacitance will be expressed in microfarads) in our original equation:

$$X_C = \frac{.159}{Hz(\mu F)}$$

Then, we multiply both the numerator and the denominator by 10^6:

$$X_C = \frac{.159}{Hz(\mu F)} \times \frac{10^6}{10^6}$$

The microfarad and the 10^6 in the denominator cancel, and our final equation becomes:

$$X_C = \frac{159(10^3)}{Hz(F)} = \frac{159(10^3)}{FC}$$

AC ELECTRONICS

Let's compute the capacitive reactance for a 10 microfarad capacitor with a 10 volt peak-to-peak signal applied at 60 Hz:

$$X_C = \frac{159(10^3)}{FC} = \frac{159,000}{(60)(10)} = \frac{159,000}{600} = 265\Omega$$

If the frequency is increased to 120 Hz, what is the capacitive reactance?

$$X_C = \frac{159(10^3)}{FC} = \frac{159,000}{(120)(10)} = \frac{159,000}{1200} = 132\Omega$$

Notice that when frequency doubles, capacitive reactance decreases by one-half.

What will happen to capacitive reactance in our original calculation if the frequency remains the same, but the peak-to-peak voltage is doubled?

$$X_C = \frac{159(10^3)}{FC} = \frac{159,000}{(60)(10)} = \frac{159,000}{600} = 265\Omega$$

Notice that the answer is the same. This is because the voltage of the input signal is *not* a factor in capacitive reactance. When you worked with resistors in DC, changing the voltage applied didn't change the resistance of the circuit. The same is true for capacitive reactance. While voltage and capacitive reactance will determine the amount of current flow in the circuit, voltage *does not* determine capacitive reactance.

Notice that throughout the previous calculations, the value of capacitance was inserted into the formula without including the "micro" metric prefix, since our formula already included that. Also note that the answer is stated in terms of ohms. This will remind you that capacitive reactance is a *steady* opposition to current flow in an AC circuit.

If the value for capacitance is stated in picofarads, you must convert picofarads into microfarads before you can insert the value into the formula. For example, if, in our original circuit, the capacitance value was 10 picofarads instead of 10 microfarads, the equation would have looked like this:

$$X_C = \frac{159(10^3)}{FC} = \frac{159,000}{(60)(0.00001)} = \frac{159,000}{0.0006} = 265M\Omega$$

As you can see, the 10 picofarad capacitor, with a much smaller capacitance, offers much more capacitive reactance.

AC CAPACITANCE

Ohm's Law in Capacitive Circuits

Because of its capacitive reactance in an AC circuit, a capacitor is just as effective as a resistor in controlling current flow. Therefore, Ohm's Law will still apply in a capacitive circuit. However, there is an important rule when using Ohm's Law in capacitive circuits:

BE CERTAIN TO WORK WITH *CAPACITIVE REACTANCE*, AND NOT CAPACITANCE.

Ohm's Law, if you'll recall, states the relationship between voltage, current, and resistance. To put it another way, Ohm's Law states the relationship between potential difference, electron movement, and electron opposition. Capacitive reactance is the opposition offered by a capacitor; capacitance is the ability to store a charge. Therefore, you must first convert all capacitance values to capacitive reactance before you can use Ohm's Law.

A purely capacitive circuit is one that has only capacitors, and no physical resistance other than the resistance of the conductive paths (which is negligible). When an AC voltage is applied to such a circuit, Ohm's Law still applies, and all of its formulas can still be used:

$$I = \frac{E}{X_C} \qquad E = I \times X_C \qquad X_C = \frac{E}{I}$$

For the sake of analysis, all voltage will be effective voltages. For example, if an AC voltage of 6.3 volts is applied to a capacitor with a reactance of 210 ohms, circuit current is:

$$I = \frac{6.3 \text{ V}}{210 \text{ }\Omega} = 0.03 \text{ A, or } 30 \text{ mA.}$$

Since you have worked with Ohm's Law extensively in previous chapters, these concepts should be familiar to you. It is important to reemphasize that all calculations involving Ohm's Law should be made using *capacitive reactance*, and not capacitance.

RC Circuits

While almost purely capacitive circuits are sometimes used in electronics, more often capacitors are combined with other components to form electronic circuits. The most commonly used circuits contain a resistor and capacitor connected in series. Despite its simplicity, this simple series RC circuit has many applications. Another commonly used capacitor circuit is the paralleled resistor-capacitor combination. While not as common as the series RC circuit, the parallel RC circuit is also found in electronic equipment. In this section you are going to investigate the operation and characteristics of both series and parallel RC circuits.

Series RC Circuits

Figure 14-6 shows a resistor and capacitor connected in series with an AC voltage source. The source voltage is a sine wave (designated E), which causes current to flow in the circuit. The capacitor constantly charges and discharges, allowing current to constantly flow in the circuit. The stored charge of the capacitor is its voltage drop, designated E_C, and the current flow through the resistor allows it to also develop a voltage, designated as E_R.

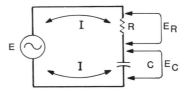

Figure 14-6 Series RC circuit.

The phase relationship between E_A, E_C, E_R, and I in this circuit is rather unique. In a purely capacitive circuit, voltage and current are 90 degrees apart, with current leading voltage. In a purely resistive circuit, voltage and current are in-phase with each other, or have 0 degrees phase difference. However, when we combine resistors and capacitors in the same circuit, the phase relationship between these values changes.

Figure 14-7 displays the phase relationships between these values. It is important to note that these signals do not represent the true amplitude of these signals, but merely the time when they occur. Current is used as a reference because current is the same throughout a series circuit. The signal for I represents both the current through the resistor and the current through the capacitor.

AC CAPACITANCE

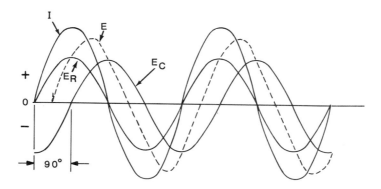

Figure 14-7 Phase relationships between current and voltage in a series RC circuit.

We know that when current flows through the resistor, it develops its voltage drop. Therefore, the voltage drop of the resistor is in-phase with current flow. However, the voltage drop of the capacitor is *not* in-phase with current flow. E_C lags (occurs at a later time) current by 90 degrees, which is the standard phase relationship between voltage and current for a capacitor. Therefore, E_R and I occur 90 degrees before E_C.

Finally, the signal for the applied voltage is obtained by adding the values for E_R and E_C at different times, plotting the results, and connecting them with a line. This allows us to see the time at which E_A occurs with respect to the other signals. However, we cannot add E_R and E_C to find the value of E_A. The method for calculating E_A will be explained later.

The signal for E_A is indicated by the dotted line. Notice that E_A is out-of-phase with both E_R and E_C. And since E_R and I are in-phase, this must mean that the applied voltage is out-of-phase with the current flowing in the circuit. However, the phase difference between these signals is *not* 90 degrees, because we are dealing with both resistance and capacitance. The phase difference must be somewhere between 0 and 90 degrees. In this example, the current leads the applied voltage by approximately 45 degrees. This phase difference is called the phase shift, or phase angle. The method for calculating phase shift will be explained later.

We previously stated that E_R and E_C could not be added together to determine the applied voltage. This is because these voltage drops occur at different times. In order to combine E_R and E_C, we must use vectors.

AC ELECTRONICS

Vector Diagrams

In the physical world, there are quantities that are expressed in specific units. Ohms, inches, and pounds are all examples of these quantities. These quantities are measured in scalar values and, in order to find the sum of a number of like quantities, you would just add them together. For example:

$$5\,\Omega + 10\,\Omega = 15\,\Omega$$

Other physical quantities have two properties; magnitude and direction. Any quantity having both magnitude and direction is called a vector. Miles traveled in a northeast direction, altitude gained in feet while traveling west, and volts at 90 degrees are all examples of vector quantities.

It is possible to add vectors, but you must take into consideration both the magnitude and the direction of the vector. In order to make this easier, a vector diagram is used to analyze AC circuits.

The basic format for vector diagramming is shown in Figure 14-8. The diagram consists of two lines (axis) perpendicular to each other. The horizontal line is the X axis and the vertical line is the Y axis. The point at which these two lines cross is the origin, and is labeled zero. The magnitude, or value, of any vector is measured outward from the origin. The direction of a vector corresponds to its phase difference in degrees.

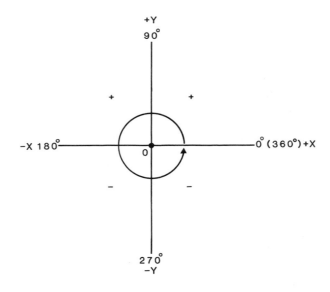

Figure 14-8 Typical AC vector diagram.

AC CAPACITANCE

In Figure 14-8, the ends of the axis are labeled 0 (360), 90, 180, and 270 degrees. When the vector is rotated, it starts at 0 degrees and travels in a counterclockwise direction until it reaches 360 degrees. This would be one complete revolution and ends at the origin. Now that we've set the basic groundwork for the vector diagram, let's look at an example.

Figure 14-9 shows the vector diagram for a series RC circuit. A current vector, I, is shown on the x axis pointing to the right. This vector represents the value of the current flowing in the series RC circuit. It is used as a reference vector for the diagram because the current value is the same at all points in a series circuit.

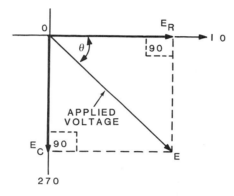

Figure 14-9 Vector diagram of a series RC circuit.

Another vector, labeled E_R, is drawn on the same line as the current vector. The length of this vector, from the origin to the point of the vector, represents the voltage dropped across the resistive portion of the circuit. This is the actual effective voltage as it would be measured with a voltmeter. The vector for E_R overlaps the current vector because they are in-phase.

The voltage across the capacitor is labeled E_C. This is the actual voltage measured across the capacitor and it is an effective value. Notice that the direction of the capacitor voltage is shifted 90 degrees from the direction of the resistor voltage. This is because there is a phase shift between the resistive and capacitive components in the circuit.

The applied voltage is the vector sum of the capacitor and resistor voltage. To graphically accomplish the vector addition, a rectangle is first formed using the resistor and capacitor voltages. This is shown by the dashed lines in Figure 14-9. The effective value of the applied voltage is the distance from the origin to the far corner of the completed rectangle. Stated another way, the value of the applied voltage is represented by the diagonal line drawn from the origin across the rectangle.

AC ELECTRONICS

The angle formed by the applied voltage vector and the resistive voltage vector represents the amount of phase shift that exists between circuit current and circuit voltage. This angle is always between 0 and 90 degrees.

You can now use the illustration created by the graphic solution of vector addition to determine the actual applied voltage. The diagonal line that represents the applied voltage splits the rectangle into two triangles. These triangles are right triangles because they each have a 90 degree angle. In addition, the length of two sides of the triangles (the capacitor and resistor voltages) are also known.

When you know the value, or length, of any two sides of a right triangle, it is possible to determine the value of the third side by using a mathematical formula called the **Pythagorean Theorem**. Appendix B, located at the end of the book, describes this theorem in detail. If you are not familiar with the Pythagorean Theorem, refer to Appendix B at this time.

In Figure 14-10, the side of the triangle representing the applied voltage is the side directly opposite the 90 degree angle. This side is called the hypotenuse and you can compute its length using the Pythagorean Theorem:

$$E^2 = (E_R)^2 + (E_C)^2$$

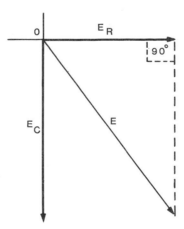

Figure 14-10 Voltage vector triangle for a series RC circuit.

This formula states that the resistor voltage squared plus the capacitor voltage squared is equal to the applied voltage squared. When you take the square root of the sum, the result will be the applied voltage.

AC CAPACITANCE

For example, let's find the applied voltage in a series RC circuit with a resistor voltage of 12 volts and a capacitor voltage of 18 volts:

$$E = \sqrt{(E_R)^2 + (E_C)^2} = \sqrt{12^2 + 18^2} = \sqrt{144 + 324} = \sqrt{468} = 21.63 \text{ V}$$

Obviously, in order to complete this calculation, you need to find the square root of 468. Finding the square root of a number is an extremely complex process. Therefore, we recommend that you use either a calculator (with a square root function) or a table of square roots.

Impedance

Impedance is the total opposition to current flow in an AC circuit. In a circuit consisting of a resistor and a capacitor, the total opposition is made up of both capacitive reactance and resistance. Because of the phase shift created by the capacitor, total impedance is not the sum of resistance and capacitive reactance. Total impedance, like total voltage, must be vectored.

The impedance of an AC circuit is expressed in ohms and is designated by the letter Z. We can define the impedance in terms of Ohm's Law just as we defined the total resistance of a DC circuit. The impedance of an AC circuit is equal to the applied voltage divided by total circuit current.

$$Z = \frac{E}{I}$$

The vector diagram for total impedance is shown in Figure 14-11. Notice that this is the same diagram as the voltage vector. The reason for this is because we are working with a series circuit. The voltage drops for the resistor and the capacitor are a result of the current flow in the circuit, and their respective opposition. Since current flow for each component is the same, their voltage drops are directly proportional to their opposition. The relation to impedance and voltage in a series circuit allows us to use the same vector diagram.

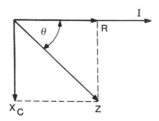

Figure 14-11 Impedance vector diagram of a series RC circuit.

AC ELECTRONICS

Current is again used as a reference. The resistance vector coincides with the current vector because the resistor voltage drop is in-phase with current. The reactance vector is drawn 90 degrees out-of-phase with the resistance vector. Completing the rectangle formed by the resistance and reactance vectors and drawing the diagonal line gives us the magnitude of the impedance. The length of the diagonal line represents the total opposition to current flow in the circuit.

Using Pythagorean's Theorem, we can write an expression for the impedance of the circuit in terms of the resistance and reactance:

$$Z = \sqrt{R^2 + X_C^2}$$

This expression says that the impedance is equal to the square root of the sum of the resistance squared and the capacitive reactance squared. Again, this formula is similar to the voltage formula, and calculations are made in precisely the same fashion. For example, what is the total impedance of a circuit with a resistance of 4 kilohms, and a capacitive reactance of 3 kilohms?

$$Z = \sqrt{R^2 + X_C^2} = \sqrt{(4k)^2 + (3k)^2} = \sqrt{16M + 9M} = \sqrt{25M} = 5k\Omega$$

Power

There is little difference between power in a DC resistive circuit and power in an AC resistive circuit. When an AC voltage is applied to a resistive circuit, current flows and electrical energy is converted into heat energy. The heat energy, or power, is dissipated in the resistance. The formulas used for calculations involving power in a DC circuit also apply to power in an AC circuit.

$$P = IE \qquad P = I^2 R \qquad P = \frac{E^2}{R}$$

You will recall that in a purely resistive AC circuit, the power curve (drawn by multiplying the voltage and current curves, and plotting the results) remained above the zero reference line. A line was drawn through the curve at the halfway point. This line represented **average power,** and could be calculated by multiplying the effective values for current and voltage. Another term often used for average power is **true power.**

AC CAPACITANCE

When you are analyzing a purely capacitive circuit, power will appear to be dissipated by the capacitor. However, the phase shift between circuit current and the applied voltage does not allow this to happen.

Figure 14-12 illustrates the power curve in a purely capacitive circuit. The voltage and current curves (90 degrees out-of-phase) are multiplied at different points, and the results are plotted. Unlike the resistive power curve, the power curve in a purely capacitive circuit is both positive and negative. What this power curve tells us is that during one half-cycle of the applied voltage, the capacitor appears to consume power. This is indicated by the positive part of the power curve. During the other half cycle, the power is negative. It is during this time that the capacitor actually acts as the supply and furnishes power to the source.

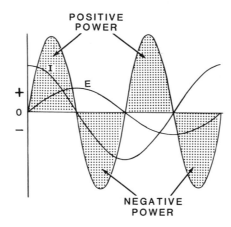

Figure 14-12 Power in a purely capacitive circuit.

When a capacitor charges, it consumes power. When a capacitor discharges, it gives power back to the circuit. Since the positive and negative power curves are equal and opposite, the total average effect is zero. This means that **no power is dissipated in a purely capacitive circuit**.

Although no power is actually dissipated in a purely capacitive circuit, there is a method for calculating the power that *appears* to be dissipated. This is called **reactive power**, and is measured in **volts-amps reactive**, or **VARS**. The same formulas are used, but X_C is used in place of R:

$$P_{VAR} = I \times E \qquad P_{VAR} = \frac{E^2}{X_C} \qquad P_{VAR} = I^2 \times X_C$$

AC ELECTRONICS

For example, what is the reactive power in a purely capacitive circuit with 10 volts rms and 2 amps rms of current?

$$P_{VAR} = I \times E = 10 \times 2 = 20 \text{ VARS}$$

Notice that effective values are still used. This answer tells us the power that *appears* to be dissipated. We know, however, that a capacitor will return any power dissipated back to the circuit.

Apparent Power

The differences between power in a resistive circuit and power in a capacitive circuit are significant. Power in an RC circuit, however, is rather simple. We know that current leads voltage by some angle less than 90 degrees. In Figure 14-13A, a power curve is obtained using the same methods as the previous power curves. Notice that the resulting curve is mostly positive, and goes negative for only a short time. Both the resistor and the capacitor consume power, as indicated by the positive portion of the curve. The negative portion of the curve is still a result of the capacitor returning its stored charge back to the circuit.

Figure 14-13 Power dissipated in a series RC circuit.

While all this sounds complex, the power dissipated in an RC circuit is found the same way as the power for any other circuit—the product of the applied voltage and circuit current. Power in an RC circuit is called apparent power, and is measured in volt-amps.

AC CAPACITANCE

While apparent power can be found by multiplying the applied voltage and circuit current, it can also be found if the power for each component is known. Because of the phase shift within the circuit, resistive and capacitive power occur at different times. Therefore, their values cannot be added together to find the total power. Instead, they must be vectored. Figure 14-13B shows the vector diagram for finding apparent power. It should be pointed out, however, that each of the values shown can be calculated by multiplying current and voltage. Resistive power (measured in watts) is drawn along the 0 degree line because its voltage drop is in-phase with circuit current. The wattage can be found by multiplying the resistor's voltage drop and circuit current.

Capacitive power, measured in VARS, is found by multiplying the capacitor's voltage drop and circuit current. This value is drawn 90 degrees *after* current, because of the phase shift between the capacitor's voltage drop and circuit current.

Finally, apparent power will be the vector sum of watts and VARS. However, this value can also be calculated by multiplying the applied voltage and circuit current. Keep in mind that all voltage and current values discussed are effective, or rms, values.

Power Factor

In a circuit containing both resistance and capacitance, the apparent power includes both the true power dissipated in the circuit resistance and the power that appears to be dissipated in the capacitor. The ratio of the true power to the apparent power in an AC circuit is referred to as the **power factor**, abbreviated **(PF)**. This is expressed in the following formula:

$$PF = \frac{\text{true power}}{\text{apparent power}} \text{ or } \frac{\text{watts}}{\text{volt/amps}}$$

The power factor is an excellent indication of the relative amounts of resistance and reactance in a given circuit. In a purely resistive circuit, the true power and the apparent power are equal. Therefore, the power factor will be equal to one. In a purely capacitive circuit, there is no true power dissipated, which means there is no power factor. For circuits containing both resistance and reactance, the power factor is always some value between zero and one. The greater the power factor, the more resistive the circuit. A lower power factor indicates a more reactive circuit.

AC ELECTRONICS

More importantly, the power factor is an indication of the efficiency of a circuit. In any circuit, work is performed only by the current that flows through the resistive portion of the circuit. The reactive current performs no work. In Chapter 16, you will see how the power factor of a circuit can be changed to make the circuit operate more efficiently.

In a series RC circuit, you can also determine the power factor using the values of the resistance and the reactance in the circuit. In fact, the power factor is equal to the ratio of the resistance to the total impedance of the series RC circuit:

$$PF = \frac{R}{Z}$$

Power factor can also be determined when you know the applied voltage and the voltage dropped across the resistance in the circuit. In this case:

$$PF = \frac{E_R}{E_A}$$

Phase Shift

To this point, we've stated that the degree of phase shift between current and voltage in an RC circuit is somewhere between 0 and 90 degrees. The degree of phase shift is determined by the voltage, impedance, and power vector diagrams. With trigonometry, you can take the lengths of the sides of the right triangles in the vector diagrams, and compute the phase shift.

Trigonometry is the branch of geometry that gives us methods and procedures for calculating the length of the sides in a right triangle, and the other angles. Appendix B at the end of the book introduces you to trigonometry, and briefly describes how trigonometry is used. If you are unfamiliar with trigonometry, refer to Appendix B at this time.

Briefly stated, trigonometric functions determine the ratio between two sides of a right triangle with respect to an angle within the triangle. The three most common trigonometric functions are sine, cosine, and tangent. The following formulas define these functions:

$$\text{sine} = \frac{O}{H} \qquad \text{cosine} = \frac{A}{H} \qquad \text{tangent} = \frac{O}{A}$$

In the previous formulas, **O** is the side opposite the angle, **A** is the side adjacent to the angle, and **H** is the hypotenuse, or the longest side of the triangle.

The impedance vector of a series RC circuit is shown in Figure 14-14A. The phase angle, indicated by the Greek letter theta (θ), can be found by using any of the three trigonometric functions. The function you use depends on the values known. If all three values are known, any of the three functions can be used.

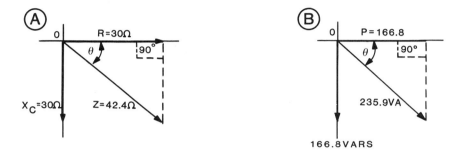

Figure 14-14 The impedance and power vectors for a series RC circuit.

In Figure 14-14A, all three values are known. We can use any of the previously mentioned trigonometric functions to determine the phase angle. The sine function is used in the following calculation:

$$\sin \theta = \frac{O}{H} = \frac{30}{42.4} = 0.707$$

Once you know the sine of the angle, use the table in Appendix C. In the table, 0.707 is the sine of 45 degrees. Therefore, the phase shift is equal to 45 degrees.

If the cosine function is used, our calculation would look like this:

$$\cos \theta = \frac{A}{H} = \frac{30}{42.4} = 0.707$$

In Appendix C, the angle with a cosine of 0.707 is also 45 degrees. In this example, cosine and sine are equal values because the opposite and adjacent sides are equal. This will not always be true.

Finding the phase angle using the tangent function would be accomplished as follows:

$$\tan \theta = \frac{O}{A} = \frac{30}{30} = 1$$

Again 45 degrees has a tangent of 1. Each function arrived at the same angle. If any of the other vector diagrams are used, the same calculations will achieve the same result, provided you use the values for the same circuit.

The power factor for a circuit can also be used to determine the phase angle of a circuit. The power vector is shown in Figure 14-14B. If you'll recall, the power factor could be calculated by dividing the true power (the power dissipated by the resistor) by the total, or apparent power. The true power is the side adjacent to the phase angle, and the apparent power is the hypotenuse. The **cosine** function expresses a ratio between these two sides, and is therefore used:

$$\cos \theta = \frac{A}{H} = \frac{166.8 \text{ W}}{235.9 \text{ VA}} = 0.707$$

Notice that cosine is the same value regardless of which vector diagram is used. 0.707 is also the power factor for the circuit.

Parallel RC Circuits

So far, your study of RC circuits has been limited to series RC circuits. It is also possible to connect RC circuits in parallel configurations. As with the series RC circuit, Ohm's Law applies to all of the components within the circuit. Also, as with the series RC circuit, vector addition is required to determine some unknown values.

However, when you are drawing vector diagrams for parallel circuits, there is one major difference. As you know, in a series circuit, current is used as a reference because it is the same as all points in the circuit. In a parallel RC circuit vector diagram you must use voltage as a reference. Remember, the voltage across all branches of a parallel circuit are equal. Let's look at a basic circuit and some vector diagrams to see exactly what differences there are.

AC CAPACITANCE

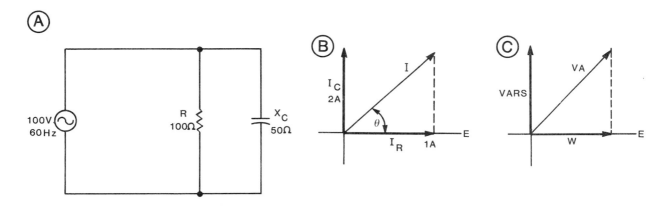

Figure 14-15 An RC parallel circuit with current and power vector diagrams.

Figure 14-15A shows an elementary RC parallel circuit. In this circuit, the voltage drop across both the resistor and the capacitor is 100 volts. Since the voltage and impedance across each branch is known, it is possible to calculate the current flow through each component using Ohm's Law:

$$I_R = \frac{E_R}{R} = \frac{100 \text{ V}}{100 \text{ }\Omega} = 1 \text{ A.}$$

$$I_C = \frac{E_C}{X_C} = \frac{100 \text{ V}}{50 \text{ }\Omega} = 2 \text{ A.}$$

The vector diagram, using these two values, is shown in Figure 14-15B. Notice that the value for I_C is drawn 90 degrees **before** I_R. Remember, current through the resistor is in-phase with the voltage across the resistor. However, current through the capacitor occurs 90 degrees before its voltage. Since voltage is used as a reference, I_C is drawn **before** I_R.

You can now find the circuit current by using either the Pythagorean Theorem or trigonometric functions. The Pythagorean Theorem method is shown here:

$$I = \sqrt{I_R^2 + I_C^2} = \sqrt{1^2 + 2^2} = \sqrt{1 + 4} = \sqrt{5} = 2.23 \text{ A}$$

You can also solve the triangle for the power vector diagram (Figure 14-15C) by using Ohm's Law, the Pythagorean Theorem, and the various trigonometric functions.

One more thing about parallel RC circuits. You may have noticed that an impedance vector diagram is not shown in Figure 14-15. This is because the impedance triangle for a parallel circuit is somewhat unique. Remember, the resistance of a parallel circuit is always less than the smallest resistive branch in the circuit. The same general principle holds true for impedance in a parallel RC circuit. The impedance is always less than the smaller of the two current opposing components. For this reason, when the impedance vector diagram is constructed, you must use the reciprocal of the resistance and capacitive reactance, and the resulting vector will be the reciprocal of the actual impedance.

As you have probably noticed, there are any number of ways to obtain the correct results when analyzing AC circuits. The key is to be extremely familiar with the trigonometric relationships, as well as Pythagorean's Theorem. Be sure that you have mastered these formulas before proceeding through this book.

AC CAPACITANCE

RC Circuit Applications

In this section you will learn some of the more important applications of capacitors in AC circuits. While capacitors are often used alone, usually they are combined with resistors or other components to form RC networks. Such networks have many practical applications, such as AC voltage dividers, filters, and phase shifters.

Voltage Divider

In Chapter 8, we introduced you to a simple resistive voltage divider. This circuit took an applied DC voltage and, through the voltage drops of the resistors, created different output voltages.

A resistive voltage divider will have the same effect on an AC signal. If you apply a 20 volt rms AC signal to a resistive voltage divider, it is possible to achieve an effective output voltage less than 20 volts rms. Since all values are in-phase in a resistive AC circuit, the output voltage will be in-phase with the input signal.

A capacitive voltage divider works much the same way. Figure 14-16 shows a simple capacitive voltage divider made by connecting two capacitors in series. The input voltage is connected across both capacitors. The output voltage is taken across C_2 only. The current flowing in the circuit produces a voltage drop across each capacitor. In this application, the output voltage is equal to the voltage across capacitor C_2.

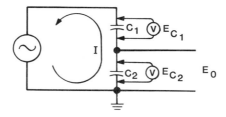

Figure 14-16 A capacitive voltage divider.

The amount of voltage developed across each capacitor depends on the current in the circuit and the capacitive reactance. According to Ohm's Law, the output voltage will be:

$$E_O = E_{C2} \quad \text{or} \quad I \times X_{C2}$$

AC ELECTRONICS

What makes the capacitive voltage divider different from the resistive voltage divider is that the frequency of the input signal affects the capacitive voltage divider. The voltage dropped across the capacitors in the circuit are in proportion to their capacitive reactance. The greater the capacitive reactance, the greater the voltage drop across that capacitor. Capacitive reactance is inversely proportional to both frequency and capacitance. Therefore, if the frequency of the input signal changes, capacitive reactance changes, which changes the voltage drops for the capacitors in the circuit. Although the frequency will affect the output voltage, there is still no phase shift between the input and output voltages.

Capacitive voltage dividers are often found in high frequency amplifier circuits. Certain types of oscillators use capacitive voltage dividers. The output voltage from a capacitive voltage divider can also be made variable by making either capacitor a variable capacitor. By changing the ratio of capacitance, the output voltage can be adjusted.

RC Filters

One of the most common applications of a capacitive circuit is the **filter**. A filter is a circuit whose primary responsibility is to produce an output voltage that changes with a change in frequency. This output voltage is supplied to a circuit that can only accept certain frequencies. A filter selects these frequencies by producing a high output voltage for the frequencies that are desired. Those frequencies which are undesired are **attenuated**. When a signal is attenuated, the voltage drop at the output is decreased to the point where it becomes unusable.

Two of the most common types of filters used in electronic circuits are the **low-pass filter** and the **high-pass filter**. A low-pass filter allows low frequency signals to pass from the input to the output with little or no attenuation. A frequency known as the cut-off frequency, F_{CO}, is the general dividing line between those frequencies that are passed and those that are attenuated.

A high-pass filter is the opposite of a low-pass filter. The high-pass filter permits frequencies above the cut-off frequency to pass, while attenuating frequencies below the cut-off point are greatly attenuated.

Simple RC networks are used as low-pass and high-pass filters. Such circuits are able to perform a frequency selective function because of the change in reactance with frequency.

AC CAPACITANCE

Low-Pass Filter

The simplest form of a low-pass filter is shown in Figure 14-17. It consists of a resistor and capacitor connected in series across an input voltage. The output voltage is taken across the capacitor. Assume that the input voltage has a fixed rms value, but that its frequency can be varied.

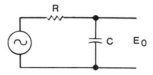

Figure 14-17 RC low-pass filter.

The best way to understand the operation of the low-pass filter is to look at the circuit as a voltage divider. The input voltage is applied across the resistor and capacitor in series. The output voltage is taken across the capacitor. The voltage division ratio depends upon the sizes of the resistance and the capacitive reactance. The value of the resistance remains constant, of course, but the value of the capacitive reactance changes as the input frequency changes. As frequency increases, capacitive reactance decreases, and vice versa.

At very low input frequencies, the capacitive reactance will be very high. If the reactance is high compared to the resistance, most of the input voltage is dropped across the capacitor. As the input frequency increases, the capacitive reactance decreases. This means that less voltage will be dropped across the capacitor and more across the resistor as the frequency increases.

For this reason, the output voltage begins to drop off as frequency increases. At very high frequencies, capacitive reactance is very low. When capacitive reactance is significantly lower than the resistor's value, very little voltage will be felt at the output.

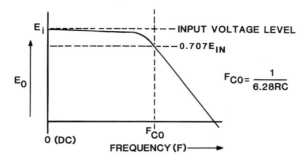

Figure 14-18 Output voltage curve for a low-pass filter.

The frequency response curve shown in Figure 14-18 illustrates this effect. This curve shows the amount of output voltage with respect to frequency. On the left-hand side of the curve, at very low frequencies, the output voltage is nearly equal to the input voltage. In fact, with a frequency of 0 Hz or DC, the capacitor offers maximum opposition and the output voltage is equal to the input voltage. As the frequency increases, the capacitive reactance begins to decrease, and the output voltage begins to drop off. At the cut-off frequency, the output voltage is equal to approximately 70.7% of the input voltage or $E_O - .707\ E_{IN}$. After the cut-off frequency is reached, the output voltage drops off at a constant rate. The cut-off frequency is a function of the resistor and the capacitor values. This is expressed by the equation:

$$F_{CO} = \frac{1}{2\pi RC} \text{ or } \frac{.159}{RC}$$

Where R is in ohms and C is in farads.

The formula can be simplified by solving for 1/6.28 and expressing C in microfarads:

$$F_{CO} = \frac{159(10^3)}{RC}$$

An important thing to note about an RC low-pass filter is that while the circuit is frequency selective, the selectivity is very gradual. In other words, the output is not sharply defined at the cut-off frequency. Higher frequencies are only attenuated, not cut out completely. In other words, the low-pass filter does pass frequencies higher than the cut-off frequency but they are lower in amplitude than those frequencies below the cut-off point. Despite this imperfection in RC low-pass filters, these circuits are still very useful. The cut-off point is defined as those frequencies that cause less than the effective voltage to be developed (coupled) to the output.

High Pass Filter

A simple RC high-pass filter is shown in Figure 14-19A. Like the low-pass filter, it consists of a resistor and a capacitor connected in series to the input voltage. In the high-pass filter, however, the output voltage is taken across the resistor. Figure 14-19B shows the frequency response curve of an RC high-pass filter.

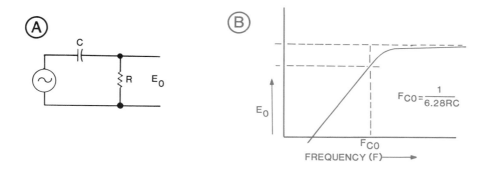

Figure 14-19 RC high-pass filter.

At very high input frequencies, the capacitive reactance will be very low. Thus, the capacitor drops very little voltage, and the resistor drops the balance. As frequency decreases, capacitive reactance increases. More and more voltage is dropped across the capacitor and less across the resistor. Thus, the output voltage also decreases. The decrease is gradual at first, but at the cut-off frequency, the attenuation becomes more pronounced and the output voltage drops at a constant rate with decreasing frequency.

Note that at high frequencies the output voltage is nearly equal to the input voltage. As the frequency decreases, the output voltage begins to decrease. At the cut-off frequency, the output voltage is approximately 70.7% of the input voltage. Below the cut-off frequency, attenuation increases and the output voltage drops accordingly.

AC ELECTRONICS

As in the low-pass filter, the cut-off frequency is a function of the resistor and capacitor values. The same expression used for computing the cut-off frequency of a low-pass filter applies to the high-pass filter.

Phase Shift Networks

RC networks are also used for phase shifting. This is where the phase of an output sine wave from a circuit is changed with respect to its input sine wave.

Phase shifting may be used for a number of reasons. Sometimes it is used to correct an unwanted phase shift that has been introduced by another component. In other applications, the phase shifting is done so that succeeding signals can be compared for magnitude and frequency.

Phase shifting networks can shift the output voltage in either a leading or lagging direction. Figure 14-20A is a schematic diagram for a leading output RC phase shifting network. Circuit current, as you know, leads the applied voltage by some phase angle between 0 and 90 degrees. The resistor will develop a voltage in-phase with circuit current. Therefore, the output voltage (taken across the resistor) will lead the applied voltage by the same phase angle. The relationship between input and output voltage is shown in Figure 14-20B.

Figure 14-20 Leading output RC phase shifting networks.

AC CAPACITANCE

Figure 14-21A illustrates a lagging output RC phase shifting network. This circuit is quite similar to the leading output network—a series RC circuit. However, now the output voltage is taken across the capacitor.

Figure 14-21 Lagging output phase shift in a series RC circuit.

The phase difference between the voltage drop of the capacitor and the applied voltage is caused by the RC time constant. In a purely capacitive circuit, the two voltages are in-phase because the RC time constant in a purely capacitive circuit is extremely small. This allows the capacitor to react almost immediately to a change in applied voltage.

In an RC circuit, however, the resistance forces a delay between a change in the applied voltage and the corresponding change in the capacitor's voltage. Therefore, the voltage drop of the capacitor **lags** (occurs later in time) the applied voltage.

The phase angle between the applied voltage and the capacitor's voltage is designated by the Greek letter **phi** (ϕ). This angle is **complementary** to the phase angle between the applied voltage and the resistor's voltage. When two angles are complementary, their sum is equal to 90 degrees.

You learned earlier that the resistor's and capacitor's voltage drops occur 90 degrees apart. If the phase angle between the applied voltage and the resistor's voltage is 35 degrees, for example, then the phase angle between the capacitor's voltage and the applied voltage must be 90 minus 35, or 55 degrees. This is expressed by the following formula:

$$\phi = 90 - \theta$$

Since the output of this circuit is taken across the capacitor, it will lag the input voltage by some phase angle between 0 and 90 degrees. The phase relationship between the input and output voltages is shown in Figure 14-21B.

These elementary RC networks are used when only small amounts of phase shift are required. When a phase shift of greater than 60 degrees is needed, other phase shifting techniques are usually used. The reason for this is that phase shifting networks also act as voltage dividers. As the phase shift across these networks approaches 90 degrees, the capacitor drops a progressively larger percentage of the applied voltage. In addition, the impedance of the circuit itself becomes a problem.

One way to get a greater phase shift is to cascade the simple RC network. Figure 14-22A is an example of a cascading leading output network, and Figure 14-22B is an example of a lagging output cascading network. The input signal is shifted by the first portion of the circuit, and that output is fed into the next circuit. If each portion of the circuit in Figure 14-22A provides a 35 degree phase shift, the final signal would be 105 degrees out-of-phase with the original input signal.

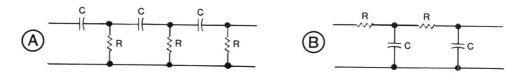

Figure 14-22 Cascaded RC phase shift networks.

Unfortunately, the amplitude of the output signal is much lower than the original input amplitude. If such is the case, the final output would then be amplified to increase the amplitude to its original value.

Another important point about these phase shift networks is that they can only be designed for one particular frequency. This is because the phase shift is determined in part by the capacitive reactance of the circuit, and the capacitive reactance is determined by the frequency of the input signal. If an RC phase shift network is designed for a particular input frequency, and that frequency changes, then the capacitive reactance for the network also changes. This changes the degree of phase shift.

AC CAPACITANCE

Chapter Self-Test

1. In a purely capacitive circuit, what is the degree of phase shift between voltage and current?

2. Which occurs first in a capacitive circuit, voltage or current? Why?

3. Define capacitive reactance.

4. What is the formula for capacitive reactance?

5. If the frequency of the input signal increases, what will happen to capacitive reactance?

6. What will happen to capacitive reactance if the amplitude (voltage) of the input signal is decreased?

7. Can the total amount of opposition in an RC circuit be found by adding the resistance and capacitive reactance values algebraically? Why?

8. How much power is dissipated in a purely capacitive circuit? Why?

9. What is the difference between vector diagrams for series RC circuits and parallel RC circuits?

10. What is the difference between a resistive voltage divider and a capacitive voltage divider?

11. What does it mean to **attenuate** a signal?

12. How is the cut-off frequency for an RC filter determined?

13. Describe the output of an RC high-pass filter as frequency changes from a very low frequency to a very high frequency.

14. Define **complementary angles**.

15. If a series RC circuit has the output taken across the capacitor, will the resulting circuit be a **leading output** phase shift network, or a **lagging output** phase shift network?

16. If you desire an output signal which leads the input signal by *greater* than 60 degrees, what type of phase shift network will work *best*?

AC ELECTRONICS

Summary

In an AC capacitive circuit, the capacitor is constantly charging and discharging, which allows current to constantly flow through the circuit.

The constant charging and discharging of a capacitor causes electrons within the dielectric material to distort their orbital paths. This is known as orbital distortion.

In a capacitive circuit, current flow will reach its maximum and minimum values before the applied voltage. This is known as a phase shift. In a purely capacitive circuit, the phase shift between voltage and current is 90 degrees.

Capacitors will offer a steady opposition to current flow in an AC circuit. This is known as capacitive reactance, and is abbreviated X_C. Capacitive reactance is affected by the amount of capacitance, and the frequency of the input signal. The formula for finding capacitive reactance is:

$$X_C = \frac{.159}{FC}$$

Capacitive reactance is measured in ohms. Once capacitive reactance is known, it may be substituted for resistance in Ohm's Law formulas.

A circuit containing capacitors and resistors is known as an RC circuit. In an RC circuit, the phase shift between the applied voltage and current flow is somewhere between 0 and 90 degrees.

Because of the phase shift between voltage and current, the voltage drops for capacitors and resistors also occur at different times. Therefore, they cannot be added together to find the total applied voltage.

A vector is any quantity having both magnitude and direction. The applied voltage for a series RC circuit can be found by vector addition. Vector addition involves creating a vector diagram, and calculating the length of the hypotenuse of a right triangle using the Pythagorean Theorem.

In a series RC circuit, current is used as a reference in vector diagrams because current is constant throughout the circuit.

AC CAPACITANCE

Impedance is defined as the total opposition to current flow offered by a circuit containing both resistive and reactive components. Total impedance for a series RC circuit can be found using vector addition.

In a purely capacitive circuit, the capacitor first consumes power, then returns the power back to the circuit. This results in no power being dissipated. This is known as reactive power, and is measured in volt-amps-reactive, or VARS. Reactive power can be found by multiplying the effective values of the applied voltage and circuit current.

In a purely resistive circuit, power dissipated is known as average power, or true power. In an RC circuit, both true power and reactive power exist. True power and reactive power can be combined, using vector addition, to find the apparent power for the circuit. Apparent power is measured in volt-amps. Apparent power for an RC circuit can also be found by multiplying the effective values of the applied voltage and circuit current.

The power factor for an RC circuit describes the ratio between true power and apparent power. It is a means of determining how efficiently an RC circuit operates. The power factor for an RC circuit is between 0 and 1. The power factor for a circuit is calculated by dividing the true power (watts) by the apparent power (volt-amps) in the following formula:

$$PF = \frac{watts}{volt\text{-}amps}$$

Trigonometry is the study of right triangles. It defines the relationship between the angles in a right triangle, and the length of the sides. Trigonometry is used to determine the phase angle in an RC circuit.

In a parallel RC circuit, voltage is constant throughout the circuit. Voltage is used as the reference in vector diagrams. Circuit current in a parallel RC circuit is the vector sum of resistive and capacitive current.

In order to find total impedance in a parallel RC circuit, the reciprocal values for resistance and capacitive reactance must by combined by vector addition.

A capacitive voltage divider operates much the same way as a resistive voltage divider. A capacitive voltage divider provides an output voltage lower in amplitude than the input voltage.

The output voltage of a capacitive voltage divider can be changed by changing the frequency of the input signal. Changing the frequency of the input changes the capacitive reactance of the components, which changes their voltage drops.

A filter is a circuit which selects a specific frequency or range of frequencies from an infinite number of frequencies.

Undesired frequencies are rejected by a filter by having the output voltage dropped below a usable level. This is known as attenuating the input. Desired frequencies are passed to the output by maintaining a high output voltage.

An RC high-pass filter allows high frequencies to pass, while rejecting lower frequencies. Any frequency below the cut-off frequency is rejected. The cut-off frequency is that frequency whose output amplitude is 70.7% of the input amplitude.

An RC low-pass filter passes frequencies below the cut-off frequency, while rejecting those above the cut-off frequency.

RC circuits are also used as phase shifting networks. A phase shift network is one that changes the phase of the output signal with respect to its input. The output can be placed before the input; this is known as a leading-output phase shift network. A lagging-output phase shift network causes the output to occur after the input.

Single phase shift networks are used if the desired phase shift is less than 60 degrees. If a phase shift of greater than 60 degrees is desired, cascading networks are used.

CHAPTER 15

AC Inductance

AC ELECTRONICS

Contents

Introduction 329

Chapter Objectives 330

Review. 331

Inductors in AC Circuits. 337

RL Circuits 342

RL Circuit Applications 353

Chapter Self-Test 357

Summary. 358

AC INDUCTANCE

Introduction

Chapter 9 introduced you to magnetism and its relationship with electronics. In Chapter 10, you applied the information you learned on magnetism, and discovered how it could be used in a component called an **inductor**. There were a number of new terms and concepts, including **self-induction, inductance,** and **henries.**

In this chapter, we will place the inductor in an AC circuit. Like the capacitor, an inductor acts differently in an AC circuit than in a DC circuit. We will begin by briefly reviewing the information on inductors we covered in Chapter 10. From there, we will discuss the operation of an inductor in an AC circuit. You will see the similarities between an inductive circuit and a capacitive circuit, as well as the differences. You will read about RL circuits, and the relationships that exist in an RL circuit. You will then learn how an RL circuit can be applied to create filters and phase shift networks.

AC ELECTRONICS

Chapter Objectives

When you have completed this chapter, you will be able to:

1. Describe the phase relationship between applied voltage and circuit current in a purely inductive circuit.

2. Describe the phase relationship between the applied voltage and CEMF in an inductive circuit.

3. Define inductive reactance.

4. List the factors that affect inductive reactance, and their relationship to inductive reactance.

5. Calculate the Q for an inductive circuit.

6. Describe the phase relationship between applied voltage, circuit current, and individual voltage drops in a series RL circuit.

7. Describe the phase relationship between applied voltage, circuit current, and individual current values in a parallel RL circuit.

8. Calculate for the applied voltage in a series RL circuit using individual voltage drops and Pythagorean's Theorem.

9. Calculate the total impedance in series and parallel RL circuits using resistance and reactance values, and Pythagorean's Theorem.

10. Calculate the total current in a parallel RL circuit using individual current values and Pythagorean's Theorem.

11. Calculate the apparent power in series and parallel RL circuits using true and reactive power values, and Pythagorean's Theorem.

12. Calculate the phase angle for series and parallel RL circuits using trigonometric functions.

13. Describe the operational characteristics for RL high-pass and low-pass filters.

14. Describe the operational characteristics for leading-output and lagging-output RL phase shift networks.

AC INDUCTANCE

Review

The following section is a brief review of the information on inductors covered in Chapter 10. This review will not be as extensive as Chapter 10, but is intended to refamiliarize you with the terms and concepts that surround inductors.

Self-Induction

When current flows through a conductor, a magnetic field is generated around the conductor. This field is not created immediately when voltage is applied to the conductor. Rather, the field builds gradually as current flow through the conductor increases. When current flow reaches its maximum value, the magnetic field around the conductor is at its maximum strength.

As the magnetic lines of force expand outward from the center of the conductor, the magnetic field causes a voltage to be induced into the conductor itself. The expansion of the magnetic lines of force with respect to the conductor represents the relative motion required to induce a voltage. The polarity of the induced voltage opposes the polarity of the voltage that causes it.

As long as the magnetic field is moving with respect to the conductor, an induced voltage is generated. As the magnetic lines of force continue to expand outward from the conductor during the rise of the current in the circuit, induced voltage is present. When the current in the circuit reaches its maximum, as determined by the applied voltage and the resistance of the conductor, the magnetic field becomes stationary. Since there is no further relative motion between the conductor and the magnetic field, there is no induced voltage. At this time, the current in the circuit is strictly a function of Ohm's Law.

When the voltage applied to the conductor is removed, current flow ceases. With no movement of electrons in the conductor, the magnetic field starts to collapse. As it collapses, the lines of force cut across the conductor and induce a voltage. Again, the collapsing lines of force cause relative motion between the conductor and the magnetic field. Therefore, a voltage is induced into the conductor. The polarity of the induced voltage is such that it tends to keep current flowing in the same direction.

The application or removal of the voltage source causes a self-induced voltage. Self-induction actually takes place for any current changes that occur. Increasing or decreasing the current in a circuit causes the magnetic lines or force to expand or collapse, and thereby cutting the conductor. The conductor cutting induces a voltage that opposes the applied voltage. The induced voltage is referred to as counter EMF, or CEMF, since it always opposes the applied voltage.

Inductors and Inductance

Inductance is the property of an electrical circuit that tends to oppose any change of current in the circuit. It is usually indicated by the letter   The conductor or wire we have been discussing exhibits the property of inductance because it opposes changes in the current flow. When the current through a conductor suddenly increases, CEMF is induced within the conductor that opposes the applied voltage. CEMF attempts to cancel the applied voltage, which temporarily holds the current to its previous level. CEMF opposes the applied voltage and therefore opposes the increase in current. The current still rises, however, because the induced voltage appears only during the time that the current is increasing. As the rate of increase is slowed, CEMF decreases in value. Once there is no longer a relative motion between the conductor and the magnetic field, no further induced voltage or additional opposition takes place.

When the current in a circuit is suddenly decreased, the magnetic lines of force collapse and induce a voltage into the conductor that opposes the applied voltage. When the applied voltage is suddenly decreased, the induced voltage maintains the current at the same level. The current will eventually decrease, however, as the magnetic field stops its collapse and no further induced voltage is generated.

As you can see, changing the current in a circuit causes CEMF to be induced into the conductor that opposes the change of current. The induced voltage tends to cause current in the circuit to remain constant. At least the circuit remains consistent for a short period of time.

The electronic component that most exhibits the property of inductance is called an inductor. The conductor or wire that we have been considering up to this point can be referred to as an inductor. While any wire or electrical conductor exhibits the property of inductance, it is normally not referred to as an inductor. Instead, an inductor is considered to be a separate and distinct type of passive electronic component. The most common inductor is a coil of wire. The term coil is interchangeable with "inductor".

Whenever a wire is wound into a coil, the inductor formed becomes more manageable. Winding the wire into a coil makes the inductor smaller and more compact. At the same time, inductance is greatly increased. By keeping the turns of wire close together, the magnetic field surrounding the wire will become more concentrated. The greater the magnetic field, the greater the induced voltage, and therefore the higher its inductance.

AC INDUCTANCE

The unit of electrical inductance is the henry, and is abbreviated H. One henry is defined as the amount of inductance that a coil has when the current, changing at the rate of one ampere per second, produces one volt of induced voltage. Inductance is a measure of how much CEMF is generated in an inductor for a specific amount of change in the current through that inductor.

The henry is a fairly large unit of inductance. While there are inductors available with an inductance of one henry or more, most inductors used in electronic circuits have a much lower inductance value. These inductance values are expressed in smaller units known as the millihenry and the microhenry. One millihenry is one thousandth of a henry. One microhenry is one millionth of a henry.

Factors Affecting Inductance

The physical characteristics of a coil determine its inductance. The amount of inductance that a coil has depends upon the number of turns in the coil, the spacing between the turns, the number of layers of windings, the type of windings, the diameter of the coil, the length of the coil and the type of core material.

The inductance of a coil is directly proportional to the number of turns of wire and the diameter of the coil. The greater the number of turns, the greater the magnetic field produced by the coil. In addition, for a given number of turns, more magnetic lines of force can be produced if the diameter of the coil is increased. Anything done to increase the magnetic field produced by the coil will increase its inductance.

The spacing between the turns also affects the inductance. Keeping the turns very close together causes the magnetic lines of force produced by each turn to add together and produce a stronger magnetic field. More space between the turns of the coil reduces the addition of the lines of force produced by each turn. This additive effect is called mutual inductance, and it can be increased or decreased by adjusting the spacing between turns.

Inductance is inversely proportional to the length of wire within the coil. That is, increasing the length decreases inductance. When the length is increased, the resistance is also increased, which reduces current flow.

The number of windings used will also affect the inductance. The greater the number of windings, the greater the inductance. The shape of the coil and the method of winding the wire also affects the inductance. Keep in mind that the closer the turns and the greater the alignment of the various layers of wire, the greater the magnetic field produced for a given amount of current; therefore, the greater the inductance.

AC ELECTRONICS

The type of material used in the core also affects the inductance. Core material refers to the type of form on which the wire is wound. Many coils are self-supporting and have no form or core. Such coils are referred to as air-core inductors. The inductance of an air-core coil is strictly a function of the factors we have just discussed. When a core is used, however, the inductance will be affected.

Most inductors are made with a core that has magnetic properties. Cores made of iron, steel, nickel, or some related alloy that can support a magnetic field and concentrate the magnetic lines of force produced by the coil. When the lines of force are concentrated, the field density increases, which increases the intensity of the magnetic field. This increases the amount of induced voltage and therefore the inductance.

It is the permeability of the core material that determines its effect on the inductance of the coil. Permeability is the ability of a material to support the magnetic lines of force. The greater the ease with which magnetic lines of force can be set up within the core materials, the higher the permeability. Air has a permeability of 1, while a magnetic material like iron has a permeability of 7,000.

Inductors in DC

Under normal conditions, an inductor has very little effect upon a DC circuit. The majority of the time, current remains at a constant steady level in a DC circuit. When current is steady, the magnetic lines of force within the coil do not move, and no CEMF is induced. The only effect an inductor has in these conditions is the small amount of resistance within the wire itself.

However, when power is initially turned on or off in a DC circuit, current is changing in value. At power on, current must move from a minimum value to its maximum value. During this time, the magnetic lines of force within the coil are expanding outward, cutting the wires within the coil. This action induces CEMF, which momentarily opposes the increase in current. As current approaches its maximum value, the rate of change becomes increasingly slower. The magnetic lines of force cut fewer wires within the same period of time, decreasing the amount of CEMF induced. Current will eventually reach its maximum value, although CEMF forces a small delay in time.

When power is turned off in a DC circuit, a similar effect occurs. Current decreases from its maximum value towards zero. As this happens, the magnetic lines of force within the coil again cut the wires within the coil, inducing CEMF. CEMF induced during a decrease in current opposes the change in current, just as CEMF induced during the increase in current opposes the change.

The difference is how CEMF opposes a decrease. As circuit current begins to decrease, CEMF opposes the decrease by introducing more current into the circuit. The CEMF induced has a potential difference which forces a small current into the circuit in the same direction as circuit current. As circuit current approaches zero, the rate of change decreases. This decreases the amount of CEMF induced, and the amount of current introduced into the circuit decreases also. Circuit current will eventually reach zero, and CEMF is no longer induced.

Inductive Time Constant

Whenever a DC voltage is applied to or removed from an inductor, it takes a finite period of time for the current to reach its maximum or minimum values. The effect is similar to the time that it takes for a capacitor to charge or discharge. The time that it takes the inductive circuit's current to rise or fall is referred to as the **constant** of the inductor.

The time constant of an inductor is defined as the time required for the current to rise to 63.2% of its maximum value or to decrease to 36.8% of its maximum value. This is illustrated in Figure 15-1. Curve A illustrates the gradual rise of current when voltage is initially applied. After the first time constant, current has risen to 63.2% of its maximum value. From there, the current rise is more gradual. After five time constants, current has reached its maximum value.

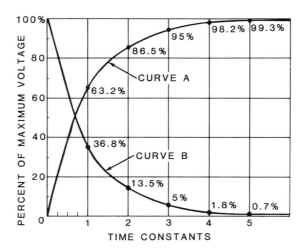

Figure 15-1 Time constant curves.

AC ELECTRONICS

In Curve B, DC power is removed, and current begins its decrease. After the first time constant, current has decreased to the point where only 36.3% of the maximum value remains. From there, the decrease in current is more gradual. Again, current reaches minimum after 5 time constants.

The value of the inductive time constant is a function of the inductance and resistance in the circuit. The time constant is directly proportional to the inductance and inversely proportional to the resistance. The following equation expresses this relationship:

$$T = \frac{L}{R}$$

As you can see, the greater the inductance, the greater the time constant. Also, the greater the induced voltage, the greater the opposition to the applied voltage. This means that the current in the circuit will take longer to rise to its final or maximum value. The larger the resistance, the smaller the current. Since current is reduced, the strength of the magnetic field and the induced voltage is also reduced.

In this expression, T is in seconds, L is in henrys, and R is in ohms. For example, a 1.5 henry inductor has a resistance of 100 ohms, the time constant is:

$$T = \frac{1.5}{100\,\Omega} \quad \text{or} \quad 0.015\ \text{s} \quad \text{or} \quad 15\ \text{ms}$$

This means that it takes 15 milliseconds for the current to rise to 63.2% of its maximum value or to fall to a value of 36.8% of the maximum circuit current. It takes approximately 5 time constants for the current to rise to the maximum value from zero or drop from its maximum value to zero. Note that the 5 time constants it takes the inductor to completely change between maximum and minimum is the same as the capacitor took to completely charge and discharge.

AC INDUCTANCE

Inductors in AC Circuits

Now that we've reviewed the basic characteristics of an inductor, and its operation in a DC circuit, let's examine the operation of an inductor in an AC circuit.

In an AC circuit, the applied voltage is constantly changing in both magnitude and polarity. Current, like voltage, will change in both magnitude and polarity. This changing current causes the magnetic field within an inductor to be constantly moving across the coils, inducing CEMF. CEMF, as you know, opposes a change in current. Since CEMF is constantly being induced, there is a steady opposition to a change in current flow. It also allows the inductor to develop a sinusoidal voltage drop.

You may have noticed that an inductor in AC acts similar to a capacitor, in that there is constant opposition, and current constantly flows. Therefore, an inductor has a reactance value, and forces a phase shift between the applied voltage and circuit current.

Current-Voltage Relationship

Figure 15-2 shows the time relationship between current, voltage, and CEMF in an inductive circuit. The phase relationship between current and CEMF is 90 degrees. The concept of self-induction states that the amount of voltage induced into a conductor is proportional to the rate of change of the magnetic field. The faster the magnetic field expands or collapses, the higher the induced voltage. As you can see in Figure 15-2, the most rapid change in current occurs at points A and B. Here, the current is changing from positive to negative or from negative to positive. Therefore, the rate of change of current is highest at these points. It is at these points that the induced voltage is highest. The positive and negative peaks of the CEMF waveform occur at the zero crossing points of the current waveform.

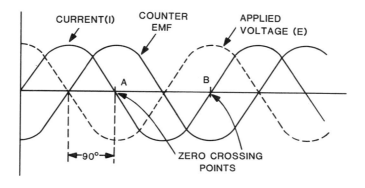

Figure 15-2 Current-voltage relationship in a purely inductive circuit.

AC ELECTRONICS

The rate of change of current at the positive and negative peaks of the current waveform, on the other hand, is zero. As a result, the induced voltage at these points is also zero.

These points illustrate the phase difference between CEMF and current. When current is at minimum (rapid change), CEMF is maximum. When current is at its peaks (minimum change), CEMF is minimum. CEMF and current, therefore, are 90 degrees out-of-phase.

The phase relationship between CEMF and the applied voltage is somewhat different. CEMF is in direct opposition with the applied voltage, and is exactly opposite in-phase to the applied voltage. We say that the applied voltage and CEMF are 180 degrees out-of-phase with one another. As you can see from Figure 15-2, the induced and applied voltages are exactly opposite one another. When one is at its positive peak, the other is at its negative peak.

CEMF is out-of-phase with both the applied voltage and circuit current. However, the phase difference between CEMF and voltage is 180 degrees, yet only 90 degrees for CEMF and current. The constant opposition CEMF offers in an inductive circuit forces a delay between the applied voltage and circuit current. Therefore, a phase shift exists between the applied voltage and circuit current. As illustrated in Figure 15-2, the phase shift is 90 degrees. When voltage is maximum, CEMF is also maximum in the opposite direction. CEMF offers maximum opposition to current, and no current flows. As the applied voltage decreases, CEMF also decreases, and current begins to increase. Therefore, when voltage is at its peak value, current is at its zero value, and vice-versa. In a purely inductive circuit, voltage leads current by 90 degrees.

This phase difference between voltage and current in an inductive circuit is opposite to the phase shift in a capacitive circuit. In an inductive circuit, voltage leads current. In a capacitive circuit, current leads voltage. The memory aid we introduced in Chapter 14 may help you remember this difference:

    ELI the ICE man.

This phrase states that voltage (E) leads current (I) in an inductive circuit (L), while current leads voltage in a capacitive circuit (C).

AC INDUCTANCE

Inductive Reactance

The CEMF induced into an inductor by a varying current opposes the applied voltage. As a result, the total effective voltage in the circuit is the difference between the applied voltage and the induced voltage. Because the induced voltage is less than the applied voltage, the effect of the inductance is to minimize or reduce current flow. The greater the inductance, the greater the CEMF. The opposition to current flow by an inductor in an AC circuit is called inductive reactance, and is abbreviated X_L. Like resistance and capacitive reactance, inductive reactance is measured in ohms.

Factors Affecting Inductive Reactance

The ability of an inductor to induce CEMF is a determining factor for inductive reactance. In other words, the amount of inductance in a circuit determines, in part, the amount of inductive reactance. If an inductor has a high amount of inductance, more CEMF can be induced. If more CEMF is induced, the opposition to current flow in the circuit increases. If a circuit has a lower amount of inductance, less CEMF is induced which allows more current flow in the circuit. We can conclude, therefore, that the amount of inductive reactance is directly proportional to inductance.

Inductive reactance is also directly proportional to the frequency of the input signal. The frequency determines the rate of change for current. The rate of change for current determines the rate at which the magnetic field within the inductor cuts the coils. As the frequency increases, the magnetic field cuts across the coils more times per second. Thus, more CEMF is induced, which increases inductive reactance. Likewise, if frequency decreases, inductive reactance decreases.

Inductive reactance is directly proportional to both inductance and input frequency. The formula for calculating inductive reactance in a circuit shows this relationship:

$$X_L = 2\pi FL$$

Multiply pi by 2, and the result is 6.28. We can substitute this into our formula:

$$X_L = 6.28\, FL.$$

In this formula, X_L is in ohms, F is the frequency in Hz, and L is the inductance in henrys. For example, to compute the inductive reactance of a 25 millihenry coil at 400 Hz, you would make the following calculations:

$$X_L = 6.28\, FL = 6.28(400)(25m) = 2512(25m) = 62800\, m\Omega \quad \text{or} \quad 62.8\, \Omega$$

AC ELECTRONICS

Quality of Merit (Q)

An inductor is a reactive component like a capacitor. However, unlike a capacitor, an inductor also has a certain amount of physical resistance. This resistance will dissipate energy like a normal resistor. If an inductor has a high amount of resistance, it will not function very well as a reactive component. An inductor with a low amount of resistance, however, will perform very well as a reactive component. The more an inductor behaves as a reactive component, the more *efficient* the inductor. We can define this efficiency by determining the inductor's quality of merit, or Q. The Q for an inductor is found using the following formula:

$$Q = \frac{X_L}{R}$$

As you can see, the Q of an inductor is a function of an inductor's reactance and its physical resistance. The Q value cannot be interpreted for a single inductor, but is used as a means of comparing one inductor to another. An inductor with a high Q value will be more efficient than an inductor with a low Q value. Generally speaking, an efficient inductor is one with a Q of 20 or higher.

Since Q is directly proportional to the inductive reactance, Q increases as frequency increases. The reactance is essentially a constant at low frequencies. Due to certain electrical characteristics, the resistance of an inductor increases as frequency increases. However, the resistance does not increase as rapidly as inductive reactance. Therefore, Q increases with frequency.

Ohm's Law For Inductive Circuits

Ohm's Law applies equally to inductive AC circuits as it does to resistive or capacitive circuits. The current flowing in an inductive AC circuit is directly proportional to the applied voltage and inversely proportional to the inductive reactance. This relationship is represented mathematically by the expression:

$$I = \frac{E}{X_L}$$

In this expression, current is in amperes, voltage is in volts, and reactance is in ohms. Increasing the voltage or decreasing the reactance causes an increase in current. Decreasing the applied voltage or increasing the reactance causes a decrease in current.

This relationship between the current, voltage, and reactance is illustrated in the following example.

What is the current flowing in a 2.5 millihenry choke when a 10 volt, 100 kHz sine wave is applied? To solve this problem, the inductive reactance must first be computed:

$$X_L = 6.28 \, FL = 6.28(100{,}000)(.0025) = 1570 \, \Omega$$

Now that the inductive reactance and the applied voltage are known, Ohm's Law can be used to compute the circuit current.

$$I = \frac{10 \text{ V}}{1570 \, \Omega} \text{ or } 0.00637 \text{ A or } 6.37 \text{ mA}.$$

Notice that the inductive reactance for the circuit must be found before current can be determined. Ohm's Law defines the relationship between potential difference (volts), electron movement (current), and opposition to current flow. In a DC resistive circuit, opposition to current is resistance. In a purely inductive circuit, opposition to current is caused by inductive reactance. Therefore, when using Ohm's Law in an inductive circuit:

BE CERTAIN TO WORK WITH *INDUCTIVE REACTANCE*, NOT INDUCTANCE.

RL Circuits

The most commonly used inductive circuit is a resistor and inductor connected in different configurations. This combination is called an **RL circuit**. Even a circuit containing only an inductor can be considered an RL circuit because of the resistance of the inductor.

RL circuits are extremely versatile, and are used in a variety of applications. However, before we discuss the applications of RL circuits, you must understand the operating principles behind both series and parallel RL circuits.

As we discuss RL circuits, you will find a number of similarities to RC circuits. Capacitors and inductors are **reactive components**. A reactive component is one that reacts differently to different inputs, and dissipates no power. Both components store a voltage. A capacitor stores voltage in an electrostatic field, and an inductor stores a voltage in an electromagnetic field. Although the similarities between the two components are extensive, their operation is completely different.

Series RL Circuits

Figure 15-3 shows a resistor and inductor connected in series with an AC voltage source. In addition to the physical resistance within the circuit, the AC voltage also causes CEMF to be constantly induced, which creates inductive reactance. Current constantly flows through the circuit, which allows both the resistor and the inductor to develop a voltage. Since current is the same throughout a series circuit, the voltage drops of the components are directly proportional to their resistance and reactance values. Both voltages can be calculated using Ohm's Law.

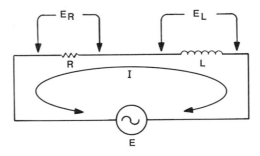

Figure 15-3 Series RL circuit.

For this discussion, we will assume the inductor to be a perfect inductance. That is, the inductor has no resistance. In this example all of the resistance in the circuit is represented by the resistor R.

The phase relationships between the voltage drops, voltage applied, and circuit current are shown in Figure 15-4. It should be pointed out that the waveforms in Figure 15-4 represent these values as they occur in *time*; not their actual peak values. The current waveform is again used as a reference, as it was with capacitors.

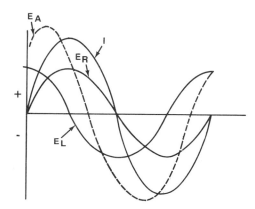

Figure 15-4 Phase relationships in an RL circuit.

The voltage drop of the resistor changes as current flow changes. Note that the waveform for E_R is in-phase with current. In any electrical circuit, the voltage drop of any resistor is in-phase with the current flow in the circuit.

Next, refer to the waveform for the inductor's voltage drop, labeled E_L. When E_L is maximum, current flow is minimum, and maximum current occurs when E_L is minimum. This defines the 90 degree phase shift between circuit current and the inductor's voltage drop. Notice also that E_L occurs before, or leads, the current waveform.

If we add the instantaneous values for E_R and E_L, we can determine the *time* of the applied voltage. This waveform is indicated by the dotted line, and is labeled E_A. Notice that E_A is out-of-phase with every other waveform shown. In a purely resistive circuit, applied voltage and circuit current are in-phase. In a purely inductive circuit, the applied voltage leads circuit current by 90 degrees. In an RL circuit, therefore, the phase shift (or **phase angle**) between the applied voltage and circuit current *must* be somewhere between 0 and 90 degrees, with voltage leading current. The phase angle between the applied voltage and circuit current is stated to be the phase angle for the circuit.

AC ELECTRONICS

Since E_R is in-phase with circuit current, the phase shift between E_R and E_A will be the same. The phase shift between E_A and E_L, however, will be the difference between 90 degrees and the phase angle for the circuit. This is because E_R and E_L are 90 degrees apart. The Greek letter **theta** (θ) is still used to identify the phase angle for the circuit, and the Greek letter **phi** (ϕ) is still used to represent the phase shift between E_A and E_L. If theta is known, then phi can be found by using the following equation:

$$\phi = 90 - \theta$$

Theta and phi will be complementary angles, since their sum is equal to 90 degrees.

Vector Diagrams

Although the waveforms for E_R and E_L were added to show E_A, this is only possible to indicate the time at which the applied voltage occurs. Because of the phase shift that exists in an RL circuit, you cannot simply add the two voltage drops to arrive at the applied voltage. Instead, they must be combined using vector addition.

Figure 15-5 shows a vector diagram of a series RL circuit. The reference vector is labeled I and represents the currents in the circuit. The voltage across the resistor is in-phase with the current flowing through it. Therefore, the resistor voltage vector, E_R, coincides with the current vector as shown. The voltage across the inductor leads the current by 90 degrees. Therefore, the inductive voltage vector, E_L, is shown 90 degrees out-of-phase (leading) with the current vector.

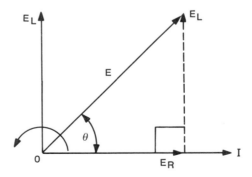

Figure 15-5 Vector diagram of a series RL circuit.

AC INDUCTANCE

Recall that we assume that each of the vectors in the diagram is rotating in the counterclockwise direction around the origin. The rotation of the vectors represents the variations of voltages and currents, in a sinusoidal manner, for a given period of time. One complete rotation represents one AC cycle. With the counterclockwise direction of rotation, you can see that the inductor voltage is ahead of (leading) the current vector. Another way of looking at this is that the current vector is lagging the inductive voltage vector by 90 degrees.

The applied voltage is the vector sum of the resistor and inductor voltages. To find this vector sum, we form a right triangle using the resistor voltages as one side and the inductor voltages as the other side. The applied voltage, E, is the hypotenuse. Note that the applied voltage leads the current by an angle that is between 0 and 90 degrees. Knowing the resistor and inductor voltages, the applied voltage can be found by solving for the hypotenuse of the right triangle. You can do this by using Pythagorean's Theorem:

$$E_A^2 = E_L^2 + E_R^2$$

We used Pythagorean's Theorem to solve these same type vector additions when we worked with capacitive circuits. The only difference is that current leads voltage in a capacitive circuit, while voltage leads current in an inductive circuit. By taking the square root of both sides in Pythagorean's Theorem, we can derive a formula for the applied voltage:

$$E_A = \sqrt{E_L^2 + E_R^2}$$

For example, a series RL circuit has a resistor voltage of 15 volts and an inductor voltage of 18 volts. What is the applied voltage?

$$E_A = \sqrt{E_L^2 + E_R^2} = \sqrt{(18)^2 + (15)^2} = \sqrt{324 + 225} = \sqrt{549} = 23.43 \text{ V}$$

Impedance

In Chapter 14, we defined impedance as the total opposition to current that is present in a circuit. Impedance takes into account any physical resistance within the circuit, as well as any reactance values. The total impedance for an RL circuit can be found two ways. First, since impedance is measured in ohms, it can be substituted for resistance in Ohm's Law. To calculate for total impedance, use the following Ohm's Law formula:

$$Z = \frac{E}{I}$$

AC ELECTRONICS

In a resistive circuit, total resistance could be found by adding the individual resistor values. The phase shift in an RL circuit does not allow us to make the same calculation in an RL circuit. Instead, the other method for finding total impedance in an RL circuit is by vector addition.

Impedance is vectored in the same manner as voltage. The value for resistance is placed along the 0 degree, or current line. The reactance value, because of the 90 degree phase shift between resistance and inductance, is placed 90 degrees in front of the current line. Total impedance is then the hypotenuse of the triangle created from the diagram. Again, Pythagorean's Theorem is used to express the formula for finding total impedance:

$$Z = R^2 + X_L^2$$

For example, what is the total impedance of a circuit with a 1.5 henry inductor and a 200 ohm resistor at 60 Hz? Before we can calculate total impedance, we must convert inductance to inductive reactance:

$$X_L = 6.28FL = 6.28(1.5)(60) = 565.2 \ \Omega$$

From there, we can insert the resistance and inductive reactance into our impedance formulas:

$$Z = \sqrt{R^2 + X_L^2} = \sqrt{200^2 + (565.2)^2} = \sqrt{40000 + 319451} = \sqrt{359451} = 599.5 \ \Omega$$

Power

Power in an RL circuit is much the same as power in an RC circuit. First, the resistor in the circuit dissipates true, or average power. This value can be obtained by inserting the resistor's voltage drop, resistance value, or current flow into any of the three standard power formulas:

$$P = IE \qquad P = I^2R \qquad P = \frac{E^2}{R}$$

AC INDUCTANCE

The inductor, like the capacitor, is a reactive component that does not dissipate power in its pure form. The inductor alternately stores energy in the form of a magnetic field and then releases it. During one-half cycle of AC operation, storage of electrical energy occurs in the magnetic field built around the inductor. When the magnetic field collapses, CEMF opposes the decrease in current by inducing a voltage into itself, which forces current flow in the circuit. The consumption and release of energy cancel one another. Therefore, the total average power dissipated by the inductive reactance is zero. Figure 15-6 shows the power consumption in a purely inductive circuit.

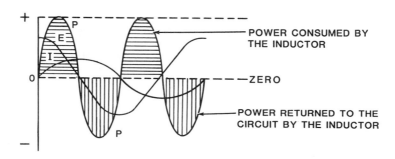

Figure 15-6 Power dissipation in a purely inductive circuit.

Unlike a capacitor, however, an inductor has resistance. For that reason, an inductor actually dissipates some power. It is the resistive portion of the inductor that causes the power to be dissipated. This is called true power. Figure 15-7 shows the current, voltage, and power curves of an RL circuit. The power curve, above the zero line, represents the power dissipated in the resistance (true power) and the power consumed by the inductance. The power curve below the zero line is the power returned to the circuit by the inductance.

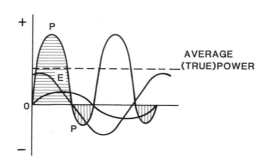

Figure 15-7 Power dissipated in an RL circuit.

AC ELECTRONICS

Figure 15-8 shows a power vector diagram for a series RL circuit. True power (the power dissipated by the resistive portion of the circuit), is given in watts. True power is drawn at 0 degrees. The power dissipated by a resistor is found by using the voltage drop of the resistor and the current flow through the resistor. Both values occur in-phase with circuit current.

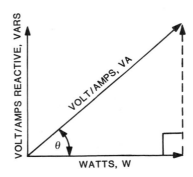

Figure 15-8 Power vector diagram.

The volt/amps reactive, or VARS, is the power in the inductive portion of the circuit that is both produced and used by the inductor. This power is drawn 90 degrees ahead of the resistive power, since the voltage drop of the inductor occurs 90 degrees ahead of the resistor's voltage drop.

The vector sum of true power in watts and the volt/amps reactive is the apparent power measured in volts/amps. The apparent power is indicated as the hypotenuse of the power triangle. While volt-amps can be found by combining true power and reactive power in a vector sum, it can also be found in an RL circuit by inserting applied voltage, circuit current, or total impedance into any of the three standard power formulas.

Phase Shift

The vector diagrams for voltage, impedance, and power for a series RL circuit all have one thing in common. The angle between the 0 degree line and the hypotenuse of the triangle are the same for each vector. In the voltage vector, the hypotenuse represents the applied voltage. Therefore the angle between the 0 degree line, which represents current, and the applied voltage is the phase angle for the circuit.

AC INDUCTANCE

In order to compute the phase angle for a series RL circuit, you must again use trigonometry. The trigonometric function you use depends on which values are known in the vector diagram. If you are using the voltage vector, for example, and you have computed the applied voltage, then you can use the sine, cosine, or tangent functions. For the angle θ, sine, cosine, and tangent use the following sides:

$$\text{sine } \theta = \frac{O}{H} \qquad \text{cosine } \theta = \frac{A}{H} \qquad \text{tangent } \theta = \frac{O}{A}$$

Where **O** is the side **opposite** the angle, **A** is the side **adjacent** to the angle, and **H** is the **hypotenuse**.

Figure 15-9 shows the voltage vector for a series RL circuit. The voltage drop of the resistor (E_R) is 30 volts, and the voltage drop of the inductor (E_L) is 40 volts. Let's assume at this point that the applied voltage has not been calculated. The dotted line in Figure 15-9 is the values of E_L. For the angle θ, which is the phase angle for the circuit, we know the length of the opposite and adjacent sides. We can therefore use the tangent function:

$$\tan \theta = \frac{O}{A} = \frac{40}{30} = 1.333$$

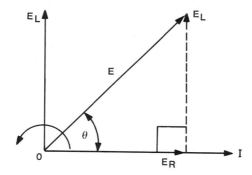

Figure 15-9 Voltage vector diagram.

AC ELECTRONICS

Appendix C shows us that the angle with a tangent of 1.333 is approximately 53 degrees. Therefore, the phase angle between the applied voltage and circuit current for this circuit is 53 degrees.

The applied voltage can be found for this circuit using Pythagorean's Theorem:

$$E_A = \sqrt{R_R^2 + E_L^2} = \sqrt{(30)^2 + (40)^2} = \sqrt{900 + 1600} = \sqrt{2500} = 50 \text{ V}$$

Once we find the applied voltage, we now know the length of the hypotenuse. This is also true for the impedance and power vectors. Once the vector sum is calculated, the length of the hypotenuse is known. We can now use either the sine or the cosine function to compute the phase angle:

$$\sin \theta = \frac{O}{H} = \frac{40}{50} = 0.800$$

The phase angle we computed with the tangent function, 53 degrees, also has a sine of 0.800.

$$\cos \theta = \frac{A}{H} = \frac{30}{50} = 0.600$$

Again, 53 degrees has a cosine of 0.600.

The examples we illustrated indicate that any trigonometric function can be used to calculate the phase angle of an RL circuit. The trigonometric function you use depends on which sides are known. The same procedures can be used for any vector diagram.

Parallel RL Circuits

Another type of inductive circuit is the parallel RL circuit. A basic RL parallel circuit is shown in Figure 15-10A. Notice that this circuit is no more than an inductor and a resistor connected in parallel. Although the coil itself has some internal resistance, this resistance will not be considered in the following discussion.

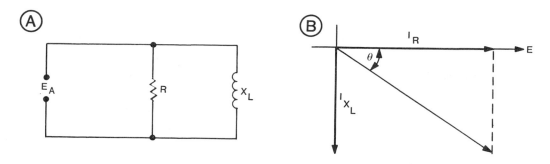

Figure 15-10 Parallel RL circuit with current vector diagram.

As with all parallel circuits, the applied voltage is dropped across each of the branches of a parallel circuit. In addition, the resistance of the entire circuit is always less than the resistance of the smallest branch in the circuit. These are important points in analyzing a parallel RL circuit.

First, since the voltage across each of the branches is equal to the source voltage, voltage is used as the reference when constructing a vector diagram for parallel RL circuits. Although the voltage across each branch is the same, the current through each branch varies according to the amount of opposition to current flow in that branch. Also, current through the inductive branch lags the current through the resistive branch. The inductive current will lag by an angle of less than 90 degrees. Figure 15-10B shows a typical vector diagram for a parallel RL circuit.

The impedance of a parallel circuit is always less than either the resistance or the inductive reactance of the circuit. Therefore, it is not possible to construct an impedance triangle, unless you use the reciprocal values for resistance, inductive reactance, and impedance.

As an example, assume that you have an RL circuit where the resistance is 50 ohms and the inductive reactance is 50 ohms. Pythagorean's Theorem can be used to compute total impedance, provided reciprocal values are used:

$$\frac{1}{Z} = \sqrt{\left(\frac{1}{R}\right)^2 + \left(\frac{1}{X_L}\right)^2} = \sqrt{\left(\frac{1}{50}\right)^2 + \left(\frac{1}{50}\right)^2} = \sqrt{(0.02)^2 + (0.02)^2}$$

$$= \sqrt{0.0004 + 0.0004} = \sqrt{0.0008} = 0.028$$

If $1/Z$ is equal to 0.028, then $Z = 35.35$ ohms.

Total impedance can also be found using vector addition. Again, reciprocal values must be used. The impedance vector diagram is shown in Figure 15-11.

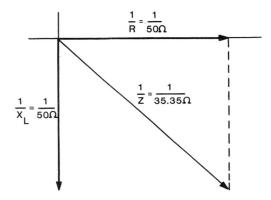

Figure 15-11 Impedance vector of a parallel RL circuit.

You may recognize these methods from Chapter 14. In fact, everything discussed thus far for the RL parallel circuit is a repetition of the formulas presented for the RC parallel circuits. This is because the mathematical theory is identical. All of the Ohm's Law and vector diagram relationships that apply to RC circuits also apply to the RL circuits. The difference is that current leads voltage in an RC circuit, while voltage leads current in an RL circuit.

You may find it easier to solve for individual branch currents and plot the vector relationships in the form of current vectors. Remember that current is used as the reference in series circuits, and voltage is used as the reference in parallel circuits. The constant values are always used as the reference.

RL Circuit Applications

Inductive circuits, like capacitive circuits, are widely used in electronics. The reactive effect of an inductor in an AC circuit makes the inductor valuable in filtering and phase shift applications. However, the inductor is less widely used than the capacitor in such applications. The reasons for this is that inductors are larger, heavier and more expensive. They also come in a narrower range of standard values and, because of their internal resistance, dissipate power. This makes the inductor far less valuable as a reactive component than the capacitor in AC circuits.

The greatest advantage of the inductor in AC applications is that it can produce a reactive effect while completing the DC circuit path. The capacitor produces a reactive effect, but *blocks* DC current. Often a complete DC current path is required in addition to the reactive effect.

Inductive Filters

Series RL networks are used as simple low and high-pass filters in the same way that series RC circuits are used. Figure 15-12 shows the two basic types of RL filter circuits. These circuits are simple RL series circuits.

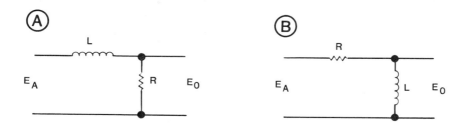

Figure 15-12 Single section RL filters.
    A. Low-pass with lagging phase shift.
    B. High-pass with leading phase shift.

The circuit in Figure 15-12A is a low-pass filter. The input signal is connected in series with the resistor and the inductor, and the output voltage is taken across the resistor. At low frequencies, the reactance of the inductor is low, and very little voltage is dropped across it. Most of the voltage is dropped across the output resistor. As the input frequency increases, inductive reactance increases. As inductive reactance increases, more of the input voltage is dropped across the inductor. Therefore less voltage appears across the output resistor. Low frequencies are passed with little or no attenuation, while high frequencies are greatly reduced in amplitude by the inductor's reactance.

AC ELECTRONICS

The circuit in Figure 15-12B is a high-pass filter. Again the input signal is connected in series with a resistor and an inductor, but in this circuit, the output voltage is taken across the inductor. At very high frequencies, the inductive reactance is very high compared to the resistance. Most of the input voltage is developed across the inductor and the output terminals. As the input frequency decreases, the inductive reactance decreases. Less voltage is dropped across the inductor and more across the resistor. The lower the frequency the lower the inductive reactance and the lower the output voltage. High frequencies are passed with little or no attenuation. Low frequencies are greatly attenuated due to the reduced inductive reactance.

These simple low-pass and high-pass RL filter circuits produce the same result as equivalent RC circuits. However, they are less desirable because the inductors are larger and more expensive than capacitors. RC circuits are preferred for filter networks when DC current paths are not required.

The cut-off frequency (F_{CO}) for an RL filter network is:

$$F_{CO} = \frac{R}{2\pi L} = \frac{R}{6.28L} = \frac{.159R}{L}$$

The cut-off frequency is that frequency above or below which frequencies are passed or attenuated. In the expression above, R is in ohms, L is in henrys and F is in Hz. At the cut-off frequency, X_L equals R and the phase shift is 45 degrees. At the cut-off frequency, the output voltage (E_O) is approximately 70 percent (0.707) of the input voltage (E_{IN}).

$$E_O = 0.707\ E_{IN}$$

For example, when the input voltage is 8 volts, the output voltage at the cut-off frequency will be 8 volts x 0.707 = 5.656 volts.

Inductive Phase Shifters

Because an inductor causes the current in a circuit to lag the applied voltage, inductive circuits can be used for phase shifting. The simple series RL circuits shown in Figure 15-12 can also serve as phase shifters. With these circuits, a phase shift between 0 and 90 degrees can be obtained. In the circuit shown in Figure 15-12A, the output voltage lags the input voltage. The phase angle is determined by the inductive reactance and the resistance values.

The circuit in Figure 15-12B produces a leading phase shift. The output leads the input by a phase angle between 0 and 90 degrees. As the amount of phase shift produced by the circuit approaches 90 degrees, the output voltage becomes greatly attenuated. Theoretically, with a 90 degree phase shift the output is zero. For that reason simple RL phase shift circuits are used to provide phase shifts of approximately 60 degrees or less. The greater the phase shift, the greater the need for amplification, in the circuit, to restore the amplitude of the output signal.

The amount of phase shift produced by these circuits can be calculated using trigonometric functions. A typical vector diagram is shown in Figure 15-13. As you know, the inductive values (X_L and E_L) are drawn 90 degrees before the resistor values (R and E_R). The resistor values are drawn on the zero line because they are in-phase with circuit current. Since the voltage values for both components are in-phase with their resistance and reactance values, we can compute the phase angle using either voltage values or resistance and reactance values.

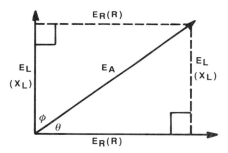

Figure 15-13 Vector diagram for computing phase shift.

The output voltage in the circuit shown back in Figure 15-12A is taken across the resistor and will lag the input voltage. The phase angle for this circuit is equal to the angle between the applied voltage and E_R. Notice that this is also the angle between the resistance side of the triangle and E_A, or the hypotenuse. This angle is labeled θ. We can compute the phase angle without knowing total impedance, or the individual voltage drops. The tangent function is the ratio between the opposite side (X_L) and the adjacent side (R). This is equal to:

$$\tan\theta = \frac{O}{A} = \frac{E_L}{E_R} = \frac{X_L}{R}$$

Referring back to Figure 15-12B, the output voltage is taken across the inductor, and will lead the input voltage. The phase angle for this circuit is equal to the phase shift between E_L and the applied voltage. This phase angle is indicated by the Greek letter phi (ϕ).

The phase angle is again found using the tangent function:

$$\tan \phi = \frac{O}{A} = \frac{E_R}{E_L} = \frac{R}{X_L}$$

The tangent function is used for both calculations because we are assuming that the only known values are applied voltage, resistance, and inductive reactance. The resistance and reactance values provide two sides of the triangle, which allows us to use tangent to compute the phase angle.

To obtain phase shifts greater than 90 degrees, RL phase shifting networks can be cascaded. Although large degrees of phase shift are obtained, the attenuation of the cascaded circuit is considerable. Some type of amplification is generally required to offset the loss in the circuit.

As with RL filters, RL phase shifters are less desirable than RC phase shifters. Inductors are larger, more expensive and have greater losses than capacitors. RC networks are preferred when DC current paths are not required.

AC INDUCTANCE

Chapter Self-Test

1. When is CEMF induced in an AC inductive circuit?

2. What is the phase shift between the applied voltage and circuit current in a purely inductive circuit?

3. What is the phase difference between the applied voltage and CEMF in an inductive circuit?

4. Which occurs first in an inductive circuit—current or voltage? Why?

5. Define inductive reactance.

6. List the factors that affect inductive reactance, and state whether they are directly or inversely proportional.

7. What is the formula for inductive reactance?

8. Can the effective values for the voltage drop of a resistor and the voltage drop of an inductor in a series RL circuit be added together to find the applied voltage? Why?

9. Define reactive power as it applies to a purely inductive circuit.

10. In a series RL circuit, resistance and total impedance are known values. Which trigonometric function may be used to determine the phase shift for the circuit?

11. Why are vector diagrams for parallel RL circuits different than vector diagrams for a series RL circuit?

12. An AC voltage is connected in series with a resistor and an inductor. An output voltage is taken across the inductor. Would this circuit operate as a high-pass or low-pass filter?

13. Will the circuit described in Question 12 operate as a leading-output or lagging-output phase shift network?

Summary

When an inductor is placed in an AC circuit, the sinusoidal voltage causes a sinusoidal current flow. Current is constantly changing, which induces a constant CEMF. The inductor also develops a sinusoidal voltage drop.

CEMF constantly opposes the applied voltage. CEMF and the applied voltage are 180 degrees out-of-phase.

Constant CEMF causes a delay in current in an inductive circuit. In a purely inductive circuit, the applied voltage leads circuit current by 90 degrees.

CEMF being constantly induced presents a steady opposition to current flow. This opposition is referred to as inductive reactance. The symbol for inductive reactance is X_L. Inductive reactance is measured in ohms.

Frequency is directly proportional to inductive reactance. If frequency changes, the rate at which the magnetic lines of force cut the coils within an inductor, thereby directly affecting the amount of CEMF being induced.

Inductance is also directly proportional to inductive reactance. Inductance is the ability to induce CEMF—if inductance changes, inductive reactances changes in the same manner.

The formula for inductive reactance is:

$$X_L = 2\pi FL$$

Inductive reactance may be substituted for resistance in Ohm's Law.

When a resistor is placed in the same circuit as an inductor, the result is an RL circuit. In an RL circuit, the applied voltage leads circuit current by some phase angle between 0 and 90 degrees. The voltage drop of the resistor and the voltage drop of the inductor are 90 degrees out-of-phase.

The voltage drops of the components in a series RL circuit cannot be added together to find the applied voltage because of the phase difference between the voltage drops. The applied voltage may be calculated using vector addition and the Pythagorean Theorem. Total impedance for a series RL circuit must also be found using vector addition and the Pythagorean Theorem.

AC INDUCTANCE

An inductor, because of its physical resistance, dissipates some power. An inductor can be compared to other inductors with regards to how efficient the inductor performs. This efficiency is a ratio of power stored (which is proportional to inductive reactance) and power dissipated (which is proportional to the resistance within the inductor). The resulting ratio is called the inductor's quality of merit, or Q value. The formula for finding the Q value for an inductor is:

$$Q = \frac{X_L}{R}$$

In the vector diagram for a series RL circuit, current is used as a reference because current is the same throughout a series circuit.

The apparent power in an RL circuit can be found by using applied voltage and circuit current in the standard power formula. Apparent power is also the vector sum of true power and reactive power in an RL circuit.

The phase angle between the applied voltage and circuit current is referred to as the phase angle for the circuit. This phase angle may be found using two known similar values, and trigonometry. The trigonometric function that is used depends on the values known, and their relation to the phase angle—opposite, adjacent, or hypotenuse.

In vector diagrams for parallel RL circuits, voltage is used as the reference because each component feels the same voltage. Circuit current for a parallel circuit can be found by using vector addition and Pythagorean's Theorem.

Total impedance for a parallel circuit can be found by using the reciprocals of the resistive and reactive values, and combining these reciprocals using vector addition and Pythagorean's Theorem.

Inductive filters operate much like capacitive filters. In a series RL circuit, a high-pass filter is obtained by taking the output across the inductor. An RL high-pass filter attenuates low frequencies, while allowing frequencies above the cut-off to pass.

An RL low-pass filter allows frequencies below the cut-off to pass, while attenuating frequencies above the cut-off frequency. The cut-off frequency is defined as that frequency which produces 70.7% of the input voltage at the output terminals.

AC ELECTRONICS

Series RL circuits are also used as phase shift networks. If the output is taken across the inductor, a leading-output phase shift network is obtained. If the output is taken across the resistor, the result is a lagging-output phase shift network.

Single stage RL phase shift networks are used if the amount of phase shift is less than 60 degrees. If the desired phase shift is above 60 degrees, phase shift networks are cascaded. However, cascading networks cause severe attenuation, so the output of a cascading phase shift network is usually applied to an amplifier to increase the amplitude of the output voltage.

CHAPTER 16

RLC Circuits

AC ELECTRONICS

Contents

Introduction	363
Chapter Objectives	364
Simple RLC Circuits	365
Resonance	371
Series Resonance	373
Parallel Resonance	383
LC Filters	392
Chapter Self-Test	396
Summary	397

RLC CIRCUITS

Introduction

Capacitors and inductors operate in AC circuits very similarly. They both react to the input signal, and change their characteristics according to changes in the input signal. However, although they essentially operate the same, they are quite opposite in their reactions. Voltage leads current in an inductive circuit, while current leads voltage in a capacitive circuit. Frequency is directly proportional to inductive reactance, while frequency is inversely proportional to capacitive reactance.

This similar yet opposite relationship between inductors and capacitors make circuits containing both components operate in a very unique manner. An RLC circuit is made up of resistors, inductors, and capacitors. This chapter introduces you to the basic RLC circuit, and shows the relationships between voltage current, and impedance. You will then learn about resonance; the single most important characteristic of an RLC circuit. You will see how resonance makes series RLC circuits operate differently than parallel RLC circuits. You will also learn how these circuits operate as filters.

AC ELECTRONICS

Chapter Objectives

When you have completed this chapter, you will be able to:

1. Define the following terms: resonance, resonant frequency, half-power point, flywheel effect, tank circuit, dampening, swamping resistor, band-pass filter, and band-stop filter.

2. Calculate the impedance, current, voltage, power factor, and phase angle for series and parallel RLC circuits.

3. State the characteristics of a series resonant RLC circuit.

4. State the characteristics of a parallel resonant RLC circuit.

5. Determine the Q of a series or parallel circuit at its resonant frequency.

6. Identify the basic types of resonant filters for both series and parallel resonant circuits.

7. Name the type of filter that would produce a given response curve.

8. Calculate the resonant frequency of an RLC circuit.

9. Determine the bandwith of a series or parallel resonant circuit.

Simple RLC Circuits

An RLC circuit has a number of unique characteristics and applications. However, before we begin to describe these characteristics, we must first analyze a simple RLC circuit to determine the relationships between voltage, current, impedance, and power. The unique operation of an RLC circuit will be very difficult to explain and understand if these basic relationships are not described.

Series RLC Circuits

Figure 16-1 shows a resistor, inductor, and capacitor connected in series with an AC generator. Because the values of capacitance, inductance, and frequency are given, the reactances in the circuit can be computed using the formulas for capacitive and inductive reactance.

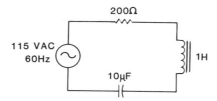

Figure 16-1 Series RLC circuit.

The inductive reactance is:

$$X_L = 2\pi FL = 6.28(60 \text{ Hz})(1 \text{ H}) = 377 \text{ }\Omega$$

The capacitive reactance is:

$$X_C = \frac{1}{2\pi FC} = \frac{159(10^3)}{(60 \text{ Hz})(10)} = \frac{159(10^3)}{600} = 265 \text{ }\Omega$$

The reactance values combine with the resistance value to form an impedance to current flow. Recall that there is a general formula for impedance:

$$Z = \sqrt{R^2 + X^2}$$

AC ELECTRONICS

When the reactance is caused by capacitance, the formula becomes:

$$Z = \sqrt{R^2 + X_C^2}$$

When inductive reactance is involved, the formula becomes:

$$Z = \sqrt{R^2 + X_L^2}$$

The circuit shown in Figure 16-1 contains both capacitive and inductive reactance. Thus, a formula is needed which contains both X_C and X_L. A vector diagram of the reactances and resistance in the circuit will help us to develop such an equation.

Figure 16-2A shows the vector diagram for this circuit. Following the conventions established earlier, R is plotted at zero degrees, X_L is plotted at +90 degrees, and X_C is plotted at −90 degrees. Notice that this places X_L 180 degrees out-of-phase with X_C. For this reason, X_L and X_C tend to cancel.

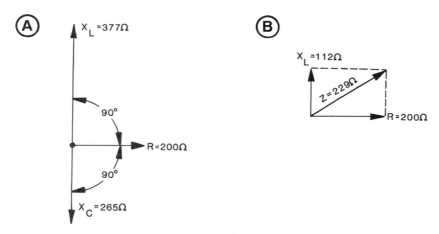

Figure 16-2 Vector diagrams for a series RLC circuit.

However, because X_L is greater than X_C, the resultant reactance is inductive. Because X_L is 377 ohms and X_C is 265 ohms, X_L completely cancels X_C and still has a value of 112 ohms. Thus, when a circuit has both capacitive and inductive reactance, the net reactance is the difference between the two. In Figure 16-2B, the resulting vector is shown. Notice how this vector is found with X_L having a value of 112 ohms, and the vector is purely inductive. The complete formula for the impedance in this series AC circuit is:

$$Z = \sqrt{R^2 + (X_L - X_C)^2} = \sqrt{(200)^2 + (377 - 265)^2} = \sqrt{(200)^2 + (112)^2}$$
$$= \sqrt{40{,}000 + 12{,}544} = \sqrt{52{,}544} = 229 \; \Omega$$

RLC CIRCUITS

The net result is the same as when an inductor having an X_L of 112 ohms is connected in series with a resistance of 200 ohms.

The previous formula for impedance is used when X_L is greater than X_C. Obviously, there will be cases where X_C is larger than X_L as shown in Figure 16-3. Here, X_L is 10 ohms, X_C is 40 ohms, and R is 40 ohms.

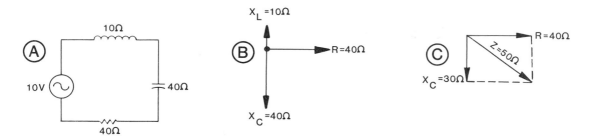

Figure 16-3 Circuit and vectors in which X_C is greater than X_L.

Figure 16-3B shows the vector diagram for this circuit. Notice that in this case X_C more than cancels X_L. Thus, the net reactance is found by subtracting X_L from X_C. The resulting vector is shown in Figure 16-3C. This vector is found with an X_C value of 30 ohms, and the vector is purely capacitive. This is shown in Figure 16-3C and results in a capacitive reactance of 30 ohms. When X_C is larger than X_L, the formula for impedance becomes:

$$Z = \sqrt{R^2 + (X_C - X_L)^2}$$

Using this formula, we find that the total impedance of this circuit is:

$$Z = \sqrt{R^2 + (X_C - X_L)^2} = \sqrt{(40)^2 + (40 - 10)^2} = \sqrt{(40)^2 + (30)^2} = \sqrt{1600 + 900}$$
$$= \sqrt{2500} = 50 \, \Omega$$

Once the impedance of the circuit is known we can determine other circuit values, such as current, the voltage dropped by each component, and the power factor. For example, the current in Figure 16-3A is:

$$I = \frac{E}{R} = \frac{10V}{50 \, \Omega} = 0.2 \, A$$

This allows us to find the voltage drop across each component:

$$E_R = I(R) = 0.2 \text{ A} \times 40 \text{ }\Omega = 8 \text{ V.}$$

$$E_L = I(X_L) = 0.2 \text{ A} \times 10 \text{ }\Omega = 2 \text{ V.}$$

$$E_C = I(X_C) = 0.2 \text{ A} \times 40 \text{ }\Omega = 8 \text{ V.}$$

The power factor can be determined using the formula:

$$PF = \frac{R}{Z} = \frac{40 \text{ }\Omega}{50 \text{ }\Omega} = 0.80$$

Since the power factor is also equal to the cosine of the angle, we can find the angle using the chart in Appendix E. We find that the angle is about 36.5 degrees. The vector shown in Figure 16-3C shows that this is a negative angle. This means that the circuit reactance is capacitive.

We can compute apparent power in volt-amps by multiplying the applied voltage times the current:

$$VA = IE = 10V \times 0.2A = 2 \text{ VA.}$$

The true power is less, since only the resistor dissipates power:

$$P = I^2R = (0.2A)^2 \times 40 \text{ }\Omega = 0.04A \times 40 \text{ }\Omega = 1.6W.$$

Parallel RLC Circuits

Figure 16-4 shows a parallel RLC circuit. Because the three components are in parallel, the same voltage is applied across each. It is easier to work with currents in parallel circuits. Recall that when a reactance is in parallel with a resistance, the total current is the vector sum of the two branch currents.

$$I_T = \sqrt{I_R^2 + I_X^2}$$

In this formula, I_X may be the current through either a capacitor or an inductor.

RLC CIRCUITS

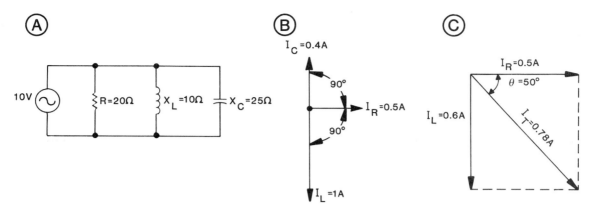

Figure 16-4 Parallel RLC circuit.

The first step in finding total current is finding the current for each branch. Since the applied voltage and all ohmic values are given, this is not difficult:

$$I_R = \frac{E}{R} = \frac{10V}{20\ \Omega} = 0.5A.$$

$$I_L = \frac{E}{X_L} = \frac{10V}{10\ \Omega} = 1A.$$

$$I_C = \frac{E}{X_C} = \frac{10V}{25\ \Omega} = 0.4A.$$

Figure 16-4B shows the branch currents plotted as vectors. I_R is in-phase with the applied voltage and is plotted at zero degrees. I_L lags the applied voltage by 90 degrees, and I_C leads the applied voltage by 90 degrees. Notice that I_C is 180 degrees out-of-phase with I_L. As far as the source current is concerned, these two currents tend to cancel.

I_L is larger than I_C. Thus, the total current through the two reactive components is I_L minus I_C, which equals 0.6 amperes. Because the reactive currents subtract, the formula for total current is written:

$$I_T = \sqrt{I_R^2 + (I_L - I_C)^2}$$

AC ELECTRONICS

This final formula can be used to find total current:

$$I_T = \sqrt{I_R^2 + (I_L - I_C)^2} = \sqrt{(0.5)^2 + (1 - 0.4)^2} = \sqrt{(0.5)^2 + (0.6)^2} = \sqrt{0.25 + 0.36}$$
$$= \sqrt{0.61} = 0.78 \text{ A}.$$

This is further illustrated by the vector shown in Figure 16-4C. The total current is shown as the vector sum of I_R and I_L. The value of I_L is the resultant current after the value of I_C is subtracted from the original value of I_L.

Once we know the total current in the circuit, we can compute other values. For example, the impedance of the circuit is:

$$Z = \frac{E}{I} = \frac{10V}{0.78A} = 12.8 \, \Omega$$

This is the impedance of the three components in parallel.

Also, the phase angle (θ) can be found using the formula:

$$\tan \theta = \frac{I_X}{I_R} = \frac{0.6}{0.5} = 1.2$$

In Appendix C you will find that a tangent of 1.2 corresponds to an angle of approximately 50 degrees. However, as shown in Figure 16-4C, the angle is negative because the circuit is inductive (ELI) and the plot is for current. Consequently, the total current lags the applied voltage by about 50 degrees.

RLC CIRCUITS

Resonance

In Figure 16-5A, an inductor and a capacitor are connected in parallel. When a voltage is applied between terminals A and B, the operation of this circuit depends on the frequency of the applied voltage. When the voltage is DC, the capacitor acts like an open, and the inductor acts like a short. Effectively, X_C is infinite and X_L is zero.

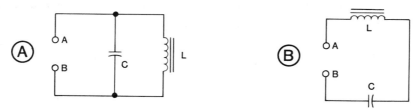

Figure 16-5 LC circuits.

Now let's assume that the applied voltage is not DC but is a very low frequency AC. In this case, X_L has a very low value and X_C is quite high. The exact values will depend on the values of C, L, and the applied frequency.

When the frequency is gradually increased, X_L gradually increases and X_C gradually decreases. As the frequency is increased, a point is reached where the value of X_L is the same as the value of X_C. That is, for every combination of L and C, there is some frequency at which X_L equals X_C. This is true whether the two components are connected in parallel as shown in Figure 16-5A, or connected in series as shown in Figure 16-5B. The condition where X_L is equal to X_C is called resonance. Also, the frequency at which X_L is equal to X_C is called the resonant frequency and is abbreviated F_O.

We can calculate the inductive and capacitive reactances for an RLC circuit using our reactance formulas. Since the resonant frequency is that frequency where X_L equals X_C, we can combine these two formulas and create a formula for F_O. The resonant frequency is the frequency where:

$$2 \pi FL = \frac{1}{2 \pi FC}$$

AC ELECTRONICS

By solving for F using algebra, the formula for F_O is found. We won't show you the step-by-step method by which this is accomplished since the calculations required are rather complex. The final formula for F_O is:

$$F_O = \frac{1}{2\pi\sqrt{LC}} \text{ or } \frac{.159}{\sqrt{LC}}$$

We can apply this formula to any LC circuit to find its resonant frequency. For example, let's assume that a 400 millihenry inductor is connected in series with a 158 microfarad capacitor. What is the resonant frequency?

$$F_O = \frac{.159}{\sqrt{LC}} = \frac{.159}{\sqrt{.4 \times .000158}} = \frac{.159}{\sqrt{.0000632}} = \frac{.159}{.00795} = 20 \text{ Hz}$$

This formula works equally well for both series and parallel LC circuits. That is, for given values of L and C, the resonant frequency is the same regardless of how L and C are connected.

RLC CIRCUITS

Series Resonance

Figure 16-6A shows a series circuit consisting of a capacitor, an inductor, and a resistor. The values of L and C are not given, nor is the frequency of the applied signal. Nevertheless, we know that the circuit is at the resonant condition because X_L is equal to X_C.

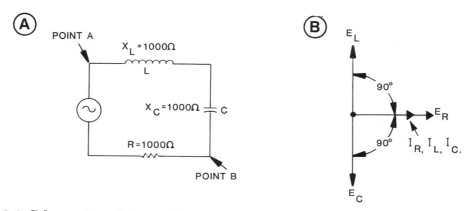

Figure 16-6 Schematic and vector diagram of a series resonant circuit.

The applied AC signal forces current to flow through the series circuit. Because the components are in series, the same current flows through all of the components.

As current flows through the circuit, a voltage is developed across each component. Because $R = X_L = X_C$, the voltages developed across each component are equal. That is, $E_R = E_L = E_C$. However, this is true only of the *magnitude* of the voltage. The phase of the voltage is different across each component. The voltage across R is in-phase with circuit current. The voltage across the inductor leads the current by 90 degrees. The voltage across the capacitor lags behind the current by 90 degrees.

Remember the memory phrase for current and voltage relationships in inductive and capacitive circuits:

> **ELI** the **ICE** man.

The best way to visualize these phase relationships is with vector diagrams. Figure 16-6B shows the vector diagram of the voltage drops and current. I_R, I_L, I_C, and E_R are all in-phase with the applied current. Consequently, they are shown with a phase angle of zero and are used as the reference for the relationship plots. E_L is drawn at +90 degrees and E_C is drawn at −90 degrees. Thus, E_L is 180 degrees out-of-phase with E_C.

373

AC ELECTRONICS

This illustrates one of the most important characteristics of the series resonant circuit. At resonance, the voltages across the capacitor and inductor are of equal magnitude but they are 180 degrees out-of-phase. Consequently, E_L exactly cancels E_C. Thus, as far as the source is concerned, the sum of these two voltages is zero. Therefore, a voltmeter connected between points A and B in Figure 16-6A measures zero volts.

When the combined voltage drops across X_L and X_C equal zero, then all of the applied voltage must be dropped across R. For example, when the source voltage is 10 volts, E_R is 10 volts.

However, since the same current must flow through C, a voltmeter placed across C also measures 10 volts. Moreover, a voltmeter placed across L measures 10 volts. At first this seems to violate Kirchhoff's Voltage Law (the sum of the voltage drops around a closed loop must equal zero). Because the 10 volts across C is *canceled* by the 10 volts across L, Kirchhoff's Law remains true for this circuit. As Figure 16-6B shows, the vector sum of E_L and E_C is zero volts.

Since the combination of L and C produces no voltage drop, their total reactance or impedance must be zero. That is, the source sees the LC combination as a perfect conductor having 0 ohms of impedance. Thus, the only opposition to current flow in the circuit is the resistance of R. At resonance the source sees no capacitance and no inductance, only resistance. This is proven by applying the impedance formula:

$$Z = \sqrt{R^2 + (X_L - X_C)^2}$$

When both inductive and capacitive reactance are the same value, $X_L - X_C = 0$. Therefore:

$$Z = \sqrt{R^2 + 0} = \sqrt{R^2} = R \text{ or } Z = R$$

At resonance, total impedance is equal to R. This means that current and voltage, as seen by the source, are in-phase. Therefore the power factor is 1.

It is important to emphasize that these conditions occur only at resonance. When the applied frequency is above or below resonance, X_L does not exactly equal X_C. Consequently, the voltage drops across L and C do not completely cancel. In this case, there is some resultant value of reactance.

The third unusual thing which happens in series resonant circuits is the most mysterious of all. It is not evident in Figure 16-6A because the value of R is equal to the value of X_L and X_C. However, look at the circuit shown in Figure 16-7. Here the value of R is much less than that of X_L or X_C. The circuit is still at resonance because X_L equals X_C.

RLC CIRCUITS

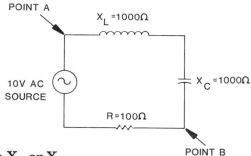

Figure 16-7 R is less than X_L or X_C.

We have seen that the current in a resonant circuit is determined solely by the applied voltage and the value of the series resistance. In Figure 16-7 circuit current is:

$$I = \frac{E}{R} = \frac{10V}{100\ \Omega} = 0.1\ A.$$

The same current flows through C and L. Therefore, the voltage drop across C must be 100 volts, as well as the voltage drop across L. Notice that the voltage across L or C is actually ten times higher than the voltage being applied to the circuit.

Once again, this may seem to violate Kirchhoff's Voltage Law. However, it does not since the 100 volts across L is canceled out by the 100 volts across C. A voltmeter connected from point A to B measures zero volts. But, if the meter is connected across L or C separately, it measures the 100 volts.

The ability to produce a voltage higher than the applied voltage is one of the most remarkable characteristics of the series resonant circuit. This is possible because of the ability of the inductor and the capacitor to store energy.

The inductor stores energy in its magnetic field. The capacitor stores energy in its dielectric. This storage occurs at resonance any time the value of R is lower than the value of X_L or X_C. The lower the value of resistance, when compared to reactance, the higher the voltage developed across the reactance. If all series resistance could be eliminated entirely, the current in the circuit would theoretically rise to an infinitely high value. The voltage across the inductor and the capacitor would also become infinitely high.

AC ELECTRONICS

In practice, of course, some series resistance always exists. The AC source and the connecting wires always have some value of series resistance. However, in LC circuits the largest source of series resistance is usually the inductor. Most inductors are wound from lengths of very small wire. Thus, they have a relatively large value of series resistance (when compared to source resistance or conductor wire). This resistance reduces the resonant current even when physical resistors are not used.

To summarize, the series resonant circuit has several important characteristics. These are listed below:

1. The impedance across the circuit is low and is equal to the series resistance.

2. The current flow is high and is limited only by the wire's resistance.

3. The applied voltage is dropped by the series resistance.

4. The voltage across the inductor or capacitor is equal to the current times the reactance. The voltage may be higher than the applied voltage.

5. The circuit acts resistive. The source current and voltage are in-phase and the power factor is 1.

Q and Bandwidth in Series Resonant Circuits

The series resonant circuit has two characteristics which we have not yet discussed. These are **Q** and **bandwidth**. Q was introduced in Chapter 15 as a method for stating the efficiency of inductors. However, the term Q as used with resonant circuits has additional aspects which must be understood. Bandwidth was also mentioned earlier in connection with filters. Here, we will discuss both Q and bandwidth in more detail.

Q In Series Resonant Circuits

One of the most important characteristics of a resonant circuit is its Q. Other names for Q are quality figure, quality of merit, and magnification factor. The Q factor is defined as *the ratio of the reactance at resonance,* to the series AC resistance. That is:

$$Q = \frac{X_L}{R}$$

RLC CIRCUITS

Since at resonance $X_L = X_C$;

$$Q = \frac{X_C}{R}$$

Normally, the reactance of a circuit in terms of Q is expressed as an inductive property which is X_L. The Q value for inductors compared the efficiency of one inductor to the efficiency of another. Although Q holds different meanings for resonant circuits, it is still not a value of any kind.

In a series resonant circuit, Q is the magnification factor that determines how much the voltage across L or C is increased above the applied voltage. For example, a 1 V peak-to-peak AC signal is applied to a series resonant circuit with a Q of 10. The voltage across L or C, at resonance, is 10 V peak-to-peak. Thus the applied voltage, E_{IN}, is magnified by the Q factor. Expressed as an equation:

$$E_L = Q \times E_{IN}$$

and:

$$E_C = Q \times E_{IN}$$

When the applied voltage and E_L or E_C are known, the Q factor is calculated using the formula:

$$Q = \frac{E_L}{E_{IN}} \text{ or } Q = \frac{E_C}{E_{IN}}$$

Q can be determined by measuring E_{IN} and E_L or E_C and using the previous formula. For example, E_{IN} is measured with an AC voltmeter and found to be 0.1 volt. E_L is measured and found to be 15 volts. The Q of the circuit must be:

$$Q = \frac{E_L}{E_{IN}} = \frac{15V}{0.1V} = 150.$$

This method of determining Q generally gives more accurate results than other formulas. The reason for this is that the AC resistance of the circuit is difficult to determine. The oscilloscope is also used to measure voltage when peak-to-peak values are required.

AC ELECTRONICS

Generally, the largest single factor that makes up the series resistance is the AC resistance of the inductor. A circuit's AC resistance can be much higher than its DC resistance. The DC resistance can be measured with an ohmmeter. The AC resistance of a component is difficult to measure. AC resistance (impedance) is usually determined by measuring the AC voltage across and the current through a component. Then the AC resistance is calculated using Ohm's Law.

An inductor has a Q factor of its own. When the only series resistance in a series resonant circuit is the inductor, the Q of the circuit is the same as the inductor. This is the highest Q that a resonant circuit can have. When additional series resistance is added, the Q of the circuit is lower than the Q of the inductor.

Bandwidth and Q

Resonant circuits are selective circuits. They respond more favorably to their resonant frequency, F_O, than to other frequencies. While the effects are greatest at F_O, the same effects exist to a smaller extent at frequencies slightly above and below F_O. Thus, a resonant circuit actually responds to a band of frequencies. The width of this band of frequencies is called the **bandwidth** of the resonant circuit. The bandwidth are the frequencies that have amplitudes from 0.707 of the peak value to the peak value. Remember, the sine wave increases from zero to a peak and back to zero. The bandwidth is the frequencies between the 0.707 increasing and 0.707 decreasing points of the output waveform.

Measuring Bandwidth

Figure 16-8 illustrates how bandwidth is measured. This graph shows the current passed by a series resonant circuit at various frequencies below, at, and above the resonant frequency. Naturally, maximum current flows at the resonant frequency. In this example, F_O is 1000 Hz and the maximum current is 10 milliamps.

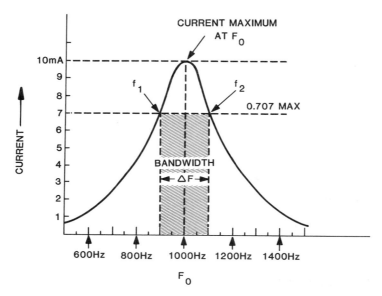

Figure 16-8 Bandwidth is measured between half-power points.

The bandwidth of the series resonant circuit is generally considered to include that group of frequencies with a response 70.7% of maximum. In our example, frequencies which produce a current of 7.07 mA or higher are considered within the bandwidth. This band of frequencies extends from 900 Hz to 1100 Hz.

The **bandwidth** is defined as the width of this band of frequencies. Consequently the bandwidth (BW) is 1100 Hz − 900 Hz = 200 Hz. The bandwidth is that band of frequencies which produce a response of 70.7% of maximum current.

Half-Power Points

You may wonder why the 70.7% points were chosen to indicate the bandwidth. Actually this is a very convenient point to use because it represents the point at which the power in the circuit is exactly one-half the maximum value. Thus, the points marked f_1 and f_2 in Figure 16-8 are referred to as **half-power points**.

AC ELECTRONICS

An example demonstrates that the power in a circuit drops to one-half when the current drops to 70.7%. Consider a circuit in which the resistance is 2 kilohms and the current is 10 mA. The power is:

$$P = I^2R = (0.01\ A)^2 \times 2k\ \Omega = 0.0001\ A \times 2k\ \Omega = 0.2\ W.$$

Now let's assume that the current drops to 70.7% of maximum or to 7.07 milliamps The power drops to:

$$P = I^2R = (0.00707\ A)^2 \times 2k\ \Omega = 0.00005\ A \times 2k\ \Omega = 0.1\ W.$$

This is one-half the previous power. Thus, reducing the current to 70.7% reduces the power to 50%.

Bandwidth Equals F_0/Q

With the bandwidth measured between the half-power points, an interesting relationship exists between the bandwidth, the resonant frequency, and the value of Q. This relationship is expressed by the equation:

$$BW = \frac{F_O}{Q}$$

This states that the bandwidth is equal to the resonant frequency divided by Q.

For the example shown in Figure 16-8, the Q is 5 because:

$$BW = \frac{F_O}{Q} = \frac{1.000\ Hz}{5} = 200\ HZ$$

The equation states that the bandwidth is directly proportional to the resonant frequency, but inversely proportional to Q.

The f_2 point is defined as the upper frequency limit of the bandwidth. The f_1 point is defined as the lower frequency limit of the bandwidth. Therefore, the bandwidth is all of the frequencies from the f_1 point to the f_2 point.

Figure 16-9 illustrates the effects of resistance on bandwidth and Q. In the circuit in Figure 16-9A, the resonant frequency is 100 kHz and the Q is 50. The bandwidth for this circuit, therefore, is:

$$BW = \frac{F_O}{Q} = \frac{100\ kHz}{50} = 2\ kHz$$

RLC CIRCUITS

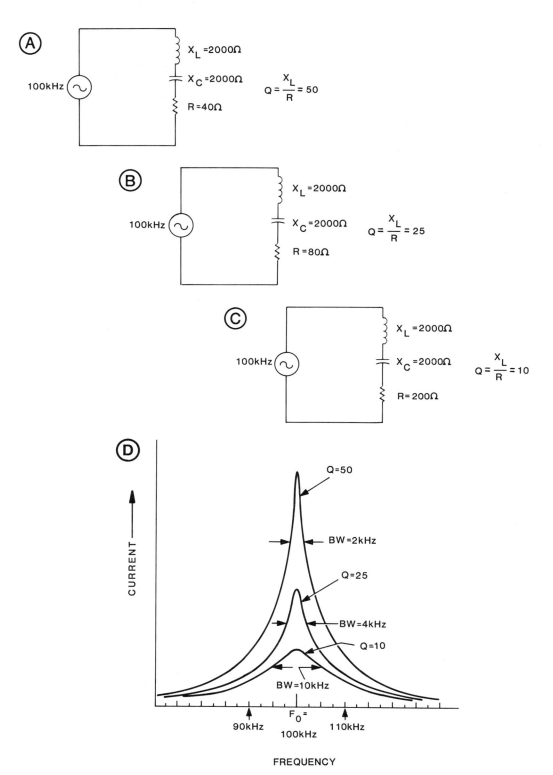

Figure 16-9 The effects of resistance on bandwidth and Q.

Figure 16-9D shows the curve for the circuit in Figure 16-9A. Notice how with a low resistance, Q is high, and the bandwidth is extremely narrow. When the Q is reduced to 25, as shown in Figure 16-9B, the current is lower and the curve is wider. The bandwidth increases to:

$$BW = \frac{F_O}{Q} = \frac{100\text{ kHz}}{25} = 4\text{ kHz}$$

And finally, the Q is reduced to 10 by the increased resistance in Figure 16-9C. The curve for circuit C is even wider than the curves for circuits A and B. The bandwidth for this circuit is:

$$BW = \frac{F_O}{Q} = \frac{100\text{ kHz}}{10} = 10\text{ kHz}$$

The curves shown in Figure 16-9D illustrate that the bandwidth increases as the value of Q decreases. When the value of Q is high, the current in the circuit is relatively high. The resonant circuit responds to a very narrow band of frequencies.

Also notice that the change in resistance in the circuit did *not* affect the resonant frequency—only Q and bandwidth.

RLC CIRCUITS

Parallel Resonance

Up to now we have considered only those resonant circuits in which the capacitor is in series with the inductor. In this section, we will consider another type of resonant circuit called a parallel resonant circuit. When the capacitor is placed in parallel with the inductor, the characteristics of the resonant circuit change completely.

Ideal Circuit

A parallel resonant circuit is shown in Figure 16-10A. We know that the circuit is in the resonant condition because X_L is equal to X_C. Thus, the applied AC signal is at the proper frequency to cause the circuit to resonate.

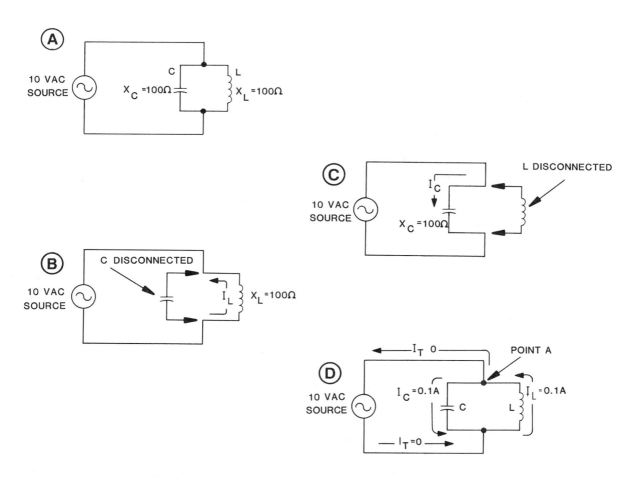

Figure 16-10 Parallel resonant circuit.

AC ELECTRONICS

To simplify the explanation of this circuit, let's initially assume that L and C are ideal components so that there is no resistance in the circuit. Of course, in practical circuits there is always some resistance, but for now, let's see what would happen in an ideal parallel resonant circuit.

When the capacitor is temporarily disconnected from the circuit as shown in Figure 16-10B, the current through the inductor can be determined using Ohm's Law:

$$I_L = \frac{E}{X_L} = \frac{10V}{100\,\Omega} = 0.1\,A.$$

The current must be supplied by an AC source. Since L is a pure inductor, I_L must lag the applied voltage by 90 degrees (ELI).

When the capacitor is reconnected and the inductor is disconnected as shown in Figure 16-10C, the current through the capacitor can be determined using Ohm's Law:

$$I_C = \frac{E}{X_C} = \frac{10V}{100\,\Omega} = 0.1\,A.$$

In a capacitor the current leads the voltage by 90 degrees (ICE). Thus, I_C must lead the applied voltage by 90 degrees. Now, we will consider the operation of the complete circuit. When I_C leads the applied voltage by 90 degrees, and I_L lags the applied voltage by 90 degrees, then I_C must be 180 degrees out-of-phase with I_L. This means that when the current is flowing in one direction through L, an equal current must be flowing in the opposite direction through C.

For example, at the instant when 0.1 amp is flowing up through L, 0.1 amp must be flowing down through C. By applying Kirchhoff's Current Law to point A in Figure 16-10D, we discover that there is no current flowing into or out of the source. That is, the same current that flows up through L, also flows down through C and no current flows to or from the source. On the next alternation, the current flows up through C and down through L but still no current flows in the external circuit.

In the ideal parallel resonant circuit, the source voltage is required only to start the oscillation. Once started, the source can be disconnected, and the current through the inductor and the capacitor will continue indefinitely. As mentioned earlier, this is true only if there are no losses in the circuit.

Now let's consider how the parallel resonant circuit appears to the AC source. The AC voltage is applied across the LC combination and yet no current flows to or from the source. Consequently, as far as the source is concerned, the circuit appears to be open. That is, it appears to have infinite impedance.

RLC CIRCUITS

Flywheel Effect

The ability of a parallel resonant circuit to sustain oscillation after the source voltage is removed is called the **flywheel effect**. It gets this name because the action is similar to that of a mechanical flywheel. Once the mechanical flywheel is started it tends to keep going until stopped by friction or some outside force. The flywheel effect of the parallel resonant circuit is illustrated in Figure 16-11.

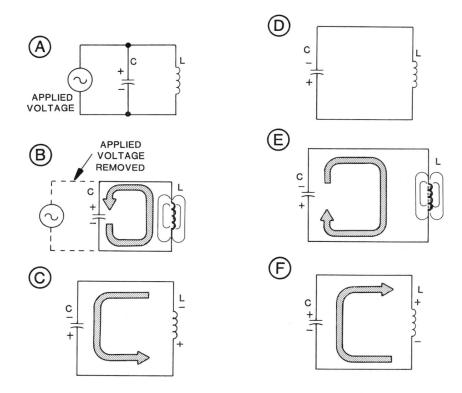

Figure 16-11 The flywheel effect.

Initially energy is supplied to the circuit by an AC source. Once started, the energy is alternately stored by the capacitor and then by the inductor. Let's pick up the action at the point where C is fully charged as shown in Figure 16-11A.

The applied voltage can be removed since it has supplied the necessary starting energy. As shown in Figure 16-11B, the capacitor begins to discharge through L. As current flows through L, the magnetic field around the inductor expands with the increasing current.

AC ELECTRONICS

When C is completely discharged, the current through L momentarily stops. Consequently, the magnetic field around L starts to collapse, inducing an EMF with the polarity shown in Figure 16-11C. This keeps the current flowing in the same direction and charges C to the polarity shown. Notice that the plates of the capacitor have switched polarities. This is due to the opposite direction of current flow through the capacitor.

When the magnetic field has collapsed, the condition shown in Figure 16-11D exists and the capacitor is again fully charged. At the next instant, the capacitor begins to discharge, sending current in the opposite direction. This causes a magnetic field of the opposite polarity to expand around L.

When C is again completely discharged, the current through L stops and the magnetic field collapses. As shown in Figure 16-11F, an EMF is induced which keeps current flowing in the same direction. Thus, C is again charged to its initial polarity.

At this point the cycle repeats itself. As you can see, the energy is interchanged between the capacitor and the inductor. Initially, the energy is stored as an electrostatic field in the capacitor's dielectric. Then it is stored as a magnetic field around the inductor. Theoretically, neither the capacitor nor the inductor dissipates energy. Therefore, the oscillations would continue indefinitely. Remember, we assumed a pure inductor, a pure capacitor, and no resistive losses in the circuit. A circuit of this type stores energy, and is commonly called a **tank circuit**.

Practical Tank Circuits

The ideal tank circuit is one which has no resistance or losses of any kind. Unfortunately, such a tank circuit does not exist. The capacitor, the inductor, and the interconnecting wires all have resistance. Normally, only the resistance of the inductor is high enough to be of importance. In reality, practical tank circuits must be analyzed as if there is a resistor in series with the inductor.

Unlike reactance, resistance dissipates power. As the current oscillates between the inductor and the capacitor, the resistor dissipates some of the power in the form of heat. Consequently, energy is removed from the tank circuit during each cycle of operation. For this reason, in a practical tank circuit, the oscillations will die out (cease) when the AC source is disconnected. Figure 16-12 shows how the voltage waveform across the tank circuit would look if the source was disconnected. Each cycle gets progressively weaker (lower amplitude) as the resistance dissipates the energy stored in the circuit. This waveform produced is called a **damped** sine wave.

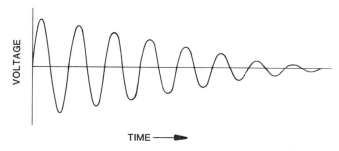

Figure 16-12 Damped sine wave.

Because energy is lost in the circuit, it must be supplied from the source to make up for the loss. The AC source provides just enough energy to make up for the loss caused by resistance. The result is that some current flows from the AC source to the tank circuit. If the resistance value in the tank is high, more power will be dissipated, and the current drawn by the tank from the source will also be higher.

This may seem to contradict Ohm's Law since current is inversely proportional to resistance. The resistance limits the circulating current within the tank. However, in doing so, it consumes power. This power must be supplied by the AC source. Therefore, the current from the source must increase.

Remember, a source must be able to supply the power to drive the load. The load demands the current needed to drop the voltage around a closed loop. In a circuit without resistance there is no load. Effectively the circuit is open and there is no current flow from the source. Circulating current is the current that flows only between the inductor and capacitor.

Q in Parallel Resonant Circuits

In the series resonant circuit we found Q by dividing the applied voltage, into either E_C or E_L. This cannot work in parallel resonant circuits since E_{IN} is applied directly across both C and L. In the parallel resonant circuit, we are concerned with current rather than with voltage. Thus, in the parallel resonant circuit, Q can be determined by dividing the source current into the tank current. Recall that in a good quality tank circuit (one with a high Q), the source current is very low while the circulating current can be very high. Therefore:

$$Q = \frac{I_{TANK}}{I_{SOURCE}}$$

AC ELECTRONICS

When source current is 1 millamp and tank current is 100 millamp, Q is:

$$Q = \frac{100\text{mA}}{1\text{mA}} = 100.$$

As with the series resonant circuit, Q can also be expressed as the ratio of X_L, or X_C, to R. R is the total AC resistance within the tank, and is usually higher than the value of R that can be measured with an ohmmeter.

In a parallel resonant circuit, Q can be thought of as a magnification factor. However, in this case it is not the voltage that is being magnified, but the impedance. Because the source current is minimum at resonance, the impedance of the tank circuit is maximum at resonance. The impedance of the tank is equal to the reactance of L or C times Q. That is:

$$Z_{TANK} = X_L(Q)$$

$$Z_{TANK} = X_C(Q)$$

For example, when the value of X_L and X_C at resonance is 1 kilohm, and Q is 100, the impedance of the tank will be:

$$Z_{TANK} = X_L(Q) = 1\text{k}\Omega\,(100) = 100\text{k}\,\Omega.$$

By transposing the equation, we can develop another useful equation. That is:

$$Q = \frac{Z_{TANK}}{X_L}$$

Thus we have three equations for Q. They are:

$$Q = \frac{I_{TANK}}{I_{SOURCE}} \qquad Q = \frac{X_L}{R} \qquad Q = \frac{Z_{TANK}}{X_L}$$

Bandwidth in Parallel Resonant Circuits

Like the series resonant circuit, the parallel tank circuit responds to a band of frequencies rather than a single frequency. Figure 16-13A shows a typical response curve. Notice that this curve has the same shape as the response curve shown earlier for the series resonant circuit. However, if you examine the curve more closely you will see that it shows the impedance of the circuit rather than the current through the circuit.

RLC CIRCUITS

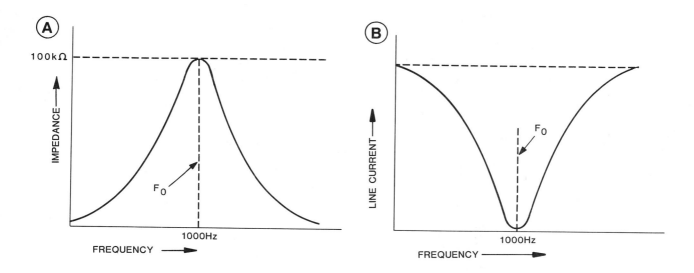

Figure 16-13 Response curves of a parallel resonant circuit.

At the resonant frequency, the impedance of a parallel resonant circuit is maximum. Above or below resonance, the tank offers a lower reactance and the circuit impedance decreases.

Remember, any time resistance (reactance) is combined in parallel, total impedance decreases. At resonance X_L and X_C are exactly equal and have opposite polarities. Effectively, they appear as an open (infinite impedance). Impedance in a parallel resonant circuit is maximum and current is minimum when the circuit is at resonance.

Because the source or line current is inversely proportional to the impedance, the current response curve has the shape shown in Figure 16-13B. Notice that the line current decreases as the resonant frequency is approached.

As with the series resonant circuit, the width of the band of frequencies to which the circuit responds is determined by the Q of the circuit. The bandwidth is determined by the formula:

$$BW = \frac{F_O}{Q}$$

Often a parallel resonant circuit will be more selective than we want. That is, it responds only to a very narrow band of frequencies. In these cases, we can increase bandwidth by connecting a relatively small valued resistor across the tank circuit as shown in Figure 16-14A. The resistor provides an alternate path for line current which increases line current.

AC ELECTRONICS

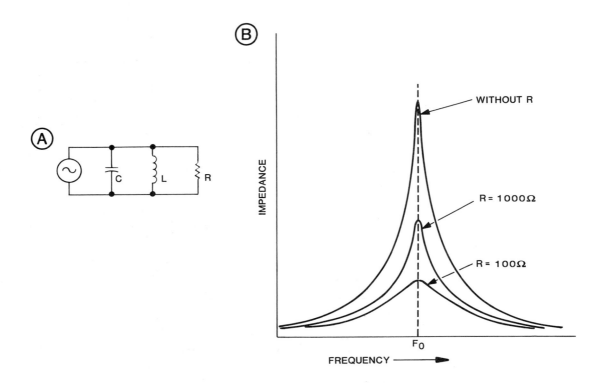

Figure 16-14 Bandwidth may be increase by loading the tank circuit.

Recall that:

$$Q = \frac{I_{TANK}}{I_{LINE}}$$

For this reason, Q is inversely proportional to line current. When the line current increases, Q decreases.

The next formula shows that bandwidth is inversely proportional to Q.

$$BW = \frac{F_O}{Q}$$

When Q decreases, bandwidth increases. Therefore, connecting a resistor across the tank circuit always increases bandwidth. This is often called loading the tank circuit. When a resistor is placed in parallel with the tank circuit, the resistor is sometimes called a shunt resistor or swamping resistor.

Figure 16-14B shows the effect of a loading resistor. Without the resistor, the impedance is extremely high and the circuit responds to a very narrow band of frequencies around the resonant frequency. When a 1000 ohm resistor is used, the additional parallel path for current reduces the circuit impedance and broadens bandwidth. A 100 ohm resistor reduces the impedance still further and widens the bandwidth even more.

AC ELECTRONICS

LC Filters

Earlier, we discussed the operation of a filter. A filter is an electronic circuit that selects a frequency (or group of frequencies) from an infinite number of frequencies. Inductors and capacitors, because of their changing voltage drops with different frequencies, allow us to pass or attenuate a frequency. When a frequency is *passed*, it provides a high output voltage. When a frequency is *attenuated*, the output voltage drops below a usable level. Thus, RL and RC circuits can be used as high-pass or low-pass filters. A high-pass filter (RC or RL) passes those frequencies above the cut-off frequency, while attenuating those below. A low-pass filter passes frequencies below the cut-off frequency, while attenuating those above. The *cutoff frequency* is that frequency which provides 70.7% of the source voltage to the output.

However, RC and RL circuits could not select a specific frequency, but only a large group of frequencies. How does a radio pick the specific station you want out of the infinite number of frequencies available?

Here is where resonant circuits become useful. Because a resonant circuit will act in a specific manner at the resonant frequency, it can be used to select the resonant frequency, while attenuating those frequencies outside the resonant frequency. These filters are known as band-pass filters. A resonant circuit can also attenuate the resonant frequency, while passing all others. These are called band-stop (or band-reject) filters.

Band-Pass Filter

A very simple band-pass filter is shown in Figure 16-15A. The filter is the series resonant circuit composed of L and C. R_L is the load to which the voltage is applied. At the resonant frequency, the series resonant circuit has a very low impedance. Thus, it drops very little of the applied voltage, E_{IN}. Most of the voltage is developed across R_L. Consequently, E_{OUT} is high at the resonant frequency.

RLC CIRCUITS

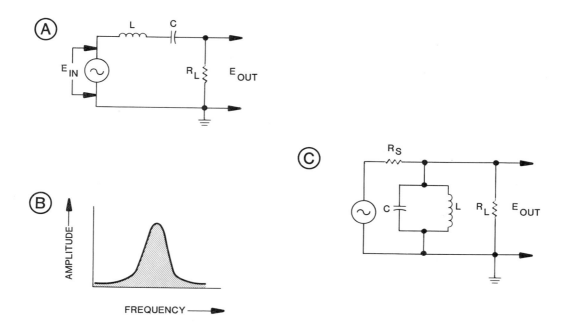

Figure 16-15 Band-pass filter.

Below the resonant frequency, the X_C of the capacitor is higher than the resistance of R_L. Consequently, most of E_{IN} is dropped across C. This leaves only a small voltage across R_L. Thus, E_{OUT} is a low voltage.

Above the resonant frequency, the X_L of the inductor is higher than R_L. Therefore, most of the voltage is dropped across the inductor and E_{OUT} is again a low voltage.

Figure 16-15B shows the response curve of a band-pass filter. At the resonant frequency of L and C, E_{OUT} is high. Above and below resonance E_{OUT} drops off quickly to a low voltage.

AC ELECTRONICS

Figure 16-15C shows that the parallel resonant circuit can also be used as a band-pass filter. Whereas the series resonant circuit is placed in series with the output, the parallel resonant circuit is placed across the output. The reason for this becomes evident when we remember the characteristics of a parallel resonant circuit.

At resonance the impedance of the tank circuit is maximum. Consequently, very little current flows through the tank circuit. Most of the current flows through R_L. The current through R_L is maximum at resonance.

Below resonance, X_L of the inductor is small when compared to the value of R_L. Thus, most of the current flows through L and very little current flows through R_L. Above resonance, most of the current flows through the capacitor again leaving little current for R_L. That is, above and below resonance, R_L is partially shorted (bypassed) by the low impedance of the tank. Thus, most of the applied voltage is dropped across R_S. However, at resonance, the impedance of the tank is high and R_L provides the path of least resistance.

Band-Stop Filter

The response of the band-stop filter is opposite that of the band-pass filter. That is, the band-stop filter stops, attenuates, or rejects the frequency to which it is tuned.

Figure 16-16A shows a simple band-stop filter. Here L and C form a parallel resonant circuit. The resonant circuit is in series with the load, R_L. At resonance, the impedance of the tank circuit is much higher than R_L. Consequently, most of E_{IN} is dropped across the tank, and very little voltage is available at the load. Above and below resonance, the resistance of R_L is higher than the impedance of the tank. Therefore, most of E_{IN} is developed across R_L. Figure 16-16B shows the response curve of the circuit. The sharpness of the curve is determined by the Q of the resonant circuit.

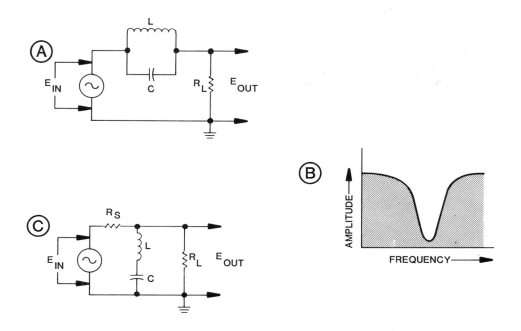

Figure 16-16 Band-stop filter.

Another circuit that produces about the same response is shown in Figure 16-16C. Here a series resonant circuit is connected across the load. At resonance, the series resonant circuit offers a very low impedance to current flow. This shorts (bypasses) most of the current around the load. Most of the applied voltage is dropped across R_S. Above and below resonance, the impedance of the filter is much higher, and R_L is no longer bypassed.

AC ELECTRONICS

Chapter Self-Test

1. What is the formula(s) for finding the total impedance in a *series* RLC circuit?

2. What is the formula(s) for finding current in a *parallel* RLC circuit?

3. Define resonance.

4. What is the formula for finding the resonant frequency of an RLC circuit?

5. Define bandwidth.

6. What is the relationship between Q, bandwidth, and the resonant frequency in a series resonant circuit?

7. In a series resonant circuit, what will happen to the bandwidth if Q decreases?

8. What is a half-power point?

9. What will happen to the bandwidth of a series resonant circuit if a resistor is placed in series with the reactive components?

10. What is a tank circuit?

11. Describe what occurs in a tank circuit when the resonant frequency is applied. What is this effect known as?

12. When a tank circuit is at resonance, what effect does the source voltage *prevent*?

13. What happens to the response curve of a parallel resonant circuit if a swamping resistor is used?

Summary

With series RLC circuits, it is easier to vector voltage or impedance. In a series RLC circuit, total impedance is calculated using the formula:

$$Z = \sqrt{R^2 + (X_L - X_C)^2}$$

When X_C is larger than X_L, the formula becomes:

$$Z = \sqrt{R^2 + (X_C - X_L)^2}$$

With parallel RLC circuits, it is easier to vector currents. The formula for finding total current is:

$$I_T = \sqrt{I_R^2 + (I_L - I_C)^2}$$

When I_C is larger than I_L, the formula is:

$$I_T = \sqrt{I_R^2 + (I_C - I_L)^2}$$

For any LC circuit, there is a frequency at which X_L is equal to X_C. This frequency is called the resonant frequency, F_O. The resonant frequency can be determined using the formula:

$$F_O = \frac{.159}{\sqrt{LC}}$$

Series resonant circuits and parallel resonant circuits behave quite differently. Although the resonant frequency is found the same way, the operation of these circuits at resonance and away from resonance are quite different. Figure 16-17 illustrates the important points for series and parallel resonant circuits, and the differences between them.

AC ELECTRONICS

SERIES RESONANT CIRCUIT	PARALLEL RESONANT CIRCUIT
Current maximum at resonance.	Line current minimum at resonance.
Impedance minimum at resonance.	Impedance maximum at resonance.
$Q = \dfrac{X_L}{R} \quad Q = \dfrac{E_C}{E_{IN}} \quad Q = \dfrac{E_L}{E_{IN}}$	$Q = \dfrac{X_L}{R} \quad Q = \dfrac{Z_{TANK}}{X_L} \quad Q = \dfrac{I_{TANK}}{I_{SOURCE}}$
Acts purely resistive at F_O.	Acts purely resistive at F_O.
At resonance, the source current and voltage are in-phase.	At resonance, the source current and voltage are in-phase.
Below resonance, the circuit acts capacitively	Below resonance, the circuit acts inductively.
Above resonance, the circuit acts inductively.	Above resonance, the circuit acts capacitively.

Figure 16-17 Comparing series and parallel resonant circuits.

At the resonant frequency, X_L is equal and opposite X_C. When the reactances are equal and opposite they cancel each other, and the only impedance in the circuit is the effective resistive component of the inductor.

When an inductor and a capacitor are connected in parallel, a tank circuit is created. Once power is applied to a tank circuit, current starts to flow between the inductor and the capacitor, without entering into the outside circuit. Power can then be disconnected. When power is disconnected, a circulating current continues to flow between the capacitor and the inductor.

The output of the LC tank circuit is referred to as the flywheel effect. The self-sustained output waveform losses some of its amplitude each cycle. This loss is caused by the resistive component of the circuit and the waveform is considered to be damped.

Tuned circuits are used as filters. Figure 16-18 summarizes the four basic types of filters discussed.

RLC CIRCUITS

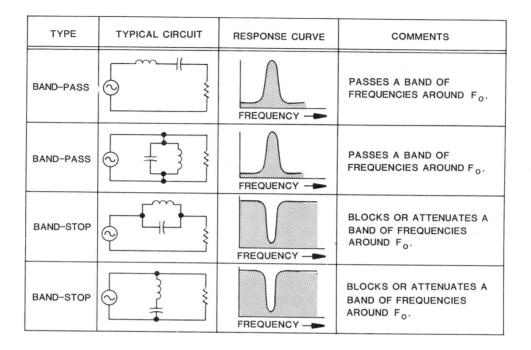

Figure 16-18 Comparing types of filters.

The power factor is equal to the cosine of the phase angle. It is the result of the inductive properties of the load. Power factor can be compensated for by varying capacitance to obtain resonance. At resonance the power factor is equal to 1 and the circuit is said to be 100% efficient. The cosine of 1 is zero degrees, which means the source and load are in-phase.

AC ELECTRONICS

CHAPTER 17

Transformers

Contents

Introduction 403

Chapter Objectives 404

Transformer Action 405

Transformer Theory 414

Transformer Ratios 418

Transformer Losses 425

Transformer Applications 429

Chapter Self-Test 434

Summary 435

TRANSFORMERS

Introduction

When we first introduced you to alternating current, we said that one of the advantages of AC was that its voltage could be increased or decreased with a transformer. This chapter covers the theory of operation and the characteristics of this component.

A **transformer** is essentially two inductors placed side-by-side. When alternating current flows through one inductor, the expanding and contracting magnetic field cuts across the windings of the other coil, thereby inducing a voltage. This is the basic action of a transformer.

Chapter 17 begins by describing in detail the action of a transformer. We will then cover how a transformer differs when a load is connected across it. We will then discuss the relationships and ratios a transformer produces with regards to voltage, current, and power. The last section of this chapter describes the number of ways power is lost within a transformer, and the different design innovations that help to prevent these losses.

AC ELECTRONICS

Chapter Objectives

When you have completed this chapter, you will be able to:

1. Define the following terms: transformer primary, transformer secondary, step-up, step-down, signal inversion, impedance matching, center-tap, auto-transformer, phase-splitting, eddy current loss, copper loss, and hysteresis.

2. Calculate turns ratio, voltage ratio, power ratio, current ratio, and impedance ratio.

3. List four types of losses that effect transformer efficiency.

4. Calculate transformer efficiency.

5. State five uses for the transformer.

6. Determine the relationship between voltage, current, and impedance ratios.

7. Name the parts of an elementary transformer.

8. State the difference between a transformer and an autotransformer.

9. Explain why the transformer is not used in DC circuits.

10. Explain why the transformer is considered a safety device.

TRANSFORMERS

Transformer Action

The **transformer** is used to transfer alternating current from one circuit to another. Normally, some characteristic of the AC signal is changed in the transformation process. For example, a low voltage AC may be **stepped-up** to a higher value, or a higher voltage may be **stepped-down** to a lower value. Often it is the current which must be changed. The transformer is also used to step-up or step-down current. However, the transformer can not step-up current and voltage at the same time, nor can it step-down these values at the same time. In this section, we will take a look at the action which allows the transformer to change voltage and current levels.

Mutual Inductance

The principle on which transformer action is based is called **mutual inductance**. When current flows through a conductor, a magnetic field builds up around the conductor. When alternating current is used, the magnetic field constantly expands and contracts. If another conductor is placed within this moving magnetic field, the magnetic field will induce a voltage into the second conductor.

The transformer is a device designed to take advantage of this principle. The two conductors are wound into coils and placed close together so that one coil is cut by the magnetic flux lines of the other. Often the coils are wound one on top of the other. When this is the case, each inductor is coated with a thin insulator material to prevent shorts.

AC ELECTRONICS

Figure 17-1 illustrates transformer action. Coil L_1 is connected to an AC voltage source. As alternating current flows through the coil, a varying magnetic field is set up. During one half-cycle, current will flow through L_1 in the direction shown. This establishes a north magnetic pole at the top of L_1. As the current increases, the field expands outward cutting the turns of wire in L_2. This induces an EMF into L_2 and, in turn, causes current to flow up through the load resistor. Thus, current in L_1 causes current to flow through L_2.

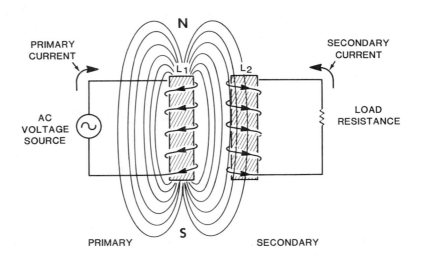

Figure 17-1 Transformer action.

Notice that the coils are wound in *opposite* directions. The current flows into the top of L_1. This induces a voltage that is negative at the top and positive at the bottom of L_1. This field induces a current into the secondary (L_2) that develops a voltage that is positive at the top and negative at the bottom. The polarity of the secondary voltage causes electrons to flow up through the load resistor. In this type transformer, the secondary voltage is 180 degrees out-of-phase with the input. This relationship is also known as signal inversion.

At the end of the first half cycle, the current through L_1 drops to 0 for an instant as the sine wave input passes through 180 degrees. As the current decreases, the field collapses back into L_1. When the current in L_1 decreases, the current through L_2 also decreases.

On the next half cycle, the direction of current through L_1 reverses. This causes a magnetic field of the opposite polarity to expand outward from L_1. Once again, the field cuts the turns of L_2 inducing EMF. However, because the polarity of the magnetic field is reversed, the polarity of the voltage induced into L_2 is also reversed. Thus, the induced EMF causes current to flow down through the load resistance.

Notice that the current in L_2 follows the current in L_1. Each time the current in L_1 reverses direction, the current in L_2 also reverses. Therefore, the alternating current in L_2 has the same frequency as the alternating current in L_1. Energy is transferred from one circuit to another even though the two circuits are electrically insulated (physically separated) from each other.

The circuit shown in Figure 17-1 is a simple transformer. The coil to which the AC voltage is applied is called the primary winding. Current in this winding is caused by the AC voltage source and is called the primary current. The coil into which current is induced is called the secondary winding, and the induced current is called the secondary current.

The amount of EMF induced into the secondary winding depends on the amount of mutual induction between the two coils. In turn, the amount of mutual induction is determined by the degree of flux linkage between the two coils. The flux linkage can be thought of as the percentage of primary flux lines which cut the secondary winding. Another expression which means approximately the same thing is the coefficient of coupling. The coefficient of coupling is a number between 0 and 1. When all the primary flux lines cut the secondary coil, the coefficient of coupling is 1. If the two coils are positioned so that some of the primary flux lines do not cut the secondary, then the coefficient of coupling is less than one.

It is impossible to create a transformer where the exact amount of primary power is transferred to the secondary winding. The two windings are separated, and the resistance of the windings will never allow 100% of the primary power to be transferred to the secondary. Thus, we can conclude that:

INPUT POWER IS ALWAYS GREATER THAN THE OUTPUT POWER.

AC ELECTRONICS

Figure 17-2 illustrates that the amount of mutual inductance depends on the flux linkage or the coefficient of coupling. In Figure 17-2A, the secondary coil, L_2, is wound directly onto the primary coil, L_1. Using this arrangement, nearly all of the flux lines produced by the primary cut the secondary windings. Therefore, the coefficient of coupling is close to one.

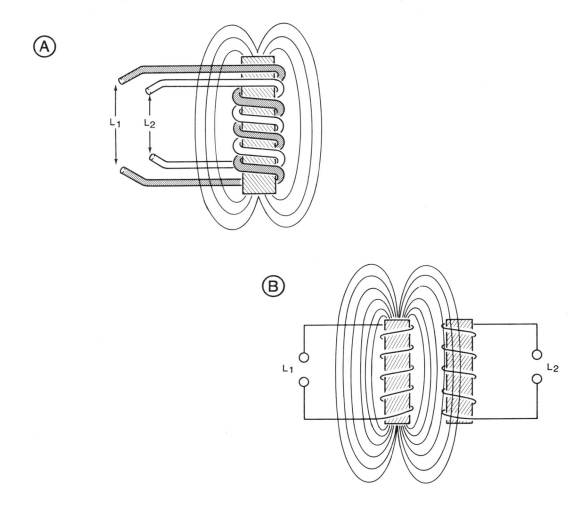

Figure 17-2A, B Mutual inductance depends on the coefficient of coupling.

In Figure 17-2B the transformer consists of two coils. Here only a few lines of flux from the primary cut the secondary. Therefore, the coefficient of coupling is low.

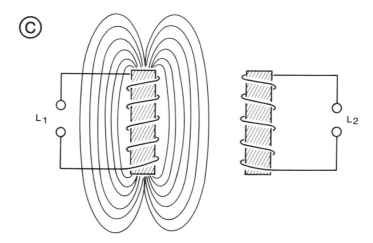

Figure 17-2C When there are no mutual inductance, the coefficient of coupling is zero.

Figure 17-2C illustrates that when the two windings are placed far enough apart, there is no flux linkage between them. In this case, there is no mutual inductance and the coefficient of coupling is zero. While this arrangement has no practical purpose, it illustrates the importance of the coefficient of coupling. Remember also that when coils are mounted perpendicular to each other the coefficient of coupling is zero.

AC ELECTRONICS

Figure 17-3 illustrates the step-by-step sequence of events that occur within a transformer. These steps are also outlined below:

Figure 17-3A: A switch is closed, connecting an EMF across the primary windings of the transformer.

Figure 17-3B: The alternating EMF causes an alternating current to flow through the coils of the primary winding.

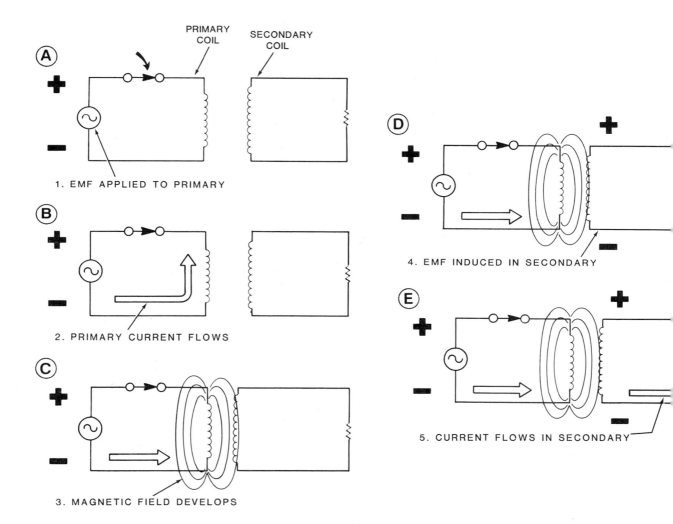

Figure 17-3 Sequence of events in transformer action.

Figure 17-3C: The current flowing through the coils causes a magnetic field around both the primary and secondary windings that constantly expands and contracts.

Figure 17-3D: The magnetic field continuously cuts across the secondary windings, inducing an EMF.

Figure 17-3E: The induced EMF across the secondary winding causes current to flow through the secondary coils, and the load.

The transformer is not used in DC circuits because there is no movement of flux lines in the circuit, except at power on or off. Without relative motion of the magnetic field, there is no induced voltage. Therefore, the transformer is considered to be an AC component.

Transformer Construction

The construction techniques for transformers vary depending on the type of transformer and the particular application. All transformers can not be used for the same application. A substation transformer used in a power distribution system may approach the size of a small house. On the other hand, a transformer used in a transistor radio may be no larger than a pencil eraser. In spite of the vast size difference, these two transformers operate on the same basic principle. Both have primary and secondary coils, where energy is coupled from the primary to the secondary by mutual inductance.

The design of the transformer is also dictated by the frequency that it must pass, the voltage and current involved, and several other factors. A power transformer may be required to handle 115 VAC 60 Hz, at 1 ampere. On the other hand, an audio transformer may work with a frequency of 455 kHz at a few millivolts and a few microamperes. Transformers are bulky, heavy, and expensive when compared to other electronic components. Therefore, they are used only in limited applications.

AC ELECTRONICS

Figure 17-4 shows the construction of an **iron-core** transformer. The iron-core transformer is rather large and heavy when compared to other electrical components. The primary winding is wound on one arm of the core, and the secondary is wound directly on top of the primary. Notice that the schematic symbol for the transformer shows two coils, with two lines between the coils. The two lines represent the iron core. The purpose of the iron core is to concentrate the lines of force into a small area. If you will recall from Chapter 9, iron has a high permeability and that permeability means "easy to magnetize."

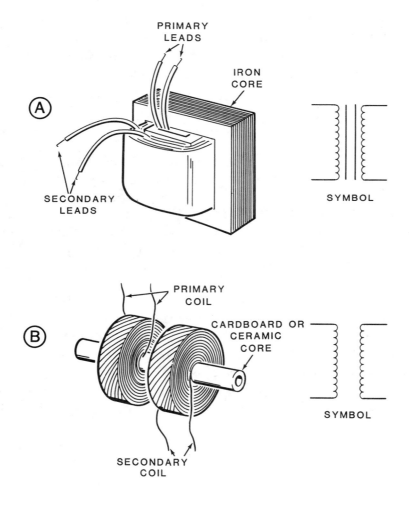

Figure 17-4 Typical transformers and their symbols.

TRANSFORMERS

The construction of the **air-core** transformer is different. It is designed to be used at much higher frequencies. Iron core losses increase as frequency increases. Transformers designed to operate at high frequencies use little or no iron in the core. Instead, a non-conductive material that has the same permeability as air is used. High frequency cores are usually either ceramic or simply small cardboard tubes. Notice that the schematic symbol for an air-core (non-conductive core) transformer is similar to the iron-core schematic symbol, but there are no lines between the windings.

Transformer Theory

In the previous section we saw how transformer action can be used to transfer an AC signal from one winding to another. However, in order for this transferred signal to be of any use, it must be connected to a circuit. Once the signal from the secondary winding of a transformer is connected to a circuit, the operation of the transformer itself changes.

Transformer With No Load

Figure 17-5A shows a transformer being operated without a load. This means that the secondary of the transformer is open. Therefore there is no secondary current. Even so, primary current exists because the primary is connected across an AC voltage source.

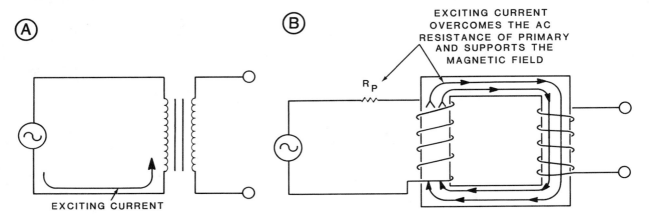

Figure 17-5 Transformer with no load.

In this case, the amount of primary current is determined by the impedance of the transformer's primary and the applied voltage. Since no power is being used in the secondary, the primary acts like an inductor. The primary of a typical iron-core transformer can have an inductance of several henrys. This tends to keep the primary current very low.

In addition to the inductance, the primary winding has a certain value of AC resistance. This limits the current even further. The small amount of primary current that flows with no load is called the **excitation current**. Figure 17-5B illustrates the two functions performed by excitation. First, it overcomes the AC resistance of the primary. In Figure 17-5B, the resistance of the inductor is shown as a separate resistor. This resistance dissipates power in the form of heat. In addition, the exciting current supports the magnetic field within the core.

TRANSFORMERS

The X_L of the primary is normally much larger than its AC resistance. Thus, the exciting current lags behind the applied voltage by almost 90 degrees. Consequently, when no current flows in the secondary, the primary of the transformer acts like an inductor.

Transformer With Load

Figure 17-6 shows a simple transformer with a load resistor connected across the secondary winding. When AC current flows in the primary, it induces a current into the secondary. Let's see how the current in the secondary affects the operation of the transformer.

In Figure 17-6A, the polarity of the applied voltage is negative at the top of the primary and positive at the bottom. This forces current to flow down through the primary winding. Using the left-hand rule, we find that the current develops a magnetic field in the direction shown.

As this magnetic field expands outward, it induces CEMF in the primary winding. This CEMF opposes the applied EMF. The applied EMF forces current to flow down through the primary, while CEMF tries to force current up through the primary. The net result is a small current which flows down through the primary.

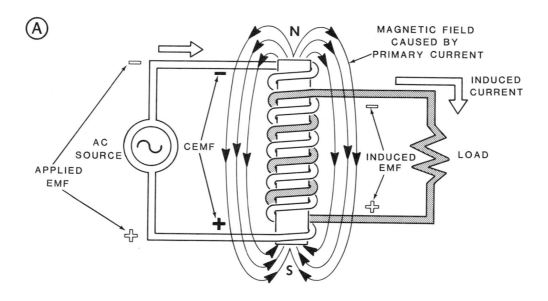

Figure 17-6A Transformer with load.

AC ELECTRONICS

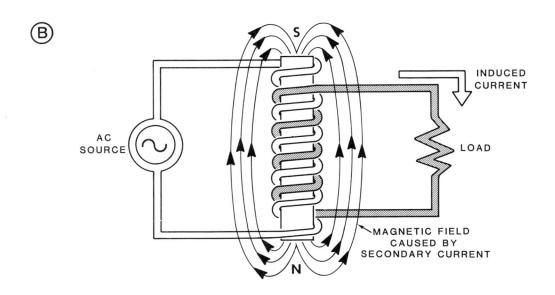

Figure 17-6B Transformer with load.

Notice that the secondary is wound directly on top of the primary. Therefore, the expanding magnetic field caused by the primary current also cuts the secondary winding. Since the secondary is wound in the same direction as the primary, the EMF induced into the secondary has the same polarity as the CEMF in the primary. Thus, the induced current in the secondary flows in the direction shown. In this configuration, there is no signal inversion between the primary and secondary windings.

The current flow in the secondary establishes a magnetic field of its own as shown in Figure 17-6B. The left-hand rule verifies that the magnetic field has the polarity shown. As the magnetic field expands, it cuts the secondary winding, inducing CEMF. This CEMF tries to force current to flow through the secondary opposite to the induced current.

The expanding flux in the secondary also cuts the primary turns. This induces yet another EMF back into the primary winding. This induced EMF is in the same direction as the CEMF of the secondary. Thus, this EMF tends to force current to flow down through the primary. If you have been keeping track of the various EMFs, you will see that the EMF induced into the primary from the secondary opposes the CEMF originally developed in the primary. Or stated another way, the current induced into the primary from the secondary aids the original primary current. This causes the primary current to increase.

This increase in primary current is caused by the expanding magnetic field of the secondary. The more current that flows in the secondary, the stronger the secondary magnetic field. This in turn, increases the primary current. Consequently, an increase in secondary current causes an increase in primary current. Later, you will see that the exact amount of increase in each depends upon the turns ratio.

The sum of the effects just described is called **mutual inductance.** The inductance is said to be mutual because the primary induces a voltage into the secondary and, simultaneously, the secondary induces a voltage back into the primary.

To be certain you have the idea, let's review this process once more.

Step 1. AC in the primary establishes a fluctuating magnetic field.

Step 2. The varying flux induces CEMF into the primary and an EMF into the secondary.

Step 3. The induced EMF causes current to flow in the secondary.

Step 4. The current in the secondary establishes a magnetic field which is opposite to the field caused by the primary current.

Step 5. The secondary flux induces an EMF back into the primary which opposes the CEMF of step 2. This decreases the primary CEMF.

Step 6. Primary current increases because the counter EMF decreases.

AC ELECTRONICS

Transformer Ratios

Transformers have many applications. They are used to step-up or step-down voltage, and step-up or step-down current. They are also used as coupling or isolation devices. In each case, we are concerned with a ratio.

In the first case, the ratio is that of an input voltage to an output voltage. In the second case, the ratio is that of a primary current to a secondary current. In the third case, the ratio is that of an input impedance to an output impedance. Each of these ratios is determined by the number of turns in the primary winding compared to the number of turns in the secondary.

Voltage Ratio

Transformers are frequently used to step-up or step-down voltages. Most electronic devices are powered by 115 VAC at 60 Hz. Some applications require higher voltages, while others can get by with much lower voltages. The transformer is used to transform the 115 VAC to whatever value is required. When the output (secondary) voltage is higher than the input (primary) voltage, the transformer is a **step-up** transformer. The amount of step-up is determined by the **turns ratio** of the transformer.

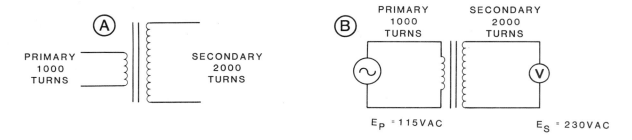

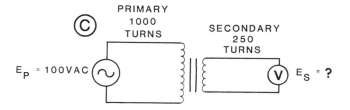

Figure 17-7 The voltage is determined by the turns ratio.

TRANSFORMERS

Figure 17-7A illustrates the turns ratio of a typical transformer. Notice that the primary consists of 1000 turns of wire, while the secondary has 2000 turns. The turns ratio can be defined as the ratio of the number of turns in the secondary, N_S, to the number of turns in the primary, N_P:

$$\text{Turns ratio} = \frac{N_S}{N_P}$$

In Figure 17-7A, the turns ratio is:

$$\text{Turns ratio} = \frac{N_S}{N_P} = \frac{2000}{1000} = 2.$$

This is expressed as a turns ratio of "2 to 1" or 2:1. When the secondary has more turns than the primary, the voltage is "stepped-up." For example, when the turns ratio is 2:1, then the secondary voltage will be twice as high as the primary voltage. Thus, the voltage ratio is equal to the turns ratio. This is expressed by the equation:

$$\frac{E_S}{E_P} = \frac{N_S}{N_P}$$

We can use either of these equations to find the secondary voltage when the turns ratio and the primary voltage are known. Figure 17-7B shows 115 VAC applied to the primary. We find the secondary voltage by rearranging the formula:

$$\frac{E_S}{E_P} = \frac{N_S}{N_P} \text{ is } E_S = \frac{N_S}{N_P} \times E_P = \frac{2000}{1000} \times 115V = 2 \times 115V = 230V$$

By choosing the proper turns ratio, the input voltage can be stepped-up to any value required. Remember that when an AC voltage is listed, it is understood to be an rms or effective value unless otherwise specified.

The transformer can also be used to step-down a voltage. To accomplish this, the secondary should have fewer turns than the primary. For example, in Figure 17-7C, the primary has 1000 turns while the secondary has 250 turns. The primary voltage is given as 100 VAC. Let's find the secondary voltage.

$$E_S = \frac{N_S}{N_P} \times E_P = \frac{250}{1000} \times 100V = 0.25 \times 100V = 25V$$

Obviously, finding the secondary voltage using this method requires that you know the number of turns for the primary and secondary winding. You will find, however, that the turns ratio also provides you with the information you need. If the turns ratio is 2:1 step-up with 100 volts on the primary, you simply multiply the input voltage by two. If the same ratio and voltage is given for a step-down transformer, you *divide* the input voltage by two. For a step-up transformer, you multiply the input voltage by the first number in the ratio. For a step-down transformer, you divide the input voltage by the turns ratio.

These equations hold true as long as the coefficient of coupling is high and the transformer losses are low. To be completely accurate, the transformer must have a coupling coefficient of 1 and an efficiency of 100%. While these conditions are impossible to achieve in practice, some transformers come very close. In this section, we will assume that an ideal transformer is used. Later, we will see that general purpose transformers fall far short of the ideal transformer. General purpose transformers have efficiency ratings between 60 and 80%.

Power Ratio

When we ignore the losses in the transformer, the power in the secondary is the same as the power in the primary. Thus, in the ideal transformer the power ratio is 1. Although the transformer can step-up voltage, it cannot step-up power. We can never take more power from the secondary than we put in at the primary. Thus, when a transformer steps-up a voltage, it steps-down the current. The output power is assumed to be the same as the input power. This is expressed by the equation:

$$P_P = P_S$$

Current Ratio

A transformer which steps-up voltage must at the same time step-down current. Otherwise, it would deliver more power in the secondary than is supplied by the primary. Since the voltage induced into the secondary is directly proportional to the number of turns, the secondary current must be inversely proportional to the number of turns.

This inverse relationship can also be shown with our knowledge of resistance and inductance. Let's say, for example, that a transformer has a 3:1 step-up ratio. This means that there are three times as many turns in the secondary winding as there are in the primary winding. Since the secondary winding has three times as many turns, there must be three times as much resistance within the wires. If there is three times as much resistance, current must decrease by as much.

However, the resistance of wire, even if it is turned thousands of times, is not very high. But the larger number of turns in the secondary also affects the inductive reactance of the secondary. If you'll recall, the inductive reactance of an inductor is directly proportional to the inductance available, and inductance is directly proportional to the number of turns. Therefore, if the number of turns increases, inductive reactance increases. In our previous example, the number of turns in the secondary was three times as many as the turns in the primary. Therefore, the secondary winding has three times as much inductive reactance, which causes three times as much opposition to current flow.

The inverse relationship between current and the turns ratio is expressed by the following formula:

$$\frac{I_P}{I_S} = \frac{N_S}{N_P}$$

AC ELECTRONICS

Figure 17-8A shows a transformer with a turns ratio of 4:1. This means that the secondary has 4 times as many turns as the primary. Thus, the voltage is stepped-up by a factor of four, from 10 volts to 40 volts. However, the current is stepped-down from 1 ampere in the primary to only 0.25 amperes in the secondary. If the current is stepped-down you divide the primary current by the first number in the turns ratio.

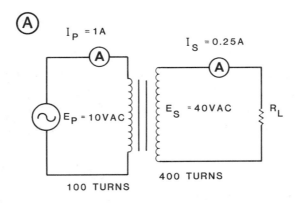

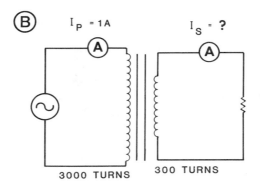

Figure 17-8 The turns ratio determines the current ratio.

A transformer can also be used to step-up current. However in doing this, it must step-down voltage. To step-up current, the primary must have more turns than the secondary, as shown in Figure 17-8B. To calculate the secondary current, the primary current must be multiplied by the first number in the turns ratio (recall that the current is being stepped-up). However, the turns ratio must first be found.

$$\text{Turns ratio} = \frac{N_S}{N_P} = \frac{300}{3000} = 0.1 \text{ or } 10:1.$$

Once the turns ratio is known, the secondary current can be found. Since the number of turns in the secondary winding is less than the number of turns in the primary winding, this must be a step-down transformer. If voltage is stepped-down, current must be stepped-up. Therefore, you multiply the primary current by the first number in the turns ratio, giving you a secondary current of 10 amperes.

Impedance Ratio

In electronics, one of the most important applications of a transformer is **impedance matching**. Maximum power is transferred from a generator to a load when the impedance of the generator matches the impedance of the load. When the impedances do not match, power is wasted.

There are many cases in electronics in which the impedance of the signal source or generator does not match the load which it must drive. For example, a stereo amplifier might be most efficient when driving a 100 ohm load. Nevertheless, the amplifier may be required to drive a 4-ohm speaker. This is a mismatch that results in wasted power and inefficient operation.

The transformer is used to correct for this mismatched impedance. The transformer can make one value of impedance appear to be another value. In the above example, a transformer is placed between the amplifier and the speaker. By choosing the proper turns ratio, the transformer can make the 4 ohm speaker appear to be a 100 ohm load to the transistor amplifier.

You have seen that the voltage or current step-up of a transformer depends on the turns ratio. The impedance matching capability of a transformer also depends on the turns ratio. However, the impedance ratio is equal to the turns ratio *squared*. The squaring occurs because voltage and current have inverse relationships with the number of turns, and impedance is a function of voltage and current. The formula is:

$$\frac{Z_P}{Z_S} = \frac{(N_P)^2}{(N_S)^2}$$

The formula can be rearranged to provide:

$$\frac{N_P}{N_S} = \sqrt{\frac{Z_P}{Z_S}}$$

Returning to our original problem, we can now find the turns ratio necessary to drive a 4 ohm speaker at maximum efficiency with a 100 ohm amplifier:

$$\frac{N_P}{N_S} = \sqrt{\frac{Z_P}{Z_S}} = \sqrt{\frac{100\ \Omega}{4\ \Omega}} = \sqrt{25} = 5.$$

This is a primary to secondary turns ratio of 5:1. Thus, when the number of primary turns is 5000, the number of secondary turns must be 1000. As you can see, any transformer with a turns ratio of 5:1 has an impedance ratio of $(5)^2$:1 or 25:1.

Impedance matching is one of the most important applications of a transformer. By choosing the proper turns ratio, transformers can match a wide range of impedances.

TRANSFORMERS

Transformer Losses

Transformers used in electronic circuits are very efficient devices. An efficiency of 60% or better is normal. All transformers have some losses. In many cases, these losses dictate the design of the transformer. Power transformers in particular are designed so that losses are minimized. The losses are separated into several categories.

Core Losses

In power transformers the largest loss occurs in the core. Even so, a much larger core loss would occur if it were not for special construction techniques. Let's examine each of these core losses.

Eddy Current Losses

The cores of power transformers are generally made of soft iron or steel. Because iron and steel are good conductors, and are ferromagnetic, a current can be induced into the core. This occurs when the core is subjected to a moving magnetic field. As we have seen, a moving magnetic field is a requirement in all transformers. Unless special precautions are taken, large circulating currents are induced into the core of the transformer. These induced circulating currents are called **eddy currents**. These eddy currents are sometimes referred to as **skin effect** because they flow along the conductor's surface.

The eddy currents produce a power loss which is proportional to the current squared ($P = I^2R$). When the eddy currents are reduced, the power loss is reduced.

Eddy currents are reduced by changing the construction of the core. In Figure 17-9A, a solid block of metal is used as the core. Because the cross sectional area of the core is large, it has very little resistance and large eddy currents flow.

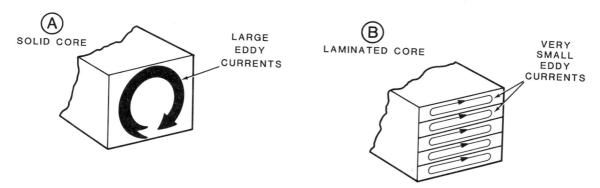

Figure 17-9 Eddy currents can be reduced by laminating the core.

AC ELECTRONICS

If thin sheets of metal are used for the core rather than a solid block of metal, these eddy currents can be reduced. Figure 17-9B illustrates this effect. The individual sheets are coated with an insulating varnish so that no current can flow between the sheets. Thus, any eddy currents produced are restricted to a single sheet of metal. Because the cross sectional area of a sheet is quite small, the resistance of each individual sheet is relatively high. This high resistance keeps the magnitude of the eddy currents low. Consequently, the power loss is much lower when the core is made of thin laminated (insulated) sheets.

The thin sheets which make up the core are called **laminations**. A laminated iron core is shown in Figure 17-10. The lamination is named according to its shape. Frequently, an iron core is constructed by combining "E" and "I" laminations as shown. Remember, the purpose of laminating a core is to reduce the power loss caused by eddy currents.

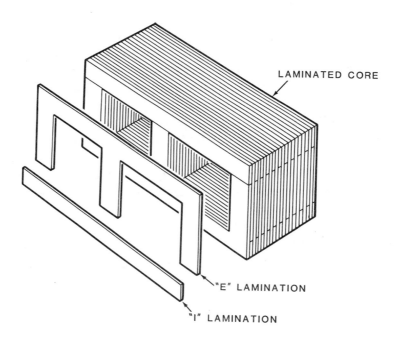

Figure 17-10 Laminated core construction.

TRANSFORMERS

Hysteresis

Another type of loss that occurs in the core of a transformer is called hysteresis. When iron is not magnetized, its magnetic domains are arranged in a random pattern. However, when a magnetizing force is applied, the domains are lined up by the magnetic field. When the magnetic field reverses, the domains must also reverse their alignment.

In a transformer, the magnetic field reverses direction many times each second. The number of reversals per second is determined by the frequency of the applied AC source. For example, with a 60 hertz source the magnetic field reverses 120 times per second. Remember the magnetic field reverses twice each cycle. When reversing direction, the domains must overcome friction and inertia. In doing this a certain amount of power is dissipated in the form of heat. This power loss is referred to as the hysteresis loss.

The terms friction and inertia refer to the basic laws of physics that state "a body at rest tends to remain at rest" and "a body in motion tends to remain in motion". Therefore, anytime you have relative motion there is friction and inertia. They are a form of resistance to motion and they do cause power to be dissipated.

Some materials such as soft iron normally have high hysteresis losses. The hysteresis loss in steel is lower than soft iron. Some large power transformers use a special type of metal called silicon steel because it has a very low hysteresis loss.

The amount of loss caused by hysteresis increases as frequency increases. An iron-core transformer that has a small hysteresis loss at low frequencies may have a larger hysteresis loss at higher frequencies. Hysteresis loss is directly proportional to frequency.

Copper Loss

Another type of loss present in all transformers is called copper loss. This loss is caused by the resistance of the copper wire in the primary and secondary windings. A transformer winding can consist of hundreds of turns of fine copper wire. Because of the length of the wire and its small cross sectional area, its resistance can be quite high. As current flows through this resistance, some power is dissipated in the form of heat. The amount of power dissipated in the form of heat is determined by the formula $P = I^2R$. Thus, another name for copper loss is the I^2R loss.

The amount of copper loss is proportional to the current squared. When the current through a transformer doubles, the copper loss increases by a factor of 4.

AC ELECTRONICS

Copper loss can be reduced by increasing the size (reducing the wire gauge number) of the copper wire used for the windings. The larger the diameter of the wire, the smaller its resistance. As you can see by the formula, power dissipated and resistance are directly proportional. Decreasing resistance decreases the copper loss. Another method of reducing copper loss is to keep the current in the transformer as low as possible.

External Induction Loss

As the magnetic field expands and contracts around the transformer it often cuts an external conductor of some kind. When a current is induced into an external conductor, some power is taken from the transformer circuit. In most cases, the power lost by external induction is so small that it can be ignored. However, the voltage induced into outside circuits sometimes causes problems. For example, in a sensitive amplifier circuit, the unwanted induced voltage from a transformer may interfere with the signal being amplified.

Interference caused by magnetic induction can be reduced by shielding. Often sensitive circuits are placed inside a metal shield which will prevent stray magnetic fields from reaching the circuits. Also, transformers themselves are often placed in thin metal housings to prevent magnetic fields from escaping. These shields are called permeability shields. Remember, magnetic flux lines can not be blocked. However, they can be rerouted.

Transformer Efficiency

Because of transformer losses, more power is applied to the primary of the transformer than is available for use in the secondary. All transformers have power losses. Hence the **efficiency** of a transformer is always less than 100%.

The efficiency of a transformer is the ratio of output power to input power. For example, when a transformer has an input power of 100 watts and an output power of 90 watts, its efficiency is:

$$\text{eff} = \frac{\text{output power}}{\text{input power}} = \frac{90W}{100W} = 0.9 \times 100 = 90\%$$

Efficiency is normally stated in terms of a percentage (%). Thus, the decimal fraction must be multiplied by 100 to convert to a percentage. The input power is the primary power, and the output power is the secondary power.

Transformer Applications

In electronic devices the transformer is used to step-up or step-down voltage. Many electronic devices require 115 VAC for power. Most of the devices have a power transformer which steps the voltage up or down, as required. In transistorized equipment, the AC line voltage is usually stepped-down and then changed to DC by rectification. In older vacuum-tube equipment, the line voltage is usually stepped-up and then rectified. Vacuum tubes require higher DC voltages than transistors. As you can see, the transformer can be used to make the line voltage compatible with both types of equipment.

Transformers are used as impedance-matching devices. They can match the impedance of one circuit to that of another. The impedances must be matched for maximum power to be transferred from one circuit to another circuit.

Each of these are basic applications of a transformer. However, there are additional applications which should be discussed.

Phase Shifting

Depending on how the transformer is wound, it provides either a 180 degree phase shift or zero phase shift. This means that the voltage in the secondary is either in-phase or 180 degrees out-of-phase with the voltage in the primary.

In some applications the phase shift is unimportant, such as when AC is rectified to DC. In other applications it is extremely important. Figure 17-11A shows a transformer in which the input signal is in-phase with the output signal.

AC ELECTRONICS

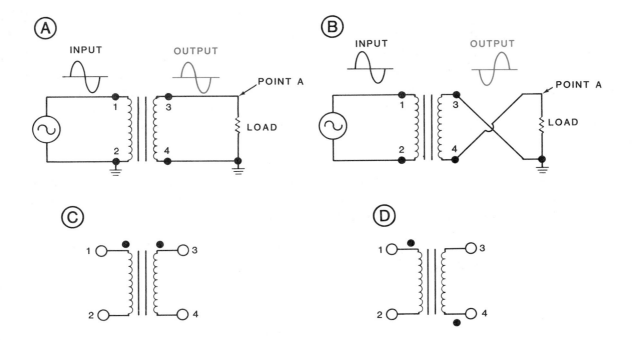

Figure 17-11 Phase relationships in transformers.

The voltage at point A with respect to ground has the same phase as the voltage at pin 1 with respect to ground. When a 180 degrees phase shift between input and output is needed, you can reverse the secondary's leads when connecting them to the load. Figure 17-11B shows the leads reversed. Notice that this places ground at pin 3 of the transformer. The voltage at point A is now 180 degrees out-of-phase with the input voltage.

The phase relationship between the windings of the transformer are indicated on schematic diagrams by dots as shown in Figure 17-11C. The ends of the windings marked by dots are in-phase. Thus, the voltage at pin 3 is in-phase with the voltage at pin 1. Figure 17-11D shows a transformer that is wound differently. Here the voltage at pin 4 is in-phase with the voltage at pin 1. Stated another way, the voltage at pin 3 is 180 degrees out-of-phase with the voltage at pin 1.

Phase Splitting

Some circuits require two AC signals of equal amplitude, but 180 degrees out-of-phase. A transformer can be used to provide the two out-of-phase signals. Figure 17-12 shows a transformer where the secondary has a **center-tap**. This means that the center of the secondary is connected to a terminal—in this case the terminal is labeled pin 4 and it is grounded. The dots indicate that the voltage at pin 3 is in-phase with the voltage at pin 1. Ignoring the center-tap, the voltage at pin 5 must be 180 degrees out-of-phase with the voltage at pins 1 and 3. When the center-tap is grounded, two signals which are equal in amplitude and 180 degrees out-of-phase exist at the opposite ends of the secondary winding.

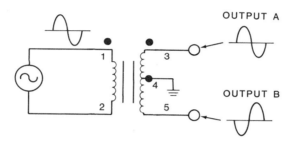

Figure 17-12 A center-tapped secondary produces two signals which are 180 degrees out-of-phase.

Isolation

Another purpose of a transformer is to isolate one circuit from another. An AC device that does not use a power transformer often has a metal chassis that connects to one side of the AC power line. Anyone touching this chassis and ground at the same time can receive an electrical shock. When a power transformer is used, the chassis is isolated from the input AC line. Separating the circuit from the input power greatly reduces the possibility of accidental shock.

Autotransformer

The **autotransformer** is a special type of transformer. In the autotransformer there is no isolation between the primary and the secondary windings. A single continuous coil is wound on a core. Part of this coil is used as the primary, while another part is used as the secondary. Usually, part of the coil is used as both primary and secondary.

AC ELECTRONICS

Figure 17-13A shows the autotransformer being used to step-down the applied voltage. Here, the entire winding serves as the primary. The lower half of the coil is also used as the secondary winding. Because there are fewer turns in the secondary than in the primary, the voltage is stepped-down and the current is stepped-up.

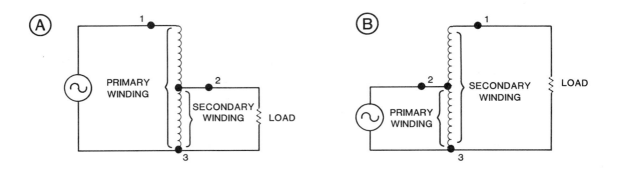

Figure 17-13 The autotransformer.

Figure 17-13B shows that the transformer can be turned around and used to step-up voltage. Here, the lower half of the coil is used as the primary while the entire coil is used as the secondary. Since the secondary has more turns, the transformer steps-up voltage and steps-down current.

Figure 17-14 compares the autotransformer with a conventional transformer. The conventional transformer in Figure 17-14A steps the applied voltage down from 120 VAC to 20 VAC. This requires a turn ratio of 6:1. Ignoring losses the current is stepped-up from 1 amp to 6 amps.

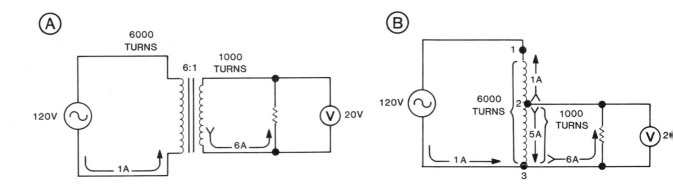

Figure 17-14 Comparing the autotransformer to a conventional transformer.

Figure 17-14B shows how the same job can be accomplished with an autotransformer. The tap at pin 2 is placed at 1000 turns. Thus, the primary, in this case the entire coil, has 6000 turns. Only 1000 of the turns are used for the secondary. Notice that 6 amperes flow in the load, but only 5 amperes flow through the secondary winding. The reason for this is that the current in the secondary winding flows in the opposite direction to the current flowing in the primary. Thus, the 1 ampere primary current subtracts from the 6 amperes of secondary current.

This illustrates the advantages of the autotransformer. First notice that fewer turns of wire are required in the autotransformer. Also, since the current in the secondary winding is lower, the copper loss is lower. In many cases, the autotransformer is also easier to construct and therefore less expensive. Its main disadvantage is that the secondary is not isolated from the primary. It is usually used in electronic circuits that have low current and high frequency requirements.

A special type of autotransformer is shown in Figure 17-15. The load is connected between the movable arm and the bottom of the coil. By moving the arm up or down the turns ratio can be changed. This causes a corresponding change in voltage across the load. The output voltage can be varied from almost zero to over 130 VAC. This device is called a **variable** autotransformer.

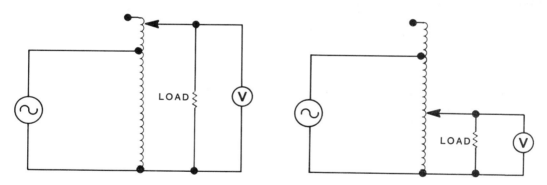

Figure 17-15 The variable transformer.

The autotransformer gets its name because it is usually used in the audio frequency range. The audio frequency range is considered to extend from approximately 10 hertz to 25,000 hertz.

Because of their inductive reactance, transformers see limited use in high frequency circuits. An exception is the special tuned tanks used in IF and RF applications. These special signal coupling transformers are used in extremely low current applications. This coupling requirement is usually in the microampere range.

AC ELECTRONICS

Chapter Self-Test

1. What are the two windings of a transformer called?

2. Which type of transformer—iron-core or air-core—is designed to work at low frequencies?

3. Ignoring losses, the voltage ratio between the primary and secondary is directly proportional to what?

4. What is the relationship between primary power and secondary power in a transformer?

5. What is meant by the coefficient of coupling?

6. List the 4 common losses associated with transformers and explain the main cause of each.

7. List 4 common applications for transformers.

8. Why is the efficiency of a transformer always less than one?

9. Name the application in which polarity inversion between the primary and secondary is not a consideration.

10. Explain why a transformer is considered a safety device.

11. What is a center-tap, and what function does it perform?

TRANSFORMERS

Summary

The transformer is a device that couples AC signals from one circuit to another.

During the transform process, the voltage and the current can be stepped-up or stepped-down. When voltage is stepped-up, current is stepped-down. When current is stepped-up, voltage is stepped down. In an ideal transformer power in the primary is equal to power in the secondary. However, frequency is never changed.

A transformer consists of two or more windings wound on a core. The winding to which an input signal is applied is called the primary. The winding from which an output signal is taken is called the secondary. The core material is usually air, steel, soft iron, or silicon steel. Metal cores have a high permeability that increases magnetic flux density.

Transformer action is based on electromagnetic mutual inductance. Current in the primary produces a magnetic field which induces an EMF into the secondary. In turn, secondary current produces a magnetic field which induces an EMF into the primary. This EMF opposes the counter EMF of the primary. Consequently, an increase in secondary current causes an increase in primary current. When the secondary is open an excitation current still flows in the primary winding.

The voltage ratio, the current ratio, and the impedance ratio of a transformer is determined by the turns ratio. The turns ratio is the number of turns in the primary compared to the number of turns in the secondary. Voltage and impedance are directly proportional to the turns ratio and current is inversely proportional to the turns ratio.

The formulas using turns ratios assumes that the transformer is ideal (has no losses). This is the same as assuming that the transformer is 100% efficient. Since all transformers have losses, their efficiencies are less than a 100%. However, 90% efficient is considered to be the norm.

A transformer can be used to match the impedance of one circuit to that of another. This is important since maximum power is transferred only when the impedance of the source matches that of the load.

Transformer losses consist primarily of core losses and copper losses. Core losses are caused by eddy currents and hysteresis.

Many transformer cores are laminated to reduce eddy current losses. Copper losses are caused by the AC resistance of the transformer's windings. Hysteresis losses are kept to a minimum by using small values of current.

Transformers are also used to provide a 180 degree phase shift between two signals. To isolate one circuit from another, to isolate an operator from the line voltage, and to block DC while passing AC signals.

An autotransformer is a special type of transformer which consists of a continuous winding on a core. It may be used like any other transformer except that it does not provide isolation between the input and output. Autotransformers are usually used in the audio frequency range which is between 10 hertz and 25 kilohertz.

An isolation transformer is a special purpose transformer that is used to isolate one circuit from another. Frequently, a 1:1 isolation transformer is used by technicians to isolate a transformerless chassis from the AC line.

THE PHYSICS OF SEMICONDUCTORS
JUNCTION DIODES
ZENER DIODES
BIPOLAR TRANSISTORS
FIELD EFFECT TRANSISTORS
CONTROL DEVICES
LIGHT-SENSITIVE DEVICES

SECTION III

The largest advances in electronics technology are the result of one simple action—the ability of certain elements to both oppose *and* conduct current. You've learned in previous chapters how materials known as semiconductors are in a special category. The third section of this book deals with that special category. We'll begin by describing how electrons move within these materials, and how these materials can be altered through the injection of impurities. From there, you will learn about a number of electronic components and how they operate within an electrical circuit. You'll learn such components as junction diodes, zener diodes, bipolar transistors, field effect transistors, SCRs, UJTs, TRIACs, and photocells. The one thing all of these components have in common is that they are built through the knowledge of how a semiconductor material operates. If some of these components already sound familiar to you, it's because their use is very widespread throughout the electronics careers.

SEMICONDUCTORS

CHAPTER 18

The Physics of Semiconductors

SEMICONDUCTORS

Contents

Introduction 439

Chapter Objectives 440

Semiconductor Materials 441

Semiconductor Atoms and Crystals 444

Conduction in Intrinsic Germanium and Silicon 447

Conduction in Doped Germanium and Silicon 450

Chapter Self-Test 454

Summary 455

THE PHYSICS OF SEMICONDUCTORS

Introduction

Throughout our discussions about conductors and insulators, we've stated that those elements with four valence electrons are neither good conductors nor good insulators. Elements of this type fall into a special category known as **semiconductors**.

The third section of this book contains six chapters. Each chapter discusses those components that function through the unique abilities of semiconductors. This chapter introduces you to the atomic structure of semiconductors, and how semiconductors operate. This chapter also covers methods of changing the structure of semiconductor materials, and how these changed materials operate. The information in this chapter is vital in understanding how semiconductor components operate.

SEMICONDUCTORS

Chapter Objectives

When you have completed this chapter, you will be able to:

1. Identify three common semiconductor devices.

2. Describe some of the uses of semiconductor devices in electronic equipment.

3. List at least five advantages semiconductor devices have over other components having similar capabilities.

4. Identify the two most commonly used semiconductor materials and identify their majority and minority current carriers.

5. Describe the crystal lattice structure of semiconductors.

6. Define the electrical characteristics of semiconductors.

7. Describe the difference between intrinsic and doped semiconductors.

8. Define the term "hole" as applied to semiconductors.

9. Recognize the difference between majority carriers and minority carriers in a doped semiconductor.

10. Explain the terms trivalent and pentavalent.

Semiconductor Materials

The term **semiconductor** is used to describe any material that has characteristics which fall between those of insulators and conductors. In other words, a semiconductor will not pass current as readily as a conductor, nor will it block current as effectively as an insulator. Some semiconductor materials are actually pure elements which are found in the periodic table of elements, while other semiconductors may be classified as compounds.

The semiconductor elements that are suited to the greatest variety of electronic applications are **germanium** and **silicon**. Germanium is a brittle, grayish-white element that was discovered in 1886. This material is generally processed from the ash of certain types of coals, in the form of germanium dioxide powder. This powder is then reduced to pure germanium in solid form.

Silicon is a non-metallic element which was discovered in 1823. This material is found extensively in the earth's crust. A white or sometimes colorless compound known as silicon dioxide (also called silica) occurs abundantly in sand, quartz, agate, and flint. These silicon compounds can be chemically reduced to obtain pure silicon in a solid form. These two materials have atomic structures which may be easily altered to obtain specific electrical characteristics. Once the pure material is available, it must be suitably modified to give it the qualities necessary to construct a semiconductor device for a specific application.

Semiconductor Components

Semiconductors are the basic building materials which are used to construct some very important electronic components. These semiconductor components are in turn used to construct electronic circuits and equipment. The three most commonly used semiconductor devices are **diodes, transistors,** and **integrated circuits**. However, other special components are also available.

The primary function of semiconductor devices is to control currents or voltages in such a way as to produce a desired end result. For example, diodes can be used as rectifiers to produce pulsating DC from AC. A transistor can be used as a variable resistance to vary the current in a heating element. Or an integrated circuit can be used to amplify and demodulate a radio signal.

SEMICONDUCTORS

Semiconductor devices are extremely small, lightweight components which consume only a small amount of power, and are highly efficient and reliable. The vacuum tubes that were once widely used in electronic equipment have been almost completely replaced by the newer and better semiconductor devices. Let's consider some of the specific reasons for this significant transition from the use of vacuum tubes to semiconductor components in electronic equipment.

Advantages

Components which are made of semiconductor materials are often referred to as **solid-state components** because they are made from solid materials. Because of this solid-state construction, these components are more rugged than vacuum tubes, which are made of a combination of glass, metal, and ceramic materials. Because of this ruggedness, semiconductor devices are able to operate under extremely hazardous environmental conditions. This ruggedness is responsible for the reliability of solid-state devices.

The solid-state construction also eliminates the need for filaments or heaters, found in all vacuum tubes. This means that additional power is not required to operate the filaments, making component operation cooler and more efficient. By eliminating the filaments, a prime source of trouble is also avoided because the filaments generally have a limited life expectancy. The absence of filaments also means that a warm-up period is not required before the device can operate properly. In other words, the solid-state component operates the instant it receives electrical power.

Solid-state components are also able to operate with very low voltages (between 1 and 25 volts), while vacuum tubes usually require an operating voltage of 100 volts or more. This means that most solid-state components use less power than vacuum tubes, which makes them more suitable for use in portable, battery-powered equipment. The lower voltages are also much safer. Pocket-size radios, hand-held calculators, and small television receivers are typical examples of devices which take advantage of highly efficient, power saving components.

The small size of the solid-state component also makes it suitable for use in portable electronic equipment. Although equipment of this type can be constructed with vacuum tubes, such equipment would be much larger and heavier. A typical transistor is only a fraction of an inch high and wide, while a vacuum tube of comparable performance may be an inch or more wide and several inches high. This smaller size also means a significant weight savings.

Solid-state components are much less expensive than comparable vacuum tube components. The very nature of a solid-state component makes it suitable for production in mass quantities which reduces cost. In fact, large numbers of solid-state components can be constructed as easily and quickly as a single component.

The most sophisticated semiconductor devices are integrated circuits, or IC's. These are complete circuits where all of the components are constructed with semiconductor materials in a single miniature package. These devices not only replace individual electronic circuits, but also complete pieces of equipment or entire systems. Entire computers and radio receivers can be constructed as a single device no larger than a typical transistor. Integrated circuits have taken us one step farther in improving electronic equipment through the use of semiconductor materials. All electronic equipment has benefited from solid-state components and particularly from the development of integrated circuits.

Disadvantages

Although solid-state components have many advantages over the vacuum tubes that were once widely used, they also have several inherent disadvantages. First, solid-state components are highly susceptible to changes in temperature and can be damaged if they are operated at extremely high temperatures. Additional components are often required simply for the purpose of stabilizing solid-state circuits so that they will operate over a wide temperature range. Solid-state components may be easily damaged by exceeding their power dissipation limits, or possibly when their normal operating voltages are reversed. In comparison, vacuum tube components are not nearly as sensitive to temperature changes or improper operating voltage.

There are still a few areas where semiconductor devices cannot replace tubes. This is particularly true in high power, ultra high radio frequency applications. However, as semiconductor technology develops, these limitations are being overcome.

Despite the several disadvantages just mentioned, solid-state components are still the most efficient and reliable devices to be found. They are used in all new equipment designs and new applications are constantly being found for these devices in the military, industrial, and consumer fields.

Semiconductors have had a profound effect on the design and application of electronic equipment. Not only have they greatly improved existing equipment and techniques, but they have permitted us to do things that were not previously possible. Semiconductors have revolutionized the electronic industry and continue to show even greater potential. Your work in electronics will always involve semiconductor devices.

SEMICONDUCTORS

Semiconductor Atoms and Crystals

From your previous studies, you know that metals such as copper and aluminum are used to carry current in an electrical circuit. You learned that these metals are classified as conductors because they offer minimum opposition to current flow. You also learned that materials such as glass, rubber, and ceramic opposes the flow of electrical current and are classified as insulators. You will now examine the atomic structure of germanium and silicon, the two most commonly used semiconductor elements.

Semiconductor Atoms

Before we examine the structure of germanium and silicon, we should review some important rules which pertain to the number and placement of the electrons that revolve around the nucleus of all atoms.

Atoms contain three basic components: protons, neutrons, and electrons. The protons and neutrons are located in the nucleus or center of the atom, while the electrons revolve around the nucleus in orbits. The atom of each particular element will have a specific number of protons in its nucleus and an equal number of electrons, in orbit, if the atom is neutral (has no charge).

The exact manner in which the electrons are arranged around the nucleus is extremely important in determining the electrical characteristics of the element. Generally, each electron has its own orbit, but certain orbits are grouped together to form a shell. For all of the elements that are known to exist, there can only be a maximum of seven shells.

The outermost shell of a particular atom is called the **valence shell** and the electrons that orbit within this shell are referred to as **valence electrons**.

In any particular atom, the outer shell can never hold more than 8 electrons. When exactly 8 electrons are present in the outer shell, the atom is considered to be completely stable and it will neither give up or accept electrons easily. Elements which have atoms of this type are neon and argon. These elements are classified as inert gases and they resist any sort of electrical or chemical activity.

When an atom has 5 or more electrons in its outer shell, it tries to fill its shell so that it can reach a stable condition. Elements of this type make good insulators, because the individual atoms try to acquire electrons instead of giving them up. Therefore, the free movement of electrons from one atom to the next is inhibited. When an atom has less than 4 electrons in its outer shell it tends to give up these valence electrons easily. Elements which have atoms of this type make good conductors because they contain a large number of free-moving electrons, which can randomly drift from one atom to the next.

THE PHYSICS OF SEMICONDUCTORS

When an atom contains exactly 4 electrons in its outer shell, it does not readily give up or accept electrons. Elements which contain atoms of this type do not make good insulators or conductors and are, therefore, referred to as semiconductors. The element carbon is a typical example of a semiconductor material.

The two semiconductor materials most commonly used in the construction of semiconductor components are germanium and silicon. Both of these materials are made up of atoms which have 4 electrons in their outermost or valence shells. A single germanium atom is shown in Figure 18-1A. Notice that the germanium atom has four shells and the distribution of the electrons from the first shell to the outer shell is 2, 8, 18 and 4. Therefore, a total of 32 electrons rotate around the nucleus of the atom, with 32 protons within the nucleus.

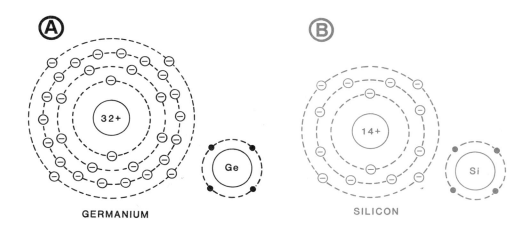

Figure 18-1 Complex and simplified atom diagrams for germanium and silicon.

A single silicon atom is shown in Figure 18-1B. Notice that this atom has only three shells and the distribution of the electrons from the first shell to the valence shell is 2, 8, and 4. The atom has a total of 14 electrons revolving around its nucleus, with 14 protons within its nucleus.

To simplify the discussions and illustrations which follow, we will use diagrams showing only the four valence electrons surrounding a central core. These simplified diagrams are shown next to the complete versions for each element. Notice that the core of the germanium atom is identified by the symbol Ge and the core of the silicon atom is identified by the symbol Si. In each case the nucleus is surrounded by four electrons.

Semiconductor Crystals

The individual atoms within a semiconductor material such as germanium are arranged as shown in Figure 18-2. Each atom shares its four valence electrons with four neighboring atoms as shown. This sharing of electrons creates a bond which holds the atoms together. This electron-pair bond is commonly referred to as a **covalent bond**, and it occurs because each of the atoms in the structure tries to take on additional electrons in order to fill its valence shell with a total of eight electrons. The end result is a structure which has a lattice-like appearance and is often referred to as a **crystal lattice**. Silicon atoms combine in the same manner as germanium atoms to form the same type of crystalline structure. The silicon atoms maintain covalent bonds just like the germanium atoms.

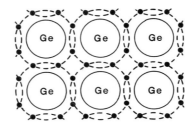

Figure 18-2 Simplified diagram of germanium crystal structure.

The germanium and silicon crystal lattices just described are free from impurities and therefore represent pure or ideal materials. Such crystals are often referred to as **intrinsic materials**. The construction of solid-state components such as transistors depends on the use of these pure or intrinsic semiconductor materials.

Conduction in Intrinsic Germanium and Silicon

Due to the crystalline structure of pure semiconductor materials such as germanium and silicon, each nucleus within the material sees eight valence electrons even though each atom actually has only four. Therefore, each atom tends to be stable and will not easily give up or accept electrons. However, this does not mean that pure semiconductors must, under all conditions, resist any sort of electrical activity. The reason for this is that another factor—*temperature*—must be considered.

Low Temperature Characteristics

At extremely low temperatures, the valence electrons are held tightly to their parent atoms (in covalent bonds), and are not allowed to drift through the crystalline structures of either semiconductor material. Since the valence electrons cannot drift from one atom to the next, the material cannot support current flow at this time. Therefore, at extremely low temperatures, pure germanium and silicon crystals function as insulators.

High Temperature Characteristics

As the temperature of a germanium or silicon crystal is increased, the valence electrons within the material become agitated and some of them will occasionally break away from the covalent bonds. Therefore, a small number of electrons will be free to drift from one atom to the next in a random manner. These free-moving electrons or free electrons are able to support a small amount of electrical current if a voltage is applied to the semiconductor material. In other words, as the temperature of the semiconductor material increases, the material begins to acquire the characteristics of a conductor. For all practical purposes, however, enough heat energy is available even at room temperature to produce a small number of free electrons which can support a small amount of current. Only when the semiconductor materials are exposed to extremely high temperatures, can a point be reached where they will conduct current as well as an ordinary conductor.

Holes

To understand exactly why a semiconductor is able to allow current to flow, we must take a closer look at the internal structure of the material. When an electron breaks away from a covalent bond, an open space or vacancy exists in the bond. The space that was previously occupied by the electron is generally referred to as a hole. A hole simply represents the absence of an electron. Since an electron has a negative charge, the hole represents the absence of a negative charge. This means that the hole has the characteristics of a positively charged particle. Each time an electron breaks away from a covalent bond, a hole is created. Each corresponding electron and hole is referred to as an electron-hole pair. A typical electron-hole pair is shown in Figure 18-3. Notice that the hole is represented by a plus sign, indicating a positive charge. The electron is simply shown as a dot although it does have a negative charge. The semiconductor material shown could be either germanium or silicon as indicated.

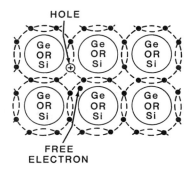

Figure 18-3 An electron-hole pair in a semiconductor material.

The number of electron-hole pairs produced within a semiconductor material increases with temperature. Even at room temperature, a small number of electron-hole pairs exist. Some of the free electrons drift randomly, and holes absorb some of these electrons. This means that some electrons will simply jump from one shell to another shell containing a hole. If an electron jumps from one shell to fill in a hole, another hole is created where the electron leaves the shell. The hole, therefore, appears to move in the *opposite direction* of the electron. If another electron moves into the hole that was just created, another hole is produced, and the previous hole appears to move randomly through a pure semiconductor material. Thus, the terms hole flow and electron flow.

THE PHYSICS OF SEMICONDUCTORS

Current Flow

When a pure semiconductor material such as germanium or silicon is subjected to a voltage as shown in Figure 18-4, the negatively charged free electrons are attracted to the positive terminal of the voltage source. The positive holes that are created by the free electrons drift toward the negative terminal of the voltage source. As the free electrons flow into the positive terminal of the voltage source, an equal number of electrons leave the negative terminal of the voltage source. These electrons are injected into the left side of the semiconductor material where many of these electrons are captured or absorbed by holes. As the holes and electrons recombine in this manner, the holes cease to exist. Therefore the holes constantly drift to the left and disappear, while the electrons flow to the right where they are drawn out of the material and into the positive terminal of the voltage source.

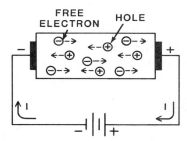

Figure 18-4 Current flow in semiconductor material.

It is important to remember that current flow in a semiconductor material consists of both electrons and holes. The holes function like positively charged particles while the electrons are actually negatively charged particles. The holes and electrons flow in opposite directions and the number of electron-hole pairs produced within a material increase as the temperature of the material increases. Since the amount of current flowing in a semiconductor is determined by the number of electron-hole pairs in the material, the ability of a semiconductor material to pass current increases as the temperature of the material increases.

It is also important to note that current flow in a semiconductor is somewhat different than current flow in a conductor. When we consider current flow in a semiconductor, we must consider the movement of holes as well as electrons. However, in a conductor we are concerned only with the number of free-electrons that are available.

Conduction in Doped Germanium and Silicon

Pure semiconductor material contains only a small number of electrons and holes at room temperature and therefore conduct very little current. However, the conductivity of these materials can be increased considerably by a process known as doping. Pure semiconductor materials such as germanium and silicon are doped by adding other materials, called impurities, to them when they are produced. Basically there are two types of impurities that are added to germanium and silicon crystals. One type of impurity is referred to as a pentavalent material because it is made up of atoms which have five valence electrons. The second type of impurity is referred to as a trivalent material because each of the atoms in this material has three valence electrons.

N-Type Semiconductors

When a pure semiconductor material is doped with a pentavalent element, such as arsenic (As), some of the atoms in the crystal lattice structure of the semiconductor are replaced by arsenic atoms. As a result, the crystal lattice of the semiconductor is like that shown in Figure 18-5. The arsenic atom has replaced one of the semiconductor atoms and is sharing four of its valence electrons with adjacent atoms in a covalent bond. However, the fifth electron is not part of a covalent bond and can be easily freed from the atom.

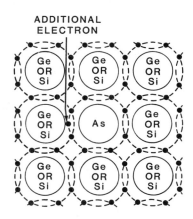

Figure 18-5 Semiconductor material doped with arsenic.

THE PHYSICS OF SEMICONDUCTORS

This extra arsenic atom is called a **donor atom** because it donates a free electron to the crystal lattice. Actually, there is a large number of donor atoms in the crystal lattice. Consequently, there are many free electrons in the semiconductor. When doping of this type occurs, **N-type material** is created. This material is referred to as N-type because of the large number of free electrons (negatively charged particles) within the structure.

If a voltage is applied to an N-type semiconductor, the free electrons contributed by the donor atoms will flow toward the positive terminal of the battery. This is shown in Figure 18-6. However, some additional free electrons will also flow toward the positive terminal. These additional free electrons are produced as electrons break away from their covalent bonds to create electron-hole pairs. This is identical to the action which takes place in a pure semiconductor material. The corresponding holes which are produced then move toward the negative terminal.

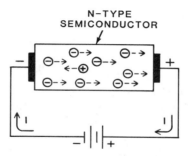

Figure 18-6 Conduction in an N-type semiconductor.

At normal room temperature the number of free electrons provided by the donor atoms will greatly exceed the number of holes and electrons that are produced by the breaking of covalent bonds. This means that the number of electrons flowing in the N-type semiconductor will greatly exceed the number of holes. The electrons, being in the majority, are therefore referred to as **majority carriers** while the holes, which are in the minority, are referred to as **minority carriers**.

P-Type Semiconductors

Doping pure semiconductor material with a trivalent element, such as gallium (Ga), causes some of the semiconductor atoms to be displaced by trivalent atoms. Each trivalent atom shares its valence electrons with three adjacent atoms in the semiconductor crystal lattice structure. However, the fourth atom does not share a covalent bond because of the missing electron. This results in a hole in the crystal lattice. This is shown in Figure 18-7.

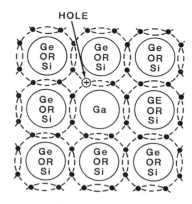

Figure 18-7 Semiconductor material doped with gallium.

A large number of holes are present in the semiconductor because many trivalent atoms have been added. These holes readily accept electrons from other atoms. Because of this, these trivalent atoms are referred to as **acceptor atoms**.

When a given hole is filled by an electron from another atom, the electron leaves another hole. The holes drift from one covalent bond to another in the direction opposite to that of electron movement. Consequently, holes behave like positively charged particles.

This type of doping causes the intrinsic semiconductor material to be referred to as **P-type material**. It is called P-type material because of the existence of a large number of holes which, as stated earlier, act as positively charged particles.

If a voltage is applied to a P-type semiconductor as shown in Figure 18-8, the holes provided by the acceptor atoms move from the positive to the negative terminal. In other words, as each electron moves into a hole, another hole is created. Since electrons move toward the positive terminal, the holes move in the opposite direction toward the negative terminal.

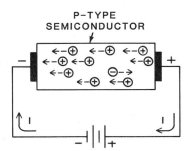

Figure 18-8 Conduction in a P-type semiconductor.

In addition to the holes provided by the acceptor atoms, many additional holes are also found in the P-type semiconductor material. These holes are produced as electrons break away from covalent bonds to create electron-hole pairs. These additional holes are also attracted toward the negative terminal, while the corresponding electrons that are produced are attracted toward the positive terminal.

Under normal conditions, the number of holes provided by the acceptor atoms greatly exceed the number of holes and electrons produced by the breaking of covalent bonds. The number of holes flowing in the P-type semiconductor will therefore greatly exceed the number of free electrons in the material. The holes are the majority carriers and the electrons are the minority carriers in P-type material.

While doping creates these materials, it is important to note that these materials *are not charged.* N-type and P-type materials are electrically neutral because each atom contains an equal number of protons and neutrons. In each material, there exists an equal number of negatively and positively charged particles. For example, if a conductor were placed across P-type and N-type materials, there would not be current flow. Notice that when we described current flow in these materials, a battery had to be connected.

Whenever intrinsic material is doped, P-type or N-type material is created. This doping can be either heavy or light. In other words, the amount of impurities injected into intrinsic semiconductor material can vary. The heavier the injection of impurities, the greater the number of majority and minority carriers. When semiconductor components are created, the amount of doping is extremely important. You will find that changing the amount of doping changes the characteristics of a component.

SEMICONDUCTORS

Chapter Self-Test

1. List four advantages of semiconductor components over vacuum tubes.

2. List the three most common semiconductor components.

3. What are the two semiconductor materials most commonly used to manufacture electronic components?

4. If a semiconductor device is made directly from pure semiconductor materials, have the materials been altered in any way?

5. Why does the nucleus within a semiconductor atom "see" eight valence electrons, although it actually contains only four? What is this known as?

6. Describe a crystal lattice.

7. What is the difference between intrinsic semiconductor materials at *low* temperatures, and intrinsic semiconductor materials at *high* temperatures?

8. What is created when an electron breaks away from a semiconductor atom?

9. Holes move in the same direction as the electrons—true or false?

10. What happens to the number of electron-hole pairs within a semiconductor as temperature increases?

11. How is a pure semiconductor doped?

12. An impurity material which is made up of atoms that have five valence electrons is called what?

13. What are the majority carriers in an N-type semiconductor?

THE PHYSICS OF SEMICONDUCTORS

Summary

Semiconductor elements have four valence electrons. The two semiconductor materials used in the manufacturing of solid-state components are the elements silicon and germanium. Silicon is more commonly used.

It is a characteristic of semiconductor atoms to readily share valence electrons with other atoms. This sharing is called a covalent bond. These covalent bonds cause the atoms to form a crystal lattice.

Pure (intrinsic) semiconductors function as insulators at low temperatures because valence electrons are held tightly in their shells. However, as temperature increases, valence electrons are able to break their covalent bonds and become free electrons, leaving holes in the crystal lattice. If a voltage is applied under these conditions, electrons can move in one direction while holes move in the opposite direction, allowing the semiconductor to act as a fairly good conductor.

Intrinsic semiconductors have limited use in electronics. However, they can be modified, by doping, to meet specific electrical requirements. Doping is the deliberate addition of an impurity material to the semiconductor element.

Doping a semiconductor with a pentavalent element (such as arsenic) adds a large number of free electrons to the crystal lattice. These electrons freely move from atom to atom in one direction within the crystal lattice when a voltage is applied.

Hole movement is also present in the crystal lattice, but the quantity of holes is much smaller than the quantity of electrons. For this reason electrons are the majority carriers of current and holes are the minority carriers in a semiconductor doped with pentavalent atoms. Semiconductors that are doped with pentavalent atoms are N-type semiconductors.

Doping a semiconductor with a trivalent element (such as gallium) produces a P-type semiconductor. The valence electrons in the outer shell of each trivalent atom form covalent bonds with three of the four adjacent tetravalent atoms in the crystal lattice. However, the absence of a fourth valence electron in the trivalent atom causes a hole in the crystal lattice. The large number of trivalent atoms in the crystal lattice causes a large number of holes, with each hole capable of accepting a free electron. However, the number of holes greatly outnumbers the free electrons. Consequently, holes are the majority current carriers and electrons are the minority carriers in P-type semiconductors.

SEMICONDUCTORS

CHAPTER 19

Junction Diodes

SEMICONDUCTORS

Contents

Introduction 459

Chapter Objectives 460

The PN Junction 461

Diode Biasing 464

Diode Characteristics 470

Chapter Self-Test 477

Summary 478

JUNCTION DIODES

Introduction

In the previous chapter, you examined the semiconductor materials that are used to construct various types of solid-state components. Now you will learn how these materials are used to construct one of the most important types of solid-state components, the semiconductor diode. Although simple in construction and operation, semiconductor diodes are widely used in many types of electronic equipment. Rectification, waveshaping, circuit protection, and logic operations are only a few of its diversified applications. This chapter is devoted to the principles of the PN junction diode.

Since transistors, integrated circuits, and other solid-state components are constructed in basically the same manner as diodes, an understanding of diode construction and operations is an essential first-step in understanding semiconductor devices.

SEMICONDUCTORS

Chapter Objectives

When you have completed this chapter, you will be able to:

1. Describe the construction of a semiconductor diode's PN junction.

2. Recognize the schematic symbol of a semiconductor diode.

3. Describe the electrical characteristics of a PN junction.

4. Identify the two parts of a diode on the schematic symbol.

6. Describe the effects of forward- and reverse-bias on a junction diode.

7. Interpret a graph of diode voltage-current characteristics.

JUNCTION DIODES

The PN Junction

Even though N-type and P-type semiconductors are electrically neutral, independent electrical charges still exist within the material. The free electrons and holes which drift throughout the semiconductors possess negative and positive charges, respectively. Since these charges are able to move, they are referred to as **mobile charges**.

But these free electrons and holes are not the only charges that exist within a semiconductor. Many of the atoms within the crystal lattice are also charged. The atoms within a semiconductor which contain extra electrons are negative ions, possessing a negative charge. Also, atoms within the semiconductor that have a deficiency of electrons are positive ions and carry a positive charge. Since these ions are within the crystal lattice structure, they are considered immobile. Within an N-type or P-type semiconductor, an equal number of mobile charges and ionic charges will exist. Since these charges are equal and opposite, the semiconductor material remains electrically neutral.

The internal structure of N-type and P-type semiconductors may be illustrated in a simplified manner as shown in Figure 19-1. The N-type semiconductor contains many donor atoms which contribute free electrons that drift through the material. The donor atoms take on positive charges and become positive ions when they release free electrons. These are represented by plus signs surrounded by small circles as shown. The free electrons which accompany the donor atoms are represented by minus signs.

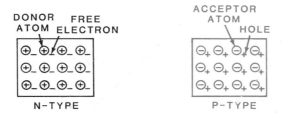

Figure 19-1 Doped semiconductors.

SEMICONDUCTORS

The P-type semiconductor contains many acceptor atoms. These atoms easily accept or absorb electrons from the semiconductor material and become negative ions. These negative ions are represented by minus signs that are surrounded by small circles as shown. The holes created by these ions act like positively charged particles and are represented by plus signs as indicated. By representing the doped semiconductors in this manner, it becomes relatively easy to analyze the action which takes place when they are combined to form diodes and other types of solid-state components.

Junction Diodes

Now let's consider the action which takes place when doped semiconductors are combined to form a diode. Basically, a diode is created by joining N-type and P-type semiconductors as shown in Figure 19-2. When these oppositely doped materials come in contact, a junction is formed where they meet. Such diodes are referred to as junction diodes and can be made of either silicon or germanium.

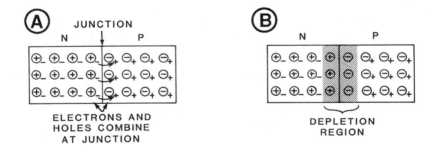

Figure 19-2 Characteristics of a PN junction.

When the junction is formed, a unique action takes place. The mobile charges in the vicinity of the junction are strongly attracted to each other and therefore drift toward the junction. The accumulated charges at each side of the junction serve to *increase* the attraction. Eventually, some of the free electrons move across the junction and fill some of the holes in the P-type material. This is shown in Figure 19-2A.

As the free electrons cross the junction, the N-type material becomes depleted of electrons in the vicinity of the junction. At the same time, the holes within the P-type material become filled and no longer exist. This means that the P-type material also becomes depleted of holes near the junction. This region near the junction where the electrons and holes become depleted is referred to as the depletion region. The depletion region extends for only a short distance on each side of the junction as shown in Figure 19-2B.

It is important to remember that free electrons and holes are actually the majority carriers for the N-type and P-type materials. Therefore, no majority carriers exist within the depletion region. Also, it is important to note that the N-type and P-type materials are no longer neutral or uncharged. In other words, the N-type material has lost free electrons, causing the positive donor atoms to outnumber the free electrons. This causes the N-type material to take on a positive charge near the junction. Meanwhile, the P-type material has lost holes, allowing the negatively charges acceptor atoms to outnumber the holes. So the P-type material takes on a *negative* charge. From all this, opposite charges now exist on each side of the junction.

The depletion region does not continue to become larger and larger until the N-type and P-type materials are completely depleted of majority carriers. Instead, the action of the electrons and holes combining at the junction tapers off very quickly. Therefore, the depletion area remains relatively small. The size of the depletion region is limited by the opposite charges which build up on each side of the junction. The negative charge which accumulates in the P-type material eventually becomes great enough to repel the free electrons and prevent them from crossing the junction. The positive charge which accumulates in the N-type material also helps to stop the flow of free electrons by attracting and holding them so that they cannot move across the junction.

Barrier Voltage

The opposite charges that build up on each side of the junction create a potential difference which limits the size of the depletion region by preventing the further combination of electrons and holes. It is referred to as the **potential barrier** or the **barrier voltage**.

The barrier voltage produced within a PN junction will usually be somewhere around several tenths of a volt. This voltage will always be higher for silicon PN junctions than for germanium PN junctions. For example, a PN junction made from doped germanium will have a typical barrier voltage of 0.3 volts, while a PN junction made from doped silicon will have a typical barrier voltage of 0.7 volts. The barrier voltage exists inside the junction diode and therefore cannot be measured directly, but its presence becomes apparent when an external voltage is applied to the diode.

Diode Biasing

Whenever diodes are used in electrical or electronic circuits, they are subjected to a variety of voltages and currents. The polarities and amplitudes of the voltages and currents must be such that proper diode action takes place. We generally refer to the voltages applied to a semiconductor diode as **bias voltages**. Let's see how these voltages affect and control the diode's operation.

Forward-Bias

In the previous discussion, you saw that the majority carriers combine at the junction to produce a depletion region. The depletion region represents an area that is void of majority carriers, but at the same time contains a number of positively charged donor atoms (ions) and negatively charged acceptor atoms (ions). These positive and negative charges are separated at the junction and effectively create a barrier voltage which opposes any further combination of majority carriers. It is important to remember that this action takes place in a PN junction diode that is not subjected to any external voltage.

When a PN junction diode is subjected to a sufficiently high external voltage, as shown in Figure 19-3, the device will function in a somewhat different manner. Notice that the negative and positive terminals of a battery are connected to the N and P sections of the diode respectively. An external resistor R is used to limit the current level to a safe value.

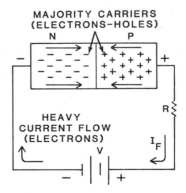

Figure 19-3 Forward-biased PN junction diode.

JUNCTION DIODES

Under these conditions, the free electrons in the N section are repelled by the negative battery terminal and forced toward the junction. There, they will neutralize the positively charged donor atoms (positive ions) in the depletion region. During this same period of time, the free electrons that had initially accumulated to create a negative charge on the P side of the junction are attracted toward the positive battery terminal. Therefore the negative charge on the P side of the junction is also neutralized. This means that the positive and negative charges which form the internal barrier voltage will be present to stop the combining of majority carriers at the junction. The PN junction diode is therefore able to support a continuous flow of current at this time. *This action will occur only if the battery voltage is greater than the barrier potential.*

Since the diode is now subjected to an external voltage, a constant supply of electrons flow into the N section of the diode. These electrons drift through the N-type material toward the junction. The movement of these electrons through the N section is sustained by the free electrons (majority carriers) that exist within this material. At the same time, the holes (majority carriers) in the P section also drift toward the junction. The electrons and holes combine at the junction and effectively disappear as they neutralize each other. However, as these electrons and holes combine and are effectively eliminated as charge carriers, new electrons and holes appear at the outer edges of the N and P sections. The majority carriers therefore continue to move toward the junction as long as the external voltage is applied. The depletion region becomes smaller since the electrons and holes neutralize each other.

It is easier to analyze the action which takes place in the P section of the diode by considering the movement of holes instead of electrons. However, it is important to realize that electrons do actually flow through the P-type material since they are attracted by the positive terminal of the battery. As the electrons leave the P section and enter the battery, holes are created at the outer edge of the P section. These holes drift toward the junction where they combine with electrons and effectively disappear.

The important point to note at this time is that electrons do flow through the entire PN junction diode when it is subjected to a sufficiently high external voltage. At this time the diode is said to be conducting current in the forward direction. Also, the diode is considered to be *forward-biased* by the external voltage. The current which flows through the forward-biased diode shown in Figure 19-3 is limited by the resistance of the P-type and N-type semiconductor materials as well as the external resistance R.

SEMICONDUCTORS

Normally the diode resistance is very low. Connecting a forward bias voltage directly to the diode will result in a current large enough to generate sufficient heat to destroy the diode. For this reason, forward-biased diodes are usually connected in series with a resistor or some other device which will limit the current to a safe level.

A forward-biased diode will conduct current as long as the external bias voltage is high enough, and the polarity is correct. For example, if the diode is constructed from germanium, a forward bias of approximately 0.3 volts will be required before the diode can begin to conduct. Silicon diodes require a forward bias of approximately 0.7 volts in order to begin conducting. The external voltage applied to the diode must be large enough to neutralize the depletion area and therefore neutralize the barrier voltage that exists across the PN junction of the diode. Once this internal voltage is overcome, the diode will conduct in the forward direction.

The polarity of the external DC bias voltage must also be correct with respect to the P and N sections of the diode. The negative terminal of the bias source must be connected to the N section and the positive terminal should be connected to the P section to achieve the forward-biased condition.

A forward-biased condition occurs any time the voltage on the N-type material is negative with respect to the voltage on the P-type material. Obviously, this can occur without connecting the diode to a battery. In Figure 19-4, +2 volts is connected to the N-type material and +6 volts is connected to the P-type material. The potential difference between these two voltages is four volts. This voltage exceeds the barrier voltage, which fulfills one of the requirements for forward bias. Second, +2 volts (connected to the N-type material) is negative *with respect* to +6 volts. Therefore, the voltage on the N-type is more negative, which fulfills our second requirement for forward bias.

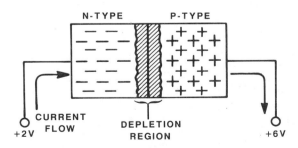

Figure 19-4 Forward biasing with two voltages of the same polarity.

JUNCTION DIODES

Once the diode is conducting a voltage will be dropped across the device. This occurs because the diode's semiconductor material has a low but finite resistance value and the current flowing through it must produce a corresponding voltage drop. As it turns out, this forward bias voltage drop is approximately equal to the barrier potential—0.3 volts for a germanium diode, and 0.7 volts for a silicon diode.

The amount of forward bias current (I_F) is a function of the applied voltage E_A, the forward voltage drop (V_F), and the external resistance R. The relationship simply involves Ohm's Law as indicated below:

$$I_F = \frac{E_A - V_F}{R}$$

For example, the forward current in a silicon diode with a bias voltage of 10 volts and an external resistor of 100 ohms is:

$$I_F = \frac{E_A - V_F}{R} = \frac{10 - 0.7}{100 \,\Omega} = \frac{9.3}{100} = .093 \text{ A or } 93 \text{ mA.}$$

Reverse-Bias

A forward-biased diode is able to conduct current in the forward direction because the external bias voltage forces the majority carriers together. These majority carriers combine at the junction of the diode and create a continuous flow of current. In order to achieve this condition, the negative terminal of the battery is connected to the N section of the diode and the positive terminal is connected to the P section. However, if the battery is reversed as shown in Figure 19-5, the diode will operate in a different manner. The negative terminal of the battery is now connected to the P section of the diode while the positive terminal is connected to the N section. The diode is now considered to be reverse-biased.

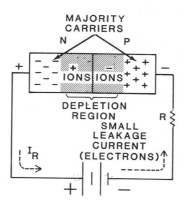

Figure 19-5 Reverse-biased PN junction diode.

SEMICONDUCTORS

Under these conditions the free electrons in the N section will be attracted toward the positive battery terminal, thus leaving a relatively large number of positively charged donor atoms (positive ions) in the vicinity of the junction as shown in Figure 19-5. In fact, the number of positive ions in the N section at times will even outnumber the positive ions that exist in an unbiased diode. This effectively increases the width of the depletion region on the N side of the junction causing the positive charge on the other side of the junction to increase.

At the same time, a number of electrons leave the negative terminal of the battery and enter the P section of the diode. These electrons fill the holes near the junction, causing the holes to *appear* as if they are moving toward the negative terminal. A large number of negatively charged acceptor atoms (negative ions) are therefore created near the junction. This effectively increases the width of the depletion region on the P side of the junction.

The overall depletion region of the diode shown in Figure 19-5 is wider than the depletion region in the unbiased diode shown earlier. This means that the opposite charges on each side of the junction are also larger and therefore create a higher barrier voltage across the junction. These opposite charges build up until the internal barrier voltage is equal and opposite to the external battery voltage. The diode *appears* to act as an open, dropping the applied voltage and not allowing current to flow. Under these conditions the holes and electrons (majority carriers) cannot support current flow and the diode effectively stops conducting.

Actually, an extremely small current will flow through the diode shown in Figure 19-5. This small current is sometimes referred to as leakage current or reverse current and is designated as I_R. It exists because of the minority carriers which are contained within the N-type and P-type materials. When the diode is reversed-biased, the minority carriers are forced toward the junction and combine. This allows them to support an extremely small current. This action closely resembles the action which takes place in the forward-biased diode, but it occurs on a much smaller scale. As temperature increases, the number of minority carriers increases, which increases leakage current.

All PN junction diodes produce a small leakage current when they are reversed-biased. This leakage current, as we stated earlier, is extremely small—normally in the microampere range for germanium diodes, and nanoamperes for silicon diodes. These currents are usually much smaller in magnitude than the usual forward current. It is important to remember that germanium diodes normally produce a higher leakage current than silicon diodes. Germanium devices are also more temperature sensitive than silicon devices. This disadvantage of germanium diodes is often offset by the desirable lower barrier potential and forward voltage drop.

JUNCTION DIODES

We can sum up the operation of a PN junction diode in this manner:

> The diode is a unidirectional electrical device since it conducts usable current in only one direction. When it is forward-biased, current flows through it freely since it acts as a very low resistance. When the diode is reverse-biased, current does not flow through it. It simply acts as an open circuit or extremely high resistance.

A junction diode can also be thought of as a polarity sensitive electrical switch. When the diode is forward-biased, the "switch" is closed, and when the diode is reverse-biased, the "switch" is open.

Diode Characteristics

Now that you have seen how a basic PN junction diode operates, you are ready to examine some of the important electrical characteristics of these devices. Since the characteristics of diodes vary considerably when they are subjected to various voltages and temperatures, it is usually best to plot the desired characteristics graphically. This makes it possible to analyze the operation of the device at a variety of points where the voltages, currents, or temperatures involved have specific and related values.

Germanium Diode

The graph in Figure 19-6 shows the amount of forward current and reverse current that flows through a typical germanium PN junction diode when the device is first forward-biased and then reverse-biased. The diode's forward and reverse bias voltages, V_F and V_R, are plotted to the right and to the left respectively on the horizontal axis of the graph. The diode's forward and reverse currents, I_F and I_R, are plotted above and below the horizontal axis respectively to form the vertical axis of the graph. The point where the horizontal axis crosses the vertical axis is the origin in the graph. This origin serves as a zero reference point for the four quantities involved. A graph like the one shown in Figure 19-6 is created by actually subjecting a diode to various forward and reverse voltages while measuring the current through the diode. However, certain precautions must be taken to insure that the diode is not damaged by excessive current or voltage.

JUNCHTION DIODES

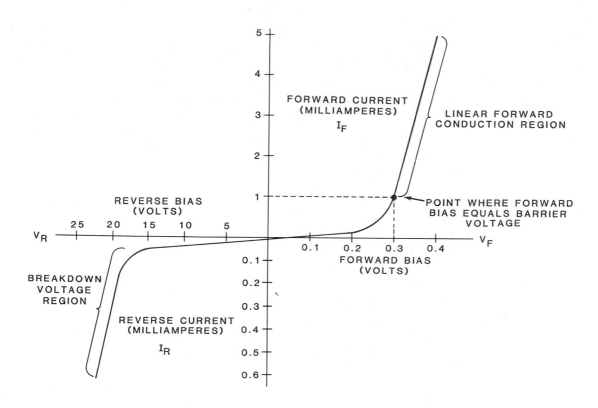

Figure 19-6 Typical germanium diode V-I characteristics.

When a large number of corresponding voltage and current values are plotted, a continuous curve is obtained. For this reason, a graph like the one shown in Figure 19-6 is often referred to as a **voltage-current (V-I) characteristic curve**. If you examine Figure 19-6 closely, you will find that the forward current and voltages are plotted on different scales than the reverse currents and voltages. This is because the forward characteristics involve low voltages and high currents, while the reverse characteristics involve relatively high voltages and low currents.

SEMICONDUCTORS

Forward Characteristics

Now that you understand the graph, let's discuss the operation of the germanium diode in detail. Notice that the forward current is extremely small and almost insignificant until the forward bias voltage across the diode increases beyond the value of 0.2 volts. Forward bias current increases much faster as the external bias voltage overcomes the diode's internal barrier voltage. Once the bias voltage exceeds the barrier voltage *(0.3 volts)*, the forward current increases very rapidly and at a linear rate because the diode than acts as a low resistance. If the forward current continues to rise, the diode would eventually be damaged by an excessive flow of current. Throughout the linear portion of the curve, the voltage across the diode is only several tenths of a volt as shown. While the forward voltage drop is not constant, it changes very little over a wide current range. A tremendous change in forward current occurs while the voltage across the diode changes only a small amount.

The point at which the bias voltage equals the barrier voltage is indicated in Figure 19-6. Notice that this point occurs when the bias voltage is equal to 0.3 volts. Also notice that the diode's forward current is equal to 1 milliampere at this time, and can increase above 5 milliamperes while the corresponding voltage across the diode remains below 0.4 volts. Figure 19-6 therefore shows that the diode's internal barrier voltage is approximately 0.3 volts. However, it is important to realize that this voltage will vary slightly from one germanium diode to the next.

Reverse Characteristics

The V-I curve in Figure 19-6 also shows that when the diode is reverse-biased, the reverse current that flows is extremely small. Notice that the reverse current increases slightly as the reverse voltage increases but remains less than 0.1 milliamperes (100 microamperes) until the reverse voltage approaches a value of 20 volts. Then the reverse current suddenly increases to a much higher value. This sudden increase in reverse current results because the reverse bias voltage becomes strong enough to tear many valence electrons from their parent atoms, increasing the number of electron-hole pairs in the N and P materials. This causes an increase in minority carriers which in turn support a higher reverse current. In other words, the junction breaks down when the reverse bias voltage approaches 20 volts.

The voltage at which the sudden change occurs is commonly referred to as the breakdown voltage. This breakdown voltage will vary from one diode to the next, because it is determined by the exact manner in which the diode is constructed. In certain cases, ordinary germanium diodes can be damaged when breakdown occurs; however, there are special diodes which are designed to operate in this region. These special devices—known as zener diodes—will be described in detail in Chapter 20.

JUNCTION DIODES

When breakdown occurs, the diode no longer offers a high resistance to the flow of reverse current and therefore cannot effectively block current in the reverse direction. For these reasons, operation in the breakdown region is avoided when an ordinary PN junction diode is being used.

If a junction diode conducts a large current when it is reverse-biased, the diode usually overheats and the junction is permanently damaged. When this happens, the diode may become shorted, open, or develop a high leakage current. The diode should be discarded at this point. An ohmmeter will usually indicate an open or shorted diode, but may not indicate a leaking diode.

Silicon Diode

While a silicon diode operates the same as the germanium diode, there are some important differences in their characteristic curves. Let's look at these differences in detail.

Forward Characteristics

The V-I curve in Figure 19-7 shows the characteristics of a typical silicon diode. Notice that the forward characteristics of this diode are basically similar to those of the germanium diode previously described; however, there is an important exception. The internal barrier voltage of the silicon diode is not overcome until the forward bias voltage is equal to approximately *0.7* volts as shown. Beyond this point, the forward current increases rapidly at a linear rate. The corresponding forward voltage across the diode increases only slightly. The exact amount of forward voltage required to overcome the barrier voltage will vary from one silicon diode to the next, but will usually be close to the 0.7 volts.

SEMICONDUCTORS

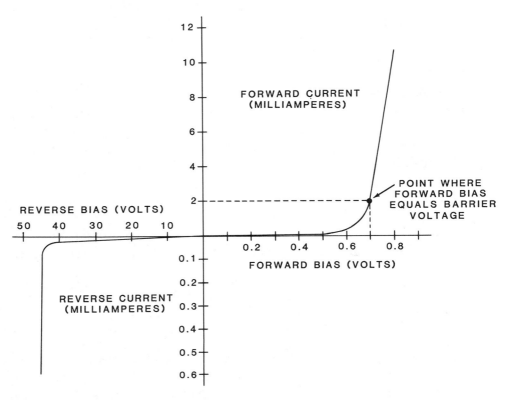

Figure 19-7 Typical silicon diode V-I characteristics.

Reverse Characteristics

The reverse characteristics of the silicon diode are also similar to those of the typical germanium diode previously described. However, the silicon diode has a much lower reverse current than the germanium diode. Notice that the reverse current remains well below 0.1 milliamperes (100 microamperes) until the breakdown voltage of the device is reached. Then, as with the germanium unit, a relatively high reverse current is allowed to flow. A breakdown voltage of 45 volts is indicated, although this voltage will vary from one silicon diode to the next. Also, the reverse currents in many silicon diodes may be in the extremely low nanoampere range, and therefore insignificant for most practical applications.

Diode Ratings

When the important characteristics of silicon and germanium diodes are compared it becomes apparent that either type can be damaged by excessive forward current. For this reason, manufacturers of these diodes usually specify the maximum forward current (I_F max) that each type can safely handle. Also, both types can be damaged by excessive reverse voltages, which cause the diode to breakdown and conduct a relatively large reverse current. To insure that the various diodes are not subjected to dangerously high reverse voltages, manufacturers of these devices usually specify the PIV (Peak-Inverse Voltage) or PRV (peak-reverse-voltage) that the diode can safely handle without damage.

Temperature Considerations

In some critical applications it is also necessary to consider the effect of temperature on a diode. In general, the diode's reverse current is the characteristic most affected by temperature. At extremely low temperatures the reverse current through a typical diode will be practically zero. But at room temperature this current will be somewhat higher, although still quite small. At extremely high temperatures an even higher reverse current will flow, which in some cases might interfere with normal diode operation. These changes in reverse current as a result of temperature changes are shown in Figure 19-8. Notice that the breakdown voltage also tends to increase as temperature increases, although this change is not great.

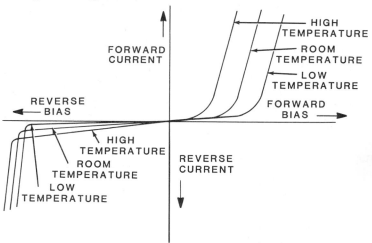

Figure 19-8 Relative changes in current due to changes in temperature.

These same relative changes occur in both types of diodes, even though the reverse currents are generally higher in the germanium types. For both germanium and silicon diodes, the reverse or leakage current doubles for approximately every 10 degrees Centigrade rise in temperature.

SEMICONDUCTORS

The forward voltage drop across a conducting diode is also affected by temperature changes. The forward voltage drop is inversely proportional to temperature. As the temperature rises, the voltage drop decreases. This effect is the same in both germanium and silicon devices.

Diode Symbols

The symbol most commonly used to represent the diode is shown in Figure 19-9A, along with a typical PN junction diode. Notice that the P-type material is represented by an **arrow** (sometimes referred to as a triangle) and the N-type material is represented by a **bar**. The arrows that are placed beside the diode and its symbol indicate the direction of forward current (I_F) or electron flow. As you discovered earlier, the forward current must flow from the N section to the P section of the diode. This means that the forward current through the symbol must flow from the bar to the arrow. In other words, *forward current flow is always against the arrow in the diode symbol.*

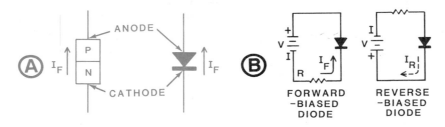

Figure 19-9 A typical junction diode, its symbol, and circuits.

The N-type material (the bar) is identified as the **cathode** of the diode, and the P-type material (the arrow) is identified as the **anode** of the diode. These two terms were once widely used to identify the two principle elements within a vacuum tube diode. However, they are now commonly used to describe the two sections of a junction diode. The cathode supplies the electrons and the anode collects the electrons.

Figure 19-9B shows how forward-biased and reverse-biased diodes are represented in schematic form. Notice that when the negative and positive terminals of the battery are connected to the cathode and anode of the diode respectively, the diode is *forward-biased*, and will conduct a relatively large forward current, (I_F). The resistor is added in series with the diode to limit this forward current to a safe value. When the battery terminals are reversed, the diode is *reverse-biased*, and only a very small reverse current (I_R) will flow through the device.

JUNCTION DIODES

Chapter Self-Test

1. How are the free electrons and holes which drift throughout a semiconductor referred?

2. How are P-type and N-type materials combined to form a PN junction diode?

3. What is the name for the region created at the junction of a diode?

4. The opposite charges which build up on each side of a PN junction create a potential difference which is known as what? What is the range of this potential?

5. When is the depletion region of a diode neutralized?

6. The action of a forward-biased diode can be compared to the action of what other component?

7. Describe the requirements for forward-biasing a diode, and describe the characteristics of a forward-biased diode.

8. Describe the requirements for reverse-biasing a diode, and describe the characteristics of a reverse-biased diode.

9. What is the term used for the small reverse current that flows through a reverse-biased diode?

10. As temperature increases, what happens to the reverse current in a diode?

11. An external source of 5 volts DC is applied to a germanium diode in series with a 1 kilohm resistor. The positive source lead is connected to the P section of the diode. How much current flows, if any?

12. Are the V-I characteristic curves of silicon and germanium diodes similar? Why?

13. List the differences between a silicon diode and a germanium diode.

14. What does PIV stand for? Where is it used?

15. In the diode symbol, what represents the P-type material?

Summary

Diodes are produced by combining N-type and P-type semiconductor material, forming a PN junction. The combination of these materials allows free electrons and holes to combine until a depletion region, containing no majority carriers, is formed in the immediate vicinity of the PN junction.

The combination of free electrons and holes causes positive charges to develop in the N-type material and negative charges to develop on the P-type material. These charges result from the ions that are left behind. Consequently, a barrier voltage develops across the junction. Eventually, the barrier voltage increases to a level that prevents further combining of electrons and holes, limiting the size of the depletion region. Germanium diodes have a typical barrier voltage of 0.3 volts, while silicon diodes have a barrier voltage of 0.7 volts.

When an external voltage is connected to a diode, it becomes possible to forward-bias the diode. In order to forward-bias a diode, two requirements must be met. First, the voltage potential connected to the N-type material must be negative in respect to the voltage potential connected to the P-type material. Second, the overall potential difference across the diode must exceed the barrier voltage of the junction—0.3 volts for a germanium diode, and 0.7 volts for a silicon diode.

If either of the above conditions are not met, the diode will not conduct. If the polarities of the external voltage are reversed, the diode becomes reverse-biased. This is because the PN junction barrier voltage increases until it is equal to and opposite that of the applied voltage. This effectively cuts off almost all current from flowing through the diode. However, a small amount of leakage, or reverse current flows by means of the minority carriers. Conduction of a forward-biased diode is much heavier than that of a reverse-biased diode. Therefore, the forward resistance of a diode is much lower than reverse resistance.

If the reverse bias voltage reaches a high enough value, the diode breaks down and conducts a relatively high reverse current. The voltage that causes this reverse current is called the breakdown voltage of the diode. Heavy reverse current produces heat that can permanently damage the diode, unless a resistor is connected in series to limit the amount of reverse current flowing through the diode. The maximum reverse voltage that a diode can withstand without damage is called the PIV, or Peak-Inverse-Voltage.

Most diode characteristics are affected by temperature. Reverse leakage current and forward voltage drop are affected the most. In a typical diode, a ten-degree centigrade rise in temperature causes reverse current to nearly double. The forward voltage drop is inversely proportional to temperature; as temperature rises, the forward voltage drop decreases.

SEMICONDUCTORS

CHAPTER 20

Zener Diodes

Contents

Introduction 483

Chapter Objectives 484

Zener Diode Characteristics 485

Current Limitations 489

Zener Diode Impedance 492

Chapter Self-Test 496

Summary 497

ZENER DIODES

Introduction

We've stated earlier that the method for making electronics work for us is to control current flow. This is accomplished by using a predetermined voltage across a specific resistance. However, when alternating current is used, the amount of voltage is constantly changing. Also, there may be times when a steady, unchanging DC voltage is required, and due to the arrangement of components, the voltage waivers. To solve this problem, we need a component that can continue to drop a specific, unchanging amount of voltage, regardless of the amount of current flowing through it.

This is the function of a zener diode. This chapter covers how a zener diode is constructed, how it operates, and its many applications. You will see the zener diode used in a wide variety of configurations within the electronics industry. For this reason, the zener diode is an extremely important component.

SEMICONDUCTORS

Chapter Objectives

When you have completed this chapter, you will be able to:

1. Explain the forward and reverse current-voltage characteristics of a typical zener diode.

2. Describe the relationship between temperature and zener diode power and dissipation.

3. Determine a zener diode's maximum safe operating current.

4. Use a zener's impedance value to determine the amount of change that can occur in its voltage.

5. Describe how the zener diode is used to provide voltage regulation.

6. Identify a zener diode on a schematic diagram.

7. Explain the terms knee of the curve and reverse breakdown voltage.

8. Compare the zener diode to a junction diode in terms of internal voltage drop.

Zener Diode Characteristics

In the last chapter, you learned that an ordinary PN junction diode breaks down and conducts a relatively high reverse current when it is subjected to a sufficiently high reverse bias voltage. This high reverse current occurs because the high reverse voltage is capable of tearing valence electrons away from their parent atoms, increasing the number of minority carriers in the N and P sections of the diode.

Ordinary PN junction diodes can be damaged if they are subjected to their breakdown voltages. This is because the high reverse currents produce more heat than the diodes can safely dissipate. However, special diodes are constructed which can operate at voltages that equal or exceed their breakdown voltage ratings. These special diodes are called **zener diodes**.

We will now examine the relationship that exists between the current flowing through a zener diode, and the voltage across the device. We will consider the action that takes place when the zener diode is forward-biased and reversed-biased, but we will be primarily concerned with the action that takes place at the point where breakdown occurs. Then we will see how a zener diode is rated according to its breakdown voltage.

Voltage-Current Characteristics

A typical zener diode V-I (voltage-current) characteristic curve is shown in Figure 20-1. Notice that the overall forward and reverse characteristics of the zener diode are similar to those of an ordinary junction diode. The primary difference is that the zener diode is specifically designed to operate with a reverse bias voltage that is high enough to cause the device to reach breakdown, and conduct a high reverse current. As shown in Figure 20-1, the zener diode's reverse current remains at a very low value until the reverse voltage is increased to a value that is sufficient to cause the zener to reach breakdown. Then the reverse current through the zener increases at an extremely rapid rate as the reverse voltage increases beyond the breakdown point. The V-I curve therefore shows that beyond the breakdown point, a very large change in reverse current is accompanied by only a very small change in reverse voltage. This action occurs because the resistance of the zener drops considerably as its reverse voltage is increased beyond the breakdown point. Once the breakdown point is reached the zener is operating in its zener region. The **zener current** may be represented by the symbol I_Z.

SEMICONDUCTORS

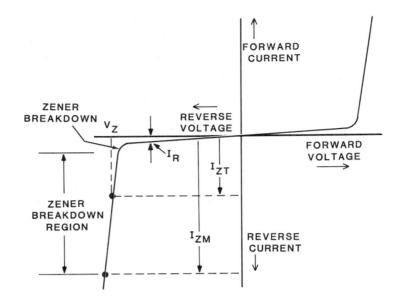

Figure 20-1 Typical V-I characteristic curve for a zener diode.

If you examine Figure 20-1 closely you will see that breakdown (also called zener breakdown) does not occur instantaneously. The curve is rounded near the breakdown point. This curve or rounded portion is often called the **knee** of the curve. When a zener diode has a knee with a very sharp edge, the zener enters the breakdown region very quickly. However, when the knee is more rounded, the breakdown region is entered at a slower rate. The importance of this characteristic will be explained later in this discussion.

Zener Voltage

The breakdown voltage of a zener diode is determined by the resistivity of the zener. This can be controlled by the various doping techniques that are used to form the device. A zener diode is manufactured to have a specific breakdown voltage rating which is often referred to as the diode's **zener voltage,** designated V_Z. Typical V_Z values may vary from several volts to several hundred volts. For example, some of the popular low voltage units have ratings of 3.3, 4.7, 5.1, 5.6, 6.2, and 9.1 volts. However, additional voltage ratings are also available.

It is important to realize that when a zener diode is rated as having a specific zener voltage (V_Z), the rated voltage does not represent the reverse voltage that is required to initially cause the diode to breakdown. The rated zener voltage is a nominal value that represents the reverse voltage across the diode when the zener current is at some specified value called the zener test current (I_{ZT}). The V-I curve in Figure 20-1 shows the relative zener voltage (V_Z) and **zener test current** (I_{ZT}) values for a typical zener diode. Notice that these values are located within the zener breakdown region of the curve. The zener test current (I_{ZT}) simply represents a typical value of reverse current that is always less than the maximum reverse current the diode can safely handle.

Like resistors and capacitors, zener diodes cannot be produced with zener breakdown voltages that are always equal to a specified value. Therefore, it is necessary to specify minimum and maximum breakdown voltage limits for each device. This is done by specifying a **breakdown voltage tolerance** for each type of diode that is manufactured. The standard zener breakdown voltage (zener voltage) tolerances are +/- 20 percent, +/- 10 percent, and +/- 5 percent. However, specially manufactured zener diodes are also available with a 1 percent tolerance. For example, a 6.8 volt 10 percent zener diode will have a zener voltage somewhere in the 6.12 to 7.48 voltage range.

Power Dissipation in Zener Diodes

Manufacturers of zener diodes also specify the maximum power dissipation of each device. Some devices are rated in milliwatts, while others are rated as high as 50 watts. Some of the most popular and widely used devices have relative ratings of 400 milliwatts, 500 milliwatts, and 1 watt. A zener diode's power dissipation rating is given for a specific operating temperature. Often the power rating is given for a temperature of 25 degrees centigrade, 50 degrees centigrade, or 75 degrees centigrade. However, the actual power that a zener diode can safely dissipate will decrease if the temperature increases above its specified level, and vice-versa.

Also, if the zener diode has axial leads, its power rating is often specified for a specific lead length. Sometimes a variety of ratings are given for various lead lengths. This is because a diode's ability to dissipate power increases as its leads are shortened. The shorter leads (when appropriately soldered in an electronic circuit) are more effective in conducting heat away from the diode's PN junction.

Power-Temperature Curves

To simplify the relationship between a zener diode's maximum power rating, its temperature, and its lead length, a power-temperature derating curve is often supplied with each type of diode manufactured. A typical curve for a zener diode is shown in Figure 20-2. Notice that three curves are shown for three different lead lengths of 1/8", 3/8", and 1". Notice also that the power dissipated decreases as temperature increases. Finally, a shorter lead length (1/8") allows the diode to dissipate more power over the same temperature spread while the longer lead length (1") reduces the overall power rating of the device. The curve also shows that the power rating of the device is effectively reduced to zero at 200 degrees centigrade.

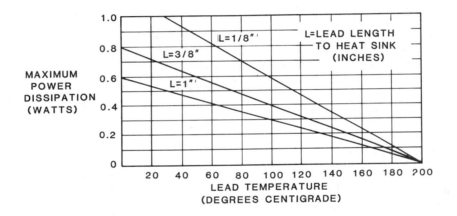

Figure 20-2 Power-temperature curve.

The temperatures indicated in Figure 20-2 are the actual temperature of the diode's leads, not the ambient or surrounding air temperature. The diode's leads are assumed to be soldered to a circuit board or component which can serve as a heat sink, which drains away the heat produced by the device.

Current Limitations

The maximum reverse current that can flow through a zener diode without exceeding the diode's power dissipation rating is commonly referred to as the maximum zener current and is represented in the V-I curve in Figure 20-1 by the symbol I_{ZM}. The I_{ZM} value of a zener diode is often specified by the manufacturer of the device. However, if I_{ZM} is not specified, it can be determined by simply dividing the power dissipation rating of the diode by its breakdown voltage (zener voltage) rating or stated mathematically:

$$I_{ZM} = \frac{\text{Power Rating}}{\text{Zener Voltage}}$$

However, it is best to play it safe and use the maximum limit of zener voltage in your calculations. For example, suppose you have a zener diode that is rated at 10 watts, and the diode has a zener voltage of 5.1 volts at +/- 10 percent tolerance. The maximum voltage limit would be equal to 5.1 volts plus 10 percent, or 5.61 volts. The maximum zener current would therefore be equal to:

$$I_{ZM} = \frac{10 \text{ W}}{5.61 \text{ V}} = 1.78 \text{ A}.$$

The V-I curve in Figure 20-1 also shows that a small reverse or leakage current (I_R) flows through the zener diode before the breakdown point is reached. Since the zener diode is normally used in its zener breakdown region, this current is usually not important. However, there are certain applications of zener diodes which require an absolute minimum leakage current before the breakdown point is reached. Manufacturers often specify the I_R value of a zener diode at a certain reverse voltage that is less than the zener voltage V_Z (often 80 percent of V_Z).

Effects of Temperature

Zener diodes also have other characteristics that must be considered in certain applications. For example, a diode's zener voltage will vary slightly as temperature changes. The amount of voltage change that takes place is usually expressed as a percentage of zener voltage (V_Z) change for each degree centigrade rise in temperature and is referred to as the zener voltage temperature coefficient. Zener diodes that have a zener breakdown voltage of 5 volts or more usually have positive zener voltage temperature coefficients. This simply means that their breakdown voltages increase as temperature increases. However, most diodes with breakdown voltages below 4 volts usually have a negative zener voltage temperature coefficient. This means that the breakdown voltage will decrease with an increase in temperature.

SEMICONDUCTORS

When the breakdown voltages are between approximately 4 and 5 volts, the zener voltage temperature coefficient may be either positive or negative. For example, a zener diode with a zener breakdown voltage of 3.9 volts might have a zener voltage temperature coefficient of −.025 percent per degree centigrade. This means that the diode's zener voltage will decrease .025 percent (or approximately .001 volt) for each degree centigrade rise in temperature.

Temperature-Compensated Zener Diodes

Special zener diodes are constructed which are temperature-compensated so that their zener voltage ratings remain almost constant with changes in temperature. These special diodes are commonly referred to as temperature-compensated zener diodes or voltage reference diodes.

A temperature-compensated diode is formed by connecting a zener in series with an ordinary PN junction diode as shown in Figure 20-3. Before we describe the operation of this device, we should point out the schematic symbol for the zener diode. Notice that the schematic symbol is much like the junction diode, with the exception that the cathode (represented by the bar on a junction diode) has two additional lines placed on the end.

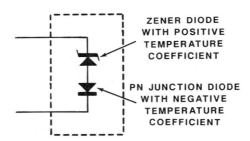

Figure 20-3 Temperature-compensated device.

Notice that the two diodes are placed with their anodes sharing a common connection. Current will flow through the reverse-biased zener and the forward-biased junction diode. When this occurs, the forward-biased diode has a voltage drop of approximately 0.6 volts, and a negative temperature coefficient. The zener diode must have a voltage rating greater than 5 volts, in order to achieve a positive temperature coefficient.

By carefully selecting the two devices so that their temperature coefficients are equal and opposite, the voltage changes effectively cancel out. Furthermore, the voltage drops across the two devices must be summed to obtain the overall voltage rating of the temperature-compensated device. For example, when a 5.6 volt zener diode is connected in series with a junction diode that has a forward voltage drop of 0.6 volts, a 6.2 volt temperature compensated device is produced. In some cases more than one junction diode may be used to obtain the necessary compensation. Typical temperature-compensated zener diodes will have temperature coefficients that range from .01 percent per degree centigrade to 0.0005 percent per degree centigrade. However, the optimum temperature stability usually occurs at or near a specific operating current which is normally specified by the manufacturer.

Zener Diode Impedance

Another important characteristic that should be considered when examining any type of zener diode is the diode's zener impedance (Z_{ZT}). This is determined by varying the zener current above and below the specified zener test current (I_{ZT}) value, and observing the corresponding change in zener voltage (V_X). This is shown in Figure 20-4. The zener impedance is equal to the change in zener voltage (ΔV_Z) divided by the change in zener current (ΔI_Z) and may vary considerably from one diode to the next. It should be noted that "change" is designated using the Greek letter delta (Δ).

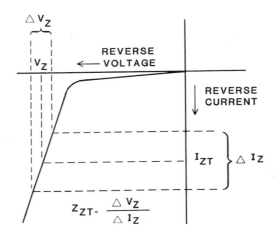

Figure 20-4 The basic method used to determine zener impedance (Z_{ZT}).

Some zener diodes with low zener voltage ratings will have a Z_{ZT} of only a few ohms. In general, the lower the zener impedance, the greater the slope of the curve in the zener breakdown region. A low Z_{ZT} therefore indicates that the zener voltage changes only slightly with changes in zener current. An ideal zener diode would not change its breakdown voltage as zener current varies and would therefore have a zener impedance of zero ohms.

The zener impedance of a diode is also useful in determining the changes in zener voltage which can occur when the diode is used at zener currents different from I_{ZT}. It is a simple matter to calculate this change in zener voltage by using a known value of zener impedance and the change or deviation in zener current. When expressed mathematically, the change in zener voltage is equal to:

$$\Delta V_Z = \Delta I_Z \times Z_{ZT}$$

This equation reads as: "change in zener voltage is equal to change in zener current times zener impedance."

The same technique used to determine Z_{ZT} may also be used to determine the impedance at the knee of the curve near the point where breakdown occurs. The impedance at the knee of the curve is commonly referred to as the **zener knee impedance** (Z_{ZK}). The zener knee impedance provides an indication of the slope or sharpness of the knee of the curve. Manufacturers of zener diodes will usually specify both the zener impedance (Z_{ZT}) and the zener knee impedance (Z_{ZK}) for each device.

Voltage Regulation

Although the zener diode may be used to perform a number of important functions, it is perhaps most widely used in applications where it is continually reversed-biased. This allows the zener to operate constantly within its breakdown region. Under these conditions, the zener diode is effectively used to provide voltage stabilization or regulation.

Voltage regulation is often required because most solid-state circuits require a mixed or constant DC power supply voltage for proper operation. If this DC voltage changes significantly, improper operation will usually result. If an AC line operated power supply is used (one that is not regulated), the DC output voltage will vary if the power line voltage changes or if the load resistance connected to the power supply changes. However, by using a zener diode regulator circuit, it is possible to compensate for these changes and maintain a constant DC output voltage.

SEMICONDUCTORS

The Basic Zener Diode Regulator

A typical zener diode regulator circuit is shown in Figure 20-5A. Notice that the diode is connected in series with a resistor, and an unregulated DC input voltage is applied. The input voltage is connected so that the zener diode is reversed-biased as shown, and the series resistor allows enough current to flow through the diode so that the device operates within its zener breakdown region. In order for this circuit to function properly, the input DC voltage must be higher than the zener breakdown voltage rating of the diode.

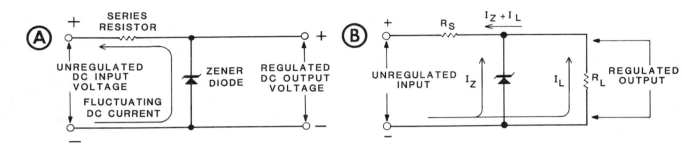

Figure 20-5 A zener diode voltage regulator circuit.

The voltage across the zener diode will then be equal to the diode's zener voltage rating and the voltage across the resistor (R_S) will be equal to the difference between the diode's zener voltage and the input DC voltage.

The input DC voltage shown in Figure 20-5 is *unregulated*. This means it is not held to a constant DC value. This voltage may periodically increase above or decrease below its specified value, causing the DC current flowing through components to fluctuate. However, since the zener diode is constantly reverse-biased, and the amount of that reverse-bias constantly exceeds the breakdown voltage, the zener diode is operating within its zener voltage region. This allows a wide range of current to flow through the zener diode, while its voltage changes only slightly. This occurs because the impedance of the zener diode drops as current increases.

Since the zener diode's voltage remains almost constant as the input voltage varies, the change in input voltage almost completely appears across the series resistor. As current fluctuates, the voltage across the zener diode will remain very close to the breakdown voltage. However, R_S feels the change in current, and the resistor's voltage changes accordingly.

ZENER DIODES

The voltage across the zener diode is used as the output voltage for the regulator circuit. The output voltage is therefore equal to the diode's zener voltage and since this voltage is held to a somewhat constant value, it is referred to as a *regulated voltage*. The output voltage of the regulator circuit can be changed by simply using a zener diode with a different voltage rating, remembering to select a series resistor that will allow the new diode to operate within its breakdown region.

Voltage regulator circuits like the one shown in Figure 20-5 are used in electronic equipment to provide constant operating voltages for various loads. However, these loads may also require operating currents as well. Furthermore, there are many cases where operating currents continually vary because load impedances are not constant. When designing a regulator circuit, it is necessary to consider the specific current or the range of currents that the regulator must supply to the load in addition to its output voltage.

This situation is shown in Figure 20-5B. Notice that the load resistor (R_L) requires a specific load current (I_L), determined by its resistance and the output voltage. The current through the zener diode (I_Z) combines with I_L and flows through the series resistor (R_S). The value of R_S must be chosen so that I_Z remains at a sufficient level to keep the diode within its breakdown region, while at the same time allow the required value of I_L to flow through the load.

If the load current increases or decreases (due to a change in load impedance), you might think that the output voltage would change accordingly. If R_L and R_S were in series *without* the zener diode installed, this might well occur. However, the zener diode prevents this from happening.

For example, when R_L increases in value, I_L decreases and the voltage across R_L tries to increase. However, the zener diode opposes this change by conducting more zener current (I_Z). With an increase in I_Z, the total current ($I_Z + I_L$) flowing through R_S remains essentially constant. The same voltage is maintained across R_S and across the parallel branch (the zener diode and R_L). If I_L were to increase, the zener would compensate by decreasing the amount of I_Z. (Keep in mind that all of this is occurring while the input voltage remains constant.)

In this way, the zener diode regulator is able to maintain a relatively constant output voltage even though changes in output current occur. The regulator circuit therefore regulates for changes in output current as well as for changes in input voltage.

SEMICONDUCTORS

Chapter Self-Test

1. True or false: zener diodes are designed to safely operate within their zener breakdown regions.

2. What is the term for the curved portion of the V-I curve near the point where zener breakdown occurs?

3. Since it is nearly impossible to manufacture zener diodes with breakdown voltages that are exactly equal to a specified value, it is necessary to indicate the minimum and maximum breakdown voltage limits. How is this accomplished?

4. How is a zener diode's power dissipation rating usually specified?

5. What will happen to the power that a zener diode can safely dissipate if temperature decreases?

6. How is the relationship between a zener diode's maximum power rating, its temperature, and its lead length expressed?

7. How is the maximum safe value of reverse current that can flow through a zener diode calculated?

8. If a zener diode has a breakdown voltage of 9.1 volts, what will its voltage temperature coefficient be?

9. How are temperature-compensated zener diodes formed?

10. What would be the *ideal* zener impedance for a diode?

11. Does a zener diode operate the same as a junction diode when they are both forward-biased? Why?

12. What is the relationship between the input DC voltage and the zener breakdown voltage in a common regulator circuit?

13. A regulator circuit compensates for what type of changes within a circuit?

ZENER DIODES

Summary

Zener diodes are designed to withstand the high reverse current that results when the diode's breakdown voltage is applied. After the breakdown voltage of the zener diode is exceeded, reverse current through the diode increases from an insignificant leakage value to a relatively high value. Once a zener diode is operating in its breakdown, or zener region, an increase in current causes the internal resistance of the diode to decrease. Conversely, if current decreases, the diode's internal resistance increases. As a result, the voltage drop across the zener diode remains essentially constant. Current through a zener diode operating in the zener region is called zener current. If zener current exceeds the specified maximum for a given diode, the diode can be permanently damaged.

You can determine the zener voltage rating (V_Z) of a given zener diode by measuring the voltage drop across the diode with a specified zener test current (I_{ZT}) through the diode. Zener test current is always greater than the reverse current that initiates zener breakdown, but less than the maximum current that the device can withstand.

Zener diodes are available for almost any zener voltage rating. However, it is impractical to manufacture a zener diode capable of dropping an exact specified voltage. Therefore, a certain amount of variation is expected. Consequently, manufacturers usually specify the amount of variation expected as a percentage of tolerance. For example, if a given zener diode is rated at 9.1 volts with a tolerance of +/- 20%, the actual voltage may range from approximately 7.3 volts to 10.9 volts.

Zener diode manufacturers also specify the maximum power that their products can safely dissipate at a given temperature. Diode power ratings increase or decrease with decreases or increases in temperature. In addition, you can increase the power rating of a zener diode with axial leads by shortening the length of the leads.

The most common application of zener diodes is as regulators in regulated DC electronic power supplies. When a zener diode is used as part of a properly designed regulator circuit, the diode maintains a constant (regulated) DC output voltage. In a typical regulator circuit, the diode is connected in series with a resistor to the unregulated output of the power supply. The load is then connected in parallel with the diode. Consequently, any variations in the unregulated output voltage caused by load current variations or line voltage changes are dropped across the resistor, while the voltage across the zener diode and the load remains constant. In addition, the sum of the zener and load currents remain fairly constant.

SEMICONDUCTORS

CHAPTER 21

Bipolar Transistors

Contents

Introduction 501

Chapter Objectives 502

Basic Transistor Action 503

Transistor Amplification 510

Transistor Amplifier Circuits 513

Chapter Self-Test 527

Summary 528

BIPOLAR TRANSISTORS

Introduction

In 1947, Bell Laboratories invented a device with two PN junctions, and discovered a method for amplifying voltage and current by varying the biasing voltages across these two junctions. This was the beginning of the bipolar transistor. The bipolar transistor has since become the single most important electronic component in use today. Its operation is essential in integrated circuits, and is used perhaps more extensively than any other component.

This chapter is designed to familiarize you with the bipolar transistor and its basic principle of operation. You will also learn how transistors are used to amplify electronic signals.

This chapter is simply an introduction to transistors. You will learn that transistors cannot function solely by themselves, but require additional circuitry around them in order to perform their function. These circuits will be explained when you enter the last section of this book.

SEMICONDUCTORS

Chapter Objectives

When you have completed this chapter, you will be able to:

1. Define the following terms: emitter, base, collector, emitter junction, collector junction, amplifier, emitter current, base current, collector current, PNP, and NPN.

2. Describe the basic physical construction of the two basic types of bipolar transistors.

3. Explain the basic principle behind bipolar transistor operation.

4. Show how a bipolar transistor should be biased for normal operation.

5. Explain the relationship between emitter current, base current, and collector current in a bipolar transistor.

BIPOLAR TRANSISTORS

Basic Transistor Action

Before we can describe how a transistor works, we must introduce you to the terms and definitions you will be using for transistors. Most of the terms here have not yet been used. However, they are vital in understanding how a transistor operates.

PNP and NPN Configurations

You are now ready to examine a solid-state component that is related to the PN junction diode. This device is called a **bipolar transistor,** although it is sometimes referred to as a **junction transistor, BJT,** or simply a **transistor.** Throughout this chapter "transistor" and "bipolar transistor" have the same meaning and are used interchangeably. There are other transistors in use today, such as the field effect transistor, or FET (which you will learn about in the next chapter), and the unijunction transistor, or UJT (which will be covered in Chapter 23). The names for these components relate to their method of operation. Once you have learned all three, you will see how their name relates to their operation.

A bipolar transistor is constructed from germanium and silicon semiconductor materials just like a PN junction diode. However, this device utilizes three alternately doped semiconductor regions instead of two as in the case of the diode. These three semiconductor regions may be arranged in two different ways. One arrangement is shown in Figure 21-1A. Notice that a section of N-type semiconductor is sandwiched between two P-type semiconductor layers to form what is commonly referred to as a **PNP type transistor.** The middle or N region is called the **base** of the transistor, and the outer or P regions are called the **emitter** and **collector** of the transistor. Normally, the base region is much thinner than the emitter and collector regions, and is lightly doped in relation to the other two. The appropriate leads must also be attached as shown to provide electrical connections to the base, emitter, and collector leads.

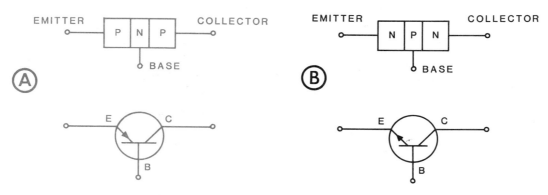

Figure 21-1 Transistors and their symbols.

SEMICONDUCTORS

The PNP transistor is usually represented in circuit diagrams or schematics by the symbol shown. The emitter, base, and collector of the transistor are identified by the letters E, B, and C, respectively; however, these letters may not always appear with the symbol. It is also important to remember that *the arrow is always located on the emitter lead*. Also, notice that on the PNP transistor, the arrow points towards the base. One way of remembering this configuration is that on a PNP transistor, the arrow "*Points iN Permanently*."

The second method of arranging the semiconductor layers in a bipolar transistor is shown in Figure 21-1B. A layer of P-type material is sandwiched between two layers of N-type material to form what is commonly known as an NPN transistor. Like the PNP transistor, this device utilizes a narrow middle region as the base. The outer layers are again referred to as the emitter and collector regions.

The symbol for the NPN transistor is very similar to the symbol for the PNP transistor. Again, the leads may not be identified by the additional letters (E, B, and C) as shown. Notice that the only difference between the PNP and NPN transistor symbols is the direction of the arrow on the emitter. Here, the arrow points away from the base, or "*Never Points iN*."

Because PNP and NPN transistors have three alternately doped semiconductor regions, these devices have two junctions of P and N type materials. In other words, each transistor has one PN junction between its emitter and base regions, and one PN junction between its base and collector regions. Therefore, bipolar transistors have three layers and two junctions, as opposed to PN junction diodes which have two layers and only one junction.

Transistor Biasing

Although bipolar transistors may be used in various ways, their basic and most important function is to provide amplification. In other words, bipolar transistors are used primarily to boost the strength or amplitude of electronic signals. It is possible to amplify current, voltage, or power. However, you will soon discover that the bipolar transistor is actually controlled by the current flowing through its terminals, and this current is controlled by varying the input voltage. Furthermore, the transistor must be *properly biased* by external voltages so that its emitter, base, and collector regions interact in the desired manner.

BIPOLAR TRANSISTORS

Biasing NPN Transistors

Like the PN junction diode, the bipolar transistor must be properly biased in order to perform a useful function. However, the bipolar transistor has two PN junctions and *both* of these junctions must be properly biased if the device is to function properly. For example, consider the NPN transistor shown in Figure 21-2A. Notice that an external voltage has been applied to the base and collector regions. The junction formed between these two regions is commonly referred to as the collector-base junction or simply the collector junction.

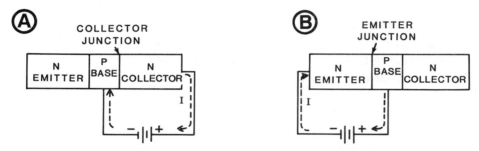

Figure 21-2 Reverse-biased collector junction.

The collector and base regions effectively form a PN junction diode even though the base is very thin and lightly doped. The external voltage is being used to *reverse-bias* the collector junction as shown. The reverse-biased collector junction will function in a manner similar to a reverse-biased PN junction diode, conducting only a very small leakage current. This extremely small current is supported by minority carriers in the N-type collector and P-type base regions. The minority carriers are holes in the N-type material and electrons in the P-type material. In this case, the number of minority carriers in both regions (the base in particular) is very small, which results in a very small leakage current. Under normal operating conditions, the collector junction of an NPN transistor is reverse-biased by connecting a positive supply voltage to its collector lead.

Now we will remove the external voltage shown in Figure 21-2A, and apply it to the emitter and base regions of the NPN transistor as shown in Figure 21-2B. The junction between these two regions is often referred to as the emitter-base junction or simply the emitter junction. The external voltage is being used to *forward-bias* the emitter junction because the N-type emitter is connected to the negative terminal, and the P-type base is connected to the positive terminal.

Although the base is extremely thin and lightly doped, the emitter junction is still able to operate like a forward-biased PN junction—the majority carriers within the two regions combine at the emitter junction. However, there is one major difference between the operation of a forward-biased emitter junction in a transistor, and a forward-biased PN junction diode. The majority carriers (electrons) within the emitter greatly exceed the majority carriers (holes) in the base. This means that the base cannot supply enough holes to combine with all of the electrons that cross the junction. However, a certain amount of forward current will still flow through the emitter junction and the external voltage source. Under normal operating conditions the emitter junction must be forward-biased.

We have now seen the action which takes place in an NPN transistor when the collector junction and the emitter junction of the device are independently subjected to the proper bias voltages. However, our analysis of transistor action is not yet complete. This is because both junctions of the transistor must be *simultaneously biased* if the device is to operate properly. In other words, the emitter junction must be forward-biased and the collector junction must be reverse-biased as shown in Figure 21-3A. When both junctions are biased in this manner, the action that takes place in the transistor is quite different. Since the emitter junction is forward-biased, electrons in the N-type emitter and the holes in the P-type base (the majority carriers) are forced to move toward the emitter junction and combine, as we stated earlier. However, because of the unique construction and biasing of the transistor, the emitter junction cannot function like a normal forward-biased PN junction.

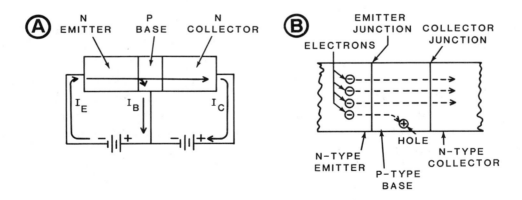

Figure 21-3 A properly biased NPN transistor showing internal and external current flow.

If you'll recall, when a PN junction is forward-biased, it conducts a forward current through the junction and the external voltage source. With a transistor, this cannot take place. The transistor has a very thin base region which is lightly doped with respect to the emitter. This means that the majority carriers (electrons in an NPN transistor) in the emitter will greatly exceed the majority carriers (holes) in the base. Therefore, most of the electrons that cross the emitter junction cannot combine with holes in the base region.

Those electrons that cannot combine within the base are attracted to the positive potential applied to the collector. In fact, most of the majority carriers entering the base from the emitter are swept across the collector junction and into the collector region. From there, they enter the positive side of the external voltage source used to reverse bias the collector junction. This action is shown in Figure 21-3B. Typically, 95 to 99 percent of the electrons supplied by the emitter flow through the collector region and into the external voltage source. This current is referred to as **collector current** and is usually designated as I_C. The remaining emitter injected electrons (1 to 5 percent) combine with holes in the base region and therefore support a small current which flows out of the base region. This relatively small current is generally referred to as **base current** and is designated as I_B. The current that flows into the emitter region is generally referred to as **emitter current** and is designated as I_E.

The relationship between I_E, I_B, and I_C can be stated mathematically using the following equation:

$$I_E = I_B + I_C$$

This equation simply states that the emitter current (I_E) is equal to the sum of the base current (I_B) and the collector current (I_C). Another way of looking at this is that the collector current is equal to the emitter current less the current lost to the base:

$$I_C = I_E - I_B$$

If the base region of the transistor was not extremely thin, the action just described could not occur. The thin base region makes it possible for a majority carrier (coming from the emitter) to move quickly into the collector region. A large base region would minimize the interaction between the emitter and collector regions, and the transistor would act more like two separately biased diodes. Furthermore, the emitter is doped much more heavily than the base region so that a large number of majority carriers will be supplied to the base and be subsequently pulled into the collector region. The collector region is also lightly doped, although not as lightly as the base. The collector is also considerably larger than either the base or emitter. This allows the collector to produce a sufficient number of minority carriers, while the number of majority carriers are reduced to a low level. A large number of majority carriers could actually interfere with collector operation because they tend to inhibit the production of minority carriers.

SEMICONDUCTORS

There are two important points which we must also consider. First, it is important to realize that the emitter-base junction of the transistor has characteristics similar to those of a PN junction diode. A barrier voltage is produced across the emitter junction of the device, and this voltage must be exceeded before forward current can flow through the junction. Throughout our previous discussion, we have assumed that the forward bias voltage exceeds this internal barrier voltage. We have also assumed that the emitter current is held to a safe value. The internal barrier voltage within a particular transistor is determined by the type of semiconductor material used to construct the device—0.3 volts for germanium transistors, and 0.7 volts for silicon transistors. This means that transistors, like PN junction diodes, will exhibit a relatively low voltage drop (0.3 or 0.7 volts) across their emitter junctions under normal operating conditions.

A second important point to consider is that the collector-base junction of the transistor must be subjected to a positive potential that is high enough to attract most of the electrons supplied by the emitter. Therefore, the reverse bias voltage applied to the collector-base junction is usually much higher than the forward bias voltage across the emitter-base junction.

Biasing PNP Transistors

A PNP transistor has an emitter made of P-type material, an N-type (lightly doped) base, and a P-type collector. In order to properly bias this transistor, we must still forward-bias the emitter junction, and reverse-bias the collector junction. However, since the polarities of the materials have switched, the connection of the external voltages must also switch.

A properly biased PNP transistor is shown in Figure 21-4A. Notice that a positive terminal is connected to the P-type emitter, and the negative terminal of battery A is connected to the base. This forward-biases the emitter junction. Also, the positive terminal of battery B is connected to the base, and the negative terminal of Battery B is connected to the collector, reverse-biasing the collector junction. The type of biasing (forward or reverse) remains the same for both types of transistors. The polarities of the voltages, however, is switched.

BIPOLAR TRANSISTORS

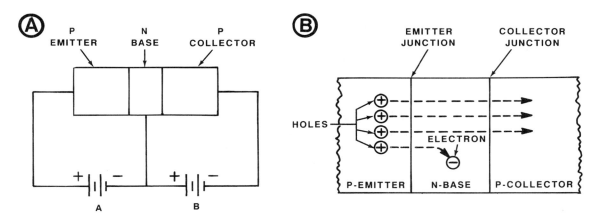

Figure 21-4 A properly biased PNP transistor showing internal current flow.

Also, once the PNP transistor is properly biased, current flow within the transistor is essentially the same. A large number of majority carriers flow from the emitter, and a very small number of these majority carriers combine with the majority carriers in the base, allowing for a small I_B. The rest of the majority carriers are attracted to the potential on the collector (remember: the potential difference on the reverse-biased collector junction is much higher than the potential on the emitter junction) and flow to the external voltage source.

The difference here is the polarity of the majority carriers. In an NPN transistor, the majority carrier for the N-type emitter was the electron. For the PNP transistor, the majority carrier for the emitter is the hole. The flow of majority carriers is the same for both the NPN and PNP transistors, but this movement is electrons for the NPN, and holes for the PNP.

This action is illustrated in Figure 21-4B. The holes (majority carriers) flow from the emitter into the lightly doped base. A small amount of these holes (anywhere from 1-5%) combine with the majority carriers (electrons) in the base. The rest of these holes are attracted to the negative potential on the reverse-biased P-type collector, cross the collector junction, and pass through the collector to the external voltage source.

Transistor Amplification

Now that we have seen the basic action that takes place in a properly biased transistor, we can demonstrate how a transistor can be used to amplify electronic signals. We will use an NPN transistor, using its electronic symbol and representing the circuit as shown in Figure 21-5. Notice that the external voltage sources that are used to forward-bias the emitter junction and reverse-bias the collector junction have also been labeled V_{EE} and V_{CC}, respectively. These two symbols are widely accepted as standard designations for the operating voltages used in this type of circuit. Furthermore, we have made V_{EE} variable so that we may vary the forward bias voltage applied to the emitter junction. The transistor has also been designated as Q_1, since the Q designation is widely used to represent transistors in electronic circuits.

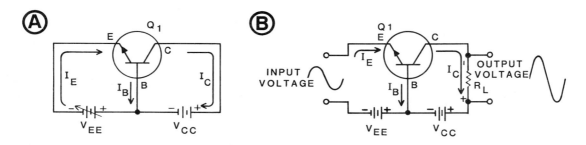

Figure 21-5 Schematic representation of a properly biased NPN transistor with input signal.

In the previous discussion on transistor action we saw that the emitter current (I_E), base current (I_B), and collector current (I_C) could be related mathematically by the expression $I_E = I_B + I_C$. We assumed that all three currents were constant. However, all three of these currents vary in a proportional manner as indicated by the mathematical expression. For example, if I_E doubled in value, then I_B and I_C would likewise double in value. The number of electrons that the emitter supplies to the base region is determined by the amount of forward bias voltage across the emitter junction. When the forward bias voltage (V_{EE}) increases, more electrons are injected into the base region, increasing I_E. Therefore, more electrons are swept into the collector, and I_C increases.

The additional forward-bias also increased the number of holes available in the base. This allows more electrons coming from the emitter to combine with these holes, increasing I_B. When the forward bias voltage decreases, the opposite effect is produced and all three currents decrease. The three currents are *directly proportional*. In other words, they all vary by the same percentage.

BIPOLAR TRANSISTORS

In order for the transistor shown in Figure 21-5A to provide amplification, the device must be capable of accepting an input signal (current or voltage) and provide an output signal that is greater in strength or amplitude. The transistor cannot perform this function if it remains connected as shown in Figure 21-5A. Instead we must make several changes to the circuit so that it appears as shown in Figure 21-5B. Notice that we have added a load resistor between the collector of transistor Q_1 and the positive terminal of V_{CC}. This resistor is used to develop an output voltage with the polarity indicated. In other words the collector current must flow through this resistor to produce a specific voltage drop.

We have also separated the connection between the emitter of transistor Q_1 and the negative terminal of V_{EE}. These open connections serve as the input of the amplifier circuit and allow an input signal to be inserted between the emitter of Q_1 and V_{EE}. Therefore, the input is not an open circuit, but is completed when an external voltage source is connected between the emitter and V_{EE}. Furthermore, the values of V_{EE} and V_{CC} have been selected to bias transistor Q_1 to ensure the values of I_E, I_B, and I_C are high enough to permit proper circuit operation.

The emitter junction of transistor Q_1 is forward-biased and has a relatively low resistance. The collector junction is reverse-biased and has a relatively high resistance. In spite of the tremendous difference in emitter junction resistance and collector junction resistance, the emitter current (I_E) is almost equal to the collector current (I_C). The value of I_C is only slightly less than I_E because a small portion of I_E flows out of the base region to become base current (I_B). The load resistor can therefore have a high resistance value without greatly restricting the value of I_C.

If the two input terminals are connected together by a piece of wire, the emitter junction feels V_{EE}, becomes forward-biased, and the transistor conducts. I_C flows through the load resistor, and a voltage is developed.

When an alternating signal is applied to the input terminals, V_{EE} varies with the input signal. At the zero reference points of the input signal, the emitter junction sees no change in V_{EE}, and the output voltage (across R_L) remains the same. However, one alternation will add to V_{EE}, and the voltage felt at the emitter junction increases. This will cause I_E to increase, which also increases I_C. The additional current flows through R_L, and the output voltage increases.

If one alternation of the input signal causes V_{EE} to increase, then the other alternation will cause V_{EE} to decrease. This decreases I_E, which in turn decreases I_C. R_L feels less current flow, and the output voltage decreases.

This is the basic action of amplification with a transistor. A change in input voltage can produce a corresponding change in output voltage. However, the output voltage change will be much greater than the input change. This is because the output voltage is developed across a high output load resistance (R_L) and the input voltage is applied to the low resistance offered by the emitter junction. A very low input voltage controls the value of I_E, which determines the value of I_C. Even though I_C remains slightly less than I_E, it is forced to flow through a higher resistance and produces a higher output voltage.

Although the example we used was an AC signal, the action just described will occur if the input signal is either a DC or an AC voltage. The important point to remember is that any change in input voltage is greatly amplified by the circuit, producing a larger (proportional) change in output voltage. This may sound very simple, but it is important to remember the *method* in which amplification is accomplished. The input signal is used to control the conduction of transistor Q_1, which controls the current through the load resistor (R_L). The transistor takes energy from the external power source V_{CC}, and applies it to the load resistor in the form of an output voltage.

Any circuit which performs this basic function is commonly referred to as a **voltage amplifier**. Such circuits are widely used in electronic equipment, since it is often necessary to raise both DC and AC voltages to higher levels so that they may be effectively used. The last section of this book covers electronic circuits, and the many types of transistor amplifier circuits are a large portion of this last section. There you will see how a transistor can be connected to different circuits which amplify currents as well as voltages. Any circuit which is used for the sole purpose of converting a low current to a higher current is commonly referred to as a **current amplifier**.

Although an NPN transistor was used in the amplifier circuit shown in Figure 21-5B, a PNP transistor will perform the same basic function. However, in the PNP circuit the bias voltages (V_{EE} and V_{CC}) must be reversed, and the respective currents (I_E, I_B, and I_C) will flow in the opposite directions.

Transistor Amplifier Circuits

As we previously stated, the bipolar transistor is primarily used as an amplifier. The method for amplification we explained earlier is more of an example than an actual circuit. A transistor can amplify in a number of different ways, and you'll discover that each different circuit has its own advantages and disadvantages.

There are three common amplifiers that are created through the use of transistors. Each arrangement uses one of the transistor's leads (emitter, base, or collector) as a reference, with the other two leads as the input and output connections. The three arrangements are known as **common-base, common-emitter,** and **common-collector.** Each of these can be constructed using an NPN or PNP transistor.

Each amplifier has different characteristics. For example, the common-emitter amplifier is used as a power amplifier. This means that the power at the output of a common-emitter is larger than the power at the input. The common-base is used primarily as a voltage amplifier. Any change in current, voltage, or power between the input and the output of an amplifier is described in terms of **gain.** For example, if you apply a 10 volt peak-to-peak signal to the input of an amplifier, and the output is also 10 volts peak-to-peak, the gain of the amplifier is equal to 1. If the output of the amplifier is larger than the input, the gain is more than one. It is not uncommon to see gain values of 250 or higher.

Gain can also be used to describe a loss. For example, if the same 10 volt peak-to-peak is applied, and the output is only 9 volts peak-to-peak, the gain would be less than one. The amount of gain for an amplifier can be calculated, but we'll save that for our section on electronic circuits.

We will now describe the operation and characteristics of each of the three common amplifiers.

SEMICONDUCTORS

Common-Base

In the common-base amplifier, the base is used as the **reference**. This means that the input signal is applied between the emitter and base leads, and the output is taken across the collector and base. The typical circuit arrangements for the common-base amplifier are shown in Figure 21-6. Figure 21-6A shows the arrangement for an NPN transistor, and Figure 21-6B shows the arrangement for a PNP transistor. Notice that the arrangement is the same for each transistor, but the biasing voltages are switched for the PNP transistor because the polarities of the emitter, base, and collector are switched. You may also recognize this arrangement as the one we used to describe basic transistor amplification.

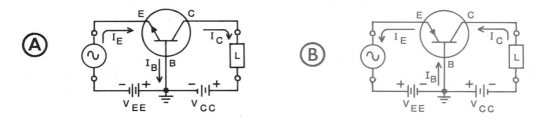

Figure 21-6 NPN and PNP common-base circuits.

In the common-base circuit, the input signal is applied to the emitter. This varies the forward-bias voltage across the emitter junction, labeled V_{EE}. The output of the circuit is taken from the collector and developed through the load, labeled L. The magnitude of the output voltage is directly proportional to the amount of collector current flowing through the load.

Each alternation of the input signal will have a different effect on V_{EE}. One alternation will *add* to V_{EE}, increasing the amount of forward-bias voltage across the emitter junction. This will increase the majority carriers in both the emitter and the base. I_E increases, and the increased number of majority carriers in the base also allows I_B to increase. The increase in I_B is very small, however. The increased emitter current also causes a directly proportional increase in collector current, I_C, and the output voltage across L increases. Therefore, a small change in voltage at the input causes a large change in output voltage, which defines amplification.

Since the input alternations are opposite in polarity, the second alternation will decrease V_{EE}. The decrease in V_{EE} will decrease emitter current, which decreases collector current, and therefore decreases the output voltage. A small change in input voltage, regardless of whether this change is an increase or a decrease, causes a large (proportional) change in output voltage.

The common-base amplifier is not used extensively in electronics. Although it is a good voltage amplifier, the amount of current at the output is slightly less than the current at the input. This is because collector current is always slightly less than emitter current. However, the common-base amplifier *does* amplify power.

Power, if you'll recall is a function of current and voltage. Although current decreases from the input to the output, and voltage increases from the input to the output, the overall amplification of voltage is larger than the loss of current. Therefore, power is increased.

Collector Characteristic Curves

In the previous chapters covering diodes, we used characteristic curves to describe the operation of diodes with various voltages applied. This method can also be used to describe the operation of transistor amplifiers. We use collector characteristic curves to describe the changes in collector current in relation to both emitter current and reverse-bias voltage. Although collector current is primarily controlled by emitter current, it can also be changed (to a lesser degree) by changing the amount of reverse-bias voltage between the collector and the base. With a characteristic curve, we can show the relationship between emitter current, collector current, and reverse-bias voltage.

In order to accurately define this relationship, a number of curves are necessary. Each curve is plotted for a specific amount of emitter current. These curves (often called a family of curves) are shown in Figure 21-7. Notice that these curves start with I_E at 0 milliamperes, and ends at 7 milliamperes. Then, as reverse voltage is adjusted, the corresponding change in collector current is plotted.

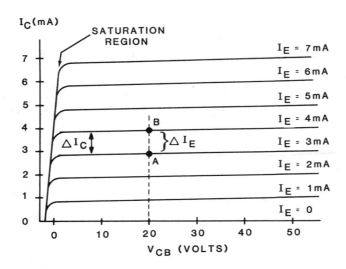

Figure 21-7 Collector characteristic curves for a common-base amplifier.

As you can see, collector current is always very close to emitter current. Initially, collector current rises drastically as reverse bias voltage (V_{CB}) increases to the first reverse-bias value, and then increases only slightly as V_{CB} increases from 0 volts to over 50 volts. What we can conclude from analyzing these characteristic curves is:

1. The largest changes in collector current come from changes in emitter current.

2. Collector current is very close (but less than) emitter current.

3. Changes in reverse bias voltage cause only slight changes in collector current.

Current Gain

The collector characteristic curves just described show the amount of current amplification a common-base amplifier can provide. We know that a common-base amplifier cannot amplify current because collector current will always be slightly less than emitter current. However, it is still common practice to describe the common-base amplifier in terms of its ability to amplify (or provide a gain) in current. This gain must be less than one, showing that the output current will be less than input current.

The current gain is determined by varying the emitter current, and observing the change in collector current. This is accomplished while keeping reverse bias voltage at a constant level. This current gain is identified by the Greek letter **alpha** (α), and is expressed mathematically as:

$$\alpha = \frac{\Delta I_C}{\Delta I_E}$$

Remember that *change* is expressed with the Greek letter delta (Δ). This equation states that the current gain of a common-base amplifier (alpha) is equal to a change in collector current divided by a change in emitter current. Since collector current is always slightly less than emitter current, alpha must always be less than one. For example, between points A and B in Figure 21-7, the emitter current varies from 3 milliamperes to 4 milliamperes. The collector current changes from approximately 2.8 milliamperes to 3.75 milliamperes. This tells us that emitter current change equals approximately 1 milliampere, while collector current change equals approximately 0.95 milliamperes. We can insert these values into our formula, and derive out alpha for this circuit:

$$\alpha = \frac{0.95 \text{ mA}}{1 \text{ mA}} = .95$$

This shows us that the gain for this particular amplifier is .95. In the case of the common-base amplifier, the current gain (input current to output current) can be expressed by alpha, although every transistor has an alpha value. The alpha value for this transistor shows us that the current gain is less than one. This means that there will be less current at the output than at the input. Manufacturers of transistors usually state the alpha value for each type of transistor they produce. Typical alpha values range anywhere from .95 to .995.

More recently, the alpha of a transistor has been referred to as the **forward current transfer ratio,** represented by the symbol h_{fb}. This representation is becoming more popular, while "alpha" is being used less.

SEMICONDUCTORS

Common-Emitter

In Figures 21-8A and 21-8B, the input signal is still applied between the emitter and the base, but now it is on the other side of the battery. This causes the changing input to be felt at the base. The output is still developed by the load, which is connected between the emitter and the collector.

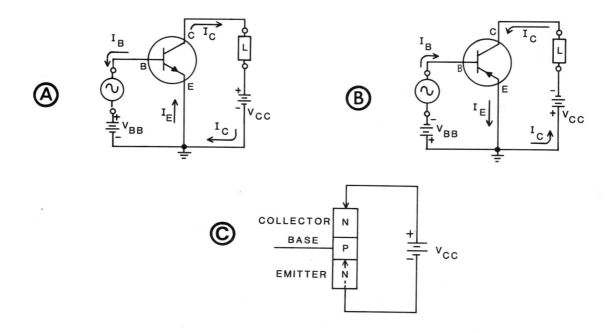

Figure 21-8 Common-emitter circuits with proper biasing.

Since the emitter is common to both the input and the output, this arrangement is known as the **common-emitter** amplifier. Although this arrangement is slightly different than the common-base, it still requires that the transistor be properly biased before it can operate correctly. Figure 21-8A shows the proper biasing for an NPN common-emitter, and Figure 21-8B shows the proper biasing for a PNP common-emitter.

Notice that the forward bias voltage is now labeled V_{BB}. This indicates that the input signal and the battery will vary base current instead of emitter current.

A second difference with this arrangement is that reverse bias voltage is not connected between the collector and the base. However, the collector junction is still reverse-biased. Figure 21-8C illustrates how this is accomplished. When V_{CC} is connected as shown, the emitter junction is forward-biased. This creates a low resistance between the emitter and the base. The positive terminal of V_{CC} makes the N-type collector positive with respect to the base. The low resistance in the emitter junction also effectively allows the negative terminal to be felt at the P-type base. This same action occurs with PNP transistors, although the polarities of the batteries and the transistor regions would be reversed.

The three currents within the circuit (I_B, I_E, and I_C) still share the same relationship as before. I_E is still the sum of I_B and I_C, and any change in one current causes a proportional change in the others. The difference with the common-emitter circuit is that the reference has changed. The transistor is connected so that the very low base current controls the much larger collector current.

As the input signal varies, it will add to V_{BB} or oppose it. This causes I_B to increase or decrease. Any change in base current forces I_E and I_C to change proportionally. This allows the small base current to change the larger collector current, resulting in current amplification. Also, the changing collector current still flows through the high output resistance of the load, which gives us voltage amplification.

The common-emitter amplifier, amplifying both current and voltage, is primarily used as a power amplifier. If you'll recall, the common-base amplifier also amplified power, but this was due to the large amplification of voltage overriding the loss in current. In the common-emitter, both current and voltage are amplified, resulting in an extremely high power amplification. The common-emitter amplifier is the only power amplifier which accomplishes its function by amplifying both current *and* voltage. The common-emitter is also the only arrangement to provide phase-shift. The output signal is 180 degrees out-of-phase with the input signal. That is, the positive alternation of input produces a negative alternation at the output.

SEMICONDUCTORS

Collector Characteristic Curves

The operating characteristics of the common-emitter amplifier can be shown using the collector characteristic curves shown in Figure 21-9. There are a number of differences between this family of curves and the family of curves for the common-base amplifier. First, the curves for the common-emitter show the difference between base current and collector current, while the common-base amplifier used emitter current and collector current.

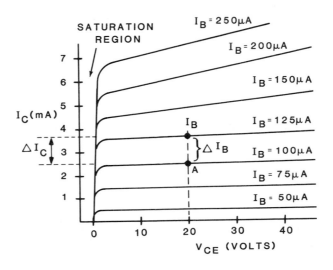

Figure 21-9 Collector characteristic curves for a common-emitter amplifier.

Second, the common-base amplifier shows these voltages with respect to collector-base voltage (V_{CB}), while the common-emitter shows its current values with respect to collector-emitter voltage (V_{CE}). The reason for this change is that in the common-emitter amplifier, V_{CE} provides reverse bias voltage, and V_{CB} provides reverse-bias voltage for the common-base amplifier. These reverse bias voltages will have a small effect on I_C, and are required in the curve to show their effect.

The curves in Figure 21-9 are plotted for various I_B values ranging from 0 to 250 microamperes. I_C is plotted vertically, and V_{CE} is plotted horizontally. Notice that I_C is measured in milliamperes, while I_B is measured in microamperes.

These curves are similar to the curves for the common-base amplifier, but they have a greater slope. This indicates that a change in V_{CE} has a greater effect on collector current in a common-emitter than V_{CB} has on collector current in a common-base.

We should also point out that the area to the left of the knee of the curve represents the saturation region. This is the region where collector current rises rapidly with a small change in V_{CE}. Transistors are normally biased beyond this region (to the right of the knee) to ensure collector current changes only slightly with a change in V_{CE}.

The curves in Figure 21-9 may be used to determine the corresponding values of I_B, I_C, and V_{CE} at any specific operating point. For example, when V_{CE} is equal to 20 volts and I_B is equal to 100 microamperes, I_C will be equal to approximately 2.5 milliamperes, as shown at Point A. The larger I_C value shows that a common-emitter amplifier is capable of providing substantial current gain.

Current Gain

As we stated earlier, the amount of current gain in a common-base amplifier is expressed by the Greek letter alpha (α), calculated by dividing the change in emitter current by the change in collector current. This is accomplished while V_{CB} remains constant. The alpha value is specifically used for the common-base because the input and output signals are applied to the emitter and collector, respectively, and alpha is a function of emitter and collector current.

Alpha will not be able to tell us much about the current gain for the common-emitter amplifier, however, because the input is applied to the *base* of the transistor. In addition, the reverse voltage is now applied to the emitter and collector (V_{CE}) instead of the collector and the base (V_{CB}).

In order to express the current gain in a common-emitter amplifier, we need to compare the amount of change in base current to the amount of change in collector current. Again, this must be compared while V_{CE} is held constant. The Greek letter beta (β) is used to express this relationship. The formula for beta is:

$$\beta = \frac{\Delta I_C}{\Delta I_B}$$

Again, the Greek letter delta (Δ) is used to express "change." This equation states that the current gain in a common-emitter amplifier (or beta) is equal to the change in collector current divided by the change in base current. Like alpha, beta is simply a number, and not a specific unit of measurement. Also, every transistor has a beta value, since every transistor has base and collector currents.

We found earlier that at Point A in Figure 21-9, I_C had a value of 2.5 milliamperes when I_B was equal to 100 microamperes and V_{CE} was at 20 volts. To find a change in these currents, we move up to Point B in Figure 21-9. V_{CE} remains at 20 volts (remember: V_{CE} must be held constant to find beta), and I_B is equal to 125 microamperes, for a change of 25 microamperes. At Point B, I_C is equal to approximately 3.6 milliamperes, which means that I_C has changed a total of 1.1 milliamperes. We can insert these values into our equation, and calculate beta:

$$\beta = \frac{1.1 \text{ mA}}{25 \text{ }\mu\text{A}} = \frac{0.0011 \text{ A}}{0.000025 \text{ A}} = 44$$

This means that the common-emitter amplifier tested in Figure 21-9 has a beta value of 44. This value could have been found at any point along the characteristic curve, provided you remained at the same V_{CE} value. Betas of 44 are common for small and medium power transistors. Beta values can range anywhere from 10 to 200.

Like alpha, manufacturers usually specify the beta value for each type of transistor they produce. Also, beta has more recently been referred to as the **forward current transfer ratio**, and is labeled h_{fe}.

Comparing Alpha and Beta

Alpha, you have learned, is a function of collector and emitter currents. Beta is a function of collector and base currents. Since emitter, collector, and base currents are all interrelated, we can use one of our two ratio values to find the other. Alpha can be used to calculate for an unknown beta, and vice-versa. The following equation may be used to find beta if alpha is known:

$$\beta = \frac{\alpha}{1 - \alpha}$$

Let's assume that a transistor has an alpha of 0.98. We can insert this value into our equation and calculate the beta value of the transistor:

$$\beta = \frac{0.98}{1 - 0.98} = \frac{0.98}{0.02} = 49$$

The equation for finding alpha using beta is somewhat different:

$$\alpha = \frac{\beta}{\beta + 1}$$

For example, if a transistor has a beta of 100, what is its alpha value?

$$\alpha = \frac{100}{100 + 1} = \frac{100}{101} = 0.99$$

When one of these values are known, it is easier to find the other value with these formulas, rather than attempt to read the characteristic curves. It is often very difficult to obtain accurate readings from these curves.

Common-Collector

The third most common type of transistor amplifier is the **common-collector**, which is also known as the **emitter-follower** amplifier. Figure 21-10 illustrates both the NPN common-collector (21-10A) and the PNP common-collector (21-10B). Notice that the emitter on each of these is on the top in the schematic instead of the bottom. It is very easy to become confused if you do not remember this.

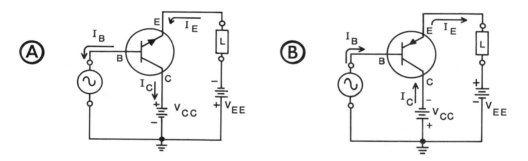

Figure 21-10 Common-collector circuits.

In the common-collector, the input signal is applied to the base of the transistor, and the output is taken across the emitter. The varying input to the base causes I_B to vary, which varies I_E and I_C in a proportional manner. Since the load is connected across the emitter, the output will feel I_E, which has the highest amount of current for any point of the transistor. A small change in I_B (which is the smallest current flow through the transistor) will cause a large change in I_E. The result is a substantial current gain.

SEMICONDUCTORS

The input voltage, however, will not be amplified. Remember, the input is taken across the base, and the output is taken across the emitter. In order to forward bias the emitter junction, the voltage on the emitter must always be slightly less than the voltage on the base. As a result, the output voltage (on the emitter) will always be slightly less than the voltage on the base (the input voltage). This results in a voltage gain of less than one, but very close to one. Common voltage gains for a common-collector amplifier are 0.97 - 0.99. The voltage at the emitter tends to follow proportionally to the voltage on the base. This is why the common-collector is referred to as an emitter-follower.

There is, on the other hand, a slight gain in power. If you'll recall the common-base amplifier, the voltage gain (greater than one) is much larger than the current loss, and this results in a gain in power. The same is true for the common-collector. The large gain in current outweighs the slight voltage loss, and the overall effect is an increase in power.

We should point out that although all three amplifiers (common-base, common-emitter, and common-collector) amplify power, that is not their primary use. The power gains in the common-base and the common-collector, which have a loss in current and voltage respectively, are small compared to the power gain in the common-emitter, which amplifies both current and voltage.

The most common function for a common-collector amplifier is in circuits which require a high input resistance and a low output resistance. The input for the common-collector is felt at the reverse-biased collector junction, which has a high resistance value. The output, however, is taken across the forward-biased emitter junction, whose resistance is extremely low. This allows the common-collector to act as an impedance-matching circuit, much like the transformers we discussed in Chapter 17.

Current Gain

Expressing the current gain in a common-collector is not as easy as with the common-emitter and the common-base. However, using these known formulas and our knowledge of the relationship between the three currents in a transistor, we can derive a mathematical formula for the common-collector's current gain. We must derive a method of expressing the change in base current as compared to the change in emitter current.

In other words, we must express this relationship:

$$\text{current gain} = \frac{\Delta I_E}{\Delta I_B}$$

We can express emitter current in terms of base current and collector current, since we know that emitter current is the sum of these values. From there, it is simple algebra:

$$\text{current gain} = \frac{\Delta(I_B + I_C)}{\Delta I_B} = \frac{\Delta I_B + \Delta I_C}{\Delta I_B} = \frac{\Delta I_B}{\Delta I_B} + \frac{\Delta I_C}{\Delta I_B} = 1 + \frac{\Delta I_C}{\Delta I_B}$$

As you know, I_C/I_B is the formula for beta (β). Therefore, we can express the current gain for a common-collector with the formula:

$$\text{current gain} = 1 + \beta$$

In other words, if a transistor has a beta of 49, the current gain—if connected in a common-collector arrangement—would be 50. However, if a transistor's beta is extremely high, the current gain in a common-collector arrangement would be equal to beta. This is because when you have an extremely high beta, the difference between emitter current and collector current is nominal.

Up to now, we have expressed beta for *all* transistors using the Greek letter delta (Δ) to indicate change. More specifically, this is referred to as **AC beta** because we are discussing the ratio between the *change* in collector and base current. However, if we were to properly bias a transistor using steady DC voltages, we could measure the values for collector and base current, and calculate beta at a steady point. This can also be accomplished for alpha. Values of this nature are referred to as **DC beta** and **DC alpha**. The DC values have the exact same formula, without the Greek letter delta (Δ) to indicate change, since there is none.

Transistor Ratings

During all of our discussions on transistor operation to this point, we assumed each transistor was working within safe operating limits. However, transistors can be damaged if they are subjected to excessively high current or voltage values. This makes it extremely important to know the maximum current or voltage each device can withstand. Manufacturers will usually specify this information for each type of transistor they produce. In addition to current and voltage, they may also specify the power and temperature limitations. These values are often referred to as **maximum transistor ratings**. The following discussion briefly describes some of the more important ratings.

SEMICONDUCTORS

Collector Breakdown Voltage

This value is similar to the breakdown voltage for a normal junction diode. This is the voltage that causes a severe increase in reverse-bias current through the collector junction. Collector breakdown voltage is often referred to as V_{CBO}.

Emitter Breakdown Voltage

Like V_{CBO}, emitter breakdown voltage (designated V_{EBO}) is the amount of reverse-bias voltage necessary to cause a sharp increase in reverse-bias current through the emitter-base junction.

Both V_{CBO} and V_{EBO} are specified at a value which will cause a specific amount of reverse current. Obviously, if either of these values are exceeded, possible damage to the transistor could result. Transistors should always be connected to external voltages well below these values.

In addition to specifying the maximum amount of reverse voltage, manufacturers also specify the maximum emitter and collector currents that each transistor can conduct without damage. Like V_{EBO} and V_{CBO}, these maximum current values should never be exceeded.

Maximum Collector Dissipation

A bipolar transistor dissipates power in the form of heat like any other component. Practically all of the power dissipated by a transistor occurs at the point of maximum resistance within the transistor—the collector junction. The amount of power dissipated can be easily calculated by multiplying I_C and V_{CE}. The maximum amount of power a transistor can dissipate is referred to as the **maximum collector dissipation rating**. These power ratings are measured at 25 degrees centigrade, and can range anywhere from 400 milliwatts to several hundred watts for high-power transistors. Temperature is specified because a change in temperature will cause a change in the operating characteristics of the transistor. Also, this rating is determined without a load connected to the transistor.

Chapter Self-Test

1. Which bipolar transistor utilizes an N-type region that is sandwiched between two P-type regions?

2. How many junctions does a bipolar transistor have?

3. Where is the arrow in the transistor's schematic symbol always located?

4. What is the junction formed between the emitter and base regions called?

5. Describe the biasing requirements for a bipolar transistor.

6. Is the collector current flowing through a transistor exactly equal to the emitter current? Why?

7. Where does all the current flowing through a transistor start?

8. Describe the relationship between emitter, base, and collector currents in a transistor.

9. Are PNP transistor bias polarities the same as the NPN transistor? Why?

10. An input signal is applied between the emitter and base of a transistor. The first alternation of the input signal causes V_{EE} to decrease. What happens to the output voltage across the load resistor, and how did it occur?

11. What are the three most common types of transistor amplifiers?

12. Define *gain*.

13. Identify the input and output terminals, as well as the primary function and amount of current gain for the common-base amplifier.

14. Identify the input and output terminals, as well as the primary function and amount of current gain for the common-emitter amplifier.

15. Identify the input and output terminals, as well as the primary function and amount of current gain for the common-collector amplifier.

16. What is another name for the common-collector amplifier? How was it derived?

17. List the many different characteristics that manufacturers will specify for a transistor.

Summary

Bipolar transistors are two junction devices that consist of three alternately doped semiconductor regions. A transistor that has a P-type region between two N-type regions is called an NPN transistor. Similarly, a transistor having an N-type region between two P-type regions is a PNP transistor.

The center region of a transistor is called the base, while the two outer regions are called the emitter and the collector. The base region is usually much thinner than the other two regions and is lightly doped. The base is P-type material in an NPN transistor, and N-type material in a PNP transistor.

The junction formed between the transistor's emitter and base is often called the emitter-base or emitter junction. Similarly, the junction between the collector and base is called the collector-base or collector junction. Under normal operating conditions, the emitter junction is forward-biased, and the collector junction is reverse-biased. This means that in an NPN transistor, the N-type emitter must be negative with respect to the P-type base, while the base is negative with respect to the N-type collector.

In a properly biased NPN transistor, electrons flow from the transistor emitter to the base. A few of these electrons leave the base as base current. However, most of the electrons pass through the base to the positive collector as collector current. Therefore, the emitter current is equal to the sum of the base and collector currents. However, when transistor circuits are being analyzed, emitter current and collector current can be considered equal because the base current is so low. Remember that in N-type material electrons are the majority carriers.

Operation of a PNP transistor is similar to that of an NPN transistor. Since holes are the majority carriers in P-type material, the operation of a PNP transistor depends on hole movement instead of electrons. Holes are confined to the internal structure of the transistor. Emitter, base, and collector currents are still produced by electrons flowing into and out of the device. A PNP transistor is properly biased when its emitter is positive with respect to the base and the collector is negative with respect to the base.

Practically all transistor applications are based on a transistor's ability to amplify weak electronic signals. Amplification is the ability to increase the amplitude of an electronic signal. The amplitude of the input signal can be expressed in terms of voltage or current. However, before the transistor can amplify, it must be connected to an external circuit in a manner that allows the input signal to control the transistor's conduction. The transistor then controls the current through an external load, such as a resistor.

BIPOLAR TRANSISTORS

A bipolar transistor can be connected in one of three basic circuit arrangements. These arrangements are referred to as the common-base, common-emitter, and common-collector amplifiers. The common-base amplifier provides voltage amplification, the common-emitter amplifier provides power amplification, and the common-collector is used to match a high resistance source to a low-resistance load.

The amount of current amplification—or gain—that a transistor can provide can be expressed as a ratio of its input and output currents. The ratio of emitter current to collector current is referred to as the transistor's alpha (α) and is always less than one. Alpha can describe the amount of gain for a common-base amplifier. The ratio of base current to collector current is known as beta (β), and can be used to describe the amount of gain in a common-emitter. Alpha and beta values can be determined using either AC (changing) values or DC (fixed) values.

SEMICONDUCTORS

CHAPTER 22

Field Effect Transistors

SEMICONDUCTORS

Contents

Introduction 533

Chapter Objectives 534

The Junction FET 535

The Insulated Gate FET 543

FET Amplifier Circuits 552

Chapter Self-Test 555

Summary 556

FIELD EFFECT TRANSISTORS

Introduction

This chapter examines another important solid-state component, commonly referred to as a field effect transistor. The field effect transistor (also called a FET) operates on a principle that is completely different from that of the conventional bipolar transistor. The FET is a three-terminal device that is capable of providing amplification. This ability to amplify allows the FET to compete with conventional bipolar transistors in many applications. A basic understanding of FET operation and construction is therefore essential if you are to have a well-rounded background in solid-state components.

Basically there are two types of field effect transistors. One type is known as a junction field effect transistor, but is commonly referred to as a junction FET or simply a JFET. The second type is known as an insulated gate field effect transistor (IGFET), although it is frequently referred to as a metal-oxide semiconductor field effect transistor (MOSFET). You will examine both of these basic FET devices, and learn how each device is constructed and how it operates.

SEMICONDUCTORS

Chapter Objectives

When you have completed this chapter, you will be able to:

1. Describe how a junction FET operates.

2. Use a FET's drain characteristic curves to determine the transconductance of the device.

3. Properly bias both N-channel and P-channel JFETs.

4. Explain the meaning of the expressions V_{GS} (off and V_P).

5. Explain the basic difference between the JFET and MOSFET.

6. Describe the difference between depletion- and enhancement-mode IGFETs.

7. Determine the proper biasing voltages for both the depletion-mode and enhancement-mode IGFETs.

8. Name the three basic FET circuit arrangements.

9. Explain the advantages and disadvantages of FETs when compared to bipolar transistors.

FIELD EFFECT TRANSISTORS

The Junction FET

The junction FET (also called a JFET) finds many applications in electronic circuits. This device is constructed from N-type and P-type semiconductor materials and is capable of amplifying electronic signals, much like the conventional bipolar transistor. However, the junction FET is constructed in a different manner than a bipolar transistor, and operates on an entirely different principle.

JFET Construction

The construction of a junction FET begins with a lightly doped semiconductor material (usually silicon) which is referred to as the substrate. The substrate simply serves as a platform on which the remaining electrodes are formed, and can be either a P-type or an N-type material. Through the use of various growth techniques, an oppositely doped region is formed within the substrate to create what is effectively a PN junction. However, it is the unique shape of this PN junction that is important.

The structure that is created by the process described above is shown in Figure 22-1. The region that is embedded in the substrate material is U-shaped so that it is flush with the upper surface of the substrate at only two points.

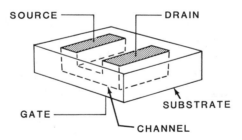

Figure 22-1 Basic construction of a junction FET.

The embedded region actually forms a channel of oppositely doped semiconductor material through the substrate. Therefore, the embedded region is generally referred to as the channel. When this channel is made from an N-type material and embedded in a P-type substrate, the entire structure is known as an N-channel junction FET. However, when a P-type channel is implanted in an N-type substrate, the device becomes a P-channel junction FET. Therefore, a junction FET is similar to a conventional bipolar transistor in that it can be constructed in two different ways. The junction FET is either an N-channel or a P-channel device, while the bipolar transistor is either an NPN or a PNP device.

SEMICONDUCTORS

The construction of a basic junction FET is completed by making three electrical connections to the device as shown in Figure 22-1. One wire or lead is attached to the substrate as shown. This connection is referred to as the **gate**. Leads are also attached to each end of the channel. These two leads are referred to as the **source** and the **drain**. In most junction FETs, the channel is geometrically symmetrical and it makes no difference which end of the channel is used as the source and which is used as the drain. Only in special types of FETs will the channel be asymmetrical; when using these special devices, the source and drain leads cannot be interchanged.

JFET Operation

Like a conventional bipolar transistor, a junction FET requires two external bias voltages for proper operation. One voltage source is normally connected between the source and drain leads so that a current is forced to flow through the channel within the device. The second voltage is applied between the gate and the source, and it is used to control the amount of current flowing through the channel.

Refer to Figure 22-2A. This figure shows a cross-sectional view of an N-channel junction FET and its required operating voltages. Notice that an external voltage source is connected between the drain (D) and the source (S) leads. This **drain-to-source voltage** is represented by the symbol V_{DS}. The voltage V_{DS} is connected so that the source is made negative with respect to the drain. This voltage causes a current to flow through the N-type channel because of the majority carriers (free electrons) within the N-type material. Current within a FET normally flows from the source to the drain. This source-to-drain current is commonly referred to as the FET's **drain current** and is represented by the symbol I_D. The channel simply appears as a resistance to the supply voltage V_{DS}.

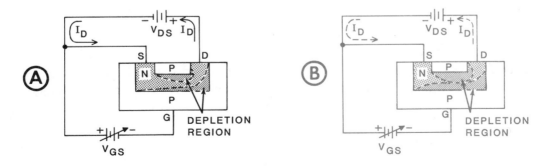

Figure 22-2 A properly biased N-channel junction FET.

FIELD EFFECT TRANSISTORS

In Figure 22-2A, a voltage is also applied between the gate (G) and the source (S) of the FET. This **gate-to-source voltage** (designated as V_{GS}) causes the P-type gate to be negative with respect to the N-type source. Since the source is effectively just one end of the N-type channel, V_{GS} effectively reverse-biases the PN junction formed by the P-type gate and the N-type channel.

This reverse bias voltage causes a depletion region (an area devoid of majority carriers) to form within the vicinity of the PN junction. This depletion region spreads inward along the length of the channel. Although two depletion regions appear to exist, only one is created. This depletion region extends around the wall of the N-type channel since all sides of the channel are in contact with the P-type substrate (which serves as the gate). Furthermore, the depletion region will be somewhat wider at the drain end of the channel than at the source end. This is because V_{DS} effectively adds to V_{GS} so that the voltage across the drain end of the PN junction is higher than the voltage across the source end of the junction.

The size of the depletion region in Figure 22-2A is controlled by voltage V_{GS}. When V_{GS} increases, the depletion region increases in size. When V_{GS} decreases, the depletion region decreases in size. Furthermore, when the depletion region increases in size, the N-type channel is effectively reduced in size (fewer free electrons are available) and less current will be able to flow through the channel. The opposite is true when the depletion region decreases in size. This means that V_{GS} may be used to effectively control the drain current (I_D) flowing through the channel. An increase in V_{GS} results in a decrease in I_D and vice-versa.

Remember, V_{GS} reverse-biases the PN junction formed by the gate and the channel. Therefore, only an extremely small (almost insignificant) leakage current flows from the gate to the source.

It is important to note that a voltage (V_{GS}) is used to control the drain current (I_D) in a junction FET. In normal operation, the voltage applied between the gate and the source (V_{GS}) serves as an input voltage to control the device. The drain current (I_D) represents the output current which can be made to flow through a load. This action is considerably different from the action that takes place in a bipolar transistor. In the bipolar transistor, an input current (not a voltage) is used to control an output current. Also, since the gate-to-source junction of the FET is reverse-biased by V_{GS}, the FET has an extremely high input resistance. This is just the opposite of a bipolar transistor, which has a forward-biased emitter-to-base junction and a relatively low input resistance. The gate-to-source voltage (V_{GS}) must never be reversed allowing the gate and the channel to become forward-biased. This would result in a relatively large current through the junction, allowing very little current to flow to the drain, which severely decreases the FET's amplifying abilities.

Gate-to-Source Cutoff Voltage

We mentioned earlier that when V_{GS} is increased, the depletion region within the FET increases in size. This allows less drain current (I_D) to flow. If V_{GS} is increased to a sufficiently high value, the depletion region is increased in size until the *entire channel* is depleted of majority carriers as shown in Figure 22-2B. This causes I_D to decrease to an extremely small value and for all practical purposes is reduced to zero. The gate-to-source voltage required to reduce I_D to zero (regardless of the value of V_{DS}) is referred to as the gate-to-source cutoff voltage and is represented by the symbol $V_{GS(off)}$. Manufacturers of FETs usually specify the applicable cutoff voltage for each type of FET that they produce.

Pinch-Off Voltage

The drain-to-source voltage (V_{DS}) also has a certain amount of control over the depletion region within the junction FET. The effect of V_{DS} can be noted in Figures 22-2A and 22-2B where the depletion region is wider near the drain than it is near the source. As explained earlier, this occurs because V_{DS} is in series with V_{GS}, causing a greater voltage to exist across the PN junction near the drain. If V_{DS} increases in value, the action becomes even more pronounced and can affect the drain current (I_D) flowing through the device.

If V_{DS} is increased from zero to higher values, I_D is also increased. However, a continued increase in V_{DS} will not result in a constant rise in I_D. Instead, a point is soon reached where I_D levels off and then increases only slightly as V_{DS} continues to rise. This action occurs because the size of the depletion region increases (especially near the drain) as V_{DS} becomes so depleted of majority carriers that it will not allow I_D to increase proportionally with V_{DS}. In other words, a point is reached where the resistance of the channel effectively begins to increase as V_{DS} increases, thus causing I_D to increase at a much slower rate.

Since I_D levels off as the depletion region expands and effectively reduces the channel width, I_D is said to be pinched-off. The value of V_{DS} required to pinch off or limit I_D is referred to as the pinch-off voltage and is represented by the symbol V_P. Manufacturers usually provide the value of V_P for a given FET for a gate-to-source voltage (V_{GS}) of zero. V_P is measured by shorting the gate and source leads. This means that I_D will increase up to its maximum possible value (when $V_{GS} = 0$) and then level off. When V_{GS} is equal to zero, the drain current flowing through the FET is often identified as I_{DSS} instead of I_D. Furthermore, manufacturers often provide the FET's I_{DSS} value when V_{DS} is equal to or greater than V_P. In this case I_{DSS} represents the FET's maximum drain current when V_{GS} is equal to zero.

FIELD EFFECT TRANSISTORS

In practice, the value of V_P (when $V_{GS} = 0$) will always be close to the value of $V_{GS(off)}$ for any given FET. In fact, these two quantities may be interchanged in any calculation in which either quantity is involved. This means that when V_{GS} is equal to or greater than V_P, the drain current (I_D) will be effectively reduced to zero. Also, when V_P is equal to the $V_{GS(off)}$ value, the drain-current (I_D) flowing through the device will be effectively pinched-off.

Drain Characteristic Curves

During the chapters covering junction and zener diodes, you saw how characteristic curves could be used to show the relationships between the biasing voltages and current flows for various components. A similar set of curves can be used to show the relationship between V_{GS}, V_{DS}, and I_D in a junction FET. In this case they are referred to as *drain* characteristic curves.

A typical set (sometimes called a family) of drain characteristic curves is shown in Figure 22-3. These curves show how I_D and V_{DS} vary in relation to each other for various values of V_{GS}. Each curve is formed by alternately adjusting V_{GS} to a specific value and then increasing V_{DS} from zero to some maximum value, while observing the change in I_D. Notice that when V_{GS} is equal to zero, I_D increases rapidly as V_{DS} increases from zero. However, I_D soon levels off at some maximum value as shown. When this point is reached, the corresponding I_{DSS} and V_P values previously described are obtained.

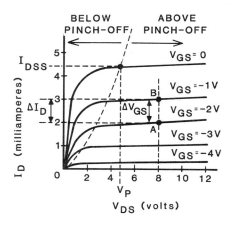

Figure 22-3 Typical drain characteristic curves for an N-channel FET.

SEMICONDUCTORS

The remaining curves in Figures 22-3 are plotted for higher values of V_{GS}. Notice that for each higher value of V_{GS}, I_D levels off at a lower value so that the corresponding pinch-off voltage (V_P) must also be lower. The dashed line that curves upward and to the right crosses each curve at the approximate point where I_D is pinched-off. When the FET is biased so that it is operating to the left of this dashed line, the device is said to be operating *below* pinch-off.

When the FET is biased to operate to the right of the dashed line, the device is operating *above* pinch-off. This region to the right of the line is often referred to as the **pinch-off region**. In most applications, the FET is biased so that it operates above pinch-off, or within the pinch-off region. Operation within the pinch-off region is assured by simply making V_{DS} higher than V_P or $V_{GS(off)}$.

As shown in Figure 22-3, the drain current (I_D) will be maximum (for any specific value of V_{DS}) when V_{GS} is equal to zero. However, I_D will decrease as V_{GS} increases in value. As explained earlier, this action results because the channel within the FET becomes depleted of majority carriers as V_{GS} increases, thus offering a higher resistance to the flow of current. Since the operation of the junction FET is controlled by varying the depletion region within the device, the junction FET is said to operate in the depletion mode.

Transconductance

Like the bipolar transistors that were discussed previously, the FET is most often used to amplify electronic signals. In the case of a bipolar transistor, this amplifying ability can be expressed mathematically as a ratio of input and output currents. In the case of the FET, a similar mathematical relationship can also be used.

The amplifying ability of a FET is measured by noting the effect that the gate-to-source voltage (V_{GS}) has on drain current (I_D). V_{GS} is varied a small amount, and the corresponding change in I_D, is observed. Then these two quantities are expressed as a mathematical ratio. This ratio is commonly referred to as the FET's **transconductance** and is expressed mathematically as:

$$gm = \frac{\Delta I_D}{\Delta V_{GS}}$$

FIELD EFFECT TRANSISTORS

This equation simply states that the transconductance (designated as **gm**) is equal to a small change in I_D divided by a corresponding change in V_{GS}. Although not shown in this equation, the FET's drain-to-source voltage (V_{DS}) must be held constant when these changes are observed. We discussed the term conductance back in the chapter on resistance. We stated that conductance was the reciprocal of resistance, and is measured in mhos (Ω). Transconductance for FETs is also the reciprocal of resistance, and is also measured in mhos.

Although the quantities I_D and V_{GS} could be measured in a test circuit, these quantities can also be determined graphically by referring to an applicable set of drain characteristic curves. To demonstrate how this is accomplished we will use the typical curves shown in Figure 22-3. We will determine the transconductance of the FET within the pinch-off region (above pinch-off) since this is the region most commonly used.

We will assume that V_{DS} remains constant at 8 volts, and V_{GS} changes from 1 volt to 2 volts as indicated at points A and B in Figure 22-3. This change in V_{GS} causes I_D to change from approximately 3 to 2 milliamperes. A total change in V_{GS} of 2 − 1 or 1 volt corresponds to a change in I_D of 3 − 2 or 1 milliampere. When these corresponding changes in I_D and V_{GS} are inserted in the equation given above, we obtain a transconductance of:

$$gm = \frac{\Delta I_D}{\Delta V_{GS}} = \frac{1mA}{1V} = \frac{0.001}{1} = 0.001 \; \Omega \text{ or } 1 \text{ m } \Omega.$$

This calculation tells us that the transconductance between points A and B in Figure 22-3 is equal to 0.001 mho, or 1 millimho. But this value will be different at different operating points on the curves, because the curves are not equally spaced within the pinch-off region.

The greater the change in I_D for a change in V_{GS}, the higher the gain of the FET. In general a high gm is a desirable FET characteristic.

Symbols

To this point we have examined only the N-channel junction FET. The operation of a P-channel FET has not been discussed because it operates in the same manner as the N-channel device, and has the same basic characteristics. The primary difference is in the manner in which the drain current (I_D) flows through the channel. In a P-channel FET, I_D is supported by the movement of holes in a P-type channel. However, these holes are still the majority carriers within the P-type channel just as the electrons are the majority carriers in the N-channel FET.

Also, the P-channel FET has a gate (or substrate) that is formed from N-type material. This is opposite to the conditions that exist within an N-channel FET. This means that the polarities of the bias voltages (V_{GS} and V_{DS}) are exactly opposite for the N-channel and P-channel devices.

An N-channel junction FET and a P-channel junction FET are shown in Figure as they would appear in a circuit diagram or schematic. The required bias voltages for each device are shown. Notice that the symbol used for each device is almost identical. The only difference is the direction of the arrow on the gate (G) lead. The N-channel junction FET symbol shown in Figure 22-4A uses an arrow that points inward. However, the P-channel junction FET symbol in Figure 22-4B uses an arrow that points outward.

Figure 22-4 Schematic representation of properly biased N-channel and P-channel junction FETs.

Also notice that the polarities of the bias voltages are exactly opposite. The N-channel FET must be biased so that its drain (D) is positive with respect to its source (S), and its gate (G) must be negative with respect to the source. The negative potential on the gate accounts for the use of the minus signs before the V_{GS} values in Figure 22-3.

The P-channel FET must be biased so its drain is negative with respect to its source, and its gate must be positive with respect to the source. The drain current (I_D) will therefore flow in a direction that is opposite to the drain in an N-channel FET. However, the majority carriers (holes) within the P-type channel always move from the source to the drain just like the majority carriers (electrons) in the N-channel device.

FIELD EFFECT TRANSISTORS

The Insulated-Gate FET

The gate and channel regions within a junction FET form a conventional PN junction, which is reverse-biased by connecting an external voltage between the FET's gate and source leads. The external voltage causes the FET to operate in the *depletion mode*, and allows the device to have an extremely high input resistance.

However, there is another type of FET which does not have a conventional PN junction, yet still must be reverse-biased. This device uses a metal gate which is electrically insulated from its semiconductor channel by a thin oxide layer and goes by two names: the insulated gate FET (IGFET), or the metal-oxide semiconductor FET (MOSFET). Unlike the junction FET, the IGFET (MOSFET) is designed to operate in one of two distinct modes. The IGFET may be either a depletion-mode device or an enhancement-mode device. You will now see how insulated-gate FETs are constructed, and learn about their important electrical characteristics.

Depletion-Mode Devices

A cross-sectional view of a depletion-mode IGFET (MOSFET) is shown in Figure 22-5A. This device is formed by implanting an N-type channel within a P-type substrate. A thin insulating layer (silicon dioxide) is then deposited on top of the device. The opposite ends of the N-type channel are left exposed so that wires or leads can be attached to the channel material. These two leads serve as the FET's source and drain. A thin metallic layer is then attached to the insulating layer so that it is directly over the N-type channel. This metal layer serves as the FET's gate, and signals are applied to this metal gate through a suitable wire or lead. An additional lead is also connected to the substrate as shown.

SEMICONDUCTORS

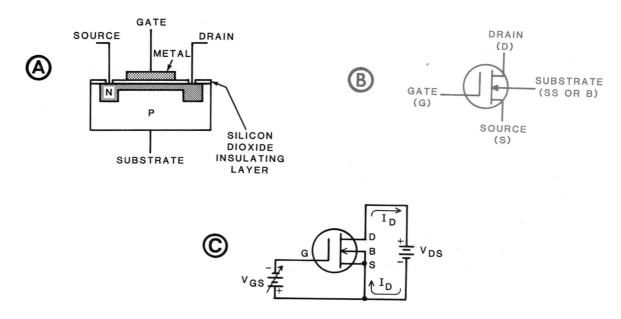

Figure 22-5 An N-channel depletion-mode IGFET with proper biasing.

It is important to note that the metal gate is insulated from the semiconductor channel by a layer of silicon dioxide. Therefore, the gate and the channel do not form a conventional PN junction. However, this insulated metal gate can be used to control the conductivity of the channel, thus allowing the device to operate in a manner similar to that of a junction FET. In other words the insulated gate can deplete the N-type channel of majority carriers (electrons) when a suitable bias voltage is applied, even though a semiconductor junction does not exist between the gate and the channel.

The insulated gate FET shown in Figure 22-5A has an N-type channel; therefore its source and drain leads are biased in the same manner as the source and drain leads of an N-channel junction FET—the drain is always made positive with respect to the source. The majority carriers (electrons) within the N-type channel allow current to flow through the channel (from the source to the drain). The source-to-drain current (normally called drain current) is in turn controlled by a gate-to-source bias voltage, just like a junction FET. When the gate-to-source voltage is equal to zero, a substantial drain current will flow through the device because a large number of majority carriers are present in the channel. If the gate is made negative with respect to the source, the channel becomes depleted of many of its majority carriers, and drain current decreases just as it would in a junction FET. If this negative gate voltage is increased to a sufficiently high value, the drain current will drop to zero just like it does in the junction FET.

There is, however, one important difference between the operation of the N-channel insulated gate FET just described and the N-channel junction FET. The gate of the N-channel insulated gate FET can be made positive with respect to the FET's source. This must never occur in an N-channel junction FET because the PN junction formed by the gate and the channel would be forward-biased.

The insulated gate FET can handle a positive gate voltage because the silicon dioxide insulating layer prevents any current from flowing through the gate lead. The FET's input resistance remains at a high value. Also, when a positive voltage is applied to the gate, more majority carriers (electrons) are drawn into the channel thus enhancing the conductivity of the channel. This is exactly opposite the action that takes place when the gate receives a negative voltage. A positive gate voltage can be used to increase the FET's drain current, while a negative gate voltage will decrease the drain current.

Since a negative gate voltage is required to *deplete* the N-type channel insulated gate FET just described, this FET is called a **depletion-mode** device. Like the junction FET described earlier, this IGFET conducts a substantial amount of drain current when its gate voltage is equal to zero. Therefore, all depletion-mode devices (either junction or insulated gate types) are said to be **normally-conducting** or **normally-on** when their gate voltages are zero.

The N-channel depletion-mode IGFET just described is often represented by the schematic symbol shown in Figure 22-5B. Notice that the gate (G) lead is separated from the source (S) and drain (D) leads. Also the arrow on the substrate (often designated as SS or B) lead points inward to represent an N-channel device. Some FETs are constructed so that the substrate is internally connected to the source lead and a separate substrate lead is not used. In such a case, the symbol can be drawn so that it shows the substrate connected to the source.

A properly biased N-channel depletion-mode IGFET is shown in Figure 22-5C. Notice that this FET is biased in the same manner as an N-channel junction FET. The drain-to-source voltage (V_{DS}) must always be applied so that the drain is positive with respect to the source as shown. However, the gate-to-source voltage (V_{GS}) can be reversed. This shows us that the gate can be made either negative or positive with respect to the source. Also notice that the substrate (B) has been externally connected to the source (S). The substrate is usually connected to the source (internally or externally) but this is not always the case. In some applications the substrate may be connected to the gate, or to other points within the FET's respective circuit. These various circuit arrangements will be described later.

SEMICONDUCTORS

The relationship between V_{GS}, V_{DS}, and I_D in an N-channel IGFET can be determined by examining the FET's drain characteristic curves. A typical set of curves are shown in Figure 22-6. The curves are formed by adjusting V_{GS} to various values, and observing the relationship between I_D and V_{DS}. You may notice a similarity between these curves and the curves for a JFET shown back in Figure 22-3. The basic difference is that positive *and* negative V_{GS} values are plotted in Figure 22-6, while only negative V_{GS} values are plotted in Figure 22-3.

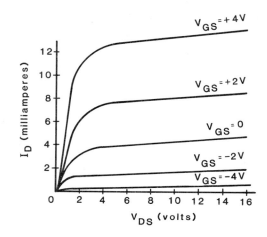

Figure 22-6 Typical drain characteristic curves for an N-channel depletion-mode IGFET.

The curves for the depletion-mode IGFET have the same general shape as the curves for the junction FET. Each curve rises rapidly, then levels off at a specific pinch-off voltage (V_P). The depletion-mode IGFET (like the junction FET) is usually operated above the pinch-off where I_D is relatively constant with changes in V_{DS}. In this region the device may be used as a highly efficient voltage amplifier. However, there are some applications where the device is operated below pinch-off. Below pinch-off, the FET's drain current (I_D) varies over a wide range and at an almost linear rate as V_{DS} changes. This means that the resistance of the device can be varied over a wide range by an input gate voltage, thus making the device useful as a voltage controlled resistor. Furthermore, the N-channel depletion mode IGFET is usually operated with a slightly negative gate-to-source bias voltage (V_{GS}), although V_{GS} may be equal to zero in some cases. This means that an AC input signal can be applied to the FET's gate (in series with V_{GS}) so that the V_{GS} varies in value.

This variance may be all negative (if V_{GS} is negative before the input), or both positive and negative (when V_{GS} is zero before the input).

Depletion-mode IGFETs may also be constructed opposite to the N-channel device shown in Figure 22.5. In this way, P-type channels which are implanted within N-type substrates. Devices constructed in this matter are referred to as **P-channel depletion-mode IGFETs**. These devices operate in the same fashion as their counterparts, but because the polarity of the materials have switched, so must many of their requirements. For instance, electrons are the majority carriers in the N-channel device, while holes are the majority carriers in the P-type device. Additionally, the drain lead must now be *negative* with respect to the source: this also being opposite to the N-channel device. With the biasing voltage being switched, current must flow in the opposite direction through the channel. But despite all of these differences, the gate may be still be negative or positive with respect to the source, and the characteristic curves retain the same shape for either device, although the polarity of the voltage and current values are opposite.

The schematic symbol for a P-channel, depletion-mode IGFET is shown in Figure 22-7. Notice that the only difference between this symbol and the symbol for an N-channel device is the direction of the arrow. In this symbol the arrow points outward to indicate that a P-type channel is used.

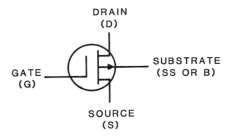

Figure 22-7 Schematic symbol for a P-channel depletion-mode IGFET.

Enhancement-Mode Devices

As explained previously, the depletion-mode IGFET is a normally-on device and therefore conducts a substantial drain current (I_D) when its gate-to-source voltage (V_{GS}) is zero. Although this type of FET is useful in many applications, there are also certain applications where a normally-*off* device is required. In other words, it is often useful to have a device that conducts negligible I_D when V_{GS} is equal to zero, but will allow a flow when a suitable V_{GS} value is applied.

SEMICONDUCTORS

An IGFET which will function as a normally-off device can be constructed as shown in Figure 22-8A. This device is similar to the depletion-mode IGFET but it does not have a conducting channel which is embedded in the substrate material. Instead, the device has source and drain regions which are diffused separately into the substrate. The device in Figure 22-8A has a P-type substrate and N-type source and drain regions; however, the exact opposite arrangement can also be used. The metal gate is separated from the substrate material by a silicon dioxide insulating layer as shown, and the various leads are attached to the device so that it has basically the same lead arrangement as the depletion-mode device.

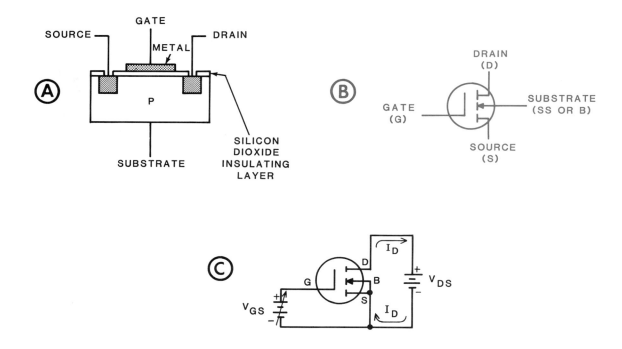

Figure 22-8 An N-channel enhancement-mode IGFET and its schematic symbol.

The device shown in Figure 22-8A must be biased so that its drain is made positive with respect to its source. With only a drain-to-source voltage (V_{DS}) applied, the FET does not conduct a drain current (I_D). This is because no conducting channel exists between the source and drain regions. However, this situation can be changed with the application of a suitable gate voltage. When the gate is made positive with respect to the source, electrons are drawn toward the gate. These electrons accumulate under the gate where they create an N-type channel which allows current to flow from source to drain. When this positive gate voltage increases, the size of the channel also increases, thus, allowing even more drain current to flow.

The action just described is similar to the action that takes place in a charging capacitor. The metallic gate and the substrate act like the upper and lower plates of a capacitor, and the insulating layer acts like a dielectric. The positive gate voltage simply causes the capacitor to charge and a negative charge builds up on the substrate side of the capacitor. The positive gate voltage effectively induces an N-type channel between the source and drain regions which sustains a drain current. Furthermore, an increase in gate voltage tends to enhance the drain current. For these reasons the device in Figure 22-8A is commonly referred to as an **N-channel, enhancement-mode IGFET (MOSFET)**.

The gate of the N-channel enhancement-mode IGFET can also be made negative with respect to its source. However, a negative gate voltage will not affect the operation of the FET since it is a normally-off (nonconducting) device. In other words, the FET's drain current is normally equal to zero and therefore cannot be further reduced by the application of a negative gate voltage.

The schematic symbol for an N-channel, enhancement-mode IGFET is shown in Figure 22-8B. Notice that this symbol is similar to the symbol used for the N-channel depletion-mode IGFET shown in Figure 22-5B. The only difference is the use of an interrupted line instead of a solid line to interconnect the source, drain, and substrate regions. The solid line is used to identify the normally-on condition of the depletion-mode device, while the interrupted line is used to identify the normally-off condition of the enhancement-mode device. In each symbol the arrow points inward to indicate that each device has an N-type channel.

A properly biased N-channel, enhancement-mode IGFET is shown in Figure 22-8C. Notice that the FET's drain is made positive with respect to its source by the drain-to-source voltage (V_{DS}). However, the gate is also made positive with respect to the source by the gate-to-source voltage (V_{GS}). Only when V_{GS} increases from zero and applies a positive voltage to the gate, will a substantial amount of drain current (I_D) flow. The substrate is normally connected to the source as shown, but in special applications, the substrate and source may be at different potentials.

The relationship between V_{GS}, V_{DS}, and I_D in a particular N-channel, enhancement-mode IGFET can be determined by examining the FET's drain characteristic curves. A typical family of curves is shown in Figure 22-9. V_{GS} is adjusted to various values and the relationship between I_D and V_{DS} is measured. These curves are similar to the curves for an N-channel, depletion-mode IGFET, but only positive values of V_{GS} are plotted instead of both positive and negative values. Notice that for each higher positive value of V_{GS}, I_D rises to a correspondingly higher value and then levels off.

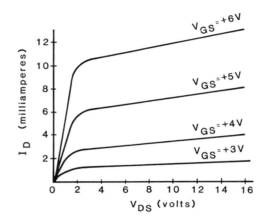

Figure 22-9 Typical drain characteristic curves for an N-channel enhancement-mode IGFET.

Although it is not apparent in Figure 22-9, V_{GS} must exceed a certain threshold voltage (usually one volt or more) before the N-type channel induced within the FET is great enough to support a usable current. Any V_{GS} value below this threshold cannot cause the FET to conduct, and the device will effectively act as if its gate voltage was equal to zero. We can then conclude that the N-channel enhancement-mode IGFET normally operates with a positive gate bias that is greater than its threshold voltage. With proper biasing the enhancement-mode device makes an excellent switch. The device can be turned on by a sufficiently high gate voltage—and turned off when the gate voltage drops below the threshold level. The inherent threshold of the device therefore provides a highly desirable region of noise immunity. This prevents low or intermediate input voltages (below threshold) from falsely triggering the device on. This characteristic makes the enhancement-mode device ideal for digital applications involving logic and switching functions. The enhancement-mode IGFET is the basic component used in many large scale digital integrated circuits.

An enhancement-mode IGFET may also be constructed in a manner that is exactly opposite the device shown in Figure 22-8A, with a P-type source and drain regions which are implanted into an N-type substrate. This type of FET must be operated with a negative gate voltage so that holes (instead of electrons) are attracted toward the gate to form a P-type channel. Such a device is referred to as a P-channel, enhancement-mode IGFET (MOSFET). The P-channel device functions in basically the same manner as the N-channel device, even though holes instead of electrons are used to support drain current through the device. The P-channel device requires bias voltages (V_{GS} and V_{DS}) that are opposite to N-channel devices. Also, the drain characteristic curves for the P-channel device have the same general shape as the curves for the N-channel device although the polarities of the voltages and currents involved are reversed.

The symbol for a P-channel, enhancement-mode IGFET is shown in Figure 22-10. This symbol closely resembles the symbol for the N-channel device. The only difference is the direction of the arrow which in this case points outward to identify the P-type channel that is induced within the device. The enhancement-mode device is usually constructed in a symmetrical manner just like the junction FET. This means that the source and drain leads can usually be reversed or interchanged.

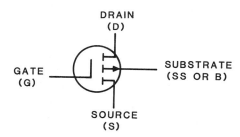

Figure 22-10 Schematic symbol for a P-channel enhancement-mode IGFET.

FET Amplifier Circuits

Like bipolar transistors, FETs are used for amplification. They also have three basic arrangements like bipolar transistors, and are named for the common element to both the input and the output signals. We will examine the common-source, the common-drain, and the common-gate amplifiers in the following discussion.

Common-Source

The most widely used FET amplifier is the **common-source** amplifier. Since the source on a FET can be compared to the emitter on a bipolar transistor, you will find that the common-source amplifier has the same characteristics as a common-emitter.

The basic common-source arrangement is shown in Figure 22-11. Notice that the input is applied between the gate and the source, and the output is taken between the source and the drain, making the source lead common to both the input and the output. The gate-to-source bias voltage is provided by V_{GG}, ensuring that a specific amount of drain current I_D will flow through the FET (before the input signal is applied), allowing it to operate within the pinch-off region. R_G is placed in series with V_{GG} and the gate to ensure the resistance at the input remains high.

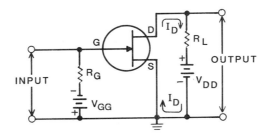

Figure 22-11 JFET common-source amplifier.

As I_D flows through R_L, the output voltage is developed. The operation of the common-source is very similar to the common-emitter amplifier used in bipolar transistors. As the input signal is applied, it alternately aids and opposes V_{GG}, which alternately increases and decreases the amount of reverse-bias voltage between the gate and the source. This causes I_D to vary, and an alternating output is developed across R_L. This provides an output voltage that is considerably higher than the input voltage.

In addition to providing voltage amplification, another desirable feature of the common-source is the difference between input and output resistance. The difference between the input resistor R_G and the output resistor R_L provides a high input resistance and a low output resistance. This can also be seen as a gain in current, although the difference between input and output resistance is not as large as with the common-emitter.

Common-Gate

In the **common-gate** amplifier (shown in Figure 22-12), an input signal is again applied between the gate and the source, and the output is taken across the drain. This may sound as if it is the same type of configuration as the common-source; however, there is one important difference. V_{SS} (the voltage that applied the reverse bias between the gate and the source) is varied by the input signal, and this variance is felt *at the source*, as opposed to the gate in the common-source amplifier.

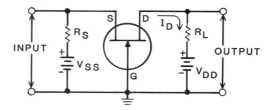

Figure 22-12 Common-gate amplifier.

As the input varies V_{SS}, this changes the amount of reverse-bias between the source and the gate, which varies I_D. The changing I_D flows through R_L, whose output voltage varies in proportion to I_D. Like the common-source, the amount of I_D traveling through R_L causes a much higher voltage to be developed at the output than the relatively small voltage which changed V_{SS} at the input. The result is voltage amplification.

If you'll recall, the common-base amplifier used a bipolar transistor, and the base was common to both the input and the output. The main function of the common-base was voltage amplification. The base in a bipolar transistor is similar to the gate with a FET. Therefore, the common-gate amplifier has much the same characteristics as the common-base amplifier.

The voltage gain in the common-gate amplifier is not as high as the voltage gain in the common-source. This is due to the values of R_S and R_L. R_S will be a lower value than R_L, because less voltage is needed to vary the amount of reverse bias between the gate and the source. However, R_L is not much larger than R_S. The resistance of R_L is partially responsible for amplifying voltage, but the majority of amplification comes from the high amount of I_D. The values of R_S and R_L are fairly close.

Common-Drain

Just as the common-source is similar to the common-emitter and the common-gate is similar to the common-base, the common-drain FET amplifier has many of the same characteristics as the common-collector. The common-drain is also known as the source-follower, since the voltage on the source follows the voltage on the gate. The schematic for a typical common-drain amplifier is shown in Figure 22-13.

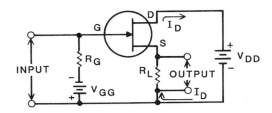

Figure 22-13 Common-drain amplifier.

The input signal is felt between the FET's gate and drain leads, and the output is taken across R_L. Although it may appear that the input is applied between the gate and the source, V_{DD} is effectively a short, as far as the input and output signals are concerned. This allows the input to be felt between the gate and the drain, and also allows R_L to develop the output signal.

The minimal voltage required to affect the reverse-bias between the gate and the drain, coupled with the high input resistance, makes the input current very low. The output, however, has a very low resistance, and varies with the high source current. This allows the common-drain amplifier to provide a substantial current gain, and is widely used in circuits which require a high-resistance input to be connected to a low-resistance load.

The three amplifiers we discussed all have N-channel FETs as the operating device. However, P-channel FETs can perform the same functions. The only changes required are the polarities of the biasing voltages.

FIELD EFFECT TRANSISTORS

Chapter Self-Test

1. The construction of a junction FET usually starts with a section of lightly doped N-type or P-type material. What is this material known as?

2. What are the three leads associated with the junction FET? Which of these leads are interchangeable?

3. The PN junction formed between the gate and source of a junction FET requires what type of biasing?

4. Can the current flowing through the channel of a FET be effectively controlled by varying the reverse bias voltage applied to the gate and source lead?

5. The FET's gate-to-source voltage (V_{GS}) determines the size of what?

6. What is the name for the mode in which a junction FET operates?

7. What will happen to I_D if there is an increase in V_{GS}?

8. Is the value of $V_{GS(off)}$ for a given FET approximately equal to the FET's V_P value? Why?

9. When the arrow in the symbol of a junction FET points inward, what type of material is the channel?

10. Can an IGFET be designed to operate in either the depletion-mode or the enhancement-mode?

11. What is the difference between a depletion-mode IGFET and a junction FET?

12. What are the majority carriers within an N-channel, depletion-mode IGFET?

13. When an N-channel, depletion-mode IGFET's gate-to-source voltage is equal to zero, will drain current flow through the device?

14. What is the primary function of a common-gate amplifier?

15. From which terminal is the output of the common-source amplifier taken?

16. In a P-channel common-drain amplifier, what will happen to drain current during the positive alternation of the input signal?

Summary

The field effect transistor can provide amplification of electronic signals like a conventional bipolar transistor. In fact, the field effect transistor (or FET) is more efficient than a bipolar transistor in certain applications. However, the FET is considered a voltage-controlled device, while a bipolar transistor is considered a current-controlled device.

The operation of a bipolar transistor depends on the movement of both majority and minority carriers—hence, the name *bipolar*. The FET, however, operates only with majority carriers. The majority carriers flow through a semiconductor channel which is implanted within a semiconductor substrate. The opposite ends of the channel are referred to as the FET's source and drain, and may be compared to the emitter and collector of a bipolar transistor. The substrate itself is used to control the movement of majority carriers through the channel and is referred to as the FET's gate. The gate of a FET may be compared to the base of a bipolar transistor. The device we just described is more appropriately referred to as a junction FET or JFET, since the substrate (gate) and the semiconductor channel form a PN junction.

Junction FETs may be constructed with an N-type channel and a P-type substrate, or vice-versa. FETs are identified by the type of material in which the channel is created. FETs are referred to as N-channel or P-channel. The gate-to-channel junction of a FET must always be reverse-biased to form a depletion region within the channel. By adjusting this reverse voltage, the size of the depletion region can vary, which determines the amount of current that flows from the source to the drain. Therefore, this reverse voltage controls the conduction of the FET.

The gate-to-channel junction is usually reverse-biased by applying a proper voltage between the gate and source leads. In addition to this voltage, a potential is also required between the source and the drain to attract majority carriers.

Since the gate-to-source voltage forms a depletion region within the channel, the FET is said to operate in the depletion-mode.

The JFET represents one basic type of FET. The second type, known as an insulated-gate FET (IGFET), is constructed in a method similar to the junction FET, but utilizes a metallic gate which is insulated from the channel. The substrate is used only to support the structure. IGFETS are available as P-channel or N-channel devices. Furthermore, they can be designed to operate in either the depletion- or enhancement-mode. The depletion-mode device operates similar to a junction FET, with the exception that the gate can be made either positive or negative with respect to the source. The depletion-mode device also conducts drain current if there is zero potential between the gate and the source; this is considered to be a normally-on device.

The enhancement-mode device does not conduct drain current when the gate-source potential is zero. This makes the enhancement-mode a normally-on device. The gate may feel either positive or negative voltage to conduct current through the gate; however, usually only one polarity of voltage will initially start the device.

Either type of FET may be connected in a common-source arrangement where the input is felt at the gate, and the output is taken across the drain. This arrangement provides high voltage gain and a substantial input resistance. FETs may also be placed in a common-gate arrangement, which provides a voltage gain that is less than the common-source. The third type of amplifier is the common-drain, where the input is placed at the gate, and the output is felt across the source. This amplifier provides a high input resistance and a low output resistance.

SEMICONDUCTORS

CHAPTER 23

Control Devices

SEMICONDUCTORS

Contents

Introduction 561

Chapter Objectives 562

Silicon Controlled Rectifiers. 563

Bi-Directional Triode Thyristors 571

Unijunction Transistors 577

Chapter Self-Test 587

Summary. 588

CONTROL DEVICES

Introduction

This chapter introduces you to four different types of semiconductor devices. Although not used as often as FETs, transistors, or diodes, each of the devices in this chapter has a specific function. They are grouped into the category of "control devices" because they normally function as electronic switches, controlling the amount of current flow within a circuit.

A term which is used to signify this type of control is "thyristor." This defines a broad range of solid-state components which are used as electronically controlled switches. Each of these devices can switch between a conducting ("*on*") state and a nonconducting ("*off*") state to effectively pass or block electrical current. Furthermore, some thyristors are capable of switching currents flowing in one direction while others can switch currents flowing in either direction.

Thyristors are widely used in applications where DC and AC power must be controlled. These devices are often used to apply a specific amount of power to a load or to completely remove it from the load. In addition, if the power is already applied to the load, these devices can also control and adjust that power. For example, a thyristor could be used to simply turn an electric motor on or off, or to adjust the speed or torque of the motor over a wide operating range.

Thyristors should not be confused with bipolar transistors or field effect transistors (FET's). Although transistors and FETs can be used as electronic switches, these devices are not as efficient, since they do not have the power handling capability of thyristors. Thyristors are devices that are used expressly for the purpose of controlling electrical power, while transistors and FETs are primarily used to provide amplification.

A variety of thyristors are now available, but many of these devices have similar or related characteristics. Most applications which involve power control are handled with a few basic components. The thyristors that are most widely used are the silicon controlled rectifier (SCR), the bi-directional triode (TRIAC), the unijunction transistor (UJT) and the programmed unijunction transistor (PUT). You will examine each of these devices in this section.

SEMICONDUCTORS

Chapter Objectives

When you have completed this chapter, you will be able to:

1. Describe the conditions necessary to turn *"on"* or turn *"off"* a silicon controlled rectifier.

2. Explain the difference between a silicon controlled rectifier's forward breakover and reverse breakdown voltage.

3. Name two applications of the silicon controlled rectifier.

4. Describe the conditions necessary to turn *"on"* or turn *"off"* a bi-directional triode thyristor.

5. Name two applications of the bidirectional triode thyristor.

6. Describe the conditions required to turn *"on"* a unijunction transistor.

7. Explain how a unijunction transistor exhibits a negative resistance once it is turned *"on."*

8. Name two applications of the unijunction transistor.

9. Describe the difference between an ordinary unijunction transistor and a programmed unijunction transistor.

10. Identify the schematic symbols for the SCR, the TRIAC, the UJT, and the PUT.

CONTROL DEVICES

Silicon Controlled Rectifiers

The silicon controlled rectifier is the most popular member of the thyristor family. This device is generally referred to as an SCR. Unlike the bipolar transistor which has two junctions and provides amplification, the SCR has three junctions and is used as a switch. As its name implies, the device is basically a rectifier which conducts current in only one direction. However, the device can be made to conduct (turn "on") or stop conducting (turn "off") and therefore provide a switching action that can be used to control electrical current.

We will first examine the basic construction and operation of an SCR, and its important electrical characteristics and basic applications.

Basic Construction and Operation

An SCR is a solid-state device which has four alternately doped semiconductor layers. The device is almost always made from silicon, but germanium has been used. The SCR's four layers are often formed by a diffusion process, but a combined diffusion-alloyed method is also used.

A simplified diagram of an SCR is shown in Figure 23-1A. As shown, the SCR's four (PNPN) layers are sandwiched together to form three junctions. However, leads are attached to only three of the four layers. These three leads are referred to as the anode, cathode, and gate.

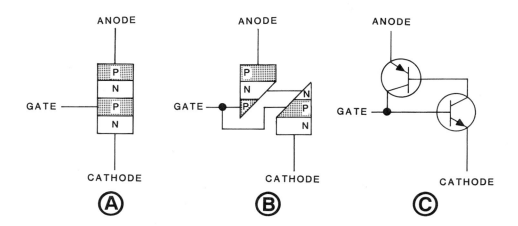

Figure 23-1 The SCR and its equivalent circuits.

The manner in which an SCR operates cannot be easily determined by examining the four-layer structure shown in Figure 23-1A. However, the SCR's four layers can be divided so that two three-layer devices are formed, as shown in Figure 23-1B. The two devices are effectively PNP and NPN transistors. When these "transistors" are interconnected (as shown), they are equivalent to the four layer device shown in Figure 23-1A. These two transistors can also be represented by their schematic symbols as shown in Figure 23-1C.

We will now bias the equivalent circuit in Figure 23-1C, just like we would bias the actual SCR shown in Figure 23-1A. First, we will make the anode of the circuit positive with respect to the cathode, but we will leave the gate open. Under these conditions the NPN transistor will not conduct, because its emitter junction will not be subjected to a forward bias voltage. Since the NPN transistor cannot conduct, it will not allow base current through the PNP, which prevents the PNP transistor from conducting. The equivalent SCR circuit will not allow current to flow from its cathode to its anode under these conditions.

If the gate of the equivalent SCR circuit is momentarily made positive with respect to the cathode, the emitter junction of the NPN transistor will become forward-biased and this transistor will conduct. This will in turn cause a base current to flow through the PNP transistor, which will cause this transistor to conduct. However, the collector current flowing through the PNP transistor now causes base current to flow through the NPN transistor. The two transistors therefore hold each other in the "on" or "conducting" state. This allows current to flow continuously from the cathode to the anode of the circuit.

It is important to note that this action takes place even though the gate voltage is applied for only a moment. This momentary gate voltage causes the circuit to switch to the "on" or conducting state, and the circuit will remain in that state even though the gate voltage is removed.

In order to switch the equivalent SCR circuit back to its "off" or "nonconducting" state, it is necessary to reduce its anode-to-cathode voltage to almost zero. This will cause both transistors to turn off and remain off until the gate voltage is again applied.

The SCR in Figure 23-1A operates just like the equivalent circuit in Figure 23-1C. In other words, the SCR can be turned "on" by a positive input gate voltage and must be turned off by reducing its anode-to-cathode voltage. When the SCR is turned "on" and is conducting a high cathode-to-anode current, the device is said to be conducting in the forward direction. If the polarity of the cathode-to-anode bias voltage was reversed, the device would conduct only a small leakage current, which would flow in the reverse direction.

CONTROL DEVICES

The SCR is usually represented by the schematic symbol shown in Figure 23-2A. Notice that this symbol is actually an ordinary diode symbol with an added gate lead. The circle surrounding the diode may or may not be used and the SCR's anode, gate, and cathode leads may or may not be identified. When the leads are identified, they are usually represented by the letters A, G, and K as shown.

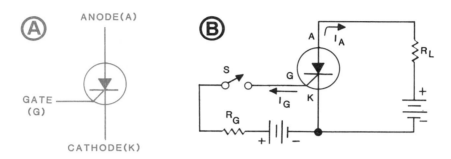

Figure 23-2 The schematic and biasing requirements of an SCR.

A properly biased SCR is shown schematically in Figure 23-2B. Notice that a switch (S) is used to apply or remove the input gate voltage which is obtained from a voltage source and resistor R_G. This resistor is used to limit the gate current (I_G) to a specific value. The SCR's anode-to-cathode voltage is provided by another voltage source, but a series load resistor (R_L) is also used to limit the SCR's cathode-to-anode current to a safe value when the device is turned "on." Without this resistor, the SCR would conduct a very high cathode-to-anode current (also referred to as anode current or I_A) and could be permanently damaged.

Voltage-Current Characteristics

A better understanding of SCR operation can be obtained by examining the voltage-current (V-I) curve shown in Figure 23-3A. This curve shows the V-I characteristics of a typical SCR. The curve is plotted by varying the SCR's cathode-to-anode voltage over a wide range while observing the SCR's anode current. The SCR is first biased in the forward direction while its gate is open as shown in Figure 23-3A. The SCR's cathode-to-anode voltage is designated as V_F at this time. The curve shows that as V_F increases from zero, the SCR conducts only a small forward current (designated as I_F) which is due to leakage. As V_F continues to increase, I_F remains very low and almost constant, but eventually a point is reached where I_F increases rapidly and V_F drops to a low value (note the horizontal dotted line). The V_F value required to trigger this sudden change is referred to as the **forward breakover voltage**.

When this voltage is reached, the SCR breaks down and conducts a high I_F which is limited only by the external resistance in series with the device. The SCR switches from "off" to "on" at this time. The drop in V_F occurs because the SCR's resistance drops to an extremely low value, and most of the source voltage appears across the series resistor.

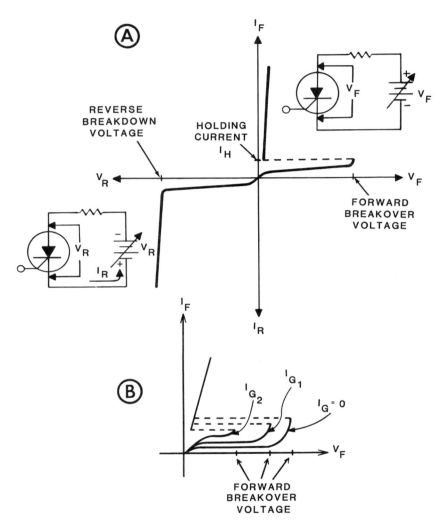

Figure 23-3 V-I characteristics of a typical SCR, with both zero and varying gate currents.

When the SCR is in the *"on"* state, only a slight increase in V_F is required to produce a tremendous increase in I_F (the curve is almost vertical and straight). Furthermore, the SCR will remain *"on"* as long as I_F remains at a substantial value. Only when I_F drops below a certain minimum value, will the SCR switch back to its *"off"* state. This minimum value of I_F which will hold the SCR in the *"on"* state is referred to as the SCR's **holding current** and is usually designated as I_H. As shown in Figure 23-3A, the I_H value is located at the point where breakover occurs (just to the left of the horizontal dotted line).

When a reverse voltage is applied to the SCR as shown in Figure 23-3A, the device functions in basically the same manner as a reverse-biased PN junction diode. As the reverse voltage (V_R) across the SCR increases from zero, only a small reverse current (I_R) will flow through the device due to leakage. This current will remain small until V_R becomes large enough to cause the SCR to breakdown. Then I_R will increase rapidly if V_R increases even slightly above the breakdown point (the curve is almost vertical and straight). The reverse voltage (V_R) required to breakdown the SCR is referred to as the SCR's **reverse breakdown voltage**. If too much reverse current is allowed to flow through the SCR after breakdown occurs, the device could be permanently damaged. However, this situation is normally avoided because the SCR is usually subjected to operating voltages which are well below its breakdown rating.

The V-I curve in Figure 23-3A shows the relationship between V_F and I_F when the SCR's gate is open. In other words, no voltage is applied to the SCR's gate and no gate current is flowing through the device. The curve in Figure 23-3 could therefore be labeled to indicate that I_G is equal to zero.

When the gate is made positive with respect to the cathode, the SCR's forward characteristics will be affected. This is because the forward-biasing of the gate allows gate current to flow. The changes that will take place in the SCR's forward characteristics are graphically represented in Figure 23-3B. In this figure, three V-I curves are plotted to show how changes in gate current (designated as I_G) affect the relationship between the SCR's forward voltage and forward current. The $I_G = 0$ curve shows the relationship between V_F and I_F when the gate current is zero and is simply a more detailed representation of the forward characteristics shown in Figure 23-3A. The I_{G1} curve is plotted for a specific but relatively low value of gate current. Notice that this curve has the same general shape as the $I_G = 0$ curve, but the forward breakover point occurs sooner (at a lower V_F value). The I_{G2} curve is plotted for a slightly higher gate current and also has the same general shape as the other two curves. However, the breakdown point occurs even sooner at this higher value of gate current.

SEMICONDUCTORS

The curves in Figure 23-3B show that the SCR's forward breakover voltage decreases as the gate current increases. In fact, the gate current could be increased to a point where the breakover voltage would be so low that the device would have characteristics that closely resemble those of an ordinary PN junction diode. The ability of the gate to control the point where breakover occurs is an advantage in many types of electronic circuits.

The curves in Figure 23-3B reveal the SCR's most important electrical characteristics. Basically these curves show that for any given gate current, a specific forward breakover voltage must be reached before the SCR can turn *"on."* However, the curves also show that for any given forward voltage across the SCR, a specific value of gate current must be reached before the device can be turned *"on."* Therefore, an SCR can be turned *"on"* only when it is subjected to the proper combination of gate current and forward voltage values.

Practical Applications

In normal operation, the SCR is subjected to forward voltages which are below its breakover voltage, and turns *"on"* by the application of a suitable gate current. This gate current is usually made high enough to insure that the SCR is switched to the *"on"* state at the proper time. However, for the SCR to be most effective, this gate current must not be constantly applied. In most circuits, a pulse is used. A pulse is a sudden increase and decrease in voltage. When this pulse is applied to the gate, the sudden increase in voltage allows gate current to flow, switching the SCR to its *"on"* state. Once the *"on"* state is reached, the holding current will keep the SCR on—the SCR will not turn off with the sudden decrease of voltage at the gate. A constant gate current is not required to trigger the SCR. This would only cause more power to be dissipated within the device. Once the SCR turns *"on,"* it can be turned off only by reducing its forward current below its respective holding current value.

The SCR is primarily used to control the application of DC or AC power to various types of loads. It can be used to open or close a circuit, or it can be used to vary the amount of power applied to a load. A very low gate current signal can control a very large load current. The SCR back in Figure 23-2B is used as a switch to apply DC power to the load resistor (R_L); but in this basic circuit, there is no effective means of turning off the SCR to remove power from the load.

However, this problem can be easily solved by simply connecting a switch across the SCR. This switch can be momentarily closed to short out the SCR and reduce its anode-to-cathode voltage to zero. This will reduce the SCR's forward current below the holding value and cause the SCR to turn off.

A more practical SCR circuit is shown in Figure 23-4. With this circuit, mechanical switches have been completely eliminated. In this high speed circuit, SCR_1 is used to control the DC power applied to load resistor R_L, and SCR_2—along with a capacitor (C) and a resistor (R_1)—are used to turn off the circuit. When a momentary gate current flows through SCR_1, it turns "on," and allows a DC voltage to be applied to R_L. This effectively grounds the left side of capacitor C and allows it to charge through resistor R_1.

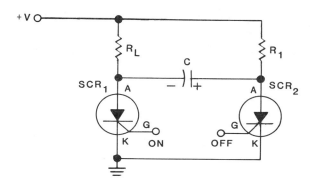

Figure 23-4 A practical DC SCR switch control.

This, in turn, causes the right-hand plate of the capacitor to become positive with respect to the left-hand plate. When a momentary gate current pulse is applied to SCR_2, it turns "on" and the right-hand plate of the capacitor is grounded, thus placing the capacitor across SCR_1. The voltage across the capacitor now causes SCR_1 to be reverse-biased. This reverse voltage causes the forward current through SCR_1 to drop below its holding value, thus, causing SCR_1 to turn off and remove power from R_L. Therefore, a momentary gate current through SCR_1 will turn on the circuit, and a momentary gate current through SCR_2 will turn off the circuit.

SEMICONDUCTORS

When using SCRs in DC switching circuits, it is often necessary to use additional components (not always additional SCRs) to provide a means of turning off the circuits. The previous example shows only one of the ways this can be accomplished.

It is also possible to use an SCR to control an AC signal. However, since an SCR is only able to control current flow in one direction, modifications must be made. One possible modification would be to *rectify* the AC signal. We have mentioned rectification in semiconductors, describing it as a method of removing one-half of an AC signal. With one alternation removed, the SCR could control the other alternation without difficulty. Another possible method would be to connect two SCRs in parallel, and opposite directions. This would allow one SCR to control one alternation, and the other SCR to control the other alternation.

CONTROL DEVICES

Bi-Directional Triode Thyristors

The silicon controlled rectifier (SCR) previously described is capable of controlling current which is flowing in one direction and is therefore a *unidirectional* device. The SCR is used in many applications which involve the control of direct currents as well as alternating currents. However, when an SCR is used in AC circuits, it can only operate with one alternation of each AC input cycle. In order to fully control an AC signal using SCRs, we would require two SCRs in parallel, and connected in opposite directions. This makes the SCR rather impractical for AC circuits.

In applications where it is necessary to achieve full control of an AC signal, it is often much easier to use a device known as a bi-directional triode thyristor. This device is more commonly referred to as a triac. The triac has basically the same switching characteristics as an SCR, with the exception that a triac exhibits these characteristics in *both* directions. This makes the triac equivalent to the two SCRs in parallel we mentioned in the previous paragraph.

Basic Construction and Operation

A simplified diagram of a triac is shown in Figure 23-5A. Notice that the device has three leads which are designated as main terminal 1, main terminal 2, and the gate. Main terminal 1 and main terminal 2 are each connected to a PN junction at opposite ends of the device. The gate is also connected to a PN junction which is at the same end as terminal 1. If you examine the entire structure closely you will see that from terminal 1 to terminal 2 you can pass through an NPNP series of layers or a PNPN series of layers. In other words, the triac is effectively a four-layer NPNP device in parallel with a four-layer PNPN device. These NPNP and PNPN current paths can be compared to the SCRs previously mentioned; the equivalent SCR circuit is shown in Figure 23-5B.

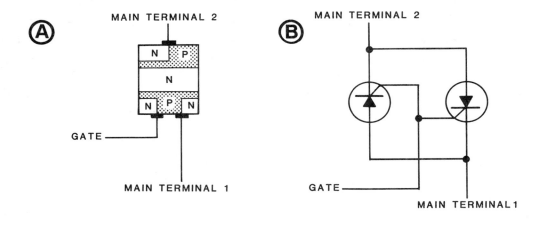

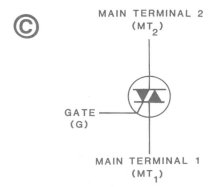

Figure 23-5 The triac, its SCR equivalent circuit, and its schematic symbol.

The triac's gate region is more complex and a detailed analysis of its operation will not be considered at this time. However, the gate is basically capable of directly or remotely triggering either of the equivalent SCRs into conduction. Notice that both of the equivalent SCR gates are tied together in Figure 23-5B to show the equivalent relationship.

The circuit in Figure 23-5B is not in all ways equivalent to the triac. This circuit is used simply to explain the basic concept involved. The primary difference is that the two equivalent SCRs would actually require different gating circuits to trigger them into conduction, but the triac is designed to respond to the currents that flow through its single gate terminal.

CONTROL DEVICES

Because of the triac's ability to control current flow in both directions, it is widely used to control the application of AC power to various types of loads or circuits. The conditions required to turn a triac *"on"* or *"off"* in either direction are similar to the conditions required to control the SCR. Both devices can be triggered to the *"on"* state by a gate current, and can be turned off by reducing their operating anode currents below their respective holding values. In the case of an SCR, current must flow in the forward direction from cathode to anode. However, the triac is designed to conduct both forward and reverse currents through its main terminals.

The schematic symbol that is commonly used to represent the triac is shown in Figure 23-5C. Notice that the symbol consists of two parallel diodes connected in opposite directions with a single gate lead attached. The device is usually placed within a circle as shown and its main terminals are sometimes identified as MT_1 and MT_2 as indicated.

Voltage-Current Characteristics

The voltage-current (V-I) characteristic curve for a typical triac is shown in Figure 23-6. This curve shows the relationship between the current flowing through its main terminals in each direction (designated as $+I_T$ and $-I_T$), and the voltage applied across its main terminals in each direction (identified as $+V$ and $-V$). Furthermore, this curve was plotted with no gate current flowing through the triac, and main terminal 1 was used as the reference point for all voltage and current values.

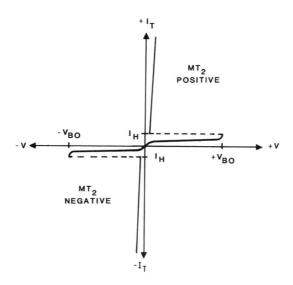

Figure 23-6 V-I characteristics of a typical triac.

Figure 23-6 shows that when main terminal 2 (MT_2) is positive with respect to MT_1 (or when the applied voltage is equal to +V), the current through the device ($+I_T$) remains at a low leakage value until +V rises above the breakover voltage ($+V_{BO}$) of the device. At this time, the triac switches from the *"off"* state to the *"on"* state, and $+I_T$ is essentially limited by the external resistance of the circuit. The triac must remain in the *"on"* state until $+I_T$ drops below a specified holding current (I_H) as shown. This is exactly what happens when an SCR is subjected to a forward voltage that exceeds its respective forward breakover voltage.

When MT_2 is negative with respect to MT_1, the triac exhibits the same basic V-I characteristics since the current through the device ($-I_T$) remains at a low leakage value until $-V$ rises above the breakover voltage ($-V_{BO}$) of the device. At this time the triac switches from the *"off"* state to the *"on"* state, and remains there until $-I_T$ drops below I_H.

The V-I curve in Figure 23-6 therefore shows that the triac exhibits the characteristics of an SCR in either direction. However, this curve does not show how the triac's gate is used to control its operation. Like the SCR previously described, the triac's breakover voltage (in either direction) can be varied by controlling the amount of gate current flowing through the device. When the gate current is increased, the breakover voltage is lowered. However, there is still one very important difference between the SCR's gating characteristics and those of a triac. The SCR always requires a *positive* gate voltage, but the triac will respond to either a *positive or negative* gate voltage. In other words, the triac's breakover voltage (in either direction) can be lowered by making its gate more positive or more negative with respect to MT_1, which is used as the reference terminal. This positive or negative gate voltage correspondingly produces a gate current that flows out of or into the gate lead. These currents then regulate the point at which the device turns *"on."*

Like the SCR, the triac is normally subjected to operating voltages that are well below its breakover voltage (in either direction). The device is turned *"on"* by subjecting it to a sufficiently high gate current, which flows into or out of its gate lead. The device is turned off by simply reducing its operating current ($+I_T$ or $-I_T$) below its respective I_H value.

CONTROL DEVICES

The triac is most sensitive when it is subjected to +V and $+I_T$ values along with a positive gate voltage. Under these conditions, the device requires the least gate current to turn "on" for any given +V value. Other combinations of operating voltages and currents result in a loss of sensitivity. To help the circuit designer determine the conditions necessary to turn "on" a specific triac, manufacturers of these components usually specify minimum or typical values of gate current (in each direction) required to turn "on" the device. These values are given for a specified operating voltage which is applied in first one direction and then the other (+V and −V). With this information, the circuit designer can insure that sufficient gate current is used to turn "on" the triac at the proper time. As with the SCR circuits, this gate current need only be applied momentarily—usually in the form of a pulse—to cause the triac to change states.

Applications

Since the triac conducts in either direction, it is ideally suited for applications where AC power must be controlled. The device can be used as an AC switch, or to control the amount of AC power applied to a load. A typical example of its use as an AC switch is shown in Figure 23-7. This circuit will apply the full input voltage across load resistor R_L when switch S is closed, or completely remove it when switch S is opened. When switch S is open, the triac cannot conduct on either the positive or the negative alternations of each AC input cycle. This is because the input voltage does not exceed the triac's breakover voltage in either direction.

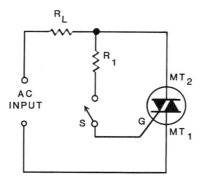

Figure 23-7 A simple AC switch.

However, when switch S is closed, resistor R_1 allows enough gate current to flow through the triac on each alternation to insure that the device turns *"on."* The triac therefore applies all of the available input power to the load, while a comparable SCR circuit (refer to Figure 23-7) is capable of supplying only half of the input power to the load. The advantage of this circuit is that the small gate current can control a high load current.

Although the triac has the ability to control current in either direction and respond to gate currents flowing in either direction, the device does have certain disadvantages when compared to an SCR. In general, triacs have lower current ratings than SCRs, and cannot compete with the SCRs in applications where extremely large currents must be controlled. Triacs are available that can handle currents (usually measured in rms values) as high as 25 amperes. By comparison, SCRs can be readily obtained with current ratings (usually expressed as average values for a half cycle) as high as 700 or 800 amperes, and some are rated even higher. Also, both devices can have peak or surge current ratings that are much higher than their respective rms or average ratings.

CONTROL DEVICES

Unijunction Transistors

We will now examine another important member of the thyristor family, known as a unijunction transistor, or simply a UJT. The UJT has physical and electrical characteristics that are quite different from those of the SCRs and triacs previously described. The UJT is actually a special type of transistor which is used as an electronic switch; not as an amplifying device. Furthermore, the UJT can be used to generate repetitive waveforms which can be used to perform many useful functions in electronic circuits.

In addition to the basic unijunction transistor, a special type of UJT is also used. This device is commonly referred to as a programmed unijunction transistor or simply a PUT. The PUT operates in basically the same manner as the UJT, but its electrical characteristics can be made to vary over a wide range.

We will examine the physical and electrical characteristics of the basic UJT, and see how it is used as an electronic switch and waveform generator. From there, we will briefly examine the special PUT and consider its advantages when compared to the UJT.

Basic Construction

As its name implies, the unijunction transistor (UJT) has only one semiconductor junction. The device basically consists of a block or bar of N-type semiconductor material, in which a small pellet of P-type material is fused. This is shown in Figure 23-8A. A lead is attached to each end of the N-type bar as shown. These two leads are referred to as base 1 and base 2. Another lead is attached to the P-type pellet; this is referred to as the emitter. These leads are commonly designated as B1, B2, and E as shown.

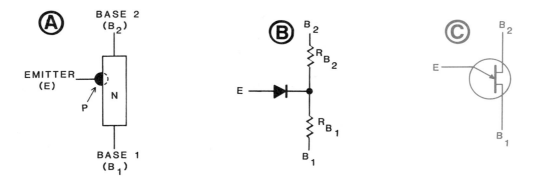

Figure 23-8 The basic UJT with its equivalent circuit and its schematic symbol.

577

SEMICONDUCTORS

The N-type bar is lightly doped and therefore has few majority carriers to support current flow. This means that the resistance between the base 1 and base 2 leads is quite high. Most UJTs exhibit an inter-base resistance (between B_1 and B_2) that is approximately 5 kilohms to 9 kilohms.

The PN junction formed between the N-type bar and P-type pellet has characteristics similar to those of a PN junction diode. Therefore, we can represent the UJT with an equivalent circuit shown in Figure 23-8B. This equivalent circuit consists of two resistors (R_{B1} and R_{B2}), and a diode which has its cathode end connected between the resistors. The operation of the UJT is much easier to understand when this equivalent circuit is used in place of the UJT's basic structure. The schematic symbol that is commonly used to represent the UJT is shown in Figure 23-8C. The arrow on the emitter (E) lead points inward to show that the UJT has a P-type emitter.

Operation

The operation of an ordinary UJT is more apparent when its equivalent circuit is analyzed. Figure 23-9 shows how the equivalent UJT circuit is biased under normal conditions. Notice that an external voltage source (V_{BB}) is connected across the B_1 and B_2 terminals so that B_2 is positive with respect to B_1. Another external voltage source (V_S) is connected across the E and B_1 terminals so that E is positive with respect to B_1. The resistor between the positive side of V_S and terminal E is used to limit the current through E to a safe level.

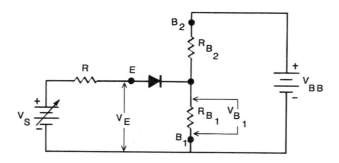

Figure 23-9 A properly biased UJT.

If voltage V_S is not high enough to cause the diode in the equivalent circuit to be forward-biased, the two resistors within the equivalent circuit (R_{B1} and R_{B2}) will allow only a small current to flow between terminals B_1 and B_2. This could be easily calculated according to Ohm's Law by dividing the total resistance between terminals B_1 and B_2 ($R_{B1} + R_{B2}$) into voltage V_{BB}. This means that voltage V_{BB} is distributed across R_{B1} and R_{B2}. The ratio of the voltage across R_{B1} (designated as V_{B1}) to the source voltage (V_{BB}) is known as the *intrinsic standoff ratio,* and is represented by the Greek letter eta (η). This relationship can be expressed mathematically as:

$$\text{intrinsic standoff ratio } (\eta) = \frac{V_{B1}}{V_{BB}}$$

Since the voltage dropped across R_{B1} and R_{B2} is proportional to their resistance values, the intrinsic standoff ratio is also equal to the ratio of R_{B1} to the total resistance between terminals B_1 and B_2 ($R_{B1} + R_{B2}$). This can be expressed mathematically as:

$$\eta = \frac{R_{B1}}{R_{B1} + R_{B2}}$$

This last equation shows that the intrinsic standoff ratio is determined by the two internal resistances. Its value is determined by the physical construction of the device, and cannot be controlled by varying V_{BB} or V_S. The intrinsic standoff ratio is specified for each type of UJT that is made and typical values will range from approximately 0.5 to 0.8. When the intrinsic standoff ratio for a particular device is known, the voltage across R_{B1} (V_{B1}) can be determined for any value of applied voltage (V_{BB}). This calculation can be made by simply transposing the first equation given to obtain:

$$V_{B1} = \eta \times V_{BB}$$

This equation simply states that the voltage across R_{B1} (V_{B1}) is equal to the intrinsic standoff ratio times V_{BB}. For example, if the UJT has an intrinsic standoff ratio of 0.5 and it is subjected to a V_{BB} of 20 volts, the voltage across R_{B1} would be equal to 0.5 times 20 or 10 volts.

The UJT functions in the manner just described as long as V_S is not high enough to forward bias the diode, allowing the diode to conduct. However, when V_S is high enough to forward bias the diode, the UJT will behave differently. In order for this to happen, V_S must be increased until the voltage between the emitter (E) and B_1 terminals (designated V_E) is high enough to forward bias the diode. The value of V_E is equal to V_S, plus the additional voltage required to turn "on" the diode. This voltage—referred to earlier in this book as the **barrier voltage**—is approximately 0.7 volts. In other words, the voltage across R_{B1} (V_{B1}) causes the diode to be reverse-biased, until V_E is approximately 0.7 volts higher than V_{B1}. Then, the barrier voltage of the diode is overcome, and the diode begins to conduct.

The value of V_E required to turn "on" the diode is called the **peak voltage**, and is usually designated as V_P. The value of V_P is determined by the source voltage V_{BB}, the intrinsic standoff ratio, and the voltage required to turn "on" the diode. This relationship can be shown mathematically as:

$$V_P = \eta (V_{BB}) + V_F$$

This equation states that V_P is equal to the product of the intrinsic standoff ratio and source voltage, plus the voltage required to turn the diode "on" (V_F). Manufacturers usually do not indicate the value of V_P for a particular UJT since it varies with V_{BB}. For example, if a UJT has an (η) of 0.6 and a V_{BB} of 10 volts, its peak voltage will be:

$$V_P = (0.6)(10) + 0.7 = 6 + 0.7 = 6.7 \text{ V}.$$

This means that the diode will turn "on" when input voltage V_E reaches 6.7 volts.

Until the V_P value is reached, the diode conducts only a very small reverse leakage current which flows through the emitter lead. However, when the V_P value is reached, the diode turns "on" and allows current to flow in the forward direction through its PN junction and through the emitter lead. This current (often designated as I_E) results because the diode turns on.

In addition to I_E beginning to flow, another action takes place. When I_E begins to flow in the forward direction, many charge carriers (holes) are injected into the lightly doped N-type bar and are swept toward terminal B_1, which is negative with respect to the emitter. These charge carriers increase the conductivity between the emitter (E) and B_1, which reduces the resistance of R_{B1}. This lower resistance causes I_E to increase even further, and more charge carriers are injected into the bar. This lowers the value of R_{B1} even further, increasing I_E even further. This action is cumulative and it starts to occur when the V_P value is reached. At this time the UJT is said to be turned *"on."* If the source voltage (V_S) is increased, the cumulative action just described becomes even more apparent. As V_S increases, I_E increases rapidly due to the decrease in resistance of R_{B1}. Furthermore, this decrease in R_{B1} causes the voltage V_E to decrease, even though I_E increases in value. The UJT therefore exhibits a **negative-resistance** characteristic after it turns on.

Negative resistance is simply the opposite of normal resistance. Normally, as current flow through a component increases, its voltage drop increases also. This is due to the resistance of the component remaining the same. However, with the UJT, the increasing current flow (I_E) causes the voltage (V_E) to *decrease*. The decrease in voltage is caused by the decrease in resistance of R_{B1}.

If V_S is increased even further, I_E would continue to increase. However, a point would eventually be reached where V_E would stop decreasing and actually begin to rise. This point marks the end of the negative resistance region. Beyond this point an increase in I_E is accompanied by a slight increase in V_E.

So far we have considered only the action that takes place between terminals E and B_1 when the UJT turns on. We have seen that R_{B1} decreases, however it is important to note that R_{B2} also decreases to a certain extent by the same charge carriers. However, the decrease in the value of R_{B2} is small when compared to the decrease in R_{B1}. This means that the total resistance of the bar ($R_{B1} + R_{B2}$) is reduced once the UJT turns on, and this results in an increase in current through these two resistors and voltage source V_{BB}. The increase in current is slight, however, because the decrease in R_{B2} is slight.

The most important action takes place between terminals E and B_1. It is this portion of the UJT that provides the most useful characteristics. The two most important features are the ability to turn on at a specific V_P value and the negative resistance characteristic that occurs for a certain period of time after the device turns on.

Voltage-Current Characteristics

The action that takes place between the E and B_1 terminals of an ordinary UJT is shown graphically in Figure 23-10. The curve in this figure shows the relationship between the current flowing through the emitter of a typical UJT (designated as I_E) and the voltage appearing across its E and B_1 terminals (designated as V_E). Figure 23-10 shows that when V_E equals zero, a small negative current ($-I_E$) flows through the emitter lead. This is a small leakage current which flows from left to right through the diode because of the relatively large voltage across R_{B1} (V_{B1}). As V_E increases, it opposes V_{B1}, and the leakage current decreases. When V_E is equal to V_{B1}, the current is reduced to zero, and any further increase in V_E results in a positive current ($+I_E$) which flows from right to left through the diode. When V_E reaches the peak voltage (V_P) value, the UJT is considered to be in the "on" state. The current that flows at this point is called the **peak current,** and is designated as I_P.

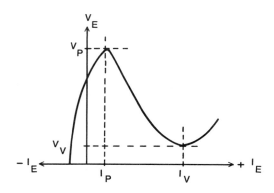

Figure 23-10 V-I characteristic between E and B_1 terminals of a typical UJT.

Beyond the V_P point, V_E decreases as $+I_E$ increases, thus, giving the device a negative resistance characteristic. This negative resistance continues until V_E starts to increase again. The point where V_E reaches its minimum value and starts to increase is called the **valley voltage** and is designated as V_V. The current flowing at this time is referred to as the **valley current,** or I_V. Beyond the V_V point, V_E increases slightly as $+I_E$ increases and the UJT no longer exhibits a negative resistance.

CONTROL DEVICES

UJT Applications

The UJT's negative resistance characteristic makes the device useful for generating repetitive signals. A circuit that is commonly used for this purpose is shown in Figure 23-11A. This circuit is commonly referred to as a **UJT relaxation oscillator**, and it is capable of generating two types of waveforms which can be used in a variety of applications.

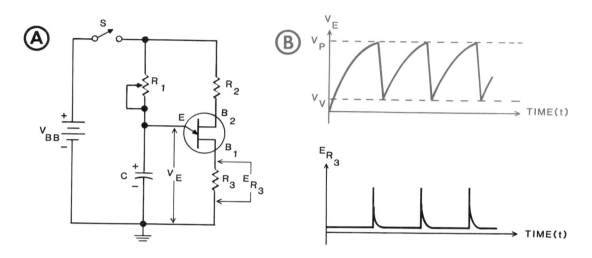

Figure 23-11 UJT relaxation oscillator.

When switch S is closed, capacitor C charges through resistor R_1 in the direction shown. When the voltage across C reaches the UJT's V_P value, the UJT turns *"on"* and the resistance between terminals E and B_1 decreases. This allows C to discharge through the UJT resistor R_3 (which has a very low value). The voltage across C quickly decreases to the UJT's V_V value and this causes the UJT to stop conducting or turn *"off"*. As soon as the UJT turns off, capacitor C starts charging again and continues until V_P is again reached. This causes the UJT to turn on, which in turn allows C to discharge until V_V is reached. This action continues with the voltage across C rising slowly to V_P and decreasing rapidly to V_V. The voltage appearing between the UJT's E and B_1 terminal fluctuates as shown in Figure 23-11B. Notice that this voltage follows a sawtooth type of pattern, and after the initial turn on, it varies only between V_V and V_P.

Each time capacitor C discharges, current is momentarily forced through R_3. These momentary pulses of current cause the voltage across R_3 (E_{R3}) to pulsate. These momentary voltage pulses are very narrow and a pulse occurs each time C discharges. Also note that the voltage never drops completely to zero. A small voltage will always be dropped across R_3 because a small current flows through the UJT (from B_1 to B_2) even when the device is in the *"off"* state.

The number of complete sawtooth waveforms or pulses produced each second (frequency of oscillation) can be controlled by adjusting R_1. If the resistance of R_1 decreases, capacitor C will charge faster, and the circuit will operate at a higher frequency. When R_1 increases in value, it takes longer for C to charge to V_P, and the frequency decreases. The frequency can also be varied by replacing capacitor C with a larger or smaller capacitor, which would change the charge rate.

The sawtooth and pulse waveforms produced by the UJT relaxation oscillator can be used to perform various functions in electronic circuits. However, the UJT is also used in applications where it is not required to oscillate continuously. For example, the UJT can function as a bistable (two-state) element, and switch from its *"on"* state to its *"off"* state when it receives an appropriate input signal. The device can also be used as a frequency divider. Another important application is its producing current pulses which are ideally suited for triggering SCRs.

Programmable UJTs

The programmable UJT, or PUT, is one of the newest members of the thyristor family. This device is not constructed like an ordinary UJT, instead it uses four semiconductor layers. However, it is capable of performing basically the same functions as the UJT. Also, the PUT has a cathode, anode, and gate, which is similar to an SCR, although they perform different functions. The schematic symbol for a PUT is shown in Figure 23-12A.

The primary difference between the ordinary UJT and the PUT is the fact that the peak voltage (V_P) of the PUT can be *controlled*, as opposed to the UJT. The PUT's anode and cathode terminals are used in much the same way as the UJT's E and B_1 terminals and the anode is always made positive with respect to the cathode. In fact, the V-I characteristics between the PUT's anode and cathode terminals are identical to the V-I characteristics between the UJT's E and B_1 terminals. Both devices exhibit the same characteristics, and both have a negative resistance region between their peak and valley points, But the V_P value of the PUT can be controlled by varying the voltage between the gate and cathode terminals, because the gate is always at a positive potential with respect to the cathode.

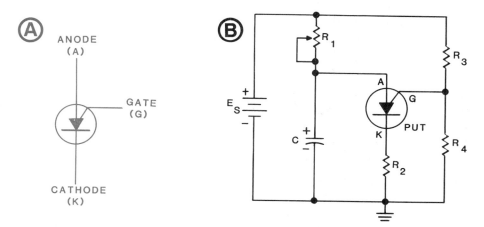

Figure 23-12 PUT schematic symbol and relaxation oscillator.

The operation of the PUT can be clearly demonstrated when it is connected in a circuit like the one shown in Figure 23-12B. This circuit is referred to as a **PUT relaxation oscillator,** and it produces the same basic waveforms shown in Figure 23-11B. Notice that the gate-to-cathode voltage (called the **gate voltage**) is obtained from resistor R_4, which is part of a voltage divider network consisting of R_3 and R_4. R_2 has a very low value, and thus does not affect the gate voltage drastically. The purpose of R_2 is to develop the output pulses. As long as the gate voltage remains constant, the PUT will remain in its "off" or *nonconducting* state until its anode-to-cathode voltage (called its **anode voltage**) exceeds the gate voltage by an amount equal to the voltage drop across a single diode (approximately 0.7 volts). At this time the V_P value is reached and the device turns on.

The anode voltage is obtained from capacitor C, which charges toward the supply voltage (E_S) through R_1. This is similar to the action that takes place in the UJT relaxation oscillator. Each time the voltage across C reaches the V_P value, the PUT turns on and allows C to discharge through its anode and cathode terminals and R_2. When the voltage across C drops to the V_V value of the PUT, the device turns off and the cycle is repeated. A sawtooth of voltage is developed across C and voltage pulses appear across R_2. As before, the frequency of oscillation can be varied by adjusting R_1 or by changing the value of C. However, in this circuit the frequency can also be varied by changing the ratio of R_3 to R_4. It is this resistance ratio that controls the PUT's gate voltage, which controls the value of V_P. If R_4 is made larger while R_3 remains constant, the gate voltage will increase and this will cause V_P to increase. The higher V_P value will make it necessary for capacitor C to charge to a higher voltage before the PUT can turn on. This will increase the time required to generate each sawtooth or pulse and decrease the frequency of operation. If R_3 increases in value while R_4 remains constant, the action will be exactly opposite and frequency will increase.

The frequency of the PUT relaxation oscillator in Figure 23-12B can be controlled by adjusting the ratio of R_3 and R_4 to control the PUT's V_P value. Although this is an extremely important consideration, it is also important to realize that resistors R_3 and R_4 control other PUT characteristics as well. The peak current (I_P) and valley current (I_V) both depend on the values of R_3 and R_4 as well as the value of the source voltage (V_S). For example, it is possible to adjust R_3 and R_4 for any specific ratio so that any specific V_P is obtained (taking into account the value of V_S). However, it is only the ratio of R_3 and R_4 that determines V_P, not the individual values of the resistors. The actual resistor values determine the PUT's I_P and I_V values for any given value of V_S.

This relationship between the external voltage divider network (R_3 and R_4) and the values of I_P and I_V is somewhat complex. However, the relationship can be simplified by relating I_P and I_V to the combined values of R_3 and R_4. R_3 and R_4 are considered to present an equivalent resistance (R_G) to the gate of the PUT which is equal to their parallel equivalent:

$$R_G = \frac{(R_3)(R_4)}{R_3 + R_4}$$

For any given value of V_S, both I_P and I_V will decrease as R_G increases. More complex equations may also be used to determine the exact values of I_P and I_V under various circuit conditions, or these values can be determined graphically when appropriate charts or curves are provided. The circuit designer may use either of these methods to calculate the required values, and in general all of the quantities just described must be considered when designing an oscillator circuit like the one shown in Figure 23-12B. The intent here is not to show a complete design procedure for a PUT oscillator, but to emphasize the PUT's important electrical characteristics.

In addition to its use in relaxation oscillators, the PUT may also be used as a triggering device for SCRs. Due to its four-layer construction the PUT is capable of supplying trigger pulses that have a higher amplitude than those obtained from ordinary UJTs, and can trigger SCRs that have higher current ratings. The PUT is also used in long duration timer circuits (which are basically identical to the circuit in Figure 23-12B), because the device has a low gate-to-anode leakage current. This is important in timer circuits, since this leakage current adds to the charging current flowing through the capacitor. In such a circuit, the capacitor must charge over a long period of time before the V_P value is reached, and an additional charging current (leakage) cannot be tolerated.

The PUT's important electrical characteristics can be controlled, thus making it more versatile than an ordinary UJT. The PUT is also more sensitive than a UJT and it responds faster. It is therefore used in place of the UJT in a number of applications, although the UJT is still preferred in certain types of circuits.

CONTROL DEVICES

Chapter Self-Test

1. How many semiconductor layers does an SCR have?

2. What is the term for the forward voltage required to turn on an SCR?

3. The minimum value of forward current required to keep an SCR in the on state is referred to as what?

4. Under normal conditions an SCR is biased so that it can be turned *"on"* with the application of what?

5. Can an SCR be used to control the amount of AC power applied to a load? What is required to achieve this?

6. When the triac turns on in either direction, it can be turned off only by reducing its main operating current ($+I_T$ or $-I_T$) below a specified current value. What is the name of that value?

7. How can a triac's breakover voltage in either direction be lowered?

8. When a triac is subjected to an operating voltage that is less than its breakover voltage rating (in either direction), can the device can be turned on by a gate current that is flowing in either direction?

9. Why can triacs be used instead of SCRs to provide full control of AC power?

10. Which lead in a UJT is attached to the P-type pellet?

11. What is the ratio of R_{B1} to the total resistance ($R_{B1} + R_{B2}$) in a UJT called? How is it labelled?

12. What is the term for the input voltage value (V_E) required to turn on a UJT?

13. Beyond the V_P point V_E decreases until it reaches a minimum value. What is this value known as? How is it designated?

14. What is the term for the special characteristic of a UJT that occurs between V_P and V_V?

15. What are the three leads for a PUT called?

16. The PUT's V_P value can be controlled by changing the voltage between which points?

SEMICONDUCTORS

Summary

The SCR is the most popular member of the thyristor family. This device has four alternately doped semiconductor layers and three leads which are referred to as the anode, cathode, and gate. This device acts like an electronic switch which can be turned on by a momentary gate voltage and turned off by reducing its operating current below a certain holding value. The SCR may be used to switch DC or AC power and it can even be used to vary the amount of AC power applied to a load. The SCR conducts current in one direction only.

The triac effectively performs the same function as two SCRs that are connected in parallel but in opposite directions. It can therefore control AC currents more efficiently than a single SCR since it operates on both the positive and negative portions of each AC input cycle. The triac must be used in conjunction with a triggering device because of its nonsymmetrical triggering characteristics. The triac conducts current in both directions, but it is not capable of the higher currents that the SCRs can handle.

Although it has characteristics that are different from those of SCRs and triacs, the unijunction transistor (UJT) is still considered to be a thyristor. This device has only one PN junction and three leads (base 1, base 2, and the emitter). The UJT's important electrical characteristics occur between its emitter and base 1 leads. The device will turn on and conduct a substantial current through these leads when the voltage across these leads reaches a certain maximum value known as the peak voltage (V_P). Beyond this point the device exhibits a negative resistance until a certain minimum voltage known as the valley voltage (V_V) is reached. These characteristics allow the UJT to generate repetitive signals when used in a relaxation oscillator circuit, or it can be used as a bistable switching element or as a triggering device.

A special type of UJT, known as a programmable UJT, or PUT, is also available. This device operates like an ordinary UJT, but its V_P value as well as other important electrical characteristics can be varied by an external control voltage. This makes the PUT more versatile than the UJT and more suitable for a broader range of applications. The PUT is widely used as a triggering device and it is often used in long duration timer circuits.

CHAPTER 24

Light-Sensitive Devices

SEMICONDUCTORS

Contents

Introduction 591

Chapter Objectives 592

Basic Principles of Light......................................593

Light-Sensing Devices 596

Light-Emitting Devices 604

Liquid Crystal Displays 609

Chapter Self-Test 612

Summary..613

LIGHT-SENSITIVE DEVICES

Introduction

In this chapter, you will examine a group of solid-state components which are capable of converting light energy into electrical energy, or electrical energy into light energy. We have grouped these components into the category of light-sensitive devices, although they are commonly referred to as **optoelectronic devices**, since their operation relies on both optic and electronic principles.

Optoelectronic devices described are divided into two basic groups: **light-sensitive** devices and **light-emitting** devices. This chapter introduces you to the most important components found within these two categories. We will not be discussing their operation as intensely as bipolar transistors or diodes, since their operation depends on light, and the theory behind measuring light, and how it interacts with other elements, is quite complex.

SEMICONDUCTORS

Chapter Objectives

When you have completed this chapter, you will be able to:

1. Describe the major characteristics of light.

4. Describe the basic function of a light-sensitive device.

5. Name four light-sensing devices and briefly describe their operation.

6. Describe the basic function of a light-emitting device.

7. Describe the basic principle of operation of the light-emitting diode.

8. Name three advantages the light-emitting diode has over an incandescent or neon lamp.

9. Explain the basic principles of operation of a liquid crystal display.

10. Identify at least one advantage of liquid crystal over other types of displays, as well as at least one disadvantage.

LIGHT-SENSITIVE DEVICES

Basic Principles of Light

The term **light** is used to identify electromagnetic radiation which is visible to the human eye. Basically, light is just one type of electromagnetic radiation, and differs from other types such as cosmic rays, gamma rays, X-rays, and radio waves only because of its frequency.

The light spectrum extends from approximately 300 gigahertz to 300,000,000 gigahertz. It is wedged midway between the high end of the radio frequency (RF) waves, which roughly extend up to 300 gigahertz, and the X-rays, which begin at roughly 300,000,000 gigahertz. Beyond the X-ray region are the gamma rays, followed by cosmic rays. To refresh your memory, one gigahertz is equal to 1,000,000,000 Hertz, or 10^9 cycles per second.

Like other types of electromagnetic radiation, light propagates (travels) through space and certain types of matter. The movement of light through space can be compared to the movement of radio waves which periodically fluctuate in intensity as they move outward from an antenna or some other radiating body. Therefore, light waves can be measured in terms of **wavelength**. As with radio waves, the wavelength of light waves is determined by the velocity at which light is moving and the frequency of its fluctuations. This relationship can be summed up with the following mathematical equation:

$$\lambda = \frac{v}{f}$$

The Greek letter **lambda** (λ) represents one complete wavelength. This equation states that one wavelength is equal to the **velocity** (v) divided by the **frequency** (f).

Light travels at an extremely high velocity. In a vacuum, its velocity is 186,000 miles per second or 3×10^{10} (30,000,000,000) centimeters per second. Its velocity is only slightly lower in air, but when it passes through certain types of matter—such as glass or water—its velocity is reduced considerably.

It is important to understand the relationship between frequency and wavelength when dealing with either radio waves or light waves. These two terms are often used interchangeably in any discussion which concerns the utilization of electromagnetic radiation. You will see both of these terms used in this course.

SEMICONDUCTORS

Within the light spectrum, only a narrow band of frequencies can actually be detected by the human eye. This narrow band of frequencies appears as various colors such as red, orange, yellow, green, blue and violet. Each color corresponds to a very narrow range of frequencies within the visible region. Above or below the visible region, light waves cannot be seen. The light waves above the visible region are referred to as ultraviolet rays, while those light rays below the visible region are commonly referred to as infrared rays. The entire light spectrum is shown in Figure 24-1 so that you can compare the three regions just described.

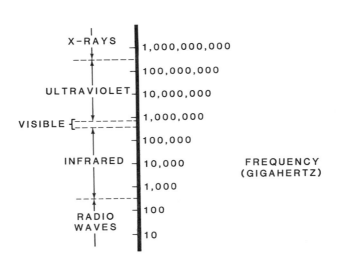

Figure 24-1 The light spectrum.

Although light is assumed to propagate as electromagnetic waves, the wave theory alone cannot completely explain all of the phenomena associated with light. For example, the wave theory may be used to explain why light bends when it flows through water or glass. However, it cannot explain the action that takes place when light strikes certain types of semiconductor materials. It is this action that forms the basis for much of the optoelectronic theory presented here. In order to explain why and how semiconductor materials are affected by light, it is necessary to assume that light has additional characteristics.

To adequately explain the operation of the optoelectronic devices in this chapter, it is necessary to consider an additional aspect of light as explained by basic quantum theory. Quantum theory acknowledges that light has wave-like characteristics, but it also states that a light wave behaves as if it consisted of many tiny particles. Each of these tiny particles represents a discrete quanta or packet of energy and is called a photon.

LIGHT-SENSITIVE DEVICES

The photons within a light wave are uncharged particles and their energy content is determined by the frequency and wavelength of the wave. The higher the frequency, the more energy each photon will contain. This means that the light waves at the upper end of the light spectrum possess more energy than the ones at the lower end of the spectrum. This same rule also applies to other types of electromagnetic radiation. For example, X-rays have a higher energy content than light waves, while light waves possess more energy than radio waves.

Therefore, light has a dual personality. It propagates through space like radio waves, but it behaves as if it contains many tiny particles. This particle-like property of light will be used in this chapter to explain the action that takes place in various types of optoelectronic components.

Light-Sensing Devices

Most of the optoelectronic devices that are now used are basically light-sensing devices. In other words, they simply respond to changes in light intensity by either changing their internal resistance or by generating a corresponding output voltage. We will now examine some of the most important light-sensitive devices.

Photoconductive Cells

The **photoconductive cell** is one of the oldest optoelectronic components. Basically it is nothing more than a light-sensitive resistor whose internal resistance changes as the light shining on it changes in intensity. The resistance of the device decreases nonlinearly with an increase in light intensity. That is, the resistance is not exactly proportional to change in light.

The two schematic symbols that can be used for a photoconductive cell are shown in Figure 24-2A. In the first symbol, arrows are shown to indicate that the device is light-sensing. The second symbol uses the Greek letter lambda (λ), which is used to represent light.

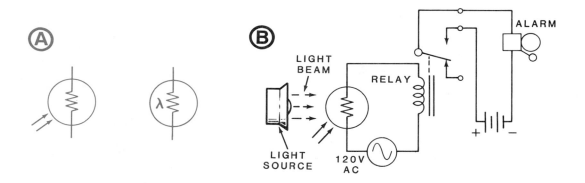

Figure 24-2 Photoconductive cell symbols, and a basic intrusion detector circuit.

LIGHT-SENSITIVE DEVICES

Photoconductive cells are more sensitive to light than other types of light-sensitive devices. The resistance of a typical cell might be as high as several hundred megohms when there is total darkness, and as low as several hundred ohms when a high degree of light strikes its surface. This represents a tremendous change in resistance for a change in light. This extreme sensitivity makes the photoconductive cell suitable for applications where light levels are low and where the changes in light intensity are slight. However, these devices do have certain disadvantages. Their greatest disadvantage is the fact that they respond slowly to changes in illumination. In fact, they have the slowest response of all light-sensitive devices. Also, they have a light memory or history effect. In other words, when the light level changes, the cell tends to remember previous illumination.

Most photoconductive cells can withstand relatively high operating voltages. Typical devices will have maximum voltage ratings of 100, 200, or 300 volts DC. However, the maximum power consumption for these devices is relatively low. Maximum power ratings of 30 milliwatts to 300 milliwatts are typical.

Photoconductive cells have many applications in electronics. For example, they are often used in devices such as intrusion detectors and automatic door openers, where it is necessary to sense the presence or absence of light. However, they may also be used in precision test instruments which can measure the intensity of light.

A simple intrusion detector circuit is shown in Figure 24-2B. The light source projects a narrow beam of light onto the cell, which allows the cell to exhibit a relatively low resistance. The cell is in series with a sensitive AC relay and its AC power source. The cell allows sufficient current to flow through the circuit and energize the relay. When an intruder breaks the light beam, the cell's resistance increases considerably, which deactivates the relay. The relay contacts close and apply power (from a separate DC source) to an alarm, which sounds a warning.

Since the photoconductive cell is constructed from a bulk material and does not have a PN junction, it is a bidirectional device. In other words, it exhibits the same resistance in either direction, and may therefore be used to control either DC or AC. Due to its bulk construction, the photoconductive cell is often referred to as a bulk photoconductor. However, you may also see it referred to as a photoresistive cell or simply a photocell.

SEMICONDUCTORS

Photovoltaic Cells

The **photovoltaic cell** is a device which directly converts light energy into electrical energy. When exposed to light, this device generates a voltage across its terminals, and this voltage increases as the light increases in intensity. The photovoltaic cell has been used for a number of years in various military and space applications. It is commonly used aboard satellites and spacecraft to convert solar energy into electrical power, which can be used to operate various types of electronic equipment. Since most of its applications generally involve the conversion of solar energy into electrical energy, this device is commonly referred to as a **solar cell**.

The basic construction for a photovoltaic cell is shown in Figure 24-3A. The existence of the P-type and N-type layers indicate that this is a semiconductor device. The metal pieces which surround the semiconductor material are used as the electrical contacts. In this case, the P-type layer is exposed to light, although these cells can also be constructed with the N-type layer exposed. The schematic symbol used to represent the photovoltaic cell is shown in Figure 24-3B. This symbol indicates that the device is equivalent to a one-cell voltage source and the positive terminal of the device is identified by a plus (+) sign.

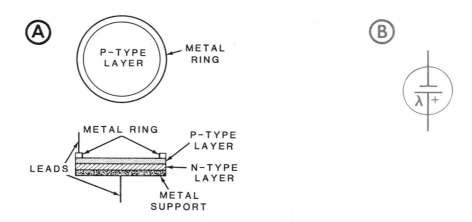

Figure 24-3 Basic silicon photovoltaic cell construction and symbols.

LIGHT-SENSITIVE DEVICES

The photovoltaic cell is designed to have a relatively large surface area which can collect as much light as possible. The cell is constructed so that light must strike the top semiconductor layer within the metal ring as shown in the top view in Figure 24-3A. Since the photovoltaic cell has a PN junction, a depletion region (an area void of majority carriers) forms in the vicinity of the junction. If the cell was forward-biased like a conventional PN junction diode, the free electrons and holes in the device would be forced to combine at the junction and forward current would flow. However, the photovoltaic cell is not used in this manner. Instead of responding to an external voltage, the device actually generates a voltage in response to light energy which strikes its surface.

In order to generate a voltage, the top layer of the photovoltaic cell must be exposed to light. The light energy striking the cell consists of many photons. These photons are actually absorbed at various depths into the semiconductor material depending on their wavelengths and energy. If a particular photon has sufficient energy when it enters the semiconductor material, it can impart much of its energy to an atom within the material.

As these photons are absorbed into the material, free electrons and holes are created, both within the depletion region and within the materials outside the depletion region. Those electrons and holes that are created outside the depletion region are drawn into it. The free electrons in the region are swept from the P-type to the N-type material, and the holes are drawn in the opposite direction. The electrons and holes flowing in this manner produce a small voltage across the PN junction, thus making the N and P regions act like the negative and positive terminals of a battery.

However, not all of the photons create free electrons and holes, and some of the electrons and holes that are created recombine. This makes the photovoltaic cell an inefficient device as far as converting light energy into electrical power, capable of producing only low output voltage. These devices usually require high light levels in order to produce useful output power. When loaded, a typical cell may provide as much as 50 or 60 milliamperes of output load current. However, by connecting a large number of cells in series or parallel, any desired voltage rating or current capability can be obtained.

SEMICONDUCTORS

When used on spacecraft or satellites, many photovoltaic cells are connected together to obtain sufficient power to operate electronic equipment. However, these devices are also used as individual components in various types of test instruments. For example, they are used in portable photographic light meters (which do not require batteries for operation). They are also used in movie projectors to detect a light beam projected through the film. The light beam is modulated (controlled) by a pattern or sound track printed near the edge of the film. In this way, the intensity of the light beam is made to vary according to the sounds (voice and music) that occurred. The photovoltaic cell simply responds to the light fluctuations and produces a corresponding output voltage which can be further amplified and used to drive a loudspeaker. This application is shown in Figure 24-4.

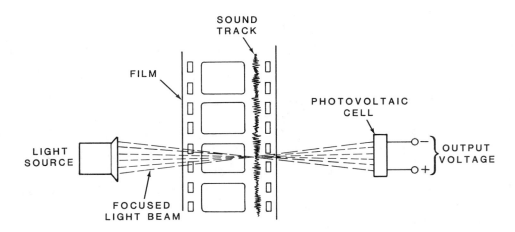

Figure 24-4 A photovoltaic cell used in a movie projector sound reproducing system.

Photodiodes

The **photodiode** is another light-sensitive device which utilizes a PN junction. Although it is constructed similar to the photo*voltaic* cell, it actually operates closer to a photo*conductive* cell. In other words, it is a device whose resistance varies with light intensity resistor.

The construction of a typical photodiode is shown in Figure 24-5A. Notice that the P-type layer is again exposed to light, while the N-type layer joins to form the PN junction. Also, the metal window and base serve as the electrical contacts.

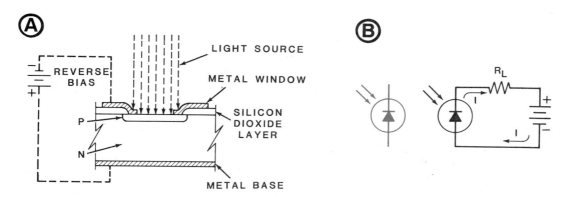

Figure 24-5 Basic construction and schematic symbol of a photodiode.

The schematic symbol for a photodiode is shown in Figure 24-5B. Like the photoconductive cell, the symbol is nothing more than a diode symbol with a circle, and arrows to indicate the device is light-sensing.

The photodiode is most commonly subjected to a reverse bias voltage. In other words, its P-type region is made negative with respect to its N-type region. Under these conditions, a wide depletion region forms around the PN junction. When photons enter this region to create electron-hole pairs, the separated electrons and holes are pulled in *opposite* directions, due to the applied reverse bias, as well as the charges that already exist on each side of the junction. The electrons are drawn toward the positive side of the bias source (the N-type region), while the holes are attracted toward the negative side of the bias voltage (the P-type region). The separated electrons and holes support a small current flow in the reverse direction through the photodiode. As the light intensity increases, more photons produce more electron-hole pairs. This further increases the conductivity of the photodiode resulting in a proportionally higher current. When a photodiode is used in this manner, it is said to be operating in the photoconductive or photocurrent mode.

To sum up, as the amount of light striking a photodiode changes, the amount of current flow through the photodiode also changes. This allows the photodiode to operate in a manner similar to a photoconductive cell.

SEMICONDUCTORS

Phototransistors

The **phototransistor** is also a PN junction device. However, it has two junctions instead of one, like the photodiode just described. The phototransistor is constructed in a manner similar to an ordinary transistor, but this device is used in basically the same way as a photodiode.

The operation of a phototransistor is easier to understand if it is represented by the equivalent circuit shown in Figure 24-6. Notice that the circuit shown contains a photodiode connected across the base and collector of a conventional NPN bipolar transistor. If the equivalent circuit is biased by an external voltage source as shown, current will flow into the emitter lead of the circuit and out of the collector lead. The amount of current flowing through the circuit is controlled by the transistor in the equivalent circuit. This transistor conducts more or less depending on the conduction of the photodiode. If light intensity increases, the diode conducts more current (its resistance decreases), thus allowing more base current to flow through the transistor. This increase in base current is relatively small, but due to the transistor's amplifying ability this small base current can be used to control the much larger collector current flowing through this device. The increase in light intensity causes a substantial increase in collector current. A decrease in light intensity would correspondingly cause a decrease in collector current.

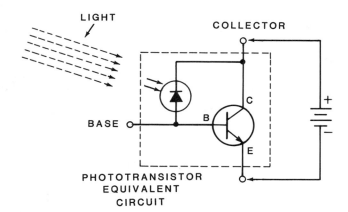

Figure 24-6 Equivalent circuit for a phototransistor.

Although the phototransistor has a base lead as well as emitter and collector leads, the base lead is used in very few applications. However, when the base is used, it is simply subjected to a bias voltage which will set the transistor's collector current to a specific value under a given set of conditions. In other words, the base may be used to adjust the phototransistor's operating point. In most applications only the emitter and collector leads are used and the device is considered to have only two terminals.

The phototransistor, like the photodiode, provides an output current that is essentially controlled by the intensity of the light striking its surface, and to a lesser degree by its operating voltage. The phototransistor is used much like the photodiode by setting its collector voltage to a specific value, and allowing the device to control its collector current in accordance with the changes in light intensity. An important difference between the phototransistor and the photodiode is in the amount of current that each device can handle. The phototransistor can produce much higher output currents than a photodiode because the phototransistor has a built-in amplifying ability. The phototransistor's higher sensitivity makes it useful for a wider range of applications than a photodiode.

Unfortunately, this higher sensitivity is offset by one important disadvantage, in that the phototransistor does not respond as quickly to changes in light intensity. This makes the phototransistor unsuitable for applications where an extremely fast response is required. Like other types of photosensitive devices, the phototransistor is used in conjunction with a light source to perform many useful functions. It can be used in place of photoconductive cells and photodiodes in many applications, and can provide an improvement in operation. Phototransistors are widely used in such applications as tachometers, photographic exposure controls, smoke and flame detectors, object counting, and mechanical positioning and moving systems. A phototransistor is often represented by the symbol shown in Figure 24-7A, and it is usually biased as shown in Figure 24-7B.

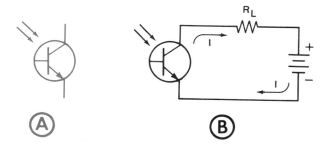

Figure 24-7 A phototransistor symbol and proper biasing.

Light-Emitting Devices

Light-emitting devices are components which produce light when they are subjected to an electrical current or voltage. In other words, they simply convert electrical energy into light energy. For many years, incandescent or neon lamps were the most popular sources of light in electrical applications. The incandescent lamp produces a considerable amount of light, but its life expectancy is relatively short. In addition, the incandescent lamp responds slowly to changes in input electrical power. Due to its slow response time, it will not faithfully vary its light intensity in accordance with rapidly charging alternating currents. The incandescent lamp, therefore, cannot be effectively used to convert high frequency electrical signals into light energy.

The neon lamp has a longer life expectancy than an incandescent lamp, and somewhat faster response to changes in input current. However, its output light intensity is much lower than that of an incandescent device. The neon lamp has been used for many years as an indicator or warning light and, in certain applications, to transmit low frequency AC signals for information in the form of light over very short distances. The neon lamp cannot be used simply for the purpose of providing illumination.

With all of their shortcomings, incandescent and neon lamps were used for many years because nothing better was available. However, in recent years a new type of light-emitting device was developed which has revolutionized the optoelectronic field. The newer device is a solid-state component, and it is physically stronger than the glass encased incandescent and neon devices. Like all semiconductor devices, it has an unlimited life expectancy. This new light-emitting device is referred to as a **light-emitting diode**, or **LED** (pronounced "el-ee-DEE"). Since it is such an important solid-state component, we will examine its operation in detail, then we will see how it is used in various applications.

LED Operation

We have seen how light energy (photons) striking a PN junction diode can impart enough energy to the atoms within the device to produce electron-hole pairs. However, the exact opposite is also possible. A PN junction diode can also emit light in response to an electric current. In this case, light energy (photons) is produced because electrons and holes are forced to recombine. When an electron and hole recombine, energy may be released in the form of a photon. The LED utilizes the principle just described. It is simply a PN junction diode that emits light through the recombination of electrons and holes when current is forced through its junction.

The manner in which this occurs is illustrated in Figure 24-8. Notice that the LED must be forward-biased so that the negative terminal of the battery will inject electrons into the N-type layer (the cathode) and these electrons will move toward the junction. Corresponding holes will appear at the P-type or anode end of the diode (actually caused by the movement of electrons), and they appear to move toward the junction. The electrons and holes merge toward the junction where they combine. If an electron possesses sufficient energy when it fills a hole, it can produce a photon of light energy. Many such combinations can result in a substantial amount of light (many photons) being radiated from the device in various directions.

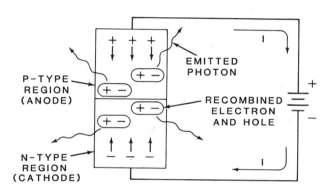

Figure 24-8 Basic operation of a light emitting diode.

At this time, you are probably wondering why the LED emits light and an ordinary diode does not. This is simply because most ordinary diodes are made from silicon, and silicon is an opaque or impenetrable material as far as light energy is concerned. Any photons that are produced in an ordinary diode simply cannot escape. LEDs are made from semiconductor materials that are semitransparent to light energy. Therefore, some of the light energy produced is able to escape from the device.

LED Characteristics

The relationship between forward current and voltage in a typical LED is shown graphically in Figure 24-9A. Notice that the forward bias must be increased to approximately 1.2 volts before any appreciable forward current flows. Then current increases rapidly for a continued increase in forward voltage. This graph effectively shows that once the LED conducts, its current can vary over a considerable range up to its maximum value, while the voltage across the LED remains at approximately 1.6 volts. Most LEDs exhibit a similar current-voltage relationship.

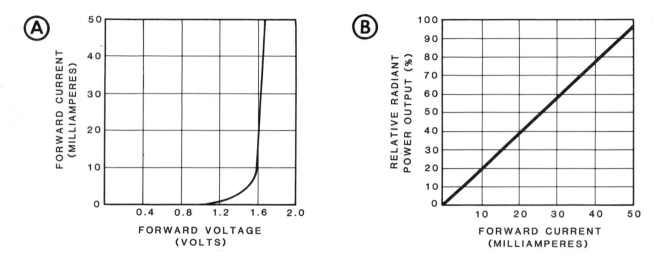

Figure 24-9 Characteristic graphs of a typical LED.

Figure 24-9B shows the relationship between the forward current and the total light output produced by a typical LED. The power is expressed only in relative terms (as a percentage). The important point to note in Figure 24-9B is that the output light increases linearly with the forward current.

Another important characteristic of LEDs is the type of material used to make the LED. Different materials will emit photons of different energy. This produces light of different wavelengths. While this may not seem important, this actually causes different LEDs to produce light of different colors. Most LEDs will emit either red or green light, while others emit an amber light. In addition, lenses must be placed on top of the LED to focus the emitted light outward. When photons are emitted, they travel in all directions, and only a small amount travel out to the human eye. If no lenses were installed, the light an LED could produce would be very difficult to see. By focusing the emitted photons, this light is intensified enough to be seen easily.

LIGHT-SENSITIVE DEVICES

As you may suspect, the amount of light produced by an LED is small when compared to an incandescent lamp. However, LEDs have several important advantages. First, they are extremely rugged. They also respond very quickly to changes in operating current, and therefore can operate at extremely high speeds. They require very low operating voltages and are compatible with integrated circuits, transistors, and other solid-state devices. They are relatively inexpensive when compared to incandescent devices. Also, they may be designed to emit a specific light color or narrow frequency range when compared to the incandescent lamp, which emits a white light that contains a broad range of light frequencies.

The disadvantages associated with LEDs (in addition to low light output) are similar to those which pertain to many types of solid-state components. They may be easily damaged by excessive voltage or current (beyond their maximum ratings) and their output light is dependent on temperature.

LED Applications

In any application of an LED, the device is seldom used alone. The LED is usually connected in series with a resistor which limits the current flowing through the LED to the desired value. To operate the LED without this current limiting resistor would be risky, since even a slight increase in operating voltage might cause an excessive amount of current to flow through the device. Some LED packages even contain built-in resistors (in chip form).

A schematic symbol that is commonly used to represent the LED is shown in Figure 24-10A, and the correct way to bias an LED is shown in Figure 24-10B. The series resistor (R_S) must have a value which will limit the forward current (I_F) to the desired value, based on the applied voltage (E) and the voltage drops across the LED. When determining the value of R_S, allowance must also be made for the internal resistance of the LED, which has a typical value of 5 ohms.

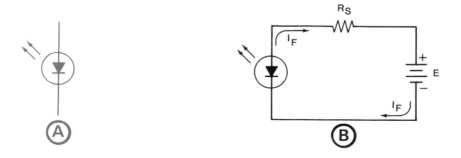

Figure 24-10 An LED symbol and a properly biased LED circuit.

SEMICONDUCTORS

LEDs and their series resistors are often used as simple indicator lights to provide on and off indications. Individual LEDs may even be arranged in specific patterns. The most common of these patterns is the **7-segment LED display**. The LEDs each illuminate one of seven segments arranged in a special pattern. The segments can be turned on or off to create the numbers 0 through 9 and certain letters.

LEDs that emit infrared light may be used in intrusion detector systems if the light is properly focused and controlled. The infrared light cannot be seen by the human eye and is very effective in this application.

Infrared LEDs are also commonly used in conjunction with light-sensitive devices, such as photodiodes or phototransistors, to form what is called an **optical coupler**. The LED and phototransistor chips are separated by a special type of light transmitting glass, and they are coupled only by the light beam produced by the LED. An electrical signal (varying current or voltage) applied to the LED's terminals (through two of the pins on the mini-DIP package) will produce changes in the light beam, which varies the conductivity of the phototransistor. When properly biased, the phototransistor will convert the varying light energy back into an electrical signal. This type of arrangement allows a signal to pass from one circuit to another, but provides a high degree of electrical isolation between the circuits. Also, the LED responds quickly to input signal changes, thus, making it possible to transmit high frequency AC signals through the optical coupler.

Another application of the LED is called an **optical limit switch**. An LED is mounted on one side of a package, and a phototransistor is mounted on the other side. Both devices are separated by a narrow slot in the package. The device may be thought of as an optical coupler with an exposed light beam. When an object (such as a dime) is inserted into the slot, the light beam is broken and the phototransistor does not provide an output signal. This condition is identified by an LED indicator lamp which is mounted above the package. When the object is removed, the light beam is restored and the indicator lamp turns on. Such a device is useful for sensing the presence of objects such as cards, tickets or tapes.

Liquid Crystal Displays

Liquid crystal displays are usually packaged as numeric or alphanumeric displays similar to those often used to package LEDs. However, the liquid crystal display **(LCD)** differs from an LED display in one important respect. LEDs *generate* light, while LCDs *control* light. Consequently, the appearance of an LCD is quite different from that of an LED. LCD characters are usually dark against a light background. A significant advantage of LCD displays is that they have very low operating voltage and current requirement. As a result, these displays have become very popular in digital wrist watches, test equipment, calculators, and other battery powered devices.

The term **liquid crystal** may seem to be contradictory because a crystalline structure is usually associated with solids. However, there is an unusual group of fluids that has a molecular structure like that of crystalline solids. The fluids in this group that are used in optoelectronic displays are **nematic** liquid crystals. Nematic liquid crystals have long molecules that are ordinarily aligned in a specific pattern. As shown in Figure 24-11A, light can readily pass through this molecular structure. Therefore, the material is transparent. However, as indicated in Figure 24-11B, if an electrical or magnetic field is applied to all or part of the liquid crystal, the molecular arrangement in that part of the crystal is subjected to the field changes. This molecular rearrangement causes scattered reflection of light with the result that the material takes on an opaque, frosty appearance.

SEMICONDUCTORS

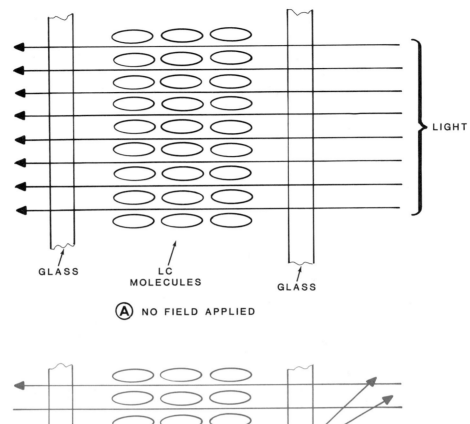

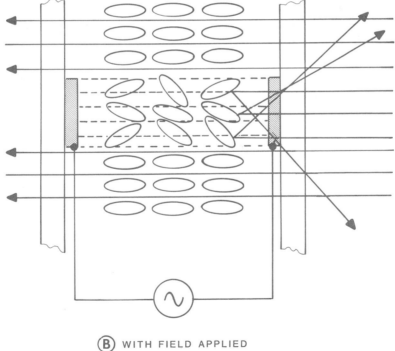

Figure 24-11 Dynamic scattering effect.

Figure 24-12 shows the construction of a typical liquid crystal display. As shown, the liquid crystal material is sealed between two layers of glass. Transparent electrodes have been deposited on each layer of glass. In addition, the molecules in the nematic liquid crystal are aligned in a uniform lattice arrangement that is perpendicular to the glass, allowing light to pass through the display.

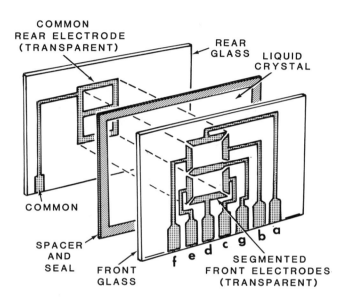

Figure 24-12 Exploded view of LCD showing construction details.

Application of a voltage to one or more front electrodes and the rear electrode produces an electrostatic field, forcing the liquid crystal molecules within the electrostatic field to rearrange themselves into a random alignment pattern. As a result, light striking this part of the display is scattered, causing it to appear opaque.

LCDs have the disadvantage that they must be illuminated by an external light source. Consequently, they cannot be read in the dark or in low light levels. In some digital wrist watches, and other similar devices, this disadvantage is partially overcome by installing a small light source in or near the display. This allows the user to switch the supply on and off as required to read the display.

The LCDs are relatively easy to read in normal room light and have the advantage over many displays—the brighter the light, the easier they are to read. Some LCDs also have the advantage of being read from either side while most visual displays can only be read from one side. This of course would depend on its mounting.

Chapter Self-Test

1. What type of radiation is light considered to be?

2. Are most of the frequencies within the light spectrum visible?

3. Light has wave-like characteristics, but it also behaves as if it consisted of many tiny particles. What are these particles known as?

4. What happens to the resistance of a photoconductive cell as light intensity decreases?

5. The photovoltaic cell directly converts light into what type of energy?

6. How is the photovoltaic cell sometimes referred?

7. How can photovoltaic cells be connected to produce a higher output voltage?

8. How does a photodiode differ in operation from a photoconductive cell?

9. Which of the phototransistor's leads are used in most applications?

10. Do incandescent lamps have a very long life expectancy?

11. Must the light-emitting diode be forward-biased to operate properly?

12. A properly biased LED emits light because of the recombination of what?

13. The light produced by an LED increases linearly with what?

14. What is normally placed in series with an LED to limit its current to the desired value?

15. What is the term for LEDs that are arranged into patterns so they will create numbers and letters?

16. Are LCDs easier to read in bright sun light than LEDs? Why?

LIGHT-SENSITIVE DEVICES

Summary

Optoelectronic devices are now widely used to perform various functions in electronic equipment. They may be divided into two general categories: light-sensing devices and light-emitting devices.

The light-sensing category includes devices such as photoconductive cells, photovoltaic cells, photodiodes, and phototransistors. These devices respond to changes in light intensity by either generating an output voltage or by changing their internal resistance.

The light-emitting category includes components such as incandescent and neon lamps, but these older components are no longer widely used. They are being replaced by a solid-state device known as a light-emitting diode or LED.

LEDs can be designed to emit light over narrowly combined regions of the light spectrum, when subjected to relatively low currents and voltages.

Infrared emitting diodes are special purpose LEDs that emit energy in the infrared region of the spectrum. They have characteristics which make them highly compatible with other types of solid-state devices such as integrated circuits and transistors. They also provide many of the same benefits that are realized with most types of solid-state devices.

There are many applications for optoelectronic devices. These components may be used individually, but they can be combined to perform a variety of functions. The LED is often combined with a photodiode or phototransistor to provide optical coupling between circuits or to detect the presence of an object. Individual LEDs are used as indicators and combinations of LEDs are used to display numbers. The number of possible applications are endless. The liquid crystal display is another numeric display device that has applications similar to those for LED displays. The LCD is very popular in applications where low power consumption is an important consideration. The true potential of these important components are just now being realized, and continued growth in the optoelectronics field is almost certain.

The LCD uses low voltage and low currents and can be easily read in bright light. In fact, the brighter the light the easier the display is to read. It also has the advantage of being visible from either side of the display. However, the LCD requires a controlled beam to form the display.

SEMICONDUCTORS

BASIC AMPLIFIERS
AMPLIFIER APPLICATIONS
OPERATIONAL AMPLIFIERS
POWER SUPPLIES
OSCILLATORS
WAVESHAPING CIRCUITS

SECTION IV

We have spent the last three sections working with and explaining the operation of a large number of electrical components, and how they work within a simple electrical circuit. All of our explanations and descriptions have been designed to lead you to Section IV—Electronic Circuits. In this section, you will combine all that you've learned, and apply it to the fundamental electronic circuits in use today. You'll learn the specific applications of amplifiers, and what must be taken into consideration in order for an amplifier to fulfill its functions. While earlier chapters may have taught you about diodes and capacitors, this section shows you how they can work together to create a power supply. You'll be introduced to your first integrated circuit—the operational amplifier. You'll see that while the op amp is comprised of several dozen components, it operates similar to a single component. You'll also learn about oscillators and waveshaping circuits, which control many equipment's operating sequence.

ELECTRONIC CIRCUITS

CHAPTER 25

Basic Amplifiers

Contents

Introduction 617

Chapter Objectives 618

The Importance of Amplifiers 619

Amplifier Biasing 623

Amplifier Configurations 632

Saturation and Cutoff 654

Amplifier Coupling 660

Chapter Self-Test 668

Summary 670

BASIC AMPLIFIERS

Introduction

In Chapters 21 and 22, you were introduced to transistors and FETs, where you learned how they produce a higher magnitude voltage or current from a smaller input. This is the process of amplification. You were then shown the three basic amplifier circuits that can be created from these components. Each of these amplifier circuits have different operating characteristics, although they basically work on the same principle. Their primary functions also differ, depending upon which terminals are used for the output and the input signal.

In this chapter, we're going to take amplifiers one step further. We'll begin this chapter by answering many of the questions that surround these circuits, such as what types of amplifiers are available, and where are they used.

We will then return to the basic transistor amplifier circuits, and review the concepts and characteristics surrounding each. We will also discuss other facets of these amplifiers that were not covered in Chapter 21. You will read about the many different ways an amplifier circuit can be biased for operation, and how one amplifier circuit can be connected, or coupled, to another.

This chapter will concentrate on transistor amplifiers, with the understanding that an equivalent amplifier circuit with a FET will have essentially the same characteristics.

ELECTRONIC CIRCUITS

Chapter Objectives

When you have completed this chapter, you will be able to:

1. Define the following terms: distortion, amplitude, thermal instability, biasing network, quiescence, switching time, buffer amplifier, saturation, cutoff, clipping, overdriven, coupling, and stage.

2. Name the two basic functions of amplifiers.

3. Identify the three basic amplifier circuit configurations.

4. Describe the most important characteristics of each basic amplifier circuit configuration.

5. Identify the basic biasing arrangements used with amplifier circuits.

6. Explain how each basic biasing circuit operates.

7. Explain the purpose of placing a resistor at the transistor's emitter.

8. Determine if a simple transistor amplifier is operating in the class A, AB, B, or C mode.

9. Identify the four basic amplifier coupling techniques and state the advantages and disadvantages of each.

BASIC AMPLIFIERS

The Importance of Amplifiers

Amplifiers are among the most widely used electronic circuits. They perform a basic but extremely important electronic function, and are used in many types of commercial, industrial, and military equipment.

We will begin by learning what an amplifier is, why it is used, and where it is used. Once you understand the important role of amplifiers in various electronic applications, you will be better prepared to understand the more detailed discussions which follow.

What is an Amplifier?

An **amplifier** is a device which is used to amplify or, in other words, increase the level or magnitude of an electronic signal. Most amplifiers are designed to faithfully reproduce the original shape of an input signal or waveform even though they increase the various instantaneous and peak values of the waveform. If an amplifier had ideal characteristics, it would amplify a signal without introducing irregularities into the signal. Such irregularities or unwanted signal variations are commonly referred to as **distortion.** Practical amplifier circuits cannot be designed so that they are 100 percent free of distortion, but the distortion can be reduced to an insignificant or acceptable level.

In certain cases, amplifiers are designed to amplify a signal but at the same time, produce certain changes in the shape of the signal. These amplifiers intentionally induce distortion into the signal to obtain a certain shape or waveform characteristic that is required for a particular application.

Most modern amplifiers use one or more semiconductor or solid-state components along with a number of associated components. The solid-state component is usually a transistor which can provide the necessary amplification. The associated components are usually resistors, capacitors, and inductors. These passive components are used to control the operation of the transistors involved, and to effectively regulate the overall characteristics, as well as the stability of the amplifier circuits.

ELECTRONIC CIRCUITS

What Type of Amplifiers are Used?

When you consider the various ways in which amplifiers are used in electronic equipment systems, you will find that there are basically only two types. They can be classified as either **voltage amplifiers** or **power amplifiers.**

A voltage amplifier is used to increase the voltage level (also known as the **amplitude**) of an input signal. For example, the amplifier may accept an input AC voltage that has a peak-to-peak value of only a few millivolts, and amplify this voltage to a level of several volts peak-to-peak, possibly higher. This output signal voltage could then be applied to another circuit, or device, which can process the signal or respond to it in some desired manner.

A power amplifier is used to increase the power level of an input signal. For example, this type of amplifier may receive an AC signal at a power level of only a few milliwatts and produce an output AC power level anywhere from several watts to several hundred watts. The power amplifier delivers a substantial amount of power to a load by accepting a relatively low input current and producing a high output current. A power amplifier usually supplies a substantial amount of power to a load, or circuit which has a relatively low resistance or impedance. A low resistance load is necessary to develop a high output current.

Although all amplifiers are basically voltage or power amplifiers, there are other ways that the basic amplifier types are grouped or classified. For example, both types can be classified according to their **circuit configuration.** When a transistor is used as the principle controlling element in an amplifier circuit, three circuit configurations are possible. These, as you learned in Chapter 21, are the **common-emitter, common-base,** and **common-collector** circuits.

Both voltage and power amplifiers may also be arranged according to their **class of operation.** Four classes of operation are commonly used and are referred to as **class A, class AB, class B,** and **class C** modes of operation. You will examine all four of these operating modes in this section.

Most amplifiers provide amplification over a specific range of frequencies. Some amplifiers are used to amplify DC signals (signals which have zero frequency) or very low frequency AC signals (usually 10 Hz or less). This type amplifier is known as a **DC amplifier.** Other amplifiers operate at progressively higher frequencies. These amplifiers may be designed to amplify a narrow or wide frequency band. Typical examples include **audio amplifiers, video amplifiers, RF amplifiers,** and **IF amplifiers.** Chapter 26 will discuss audio and video amplifiers, since they are among the most common, and easiest to understand.

BASIC AMPLIFIERS

Where are Amplifiers Used?

Amplifiers are essential in most types of electronic equipment. They are used in many of the electronic instruments and devices that you use each day. For example, amplifiers are used in radios, television sets, stereo systems, tape recorders, and telephones. In each of these devices, the amplifiers are used to raise the amplitude of an AC signal to a usable level. In a radio or television set, amplifiers are needed to increase the level of the extremely weak (usually microvolts) AC signal that is picked up by the associated antenna. These signals are amplified to a level (usually volts) which is suitable for processing.

The related picture and sound information (only sound in the radio) is then extracted from the incoming signal. This information is further amplified to a level that operates the picture tube and loudspeaker, which convert the amplified signal into the visual display and audible sound.

In a telephone, small signals must be amplified to a level which can drive a speaker which produces the various audible tones. In a tape recorder or stereo system, amplifiers are needed to amplify the weak signals that are extracted from the magnetic tape or from a record. The amplified signals can then be applied to one or more speakers.

Amplifiers are also used extensively in military applications. For example, they are used in two-way radio communications systems where it is necessary to amplify a signal (which contains voice information). The signal strength must be increased to a very high level so that it can be applied to an antenna and transmitted through space. At the receiving end, the signal must again be amplified so that it can be processed and used to drive a speaker.

Amplifiers are used in radar sets to amplify the short bursts of high frequency AC energy which is applied to an antenna and radiated into space. When this energy strikes an object, it is reflected back to the antenna, where it is picked up and again amplified before it is processed. Amplifiers play an important role in automatic fire-control systems which are used to control the large guns aboard naval vessels. Amplifiers are also used in guided missiles, satellites, and spacecraft as part of the onboard communication systems. They amplify signals obtained from sensors and instruments that are used in a variety of board experiments.

ELECTRONIC CIRCUITS

Much of the electronic equipment used to monitor or control industrial processes, contains a variety of amplifiers. Amplifiers are also used in electronic security systems. Amplifiers are the heart of most public address and intercom systems. Even our world-wide telephone communication network would not be possible without amplifiers.

As you can see, there are many applications for amplifiers in electronics. The examples just given are only a small sample. These important circuits are used in so many different applications that it is almost impossible to itemize them all. It is safe to say that amplifiers are among the most common circuits in electronics.

BASIC AMPLIFIERS

Amplifier Biasing

In Chapter 21, we described the requirements for biasing a transistor. If you'll recall, the emitter junction of a transistor must be forward-biased, and the collector junction must be reverse-biased. This allows current flow to begin at the emitter, then split; a small portion of current flows through the base, and the balance of current flows through the collector.

Once we described the biasing requirements for the transistor in Chapter 21, we then discussed how the transistor operated within a circuit. We will follow essentially the same format for this chapter. Before we describe the various amplifier circuits where a transistor is used, we will show you the various methods of biasing.

In Chapter 21, the two different biasing voltages were provided by two separate batteries. This is referred to as **double-battery biasing.** This type of arrangement, although it was the first type of biasing arrangement introduced, is too costly and also unnecessary. The required operating voltages can be provided by a single voltage source.

You will now examine some practical amplifier circuits which are biased with a single voltage source and one or more resistors or capacitors. We will start with the most basic biasing techniques and then progress to more complicated, but more useful, biasing arrangements. Much of our discussion will be based upon the information provided in Chapter 21, so it may be beneficial to briefly review Chapter 21 before proceeding any further.

Since the common-emitter circuit is used more extensively than the common-base or common-collector circuits, our discussion will be primarily centered around common-emitter biasing techniques. We will, however, show how the common-base and common-collector amplifiers can be biased in the more popular biasing techniques.

ELECTRONIC CIRCUITS

Base-Biased Circuits

The simplest method of biasing a common-emitter transistor amplifier is shown in Figure 25-1A. Notice that a single voltage source (V_{CC}) is used to provide both forward and reverse bias-voltages for the NPN transistor used in the circuit. A PNP transistor could also be used in this circuit arrangement if the polarity of V_{CC} was reversed.

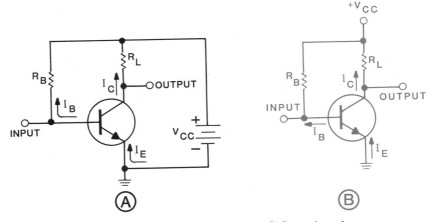

Figure 25-1 A base-biased common-emitter amplifier circuit.

Two resistors (R_B and R_L) are used to distribute the voltage in the proper manner. Resistor R_L is simply a collector load resistor which is in series with the collector. The collector current (I_C) flows through this resistor and develops a voltage across it. If R_L were not in the circuit, the voltage source would be directly connected to the transistor, which could easily damage the component. The sum of the voltage across R_L and the transistor's collector-to-emitter voltage must equal V_{CC}.

Resistor R_B is connected between the base and the positive side of V_{CC}. This resistor is in series with the base lead and it controls the amount of base current (I_B) flowing out of the base. This I_B value flows through R_B and develops a voltage. Most of the source voltage V_{CC} is dropped across R_B and the remainder (the difference) appears across the transistor's base-to-emitter junction to provide the necessary forward-bias voltage. A single voltage source is able to provide the necessary forward and reverse-bias voltages for the NPN transistor because the transistor's base and collector must both be positive with respect to the emitter. Therefore, the positive side of V_{CC} can be connected to the base and collector through R_B and R_L.

BASIC AMPLIFIERS

Since the base current (I_B) in the circuit just described is primarily controlled by resistor R_B and the voltage source V_{CC}, the circuit is often called a **base-biased** circuit. The value of R_B is usually chosen so that the I_B value will be high enough to cause a substantial I_C value to flow through RL. The I_B value may be in the microampere range, while the higher I_C value is likely to be in the milliampere range.

The input signal voltage is applied between the transistor's base and emitter (in other words, between the input terminal and ground) as shown in Figure 25-1A. The input signal voltage either aids or opposes the existing forward-bias voltage across the emitter junction. This in turn causes I_C to vary, which varies the voltage across R_L. The transistor's collector-to-emitter voltage likewise varies, and produces the output signal voltage which appears between the output terminal and ground.

The circuit in Figure 25-1A is shown again in Figure 25-1B as it would appear on a typical schematic diagram. Instead of showing voltage source V_{CC}, only the upper positive terminal is shown and is identified as $+V_{CC}$. The ground symbol identifies the negative side of V_{CC}. This type of drawing is generally used on schematic diagrams and it will be used in the discussions which follow.

The base-biased circuit is seldom used in electronic equipment because it is extremely unstable. The circuit has been described because it serves as a logical introduction to the more complex and more practical circuits which will be described later. The base-biased circuit is unstable because it cannot compensate for changes in its steady-state (no-signal) bias current. For example, temperature changes can cause the transistor's internal resistances to vary and this causes the bias currents (I_B and I_C) to change. This, in turn, causes the transistor's operating point to shift and it can also reduce the gain of the transistor. This entire process is referred to as **thermal instability,** and it is inherent in any circuit that is base-biased.

Feedback Bias

A practical transistor amplifier circuit must be able to compensate for thermal instability. Any time bias current changes, both output current and output voltage change. It is possible to compensate for these changes by feeding a portion of the unwanted output current or voltage back to the circuit's input, so that it opposes the change. When this is done, the circuit is said to be using **degenerative feedback** or **negative feedback.**

ELECTRONIC CIRCUITS

Negative feedback should not be confused with negative voltage feedback. Negative feedback means a signal is fed back from the output to the input that opposes what the circuit is trying to do. It is always degenerative, which again means opposing, and always results in a lower gain from the circuit. When you use degenerative feedback, some of the gain is sacrificed to provide thermal stability.

A common-emitter circuit which uses degenerative feedback is shown in Figure 25-2. Notice that base resistor R_B has been connected directly across the NPN transistor's base and collector leads. With this circuit arrangement, the base current flowing through R_B is determined by the voltage at the collector.

When the temperature of a transistor rises, the size of the depletion regions at both the emitter and collector junctions becomes smaller. This effectively reduces the amount of internal resistance offered by the transistor, which causes an increase in emitter, base, and collector currents. In the circuit shown in Figure 25-2, as the collector current rises, the voltage drop of R_L also rises. This causes the transistor's collector-to-emitter (output) voltage to decrease. This decrease reduces the voltage applied to R_B. This decrease in E_{RB} causes forward-bias to decrease, which reduces the increased current flow back to its normal value. Since the feedback signal is obtained from the collector of the transistor, this type of circuit is said to use **collector feedback**. A collector feedback circuit can provide a certain amount of temperature stability, but it still does not provide complete stability.

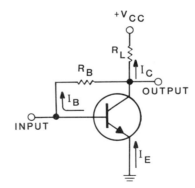

Figure 25-2 A common-emitter circuit which uses degenerative feedback.

A different type of feedback arrangement is shown in Figure 25-3A. Notice that R_B is connected directly to the positive side of V_{CC} as it was in the base-biased circuit shown in Figure 25-1. Also, an additional resistor (R_E) has been connected in series with the emitter lead. The emitter current must flow through R_E, which develops a voltage drop. A series path is created, starting from ground, and continuing through the emitter resistor, the transistor's emitter junction, the base resistor, and finally stopping at V_{CC}. The voltage across the emitter junction is very small, and can be neglected in our analysis. When an increase in temperature causes I_C to increase, I_E also increases. This causes the voltage across R_E to increase, which in turn causes less voltage to appear across R_B. A lower voltage at the base decreases forward-bias, which reduces I_C and I_E towards their normal values.

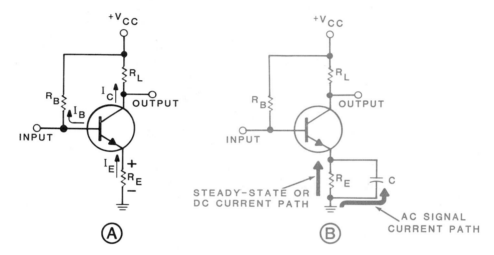

Figure 25-3 A common-emitter circuit which uses emitter feedback.

In a practical circuit, I_C and I_E may still increase just slightly since the decreasing I_B value cannot completely counteract or limit their rising values. This refers to another law of physics which states "the effect can never overcome the cause." However, the resulting change is considerably less than it would be without feedback. Since the feedback in this circuit arrangement is generated at the transistor's emitter, the circuit is said to use **emitter feedback.** Such an arrangement provides a substantial amount of temperature stability. Emitter feedback is always degenerative feedback, and it always decreases gain.

ELECTRONIC CIRCUITS

In Figure 25-3B, the same circuit is shown, with the exception that a capacitor is connected across the emitter resistor. To understand the function of this capacitor, we must first examine what happens to the voltage drop of the emitter resistor, and the problems it can cause when the capacitor is *not* used.

If you'll recall from Chapter 21, when an input signal is applied to an amplifier, the amount of forward-bias is varied. This changing forward-bias causes emitter current to constantly change, and this changing emitter current is felt through R_E. As the emitter resistor feels this changing current, its voltage drop changes in the same fashion.

This is where the problem occurs. An amplifier can only achieve its highest amount of gain when the changing input voltage causes the *greatest* change in forward-bias. If the emitter voltage increases and decreases at the same time the base voltage increases and decreases, the changing base voltage can only *slightly* affect forward-bias. Therefore, the greatest gain can only occur if the base voltage is changing while the emitter voltage remains constant.

The bypass capacitor placed across the emitter resistor performs this exact function. As emitter current attempts to change, the bypass capacitor alternately charges and discharges with these changes, keeping the voltage on the emitter at a constant level. This allows the amplifier to provide maximum gain, while at the same time, it frees the emitter resistor to change its voltage drop only to compensate for changes in temperature.

BASIC AMPLIFIERS

Voltage Divider Biasing

The common-emitter circuit which uses emitter feedback can provide a reasonable amount of temperature stability. However, the stability of this circuit can be improved even more by removing resistor R_B and installing two resistors (R_1 and R_2) as shown in Figure 25-4. These two resistors complete a series path between V_{CC} and ground. Both resistors develop voltages, whose sum equal V_{CC}. This can be thought of as a series voltage divider.

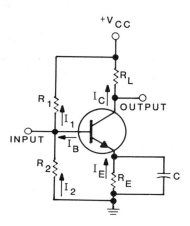

Figure 25-4 A common-emitter circuit which uses voltage divider biasing and emitter feedback.

Obviously, the voltage drop of R_2 must be lower than V_{CC}. Also, since the upper end of R_2 is connected to the transistor's base, the voltage at the base (with respect to ground) is equal to the voltage dropped across R_2. Furthermore, the base voltage is positive with respect to ground, since the current through R_2 (I_2) flows *up* through the resistor.

R_2 feels only that current which flows from ground towards V_{CC}. R_1 also feels this current, but in addition, feels that current which flows from the base of the transistor towards V_{CC}, which is the base current. Remember, the base voltage is positive with respect to the emitter, and the emitter junction is forward-biased. The effect that base current has on the voltage divider is insignificant, however, because the ohmic values for R_1 and R_2 are selected to prevent this from happening. The sole purpose for this voltage divider is to provide a steady DC voltage at the base of the transistor.

ELECTRONIC CIRCUITS

Emitter and collector currents for this circuit flow in the same manner as for the emitter feedback circuit shown in Figure 25-3. Since emitter current flows up through R_E, the emitter voltage is positive with respect to ground. Since the base voltage is also positive with respect to ground, the emitter voltage must be *less positive,* in order to properly forward-bias the emitter junction.

Voltage divider biasing is also used with common-base and common-collector circuits. For example, a common-base circuit using a voltage divider is shown in Figure 25-5A. In this circuit, resistors R_1 and R_2 are again used to keep the base at a constant positive voltage with respect to ground. Emitter resistor R_E and load resistor R_L again perform the same basic functions they performed in the common-emitter circuit. However, in this circuit, the input signal is felt at the emitter, and the output is felt at the collector. The base voltage is provided by the voltage divider, comprised of R_1 and R_2, with the voltage drop of R_2 being felt at the base. The bypass capacitor is placed at the base for the same reason as it was placed at the emitter for the common-emitter amplifier. Since the base voltage varies because of changing forward-bias, the capacitor alternately charges and discharges, keeping the base voltage as a fairly constant value.

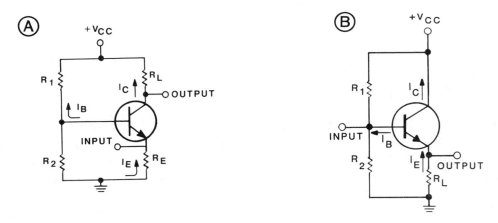

Figure 25-5 Common-base and common-collector circuits with voltage divider bias.

A common-collector circuit with voltage divider bias is shown in Figure 25-5B. This circuit is biased in basically the same manner as the common-emitter circuit, with the exception that no load resistor is used. For this circuit, the output voltage is taken across the emitter resistor. Resistor R_1 and R_2 form the voltage divider, with the voltage drop of R_2 providing the base voltage. Also, you may notice that there is no bypass capacitor. This is because there is really no reason for one for this circuit. A bypass capacitor cannot be placed at the base, otherwise the base would not feel the input signal. And placing a bypass capacitor across the emitter resistor would prevent any AC signal voltage from being developed at the output.

In conclusion, this type of biasing provides excellent temperature stability, and the bypass capacitor provides for maximum amplifier gain. For these reasons, voltage divider biasing is the most widely used type of biasing for transistor amplifiers.

Although we have examined several practical amplifier circuits which use voltage divider bias along with collector or emitter feedback, we still have considered only a few of the many biasing techniques that are used. However, this information should provide a sufficient basis of information, enabling you to understand more sophisticated circuit arrangements when you encounter them. These sophisticated biasing techniques are essentially combinations of the techniques discussed, with the addition of special purpose components to perform a specific function.

Amplifier Configurations

We will now discuss the three different amplifier configurations available with a transistor. While some of the information presented in this portion was originally presented in Chapter 21, there is also an extensive amount of new information. Also, the material discussed here will be used extensively throughout this entire last section of the textbook. Therefore, a complete and thorough understanding of this portion is essential in your studies.

We will discuss each amplifier using voltage divider biasing. Although you may see these amplifiers using the other biasing techniques, voltage divider biasing is used because it is the most common type of biasing method, and it is the easiest circuit for describing amplifier operation.

Common-Base Circuits

In the common-base amplifier, the transistor's base is common to both the input and output signals. The input signal is applied at the emitter, while the output signal is taken at the collector. Although the common-base circuit is not used as extensively as other amplifiers, it does have certain features which make it useful in a number of applications. You will find that the common-base amplifier provides a substantial voltage gain, even though there is less current at the output than at the input. For this reason, the primary function of the common-base amplifier is to operate as a voltage amplifier.

Circuit Configuration

A simple common-base circuit which uses an NPN transistor is shown in Figure 25-6A. Resistors R_1 and R_2 are the voltage divider, with V_{R2} providing the base voltage. R_E provides the emitter voltage, and the collector voltage is that portion of V_{CC} which is *not* dropped by R_L. The right side of our circuit (often called a biasing network) can be thought of as another voltage divider. Emitter current begins at ground, flows though the emitter resistor, then through the transistor, and through R_L to V_{CC}. Therefore, the sum of the voltage drops for these three components—the emitter resistor, the transistor, and the load resistor—must equal V_{CC}.

Since we are using a positive voltage source, all of the voltages felt at any point in this circuit are positive, and all currents flow from ground to V_{CC}. It may be interesting to point out that whatever polarity of the base material, V_{CC} must be of the same polarity. For the NPN transistor, the base material is positive; therefore, the voltage source is also positive.

BASIC AMPLIFIERS

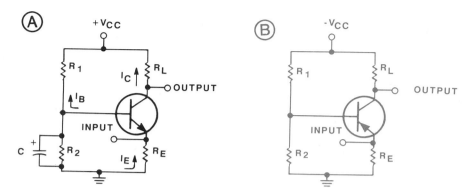

Figure 25-6 NPN and PNP common-base amplifiers.

The common-base amplifier may also be formed with a PNP transistor as shown in Figure 25-6B. Notice that this circuit is basically the same as the NPN circuit, except for the polarity of V_{CC}. With the PNP transistor, the base material is negative; therefore, the polarity of V_{CC} must also be negative. Current flow throughout this circuit is from V_{CC} to ground.

Quiescence

Since there is a voltage source, current flows in the circuit, although there is no input signal. The transistor is properly biased, due to the magnitude of the voltages felt at each terminal of the transistor. Although the base, emitter, and collector voltages are all positive, they are of different values. The voltage on the positive base material is more positive than the voltage on the emitter, thus forward-biasing the emitter junction. In addition, the voltage on the negative collector is more *positive* than the voltage on the base, thus reverse-biasing the collector junction.

Before an input signal is applied, current still flows through the transistor, as well as the voltage divider networks. Once the input signal is applied, it simply varies these currents. However, before the input is applied, the amplifier is said to be at **quiescence,** or in its **quiescent state.** Quiescence (pronounced "KWEE-es-ens" or "KWY-es-ens") simply means "at rest," or "quiet." The amplifier simply has current flow, and is waiting for an input signal to vary that current flow. This action can be compared to an idling car engine: the motor is running, but it is not doing any work. Once the car is placed in gear (or an input is applied to the transistor), its primary function is fulfilled.

Circuit Operation

Once V_{CC} has been applied to the amplifier, it is ready to accept an input signal. We will refer to the circuit in Figure 25-7 as we describe the reaction the amplifier has when the input is applied.

As we stated earlier, the input signal is applied to the emitter. As the input signal is felt at the emitter, the emitter voltage changes. If you'll recall, the emitter voltage for the NPN transistor is a positive value. When the positive alternation of the input is applied to the emitter, it makes the emitter voltage more positive. This serves to *decrease* the amount of forward-bias at the emitter junction. The decrease in forward-bias causes emitter current to also decrease.

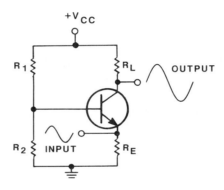

Figure 25-7 Operation of a common-base amplifier.

Since both base and collector current are dependent upon emitter current, any change in emitter current affects base and collector current in the same manner. However, since the output of the common-base amplifier is at the collector, we are primarily concerned with collector current. Therefore, when emitter current decreases, collector current also decreases.

This decrease in collector current flows through R_L, which receives its current solely from the collector. The voltage across R_L must decrease when the current flowing through it decreases. However, R_L does not provide the output voltage. The output voltage is across the transistor and the emitter resistor. We stated earlier that the sum of the voltage drops for the right side of this biasing network—the emitter resistor, the transistor, and the load resistor—must equal V_{CC}. When one voltage drop decreases, the other two must increase to compensate for the loss. When V_{RL} decreases, , the voltage drop of the transistor (V_{CE}) and V_{RE} must increase. Therefore, the output voltage also increases.

Figure 25-8 shows the input signal, and the changing values of I_E, I_C, V_{RL}, and the output signal (V_{OUT}). Refer to this figure while we summarize the effect of the positive input alternation on the common-base amplifier:

1. The positive alternation increases emitter voltage.

2. The forward-bias on the emitter junction decreases.

3. Emitter current decreases.

4. Collector current decreases.

5. The voltage drop of the load resistor decreases.

6. The voltage drop of the transistor and the emitter resistor (the output voltage) increases.

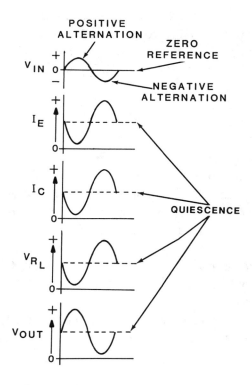

Figure 25-8   The input and output signals of a common-base circuit.

Obviously, the negative alternation affects the same values, but oppositely. The negative alternation causes the voltage on the emitter to decrease, which increases the forward-bias on the emitter junction. This causes an increase in emitter and collector currents, which increases the voltage drop of the load resistor. When V_{RL} increases, the output voltage (comprised of the transistor and resistor voltage drops) decreases.

Let's return to Figure 25-8, and compare only the input and output signals. Notice how the positive alternation of the input signal causes the output to also go positive, and the negative alternation of the input signal causes the output to go negative. In other words, the input and output signals are **in-phase** with each other. This is a trait inherent in the common-base amplifier. Notice also how all of the values in Figure 25-8 vary above and below their quiescent values, with the exception of the input signal. The output signal in particular is always above 0 volts. If a PNP transistor were used, all of the voltages in the circuit would be negative, and the output signal would vary above and below the negative quiescent value, never going above 0 volts.

Keep in mind that these changes in voltage and current for both alternations occur very quickly, usually within the period of 20 nanoseconds. The fast reaction time (sometimes referred to as **switching time**) is a characteristic trait of a transistor, and makes the transistor a very versatile component.

Current Gain

If you'll recall from Chapter 21, **gain** is defined as the change between the input and output signals of an amplifier. **Current gain,** therefore, means the change between the input and output currents. If these two currents are equal, the gain has a value of 1. If the output current is larger than the input current, the gain is greater than one. If the output current is less than the input current, the gain is less than one.

For the common-base amplifier, current flow through the emitter is the input current, while collector current serves as the output current. From our knowledge of basic transistor action, we know that collector current must always be slightly less than emitter current, although these values are normally very close. Therefore, the amount of current gain for the common-base amplifier must be less than (but very close to) 1.

BASIC AMPLIFIERS

In Chapter 21, we stated that the relationship between emitter current and collector current for any transistor is expressed using the transistor's **alpha** value. This is true for any transistor, even if the transistor is not connected in a common-base configuration. Alpha, represented by the Greek letter alpha (α), is found simply by dividing the emitter current by the collector current. This is expressed mathematically as:

$$\alpha_{DC} = \frac{I_C}{I_E}$$

A transistor's alpha value can be determined using two different methods. First, the two current values are found when the circuit has no input signal. Here, emitter and collector current are steady values, resulting solely from the two biasing voltages. When alpha is calculated in this manner, it is called **DC alpha.**

Alpha can also be calculated using the amount of change between the two currents. Here, emitter and collector currents are measured, and V_{EE} is then adjusted, and the *amount of change* in the two current are measured. Alpha is then calculated using the amount of change for the two currents. This is referred to as a transistor's **AC alpha,** and is calculated using the equation:

$$\alpha_{AC} = \frac{\Delta I_C}{\Delta I_E}$$

If you'll recall, the Greek letter delta (Δ) is used to represent change.

For any given transistor, the DC and AC alpha values are almost equal, and both are always slightly less than 1. Furthermore, the DC and AC alpha values are always determined while the transistor's collector-to-base voltage (V_{CB}), is held constant.

A transistor's alpha is often referred to as the transistor's **common-base, forward-current transfer ratio.** Also, the symbols h_{FB} and h_{fb} are sometimes used to represent the DC and AC alpha values respectively.

In most practical applications, some type of load resistance is usually connected between the transistor's collector and base so that I_C flows through the load to perform a useful function. The DC or AC alpha values provide a reasonably accurate indication of the amount of current gain that can be obtained using a common-base amplifier.

Voltage Gain

Even though the common-base amplifier has a current gain of less than 1, it can provide a substantial voltage gain. In other words, the amplitude of the output signal is considerably larger than the amplitude of the input signal.

Figure 25-9 shows the right-side of our biasing network, which will suit our purposes for this discussion. The amount of voltage gain is partly due to the ohmic values of the emitter and load resistors. For the emitter voltage to be less than the base voltage, the emitter resistor must be a very small value. Therefore, only a small input signal is required to change the amount of forward-bias on the emitter junction.

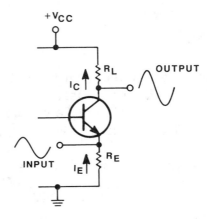

Figure 25-9   The voltage gain of a common-base amplifier.

The load resistor, on the other hand, is considerably larger than the emitter resistor. This is because the load resistor must drop a higher amount of voltage to develop that portion of V_{CC} which is not dropped by the transistor or the emitter resistor.

The load resistor changes its voltage drop with any change in collector current. Due to the characteristics of the transistor, as well as the high ohmic value of the load resistor, any small change on the emitter junction forward-bias causes a large change in collector current. This large change in collector current causes a proportional change in the voltage drop of R_L, which changes the output signal. For these reasons, the common-base amplifier is able to provide a high voltage gain.

The voltage gain (represented by A_V) for any amplifier is found by dividing the input peak-to-peak voltage by the output peak-to-peak voltage. This is done using the following formula:

$$A_V = \frac{\text{OUTPUT P-P}}{\text{INPUT P-P}}$$

For example, if a common-base amplifier produces a 10 volt p-p output signal when a 0.5 volt p-p input signal is applied, the voltage gain for this amplifier would be:

$$A_V = \frac{\text{OUTPUT P-P}}{\text{INPUT P-P}} = \frac{10V}{0.5V} = 20$$

BASIC AMPLIFIERS

For this amplifier, any input voltage is amplified 20 times at the output. This shows us that the output voltage is equal to the input voltage multiplied by the gain. Expressed as a formula:

$$\text{OUTPUT P-P} = \text{INPUT P-P} \times A_V$$

Power Gain

Since power is a function of current and voltage, the common-base amplifier has the ability to provide a power gain of greater than one. Although the current gain is less than one for the common-base it is offset by the substantial voltage gain. The power gain (A_P) is equal to the product of the current gain (alpha) and the voltage gain (A_V) or, expressed mathematically:

$$A_p = (\alpha) \times A_V$$

Since the transistor's alpha may have a typical value of 0.99 and the voltage gain may be as high as 1000, the power gain may reach a value of 990 as shown by the following equation.

$$Ap = (0.99)(1000) = 990$$

The power gain may reach a value that is almost as high as the voltage gain. Typical common-base circuits may have power gains that range from 100 or 200, to as high as 1000.

Input and Output Resistance

The amount of resistance offered at the input and output of an amplifier is a consideration which was not discussed in Chapter 21. When a signal is applied to an amplifier, the amount of resistance at the input has an effect on the amplitude of the signal. Also, the resistance at the output has an effect on the amount of current that flows through any load that is connected to the output. This consideration is similar to impedance-matching for transformers. The amount of resistance at the input and output terminals is an important consideration when connecting a signal to the amplifier, or connecting the output of an amplifier to a load.

A substantial input current (I_E) flows in a common-base circuit as a result of the forward bias applied to the transistor's emitter junction. Therefore, when the input signal is applied between the emitter and base of the common-base transistor, it sees a very low input resistance. The input resistance of a common-base transistor typically ranges from 30 to 150 ohms. Also, the input resistance values are for the common-base transistor itself. When additional components are added to the circuit, the overall input resistance may be higher or lower than when the transistor is used alone.

ELECTRONIC CIRCUITS

The output resistance of a common-base circuit is essentially the high resistance offered by the transistor's reverse-biased collector junction. This resistance appears between the transistor's collector and base regions. A typical common-base circuit might have an output resistance that is between 300 kilohms and 1 megohm. These are typical values for the transistor only. The output resistance of the overall circuit may be higher or lower when additional components are added to the circuit.

Common-Emitter Circuits

In the common-emitter circuit, the transistor's emitter lead is common to both the input and output signals. The input is applied to the base of the transistor, and the output is taken from the collector. Furthermore, the common-emitter circuit can be formed using either NPN or PNP transistors. Since it has many desirable features, the common-emitter circuit is used more extensively than any other circuit arrangement. It has the ability to amplify both current and voltage, which allows the common-emitter to function primarily as a power amplifier.

Circuit Configuration and Operation

A common-emitter circuit with voltage divider biasing is shown in Figure 25-10. This circuit uses an NPN transistor, although a PNP transistor would prove equally useful. Notice that the polarity of V_{CC} is positive, which correlates to the base material of the transistor. Resistors R_1 and R_2 serve as the voltage divider to provide the base voltage. All of the voltages in this amplifier are positive, due to the direction of the current flow from ground to V_{CC}.

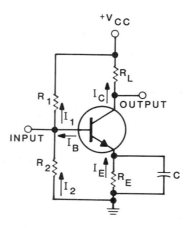

Figure 25-10 The common-emitter amplifier.

Resistor R_E develops the emitter voltage. This voltage must be less than the base voltage in order to properly forward-bias the emitter junction. For this reason, it is safe to assume that the ohmic value for R_E is smaller than R_2, the resistor which provides the base voltage.

The load resistor R_L develops that voltage which is not developed by the transistor and R_E (Recall that the right side of the biasing network can also be thought of as a voltage divider). The output voltage is that voltage which is developed by the transistor and R_E. And finally, the bypass capacitor is connected across the emitter resistor to provide maximum gain, and to allow the emitter resistor to account for thermal stability.

Like the common-base amplifier previously described, the amplifier sits at its quiescent state when there is no input signal. Emitter, base and collector currents flow because the amplifier is properly biased for operation, and the voltage source V_{CC} is applied.

The primary difference between the operation of the common-emitter and the common-base amplifiers is the location of the input signal. In the common-base, if you'll recall, the input signal was felt at the emitter. This varied the amount of forward-bias at the emitter junction by changing the voltage felt at the emitter. The common-emitter amplifier performs the same function, but the forward-bias is varied by changing the voltage on the *base*. From that point, the transistor reacts to the changing forward-bias in the same manner as the common-base.

Since the positive and negative input alternations are now applied at the base, each alternation has a different effect on forward-bias than when they were felt at the emitter. The positive input alternation, for example, decreased forward-bias when it was applied at the emitter. When this same alternation is applied to the base, it *increases* forward-bias. The input signal applied at the base also changes the voltage and current gains for the common-emitter amplifier. These differences will be discussed later.

For now, let's return to the operation of the common-emitter. When the positive input alternation is applied to the base of the amplifier, it increases the already-positive voltage on the base, which increases forward-bias. Increasing forward-bias, as you know, increases emitter current, which increases collector current. The load resistor R_L reacts to this increasing collector current by increasing its voltage drop.

Since the sum of the voltage drops of the three components on the right-side of the biasing network must equal V_{CC}, and increase in V_{RL} causes a decrease in the voltage drop of the transistor, and a decrease in V_{RE}. Since the transistor and R_E determine the output voltage, a decrease in their voltage drops means that the output voltage decreases.

ELECTRONIC CIRCUITS

Let's summarize the effect that the positive input alternation has on the common-emitter amplifier. As we go through each step, refer to Figure 25-11, which shows the input signal (V_{IN}), and the effects it has on I_B, I_C, V_{RL}, and the output signal (V_{OUT}).

1. The positive input alternation is applied at the base, increasing the base voltage.

2. As base voltage increases, the forward-bias at the emitter junction increases.

3. Emitter current increases.

4. Collector current increases.

5. The voltage drop of the load resistor increases.

6. The voltage drop of the transistor and R_E (the output voltage) decreases.

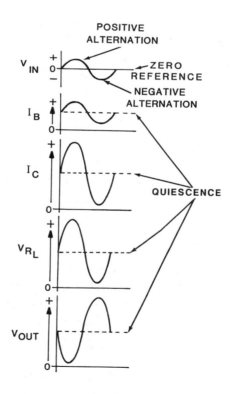

Figure 25-11 The input and output signals of a common-emitter circuit.

The negative input alternation affects the common-emitter in the exact opposite manner as the positive input alternation. The negative input alternation is felt at the base, and base voltage decreases. This decreases emitter junction forward-bias, which in turn decreases emitter current. When emitter current decreases, collector current also decreases, and this decreasing collector current is felt by R_L, whose voltage drop decreases. Since V_{RL}, V_{RE}, and V_{CE} (the voltage drop of the transistor) must equal V_{CC}, when V_{RL} decreases, V_{CE} and V_{RE} (the output voltage) increases.

Let's again refer to Figure 25-11, and compare only the input and output signals. Notice how the positive input alternation causes the output signal to develop a *negative* alternation. Likewise, the negative input alternation causes a positive output alternation. The reason for this is the effect the input has on forward-bias when it is applied to the base, as opposed to the emitter of the transistor. The input and output signals for the common-emitter amplifier are said to be out-of-phase, and their phase relationship is 180 degrees. Notice also that each value in Figure 25-11 varies above and below their quiescent values, with the exception of the input signal.

Current Gain

Since the input signal for the common-emitter amplifier is applied at the base, base current (I_B) is considered to be the input current. Output current, like the common-base, is still collector current (I_C). To determine the difference between these current values, we need only remember that emitter current splits once it enters the transistor, with a small portion going to the base, and the balance traveling through the collector. Therefore, collector current is always considerably larger than base current. This allows the common-emitter amplifier to provide a large current gain.

In Chapter 21, we introduced the term **beta**. Beta is simply a comparison of collector current to base current. Beta can be calculated using the following formula:

$$\beta_{DC} = \frac{I_C}{I_B}$$

While every transistor has a beta value, beta can be used to specifically express the amount of current gain in a common-emitter amplifier. We can express this value using two different methods. First, the amplifier can be connected with no input signal, and the biasing voltages provide a steady amount of base and collector currents. Beta is then calculated using these steady values. This is referred to as the amplifier's **DC beta value.**

ELECTRONIC CIRCUITS

The second method for finding beta is based upon the fact that a small change in base current causes a proportional change in collector current. The same circuit is used, and base current is changed by a predetermined amount. From there, the change in collector current is measured, and beta is then calculated using these values. This is referred to as the amplifier's **AC beta,** and is expressed with the following equation:

$$\beta_{AC} = \frac{\Delta I_C}{\Delta I_B}$$

Like the DC beta previously described, the AC beta is simply a ratio of output and input currents. However, the AC beta represents the gain obtained when a changing (AC) signal is applied to the common-emitter transistor, while the DC beta is determined under steady-state or no-signal conditions. DC and AC beta values are nearly the same, ranging from 10 to more than 200, with values of from 80 to 150 being typical. These values occur because collector current is always much larger than base current.

The current gain or beta of a common-emitter transistor is sometimes called the transistor's **common-emitter, forward-current transfer ratio.** The DC beta (forward current transfer ratio) is often represented by the symbol h_{FE} and the AC beta (forward-current transfer ratio) is represented by h_{fe}.

A transistor's beta is determined while V_{CE} is held constant (no load is used). However, in most practical applications, some type of load resistance is used with the transistor. The transistor's collector current must flow through this load in order to perform a useful function. Although a load resistance is normally connected in the common-emitter transistor circuit, the transistor's beta value still provides a reasonably accurate estimate of the current gain provided by the circuit. The current gain of a transistor used in the common-emitter circuit configuration is usually high.

Voltage Gain

Because of the unique configuration of the common-emitter amplifier, it can also be used to provide a voltage gain. In other words, it can produce a larger output amplitude with a smaller input voltage.

The reason behind the common-emitter's voltage gain ability is simply the location of the input signal being at the base. Since the input signal determines the voltage on the base (which is rather small), only a small change is required to change the amount of emitter junction forward-bias. Then (if you'll recall), any small change in forward-bias causes a large change in emitter and collector current. A large change in collector current results in a large change in V_{RL}, which results in a large change in the output voltage.

Voltage gains for a common-emitter amplifier, designated as A_V, can range anywhere from 2 to 200, with values of 80 to 150 being typical.

Power Gain

The input signal voltage applied to a common-emitter circuit is accompanied by a corresponding input signal current. Also, the output signal voltage produced by the circuit is accompanied by a corresponding output signal current. Therefore, the input portion of the circuit receives a certain amount of signal power and a certain amount of signal power is developed in the output portion of the circuit.

Since the common-emitter provides both current gain and voltage gain, it also provides a substantial power gain. In other words, the circuit effectively provides power amplification. This power gain, usually expressed as A_P, can be calculated using the following formula:

$$A_P = \frac{P_{OUT}}{P_{IN}}$$

The power gain of a circuit may also be determined by multiplying the current gain by the voltage gain. Remember that the current gain of a common-emitter circuit is essentially equal to its beta. The power gain of the common-emitter circuit is approximately equivalent to the transistor's beta (β) times the voltage gain (A_V) of the circuit. This relationship can be mathematically expressed as:

$$A_P = (\beta)A_V$$

For example, the power gain for a common-emitter amplifier with a beta of 100, and a voltage gain of 100, would be equal to:

$$A_P = (100)(100) = 10,000.$$

The power gain of the common-emitter circuit would therefore be equal to 10,000. This simply means that an input signal power level of 1 milliwatt (.001 watt) would be raised to an output level of 10,000 milliwatts (10 watts). Although this power gain may seem extremely high, it is a value which is often achieved or exceeded using common-emitter circuits. Such extremely high power gains are possible because the common-emitter circuit provides a substantial gain in both current and voltage.

Input Resistance

A transistor's emitter junction must be forward-biased so that a small base current (I_B) flows through the transistor's base and emitter regions. In the common-emitter circuit, the input signal voltage is applied across this forward-biased emitter junction. The emitter junction offers relatively little opposition to the flow of input signal current. For this reason, the common-emitter circuit is said to have a relatively low input resistance.

The input resistance of an amplifier is essentially an AC quantity. It is not a simple DC resistance which can be measured with an ohmmeter. The input resistance is the opposition presented to a changing (AC) input current. Furthermore, the input resistance is not a fixed value, but varies slightly as the amount of base current flowing through the emitter junction is changed.

If no additional components are connected in series with the transistor's base and emitter leads, the common-emitter circuit might have a typical input resistance that is between 500 ohms and several thousand ohms. These values effectively represent the input resistance of the transistor itself. However, when additional components are connected in the circuit, the input resistance of the overall circuit might be considerably higher or lower than the values given. Such additional components are often used to provide the proper bias voltages and currents for controlling the transistor's operation.

Output Resistance

A transistor's collector junction is reverse-biased under normal operating conditions. Although collector current can flow through the collector junction due to the unique action that occurs within the device, this reverse-biased junction still exhibits a relatively high resistance.

BASIC AMPLIFIERS

The resistance at the output terminals of a common-emitter circuit is, effectively, the high resistance offered by the collector junction. This resistance, which appears between the transistor's collector and emitter regions, is commonly referred to as the **output resistance,** or **output impedance,** of the common-emitter circuit. The output resistance is an AC quantity just like the input resistance. It represents the amount of opposition offered to an AC or changing current by the transistor.

If no load resistor is connected in series with the transistor's collector lead, the output resistance of a typical common-emitter circuit might be in the neighborhood of 40 or 50 kilohms. Under those conditions, the output resistance of the circuit is actually the output resistance of the transistor itself. When a load resistance is connected in the circuit, the output resistance of the circuit is affected. Also, the circuit's output resistance is affected by a change in the transistor's steady-state collector-to-emitter voltage (V_{CE}), whether a load resistor is used in the circuit or not. This means that the output resistance is not a fixed quantity, but varies with load resistance and operating voltage.

Common-Collector Circuits

In the common-collector circuit (also known as the emitter-follower circuit), the transistor's collector lead is common to both the input and output signals. In this arrangement, the input signal is applied to the base, and the output is taken across the emitter. The common-collector circuit has characteristics which are quite different from those of both the common-emitter and common-base circuits. The common-collector has the ability to provide a very high amount of current gain, which allows the circuit to be used as an impedance-matching device.

Circuit Configuration

A common-collector circuit with voltage divider biasing is shown in Figure 25-12. Before we point out the differences between this network and the common-base and common-emitter networks, we should point out the similarities.

ELECTRONIC CIRCUITS

We are using an NPN transistor in this circuit, and since the base material is positive, the polarity of V_{CC} is also positive. Emitter, base, and collector current, as well as current through R_1 and R_2, flow up from ground to V_{CC}. R_1 and R_2 serve as the voltage divider, with V_{R2} providing the base voltage. Although all of the voltages are positive, the voltage on the base is more positive than the voltage on the emitter, due to the relative ohmic values of R_L and R_2. This fulfills the requirement for forward-biasing the emitter junction. Also, since the collector is tied directly to $+V_{CC}$, it is more positive than the base voltage. This properly reverse-biases the collector junction.

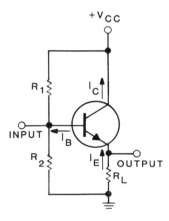

Figure 25-12 The common-collector amplifier.

As we were pointing out the similarities in completing the biasing requirements for the common-collector, we hinted at some of the differences in this configuration. While the input is applied to the base of the common-collector (similar to the common-emitter), the output is taken across the emitter resistor. Therefore, the resistor that is tied to the emitter is no longer labeled R_E, but instead, it is labeled R_L. In other words, the voltage drop of this resistor serves not only to develop the emitter voltage, but the output voltage as well.

The right side of this network is still essentially a voltage divider, but since the emitter resistor develops the output voltage, there is no need for a resistor above the collector. V_{RL} and V_{CE} (the voltage drop of the transistor) add to equal V_{CC}, and the transistor compensates for the changing voltage drop of R_L. The lack of a collector resistor is the second difference between this and other networks.

BASIC AMPLIFIERS

Finally, you may have noticed the absence of a bypass capacitor. In the common-base and the common-emitter, the bypass capacitor was connected to the common terminal to prevent that terminal from developing an unwanted AC signal. For the common-collector amplifier, the common terminal—the collector—cannot develop an AC signal, because it is tied directly to V_{CC}. If a bypass capacitor is connected to either the base or the emitter, these terminals would not be able to develop an input or output signal. Therefore, there is no need for a bypass capacitor.

As we've stated before, each amplifier sits in its **quiescent state** when there is no input signal applied. Since V_{CC} is applied, current flows, and the proper biasing voltages are developed.

Circuit Operation

We are now ready to apply an input signal to the common-collector amplifier. Figure 25-13A shows the same network as Figure 25-12, and Figure 25-13B shows the input and output signals, as well as the changing values for base and emitter current.

When the positive input alternation is applied, the base voltage (which is already positive) increases. This, as you may be able to predict, causes forward-bias to increase, which increases emitter current. So far, the common-collector is operating similar to the common-base and common-emitter amplifiers.

However, that is the extent of the similarities. The common-base and the common-emitter amplifiers had their outputs across the collector, and we were concerned with the effect the input signal had on collector current. For the common-collector amplifier, the output is taken from the emitter current, so any change in emitter current causes the same change to occur at the output. R_L feels the increase in emitter current, and its voltage drop increases. This also means that the output voltage increases.

ELECTRONIC CIRCUITS

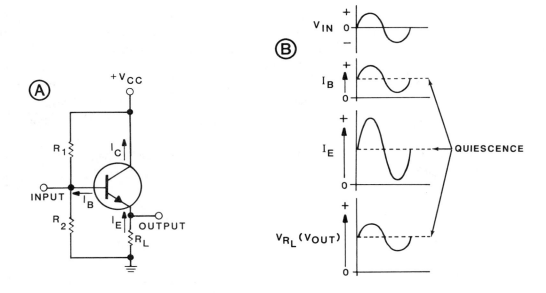

Figure 25-13 A common-collector circuit and its associated input and output signals.

Refer to Figure 25-13B as we summarize the effect of the positive input alternation:

1. The positive input alternation is applied to the base, which increases base voltage.

2. Forward-bias increases.

3. Emitter current increases.

4. R_L feels the increase, causing V_{RL} and V_{OUT} to increase.

The negative input alternation cause the exact opposite effect. When the input signal goes negative, base voltage decreases. This decreases forward-bias, which in turn causes emitter current, V_{RL}, and V_{OUT} to decrease.

As we look at Figure 25-13B, we can see that each value, with the exception of the input signal, varies above and below their quiescent values. Also, when the input and output signals are compared, we can see the in-phase relationship between these signals. The common-collector's input and output signals are said to be **in-phase** with one another.

BASIC AMPLIFIERS

Current Gain

Since the output current (I_E) is much higher than the input current (I_B) in a common-collector circuit, the circuit produces a high current gain. In fact, the current gain in a common-collector circuit is slightly higher than the current gain in a common-emitter circuit. This occurs because the output current (I_E) in the common-collector circuit is just slightly higher than the output current (I_C) in a common-emitter circuit. It is important to remember that I_E is always higher than I_C.

Back in Chapter 21, we derived a formula for determining the current gain for the common-collector amplifier. This formula is expressed as follows:

$$\text{Current Gain} = 1 + \beta$$

For example, assume that the transistor used in the common-collector circuit has a beta value of 30 (remember, *every* transistor has a beta value). This means that it has a current gain of 30 when connected in the common-emitter configuration. The current gain of this common-collector circuit would be equal to 1 + 30, or 31. This means that the amplitude or value of the output signal current would be 31 times greater than the input signal's current amplitude.

The current gain of the common-collector circuit is always slightly higher than the common-emitter's current gain. For all practical purposes, when the transistor's beta is higher than 30, the current gain of the common-collector circuit is assumed to be equal to the transistor's beta.

Voltage Gain

Due to the location of the input and output terminals, the common-collector circuit *does not* provide voltage amplification. In order to properly forward-bias the emitter junction, the base voltage must be slightly higher than emitter voltage. Since the input for the common-collector is located on the base, the voltage of the input signal must be slightly higher than the voltage on the emitter, where the output is taken. Therefore, the voltage gain for the common-collector amplifier is always less than 1, although it is very close.

You may also recall from Chapter 21 that this is how the term **emitter-follower** was derived. Because of the location of the input and output terminals, the output signal (at the emitter) *follows* the input signal (at the base).

ELECTRONIC CIRCUITS

Power Gain

Although the common-collector circuit provides a voltage gain of less than 1, it provides a large current gain. Because of the large current gain, the common-collector circuit configuration can provide a moderate power gain.

The power gain for any amplifier is equal to the voltage gain multiplied by the current gain. This holds true for the common-collector, also. We know that the current gain is equal to $1 + \beta$, and the voltage gain is less than one, although for our purposes, it equals 1. Therefore, the power gain for a common-collector is the same value as the current gain. This means that the power gain of a common-collector circuit is usually much lower than the power gain provided by either of the other two circuit configurations.

Input and Output Resistance

We stated earlier that the primary function of the common-collector amplifier is to act as an **impedance-matching** device. We also previously defined impedance-matching as the ability to connect two circuits with different resistance values in such a way that their combined resistance do not affect their separate current flows. Since this impedance-matching ability is the primary function of the common-collector, this makes the input and output resistance values extremely important.

Unfortunately, determining the input and output resistance of the common-collector is rather complicated. Because of the location of the input and output terminals, the input resistance includes the resistance of not only the forward-biased emitter junction, but the resistance of R_L as well. The output resistance is just as complicated; not only does the resistance of R_E enter into the calculation, but the resistance of the input signal source, as well as other values, must also be considered.

This makes the methods for calculating input and output resistance for the common-collector extremely complex, and we will not go into great detail with these methods. We can, however, discuss the basic premise in a way that is easily remembered.

We know that the input current for the common-collector is the base current. And base current is the smallest of all current values for a transistor. Since input current is extremely small, input resistance must be a very large value. Typical input resistances for a common-collector range anywhere from 100 to 500 kilohms. This makes the input resistance for a common-collector the highest input resistance of the three amplifiers.

The output resistance, however, is extremely small. This can be related to the high output current available from the common-collector. The output is taken across the emitter, which has the highest amount of current for a transistor. The output resistance normally ranges from 80 to 120 ohms, and is the lowest output resistance of any of the three amplifiers.

This large swing between input and output resistance is what makes the common-collector amplifier such an efficient impedance-matching device. The common-collector circuit can be connected to a high resistance signal-voltage source, without loading down the source (drawing an excessive amount of load current). Then a low resistance load can be connected to the output of the common-collector circuit without loading the output of the circuit. In this way, the common-collector serves as an intermediate circuit, which effectively matches a high resistance source to a low resistance load. By acting as a resistance or impedance matching device, the circuit allows a maximum transfer of power from the source to the load.

The common-collector circuit is used extensively to couple high to low resistances so that an efficient transfer of power results. The circuit effectively serves as a buffer between the source and the load and is often referred to as a buffer amplifier when used for this purpose. As a **buffer amplifier,** the common-collector provides isolation between the input voltage source and a low impedance load.

Saturation and Cutoff

To this point, you have seen how an input signal, when applied to an amplifier, can cause either an increase in voltage or an increase in current. However, there are limits to the capabilities of these amplifiers. For example, a common-emitter amplifier with a V_{CC} of 12 volts DC cannot amplify a 20 volt peak-to-peak input signal. When an amplifier reaches its limits in operation, it either **saturates** or **cuts off**. We will now examine each of these conditions for each amplifier, and in addition, introduce you to the different **classes of operation** for all amplifiers.

When a junction diode is forward-biased, a maximum amount of current flows through the diode, with little or no resistance. Once the diode reaches this point, a change in forward-bias voltage causes essentially no change in current flow. The diode acts like a closed switch, providing maximum current, having very little resistance, and developing no voltage.

A transistor behaves rather differently. We know that a small change in forward-bias voltage causes a large change in transistor current. However, a transistor can only be forward-biased to a certain extent—once the transistor reaches this point, increasing forward-bias cannot cause an increase in current. This is the point of **maximum forward-bias.** The emitter junction has reached a point where it achieves the smallest possible depletion region, and the highest possible current flows through the emitter, base and collector. This is the point of **saturation,** and when a transistor reaches this point, it is said to be **saturated.**

In addition to saturation, it is also possible to achieve the opposite effect. When we applied an input signal to the amplifiers in our previous discussions, the changes in voltage caused forward-bias to increase or decrease. Forward-bias was never *removed,* only increased or decreased. If the input signal ever reaches a voltage level which removes forward-bias, the transistor is no longer properly biased for conduction. All current flow through the transistor is lost, and the transistor essentially offers maximum resistance to the circuit. This can be compared to an open switch. When a switch is open, there is no current flow, and there is maximum (infinite) resistance. The transistor, in this case, is essentially turned off. When a transistor reaches this point, it has reached the point of **cutoff.**

To further define saturation and cutoff, we'll use the NPN common-emitter shown in Figure 25-14. Figure 25-14A shows the circuit before an input signal is applied. $+V_{CC}$ is applied at 12 volts, and the amplifier is biased for a quiescent output at the collector of +7 volts. For purposes of this discussion, this amplifier has a voltage gain (A_V) of 2, although this is rare.

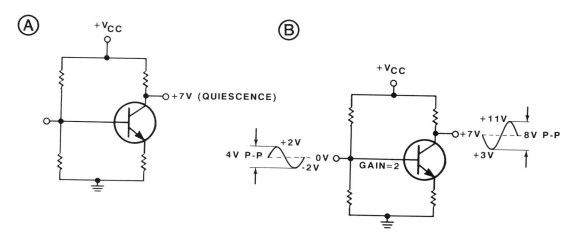

Figure 25-14 Normal operation of an NPN common-emitter.

Figure 25-14B shows the same circuit with a 4 volt peak-to-peak input signal applied. In order to determine the output peak-to-peak voltage, we simply multiply the gain by the input peak-to-peak voltage. Therefore, the output is 8 volts peak-to-peak.

This output voltage, if you'll recall the waveforms from Figures 25-8, 25-11, and 25-13B, rises above and below the quiescent value. Therefore, our 8 volt p-p output centers on +7 volts, with 4 volts going above +7, and 4 volts going below +7. The voltage levels of our output are +3 volts and +11 volts (don't forget, the common-emitter inverts the output).

Figure 25-15 shows our exact same circuit, but now the input p-p voltage has been increased to 8 volts. When the input amplitude is multiplied by the gain, our output is 16 volts p-p, and our voltage levels are -1 volt for the negative output alternation, and +15 volts for the positive alternation. However, we are limited in our voltage drops. Kirchhoff's Law tells us that our transistor can never develop a voltage higher than V_{CC}, which is +12 volts. Also, since V_{CC} is positive, we cannot develop a negative voltage. No component in a circuit can drop less than 0 volts.

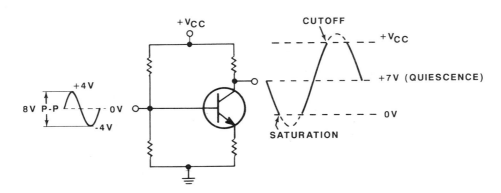

Figure 25-15 An overdriven common-emitter, with saturation and cutoff points.

In other words, our input signal is too large for our transistor, and forces the transistor into saturation and cutoff. When the positive input alternation is applied, forward-bias is increased to maximum. Our transistor acts like a closed switch, providing maximum current with minimum resistance, and developing no voltage. Our circuit is saturated, and the output voltage is 0 volts once saturation is reached.

Notice that saturation does not occur the instant the alternation is applied, but at that point where the input alternation provides maximum forward-bias. The negative output alternation, at the point it reaches 0 volts, is said to be clipped. Once the positive input alternation decreases back towards zero volts, maximum forward-bias is lost, and the transistor is no longer saturated.

When the input goes negative, forward-bias decreases, and the output voltage starts to increase because of the decrease in collector current. However, before the negative peak input is reached, the base voltage is decreased to the point where forward-bias is lost. Now the transistor is **cutoff,** and no current can flow. The output voltage rises towards V_{CC} as forward-bias decreases, and is **clipped** at V_{CC} at the point of cutoff. The transistor acts like an open component, developing V_{CC}, and removing current flow from the circuit. As the negative input moves away from its peak towards 0 volts, forward-bias is returned, and the transistor begins to conduct again. Any time an amplifier is driven to saturation and cutoff, the amplifier is being **overdriven.**

While saturation and cutoff may appear to be unwanted, amplifiers are now being created to operate efficiently at saturation and cutoff. Let's take the exact same circuit, and apply a square wave, as shown in Figure 25-16. The voltage levels for our square wave are $+V_{CC}$ and 0 volts. When the input rises to $+V_{CC}$, the circuit almost immediately saturates, and provides a 0 volt output. When the input drops to 0 volts, the circuit almost immediately cuts off, and develops $+V_{CC}$ at the output. Notice that the input signal is exactly reproduced at the output, with the exception of a 180-degree phase shift. This is known as an **inverter,** and is a very common circuit. The amplifier is biased so it can be easily saturated and cutoff. The transistor essentially acts like a switch, opening and closing to develop 0 volts and V_{CC}. Often times manufacturers will specify a transistor's **switching action,** stating the speed in which a transistor can alternate from saturation to cutoff.

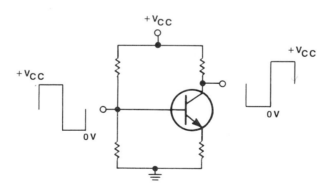

Figure 25-16 A common-emitter as an inverter.

As we move through this section on circuits, you will see the application of this switching action, as well as amplifiers biased for saturation and cutoff, quite often. This is due to the increased use of square waves instead of analog signals.

Classes of Operation

In addition to classifying amplifiers by their configuration—common-base, common-emitter, and common-collector—amplifiers may be classified by the percentage of the input signal cycle which causes the output voltage to be clipped. The four most common classes of operation are class A, class AB, class B, and class C. Since clipping is determined by input voltage, gain, and quiescent output, amplifiers are not placed into any of these categories until they are connected in the circuit where they will operate.

When we refer to the percentage of the input signal, this does not mean a voltage percentage. If you'll recall, an input signal has a cycle (or period) of 360 degrees. The different classes of operation are determined by the percentage of that 360 degree input which causes clipping at the output. It is a percentage of *time,* not a percentage of *voltage.*

A class A amplifier causes absolutely no clipping. In other words, the transistor never reaches saturation or cutoff points. This type of amplifier is preferred in audio equipment, because clipping an audio signal results in distortion. However, these circuits are constantly conducting a very high amount of current, and must be constantly monitored for overheating.

When an amplifier is connected as a class AB, the output is properly amplified for more than 50% of the input, but there will be some clipping. For example, if the positive alternation of the output is successfully reproduced, but the negative alternation is clipped, the amplifier is a class AB.

The other classes of operation are determined in a similar manner. A class B amplifier cause output clipping for 50% of the input, and a class C amplifier clips the output for more than 50% of the input. Figure 25-17 shows typical outputs for the different classes of operation.

BASIC AMPLIFIERS

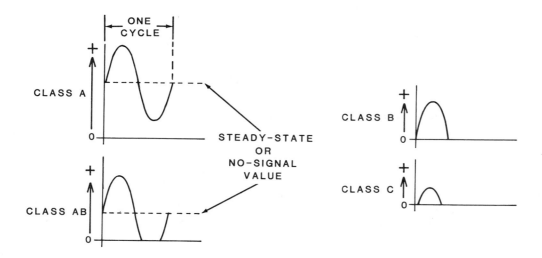

Figure 25-17 The output current for class A, class AB, class B, and class C amplifiers.

Amplifier Coupling

In some applications, one amplifier network cannot provide enough amplification. Therefore, it becomes necessary to use two or more amplifiers together to obtain a higher overall gain. When amplifiers are joined together, they are said to be **coupled**. When more than one amplifier is connected in a circuit, each separate amplifier is referred to as a *stage*. For example, if a circuit contains four amplifiers linked together, the first amplifier would be called the first stage of the circuit, and so on.

Amplifier coupling must be accomplished in such a way that the coupling will not upset or disrupt the operation of either circuit. There are four basic coupling methods which are widely used. We will now briefly examine each of these methods.

Resistance-Capacitance (RC) Coupling

The resistance-capacitance, or RC, coupling technique is one of the most widely used methods. When RC coupling is employed, the transistor's output load is a resistance, and the signal must pass from one amplifier stage to the next through a coupling capacitor.

Figure 25-18 shows how RC coupling is used to interconnect two common-emitter amplifiers with voltage-divider bias. Notice how all of the components are designated in sequence (R_1 through R_8 and C_1 through C_3). This is the method normally used to identify components when a number of components are used in a schematic diagram. Also, the transistors are designated Q_1 and Q_2. The Q designations are widely used to represent transistors on most schematic diagrams. On some schematics, the symbol T is used to designate transistors. Capacitor C_2 serves as the coupling capacitor. This capacitor couples the output signal of the first stage to the input of the second stage.

BASIC AMPLIFIERS

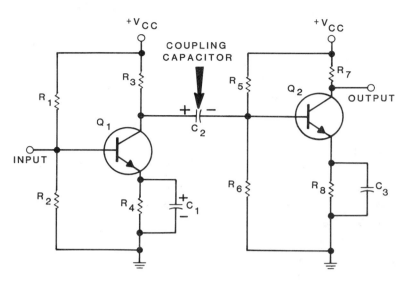

Figure 25-18 Two RC coupled common-emitter amplifiers.

Coupling capacitor C_2 must always be charged to a voltage that is equal to the voltage difference the collector of Q_1 and the base of Q_2. Furthermore, the collector of Q_1 is much more positive with respect to ground than the base of Q_2, which forces the voltage across capacitor C_2 to have the polarity shown. If the input signal to the first stage is an AC voltage, the output voltage of Q_1 will increase and decrease in accordance with the input AC signal (assuming the amplifier is operating in class A). This output voltage will always remain positive, since the output varies above and below quiescence. When the output voltage increases (goes more positive), the voltage difference between the collector of Q_1 and the base of Q_2 becomes greater, and capacitor C_2 charges to this higher voltage. When the output voltage of Q_1 decreases, the voltage difference between the collector of Q_1 and the base of Q_2 becomes less, and C_2 discharges to this lower voltage.

The action of C_2 charging and discharging allows the output of the first stage to be applied to the second stage. The varying DC voltage that appears between the collector of Q_1 and ground is actually applied across R_6, since the charge and discharge action of C_2 causes current to increase and decrease through R_6.

Notice that C_2 also acts as a block to the DC voltage connected to the collector of Q_1. C_2 provides isolation (DC block) between the two supply voltages (V_{CC} for both Q_1 and Q_2).

The ability of RC coupling to block DC allows the reference voltage to change between the output of the first stage and the input of the second stage. Let's assume the output of the first stage is referenced on +7 volts—the quiescent output of the collector. This cannot be applied to the input of the second stage, otherwise it would overdrive the transistor. The coupling capacitor effectively blocks this +7 volts, yet allows the peak-to-peak voltage to pass.

So where is the reference for the input to the second stage. Let's also assume that R_6 on the second stage has a voltage drop of +3 volts with no input signal. R_6 will feel the varying voltage from the capacitor, and current flow through R_6 will increase and decrease, as previously described. Therefore, its voltage drop will vary above and below its quiescent voltage drop of +3 volts. In other words, the coupling capacitor changed the reference for the signal from +7 volts at the first stage, to +3 volts as it enters the second stage.

In order for the coupling capacitor to efficiently transfer the AC signal to the second stage, this capacitor must offer very little opposition to the AC signal current, even at the lowest AC signal frequency which must be amplified. Therefore, a coupling capacitor usually has a high capacitance value (typically 10 microfarads or higher). With a high capacitance, its capacitive reactance will be very low. Generally, the coupling capacitor is an electrolytic type and it is necessary to observe polarity (the leads are marked + and -) when you connect this capacitor in the circuit.

RC coupled amplifiers can provide AC current and voltage amplification over a wide frequency range, but there are definite upper and lower limits. The reactance of the coupling capacitor increases as frequency decreases. This means that the lower frequency limit is determined by the size of the coupling capacitor. Using a higher capacitance value will extend the lower frequency limit. However, in most RC coupled circuits, a lower limit of a few hertz (a few cycles per second) is the best that can be obtained. The coupling capacitor simply will not pass DC and it offers tremendous opposition to extremely low frequencies.

The upper frequency limit of the RC coupled stages is primarily determined by the type of transistor used, although the various components in the circuit can also have an affect. A transistor's current gain (beta) decreases when the frequency extends beyond a certain point, and eventually a point is reached where the transistor can no longer provide a useful current gain. The decreasing current gain also causes the voltage gain to decrease proportionally.

RC coupling is used for frequencies from a few hundred hertz to frequencies in the MHz ranges.

BASIC AMPLIFIERS

Impedance Coupling

The impedance coupling technique is similar to the RC coupling method previously described. However, with this method, an inductor is used in place of the collector load resistor in the first stage, as shown in Figure 25-19.

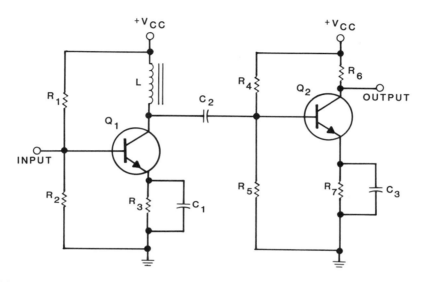

Figure 25-19 Two common-emitter amplifiers using impedance coupling.

Impedance coupling works like RC coupling since the inductor (L) performs the same basic function as a load resistor and the coupling capacitor (C_2) transfers the signal from one stage to the next. However, the inductor has a very low DC resistance across its windings. The resistance of a typical inductor may be only several hundred ohms since the inductor may be formed from low resistance wire.

Since the inductor has a low DC resistance, it drops only a small DC voltage when the collector current flows through it. This means that the inductor itself consumes only a small amount of DC power. Therefore, most of the source voltage (V_{CC}) is applied through the inductor to the collector of Q_1. However, the inductor still offers a substantial amount of inductive reactance to the AC signal. This inductive reactance can be as high as a load resistor that might otherwise be connected in the circuit. Therefore, an AC signal can be developed across the inductor the same as it would be developed across a load resistor. However, the inductor consumes less power than a resistor, increasing the overall efficiency of the circuit.

ELECTRONIC CIRCUITS

Unfortunately, the inductive reactance of the inductor does not remain constant as the signal frequency changes. Remember, inductive reactance is directly proportional to frequency. This means that the signal voltage appearing across the inductor will increase with signal frequency, thus causing the output voltage to increase as frequency increases.

The voltage gain of the first stage increases with frequency, causing the overall voltage gain of the two stages to increase. This means that the impedance-coupled amplifier stages provide a higher voltage gain as the signal frequency increases. The gain is eventually limited by the transistor's beta, just as it would be limited in the RC coupled circuit. Since the voltage gain of impedance-coupled stages tends to vary with frequency, this type of coupling is suited for applications where one signal frequency or a narrow (fixed) range of frequencies must be amplified.

Direct Coupling

When very low signal frequencies must be amplified (10 Hz or lower), the RC and impedance-coupling techniques cannot be effectively used. This is simply because the coupling capacitor cannot pass the extremely low frequencies due to an extremely high reactance value.

When low frequency signals or even DC signals must be amplified, a technique known as **direct coupling** is generally used. A typical direct-coupled circuit is shown in Figure 25-20. Notice that the base of the second stage transistor (Q_2) is connected directly to the collector of the first stage transistor (Q_1). No coupling capacitor is used between the two stages.

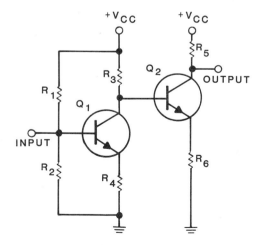

Figure 25-20 Two common-emitter stages using direct coupling.

Since the base of Q_2 obtains its voltage from the collector of Q_1, a separate voltage divider is not required to provide the base voltage for Q_2. This means that fewer components are required to construct the two common-emitter stages when they are direct coupled.

In direct coupling, R_3 must serve as both a collector load resistor for Q_1 and as a base resistor for Q_2. This resistor must limit the collector current through Q_1 to the proper value, and also control the base voltage of Q_2 so that the base current through Q_2 will have the proper value. Normally, resistor R_3 must have a high value to satisfy both of these conditions. The typical circuit values used in this circuit may be considerably different than the values used in an RC coupled amplifier.

The first stage functions like an ordinary common-emitter amplifier and produces an output signal voltage between the collector of Q_1 and ground. This signal voltage is applied directly to the base of Q_2 where it causes the base current of Q_2 to vary. This in turn, causes the collector current of Q_2 to vary and an output signal voltage is developed between the collector of Q_2 and ground. The circuit can amplify either DC signals (fixed or constant input voltages) or slowly changing AC since there are no capacitors in the signal path.

Direct-coupled amplifiers can provide a uniform current or voltage gain over a wide range of signal frequencies. These amplifiers may be used to amplify frequencies that range from zero (DC) to many thousands of hertz; however, they are particularly suited for low frequency applications.

Unfortunately, direct-coupled circuits are not as stable as RC and impedance-coupled circuits. This is because the second stage is essentially biased by the first stage. Any changes in the output current of the first stage (due to temperature variations) are effectively amplified by the second stage in much the same way that it would amplify a slowly changing input signal. This means that the biasing and quiescent voltages of the second stage can change if the circuit is not carefully designed. To obtain a circuit that has a reasonably high degree of temperature stability, it is often necessary to use expensive precision components to compensate for temperature changes.

Transformer Coupling

A technique known as transformer coupling is also occasionally used to couple two amplifier stages. A typical transformer-coupled circuit is shown in Figure 25-21. Notice that the first stage is coupled to the second stage through a transformer. The collector current of Q_1 flows through the input winding of the transformer. The transformer's output winding is connected between the base of Q_2 and the junction of the voltage divider (R_4 and R_5). This places the winding in series with the base lead. The changing signal voltage at the output of the transformer causes the base current of Q_2 to vary so that Q_2 can amplify the signal and produce an output signal voltage.

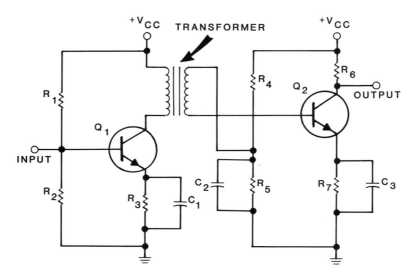

Figure 25-21 Two common-emitter stages using transformer coupling.

The transformer can efficiently transfer the signal from the high impedance output of the first stage to the low impedance input of the second stage. The transformer can effectively be used as an impedance-matching device in much the same way that a common-collector amplifier is used to match a high impedance source to a low impedance load. Optimum coupling between the two stages can be obtained by adjusting the turns ratio of the transformer's input and output windings. When the proper turns ratio is found, the low input impedance of the second stage will appear as a much higher impedance at the output of the first stage due to transformer action.

Although the transformer provides an efficient means of coupling two stages together (up to approximately 90%), it does have several disadvantages. First of all, the transformer is usually quite large and heavy, and it is also expensive when compared to the price of a resistor or capacitor. Also, the transformer cannot pass a DC signal. Only AC signals can pass through the transformer and the frequency range of these AC signals is somewhat limited, thus making the amplifier useful over a relatively narrow band of frequencies.

Chapter Self-Test

1. Most amplifier circuits are designed to faithfully reproduce the shape of a signal. If this is not accomplished, what is introduced by the circuit?

2. Amplifiers are generally considered to amplify what values?

3. List three different applications of an amplifier.

4. Which type of biasing was the first to be introduced?

5. Which type of bias does not provide thermal stability?

6. Thermal stability is accomplished using what type of feedback?

7. What is the simplest type of biasing?

8. Describe the effect of a bypass capacitor.

9. Which biasing technique uses two series resistors connected between V_{CC} and ground?

10. Define the term quiescence.

11. For the common-base amplifier, identify the following items:

    a. Input and output terminals.
    b. High or low current gain?
    c. High or low voltage gain?
    d. High or low input resistance?
    e. High or low output resistance?
    f. Phase shift between the input and output signals.
    g. Primary function?

12. Describe the method for calculating alpha.

13. What is alpha's particular significance in a common-base amplifier?

14. What is the difference between DC alpha and AC alpha?

BASIC AMPLIFIERS

15. For the common-emitter amplifier, identify the following items:

    a. Input and output terminals.
    b. High or low current gain?
    c. High or low voltage gain?
    d. High or low input resistance?
    e. High or low output resistance?
    f. Phase shift between the input and output signals.
    g. Primary function?

16. Describe the method for calculating beta.

17. What is beta's particular significance in a common-emitter amplifier?

18. For the common-collector amplifier, identify the following items:

    a. Input and output terminals.
    b. High or low current gain?
    c. High or low voltage gain?
    d. High or low input resistance?
    e. High or low output resistance?
    f. Phase shift between the input and output signals.
    g. Primary function.

19. Describe the method for calculating the current gain of a common-collector amplifier.

20. Define the term saturation, and describe the characteristics of a saturated transistor.

21. Define the term clipping.

22. What happens to an amplifier when it is overdriven?

23. List the four most common classes of operation, and describe the differences between each.

24. What are the advantages to RC coupling?

25. When transformer coupling is used, what function does the transformer fulfill, in addition to transferring the voltage from one amplifier to another?

26. Is impedance coupling more efficient than RC coupling? Why?

27. What are the preferred input frequency requirements when using DC coupling?

Summary

Amplifiers are usually classified as to primary function, circuit configuration, type of bias, and class of operation. The primary function of any amplifier circuit is to increase either voltage or power.

Amplifiers can be biased with a passive network, either in series or parallel with the amplifying component. Biasing with a resistor in the emitter circuit lowers the overall gain but stabilizes the circuit, thus preventing thermal runaway, which would damage the transistor. When a feedback signal (either from the emitter or collector) is used to prevent thermal run away, it is degenerative feedback and decreases the amplifier's gain. This type of bias is also referred to as self or dependent bias. The amount of feedback depends on how hard the circuit is conducting.

A voltage divider biasing network is connected in parallel with the transistor and allows you to select an operating point independent of the input signal. The operating point is a function of the ratio of the values of the two resistors used in the divider network. The voltage-divider biasing network is considered to be fixed or independent bias. The ratio of the resistors is preset and is independent of the circuit's operation.

The common-emitter configuration provides high current gain, high voltage gain, and high power gain. It also has a low input impedance, high output impedance, and 180 degree phase inversion. As a result, it is the most widely used configuration.

The common-base amplifier provides unity current gain and high voltage gain, but only medium power gain. It also has very low input impedance, very high output impedance, and no phase inversion.

The common-collector amplifier provides a voltage gain of 1, high current gain, and low power gain. It also has high input impedance, low output impedance, and no phase inversion. This circuit is usually used for isolation between stages or to match low impedance output circuits to high input impedance circuits. Impedance matching is required for a maximum transfer of power.

If an amplifier approaches the limits of conduction, it becomes saturated. A saturated transistor has maximum emitter, base, and collector current, and develops no voltage.

If an input signal removes the forward-bias of a transistor, the transistor is cut off. A cutoff transistor has no current, and develops V_{CC}.

BASIC AMPLIFIERS

When a transistor is saturated or cutoff, the amplifier is being overdriven. When an amplifier is overdriven, the output becomes clipped. Clipping occurs when one or both of the output peaks cannot be reached; the output either stops at V_{CC}, or 0 volts.

Amplifiers are classified according to the time of the output signal compared to the time of the input signal. The four most common classes are listed as class A, AB, B, and C.

The class A amplifier is the most basic type of amplifier. This circuit has current flow for the complete 360 degrees of each input cycle. It is characterized by a minimum of signal distortion, low output power capability, and the lowest efficiency of the four classes. However, due to its faithful reproduction of input signals, it is widely used in audio equipment. In the class A amplifier, the transistor is operated only in its linear region.

The class AB amplifier has output for more than 180 degrees, but less than 360 degrees, of the input signal. Class B operation has an output for exactly 180 degrees of the input signal. Class C amplifiers have output for less than 180 degrees of the input signal. Class C operation is the most efficient class of operation because the transistor is cutoff more than it is conducting.

Sometimes a single amplifier is not capable of providing enough gain for a specific function. It is often necessary to combine amplifiers by a process called coupling. There are basically four types of coupling networks commonly used in basic circuits.

RC coupling is the most common because it is the most flexible and the least expensive. However, it can not be used to couple DC signals because capacitors block DC currents.

Direct coupling is used for DC signal coupling. It is the most difficult coupling circuit to design, due to the interaction between the amplifier stages.

Impedance coupling has the advantage of efficiency over other forms of resistive coupling because it does not have excessive power dissipation. It presents almost no opposition to DC signals, but provides a larger impedance to develop AC signals. Its primary disadvantages are that it can't be used with DC signals, and its coupling efficiency varies with frequency.

Transformer coupling provides impedance matching and isolation. It also provides step-up, and step-down functions with an efficiency rating in the 90% range. However, it cannot be used with DC signals, it's expensive, heavy, and takes up a lot of space.

> # ELECTRONIC CIRCUITS

CHAPTER 26

Amplifier Applications

ELECTRONIC CIRCUITS

Contents

Introduction 675

Chapter Objectives 676

Audio Amplifiers 677

Video Amplifiers 693

Differential Amplifiers 704

Chapter Self-Test 712

Summary 713

AMPLIFIER APPLICATIONS

Introduction

Now that you are familiar with basic amplifier circuits, it is time to expand your knowledge by examining some amplifier circuits that are designed for specific applications. The characteristics of various types of amplifiers vary greatly, and it is necessary to understand the purpose and characteristics of each type that you will work with.

In this chapter, you will examine some of the important amplifiers that are used extensively in various types of electronic equipment. Most amplifiers are categorized by their optimum frequency range. **Audio amplifiers** are used to amplify low frequency AC signals, ranging from approximately 20 Hz to 20 kHz. You will then learn about amplifiers which operate in the frequency range of 10 Hz to 5 MHz. These amplifiers are known as **video amplifiers,** since video signals fall into this range.

In addition to audio and video amplifiers, you will also learn about the **differential amplifier,** an extremely versatile amplifier capable of operating in a number of different ways. Operational amplifiers, which you will read about in Chapter 27, use differential amplifiers quite extensively.

The operation of most of these amplifiers should already be familiar to you. Initially, the configurations are simple, involving networks you learned in Chapter 21. This chapter shows the applications of these networks, and their responses to various frequencies. You will also learn how different coupling techniques provide various circuit characteristics.

Obviously, there are a large number of amplifiers that do not fall into these two categories. Audio and video amplifiers are among the more common types of amplifier circuits, and a firm knowledge of these will enable you to work with most other amplifier circuits.

ELECTRONIC CIRCUITS

Chapter Objectives

When you have completed this chapter, you will be able to:

1. Define the following terms: preamp, cascading stages, interstage transformer, damping factor, single-ended amplifier, complementary signals, thermal runaway, flat frequency response, bass frequencies, treble frequencies, swamping resistor, differential inputs, and CMRR.

2. Explain the function and operation of audio voltage amplifiers.

3. Describe the operation of an audio single-ended power amplifier.

4. Describe the operation of a push-pull amplifier.

5. Describe the operation of the three different phase splitters.

6. Describe the operation of two different complementary amplifiers.

7. Identify the methods for controlling volume and tone in an audio amplifier.

8. Identify the additional requirements for an amplifier in order to provide amplification of video frequencies.

9. Describe the effects of junction and Miller capacitance on an amplifier.

10. Describe the functions of series peaking, shunt peaking, and combination peaking in a video amplifier.

11. Describe the operation of a differential amplifier in the single-input, single-output mode of operation.

12. Describe the operation of a differential amplifier in the single-input, differential output mode of operation.

13. Describe the operation of a differential amplifier in the differential-input, differential-output mode of operation.

14. Describe the methods in which a differential amplifier rejects common-mode inputs.

# AMPLIFIER APPLICATIONS

Audio Amplifiers

There are many applications in electronics where it is necessary to amplify AC signals that are within the audio frequency range. The circuits used to amplify these signals are generally referred to as **audio frequency (AF) amplifiers**, or simply, **audio amplifiers**. The audio frequency range extends from approximately 20 Hz to 20 kHz. This range includes the frequencies to which the human ear responds. The various frequencies produced by the human voice fall within this range. Therefore, audio amplifiers play an important role in any electronic equipment that is used to transmit, receive or process sound signals. Such equipment includes radio and television receivers, and various types of two-way radio communication systems. Also included are tape recorders, phonographs, and public address systems.

Audio amplifiers provide both voltage and power amplification. They are generally designed to operate in the class A mode, although class B amplifiers are also used in applications that require more power. In some applications, audio amplifiers must be designed to produce an absolute minimum amount of distortion. In other applications, a substantial amount of distortion may be permissible. Certain types of audio amplifier circuits contain various controls and adjustments that are used to regulate the overall gain and frequency response of the circuit.

Voltage Amplifiers

Many audio amplifiers are specially designed to amplify low-voltage audio signals and may be broadly classified as voltage amplifiers. In most cases, voltage amplifiers are used to increase the voltage level of an input signal to a value high enough to drive a power amplifier stage. When this is the case, the amplifier is usually referred to as a **preamp (pre-amplifier)**.

The preamp, or **voltage conditioning signal amplifier**, applies its high voltage output to a power amplifier. The power amplifier (usually a common-collector configuration) can then supply a high output signal current to operate a low impedance loudspeaker, or some other device requiring high power.

ELECTRONIC CIRCUITS

A simple voltage amplifier stage is shown in Figure 26-1A. Notice that the circuit is a common-emitter amplifier. These amplifiers are usually biased class A to provide minimum distortion. Let's briefly review the basic operation of the common emitter. R_1 controls the transistor's input base current and provides the necessary forward bias to operate the transistor. Emitter resistor R_3 provides emitter feedback, which improves the circuit's thermal stability. C_3 is a bypass capacitor and it prevents the AC input signal from becoming part of the degenerative feedback. The addition of the bypass capacitor increases the AC gain of the circuit. R_2 is the collector load resistor. The collector load resistor is the component across which the output signal is developed.

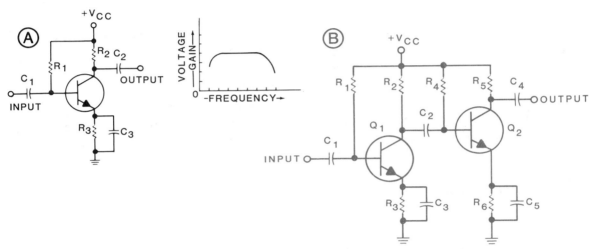

Figure 26-1 Single and double stage voltage amplifiers.

The AC input signal voltage is applied to the circuit through coupling capacitor C_1. This voltage appears between the base of the transistor and circuit ground, and it effectively controls the transistor's base-emitter current. The amplified AC output signal is developed between the collector and circuit ground. It must pass through coupling capacitor C_2 before it is applied to the next stage.

This amplifier circuit can provide substantial voltage gain over a relatively wide frequency range. A typical frequency response curve is also shown in Figure 26-1A. Notice that the voltage gain tapers off at each end of the frequency range. The decrease in gain at the lower end is due to the increase in the reactance of each capacitor. The drop-off at the upper end is due to the reduced beta of the transistor. The circuit cannot amplify DC signal voltages because coupling capacitors are used in the signal path.

AMPLIFIER APPLICATIONS

When the voltage amplifier shown in Figure 26-1A cannot provide sufficient amplification, two or more stages can be connected together to form an RC-coupled (resistance-capacitance) amplifier that has a higher overall gain. However, in any RC-coupled amplifier, the gain of each stage is affected by its load resistance. For example, when two common-emitter stages are RC coupled as shown in Figure 26-1B, the relatively low input impedance of the second stage appears to be in parallel with the collector load resistance (R_2) of the first stage, as far as the AC signal is concerned. This effectively lowers the collector load resistance and reduces the voltage gain of the first stage.

When a number of stages are RC-coupled, the gain of each stage is somewhat reduced and the overall voltage gain (which is the product of the individual stage gains) may be lower than expected. Therefore, in applications where a tremendous amount of audio signal amplification is required, a number of RC-coupled amplifier stages may be required to provide sufficient gain. When amplifiers are combined in this manner they are said to be *cascaded.*

A more efficient voltage amplifier can be formed using transformer coupling. This is shown in Figure 26-2. Coupling transformer T_1 is connected so its input (primary) winding serves as the collector load for the first stage. The output (secondary) winding is connected in series with the base lead of the second stage. The AC signal developed across the primary winding is coupled to the secondary winding through basic transformer action, where it affects the base of Q_2. The second stage is base-biased just like the first stage, with resistor R_3 and R_1 providing the proper no-signal input base current for both Q_2 and Q_1. The output of the second stage is then taken across transformer T_2.

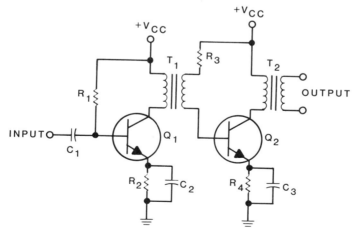

Figure 26-2 A two-stage, transformer-coupled audio amplifier.

Although only two stages are shown in Figure 26-2, additional transformer-coupled stages could be added. Because the transformer is used to connect two stages together, it is called an **interstage** transformer. Transformers are often used to provide isolation between stages, as well as to match input and output impedances. This allows a minimum of loading and a maximum transfer of power.

Greater efficiency of transformer coupling results in higher gain. The transformer can respond only to a specific range of AC frequencies without seriously effecting the signal's amplitude. Often the overall frequency response of a transformer-coupled circuit is not as wide as that of an RC-coupled circuit. However, transformer coupled circuits, because of their high gains, are widely used in audio applications. The term gain, in this case, means the gain of the amplifier component and the step-up (voltage) or step-down (current) ratio of the transformer's primary to secondary windings. Most RC-coupled and transformer-coupled voltage amplifiers operate in the class A mode and are designed to produce a minimum of signal distortion.

Power Amplifiers

Once sufficient voltage levels are available, a power amplifier is used to drive any load that requires high power. This is the final stage of amplification before the signal leaves the amplifier system. In systems requiring extremely high power, one or more stages of power amplifiers may proceed the final power amplifier.

Power amplifier circuits are designed to work into specified loads. For example, an amplifier may be designed to deliver its rated power when connected to an 8 ohm load. If it is connected to a lower impedance load, it may become overloaded. Typically, solid-state amplifiers can operate safely into loads between 4 and 16 ohms.

A power amplifier should always have a very low characteristic output impedance. For example, a circuit that is designed to drive an 8-ohm load might have an output impedance on the order of tenth's of an ohm. Low output impedance assures that load differences will not affect the voltage levels being delivered. The ratio of the load impedance (Z_L) to the output impedance (Z_{OUT}) is called **damping factor.** The damping factor (designated as **D**) can be calculated using the following formula:

$$D = \frac{Z_L}{Z_{OUT}}$$

Wire size and length, terminal connections, and fuses in the speaker line all add resistance. With minimum resistance in the line, the speaker system and wiring will provide a higher damping factor which makes the amplifier system sound clearer and sharper.

Single-Ended Power Amplifier

The power amplifier in Figure 26-3 uses a single common-emitter amplifier. This is referred to as a **single-ended** power amplifier, since only one active device (the transistor) develops the output signal. T_2 provides impedance-matching between the high collector output impedance and the low load impedance. In order to reproduce a complete waveform with minimum distortion, single-ended amplifiers are biased to operate class A.

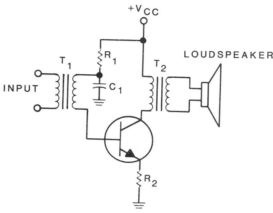

Figure 26-3 A basic audio power amplifier.

Input and output transformer coupling provides excellent gain. However, because of the resistance of the wire in the transformer and losses at high power, damping is generally low. A damping factor of 10 is typical for a transformer coupled circuit.

Because the output transformer (in this case T_2) must handle a large amount of power, output transformers are generally much larger than interstage transformers. The size of this transformer usually contributes heavily to the overall weight and cost of the amplifier system. Output transformers as a general rule are heavy, bulky, and relatively expensive.

ELECTRONIC CIRCUITS

Push-Pull Power Amplifiers

Another disadvantage of the single-ended power amplifier is that the core of the output transformer becomes magnetized. When there is no input signal, the collector of the transistor is a steady DC signal, which forces DC current to flow through the primary winding. It is this DC current which magnetizes the core. To solve this problem, the core must be large enough to withstand this current, plus the AC signal current, without saturating. This calls for a transformer that is large, heavy, and expensive.

A two-transistor circuit called a push-pull amplifier can be used to overcome this problem. With a push-pull amplifier, DC current flows into opposite ends of the primary, canceling any magnetic fields that may have been produced. Thus, the push-pull circuit can generally get by with smaller, lighter, and less expensive transformers.

Transformer Output

A push-pull power amplifier is shown in Figure 26-4. A single input is applied to two common-emitter amplifiers. However, this input cannot be the same for each transistor. In order for the push-pull to properly function, the two inputs must be exactly the same, but 180 degrees out-of-phase. In order to accomplish this, the input signal must be inverted. In Figure 26-4, a simple block inverter is shown. There are a number of ways this can be accomplished, and we will examine these inverting methods later in this chapter.

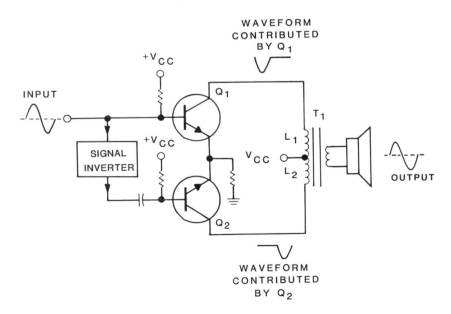

Figure 26-4 Transformer type, push-pull power amplifier.

AMPLIFIER APPLICATIONS

The inverted inputs cause the two different input alternations to be applied to each transistor at the same time. When Q_1 feels a positive alternation, Q_2 feels the negative alternation, and vice-versa. Since the two transistors are NPN-type, the positive alternation applied to the positive base properly forward-biases the emitter junction, and the transistor conducts when the negative alternation is applied, forward-bias is lost, and the transistor cuts off. When opposite alternations are applied at the same time to each transistor, this results in the two transistors alternating their conducting and cut-off times.

As a result, the two waveforms shown in Figure 26-4 are produced. For that portion which the transistor conducts, an output voltage is developed. For that portion which the transistor is cut off, the output levels off at quiescence. These output waveforms still have the same amplitude, but are 180 degrees out-of-phase.

The two outputs are applied to transformer T_1. Notice that V_{CC} splits the primary winding into two inductors, labeled L_1 and L_2. During the positive input alternation, Q_1 conducts, and the top waveform is felt by L_1. Current flows down through L_1 towards V_{CC}, and the secondary winding produces a positive alternation (remember, a transformer with no dots shown is out-of-phase). During the negative alternation, Q_2 conducts, and the bottom waveform is applied to L_2. Current flows up towards V_{CC}, and the secondary winding develops a negative alternation. This output signal is applied to the loudspeaker, which converts the electrical signal into sound waves.

Phase-Splitting

We stated earlier that in order for the push-pull amplifier to function properly, the two input signals required a phase-shift of 180 degrees. When two signals are identical in frequency and amplitude, and have a 180 degree phase-shift, they are said to be **complementary signals**. Do not confuse this definition with complementary *angles,* whose sum is 90 degrees.

Any circuit used to create complementary signals is called a **phase splitter**. Push-pull amplifiers require phase splitters to fulfill their input requirements.

One method of supplying complementary input signals is with a center tapped transformer, as shown in Figure 26-5A. With a sine wave applied to the primary of the transformer, the output signal between the center tap and point A is opposite in polarity to the signal between the center tap and point B. Equal complementary amplitudes depend on how close the ground point is to the electrical center on the transformer's secondary. The amplitude of the secondary voltage is determined by the input amplitude and the primary to secondary turns ratio of the transformer.

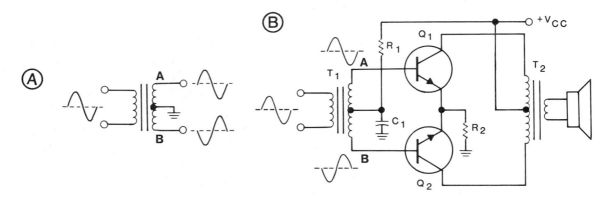

Figure 26-5   Transformer used as a phase splitter in a push-pull power amplifier.

Figure 26-5B shows how a transformer-type phase-splitter is used in the push-pull configuration. An input signal from a previous stage is applied across the primary of transformer T_1. A single bias resistor (R_1) is shared by Q_1 and Q_2. The base current path is through the secondary of transformer T_1. Capacitor C_1 connects the center tap to AC ground. The two complementary signals at points A and B are connected directly to the bases of the transistors as shown.

Two cascaded, common-emitter amplifiers, connected as shown in Figure 26-6A, will also provide complementary signals. The output signal of Q_1 is inverted compared to its input signal. Transistor Q_2 also inverts the signal, but notice that its input is taken from the output of transistor Q_1. Resistor values must be selected so that the amplifiers provide equal amplitudes.

However, complementary signals can also be obtained from a single amplifier, as shown in Figure 26-6B. Notice that one output is taken from the collector while the other is taken from the emitter. The signal phase difference between the base and collector is 180 degrees, while the phase difference between the base and emitter is 0 degrees. Therefore, the two output signal voltages are complementary.

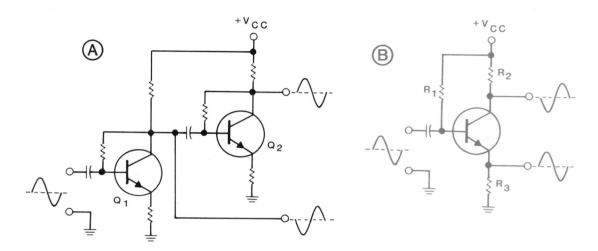

Figure 26-6 Double and single stage phase splitters.

The amplitude of the two outputs are controlled by the resistance values of resistors R_2 and R_3. When they are equal, the output signals will be equal in amplitude. In actual practice, the resistors are not equal because of the internal resistance of the transistor. To compensate for the transistor's characteristic, both the collector and the emitter resistors are variable. With both resistors variable, the outputs can be calibrated to have exactly the same outputs.

ELECTRONIC CIRCUITS

When a single transistor phase splitter is connected to a power amplifier as in Figure 26-7, the DC collector and emitter voltages of Q_1 do not match the base voltages of push-pull transistors Q_1 and Q_3. Thus, coupling capacitors C_2 and C_3 are required between the stages to offset the voltages. In addition, transistors Q_2 and Q_3 need their own bias resistors, R_3 and R_4.

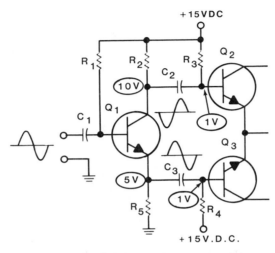

Figure 26-7 Push-pull power amplifier using a transistor as a phase splitter.

Complementary Amplifier

A different type of push-pull amplifier does not require input or output transformers. In fact, it does not even require a phase splitter. Instead, it uses the complementary nature of NPN and PNP transistors to accomplish push-pull action.

Recall that a positive voltage on the base of an NPN transistor causes it to conduct. However, the PNP transistor requires a negative base voltage to conduct. These characteristics allow us to build a push-pull amplifier using a minimum number of components. The circuit is called a **complementary amplifier** and is shown in Figure 26-8A.

AMPLIFIER APPLICATIONS

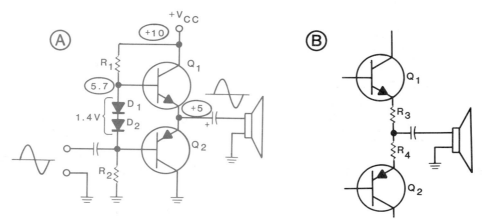

Figure 26-8 Basic complementary power amplifier.

Q_1 is an NPN transistor and Q_2 is a PNP transistor. The two transistors are connected in series between V_{CC} and ground. The two emitters are connected together. For maximum voltage swing, the voltage at the emitters must be one-half of V_{CC}. The DC emitter voltage is dependent on equal current flow through the transistors, which is dependent on the resistance values of R_1 and R_2. When each transistor is properly biased, there will be approximately 0.7 volt difference between its base and emitter, or a total of 1.4 volts, when measured from one base to the other. Diodes D_1 and D_2 help keep the voltage difference constant.

Diodes D_1 and D_2 also prevent a condition called **thermal runaway.** This condition is caused by a constantly increasing thermal feedback. As the temperature in the transistors begins to increase, it causes the current in them to increase. The increase in current causes the temperature to rise even higher. Unchecked, this thermal runaway would destroy the transistors.

The two diodes are placed close to the transistors, and have a temperature characteristic that closely matches the transistor's temperature characteristics. When the diodes sense the rising temperatures, the current in them also increases. This decreases the voltage across the diodes, which reduces the voltage difference between the transistor's bases. This decrease in base voltage causes a decrease in base current, thereby, stabilizing the current flowing through the transistors. Adding emitter resistors, as shown in Figure 26-8B, also contributes to thermal stabilization by adding degenerative feedback.

The output signal from this circuit is taken from the junction of the two emitters (or emitter resistors). Because of the DC voltage, direct speaker connection is not possible without using a DC blocking capacitor as a coupling device.

ELECTRONIC CIRCUITS

The output of the complementary power amplifier is at the emitters, making it function as a common-collector, and fulfilling the high output current requirement. However, a common-collector has a voltage gain of less than one. For this reason, it is often desirable to drive it with a class A, common-emitter amplifier, as shown in Figure 26-9. This stage provides voltage gain to the overall amplifier. However, this amplifier can also be affected by rising temperatures, since any increase in collector current is fed directly to transistors Q_2 and Q_3. Therefore, negative feedback is coupled back to the base of Q_1. The feedback path is through R_4 and R_3, from the output of the amplifier. This feedback counteracts any changes in Q_1 due to temperature changes. Operating bias for Q_1 is also supplied by this source. C_2 bypasses AC signals to ground and prevents their being coupled back to the input of Q_1.

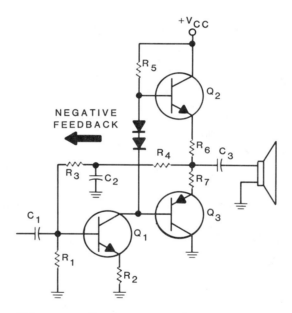

Figure 26-9 Class A amplifier precedes basic complementary power amplifier circuit.

We can take the complementary amplifier one step further. The only real problem with this amplifier is its high output impedance at low frequencies, due to the high reactance offered by the output coupling capacitor. In other words, a complementary power amplifier must be designed to eliminate coupling capacitor C_3.

Figure 26-10 shows how this is possible. Notice how the output signal from the emitters is now direct-coupled to the loudspeaker. This is accomplished with the addition of a negative V_{CC}, connected at the bottom instead of ground. The positive V_{CC} is reduced to half its former value, and $-V_{CC}$ is connected at the same, but negative, value. This produces the same potential difference as the source voltage in Figure 26-9, and allows the voltage at the emitters to be 0 volts when no input signal is applied.

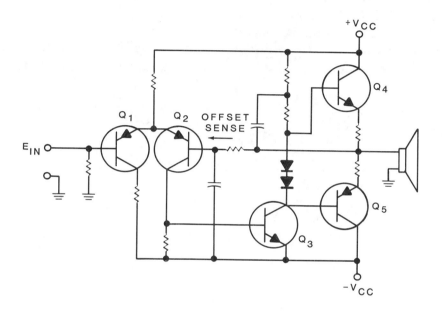

Figure 26-10 Direct coupled output power amplifier.

A differential amplifier composed of Q_1 and Q_2 is placed immediately preceding the class A amplifier, Q_3. One input accepts the DC voltage offset information from the output.

The very low output impedance offered with the elimination of the output capacitor results in a good damping factor. Also, a new characteristic has been added. Notice that all stages are DC coupled. This extends the frequency range down to DC. This type of amplifier can be used to drive loads requiring large amounts of DC power, such as DC motors.

Volume Controls

All of the amplifiers we've discussed to this point provide a fixed, or non-adjustable gain. However, due to component aging, we require a method for adjusting the overall gain of the amplifier. Increasing the gain increases the output peak-to-peak voltage, and amplitude of the output determines the volume of the signal.

The circuit in Figure 26-11A demonstrates a method for increasing the overall gain of an amplifier. Notice how the output of stage 1 is RC coupled to stage 2 through potentiometer R_2. The output signal from stage 1 passes through coupling capacitor C_1, and a portion of this output is developed by R_2. The voltage drop of R_2 then becomes the input to the second stage, passing through coupling capacitor C_2.

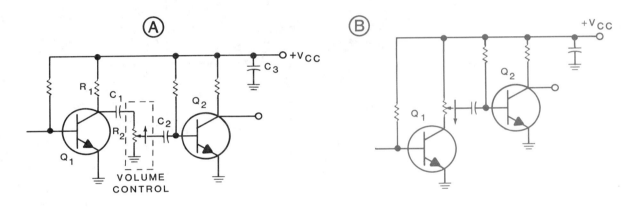

Figure 26-11 Complex and simple volume controls.

The two coupling capacitors allow the potentiometer to be adjusted without affecting the bias of either stage. The control arm of R_2 determines what portion of R_2's voltage drop is applied to the second stage. If this control is rotated counterclockwise, ground will be applied to the base of Q_2, and no output will be available. However, as the control is adjusted clockwise, the base of Q_2 feels a higher and higher AC signal. As the amplitude of this input to Q_2 increases, the output will also increase, thereby increasing the volume.

This circuit can be simplified, as shown in Figure 26-11B, with the control component serving as the collector load resistor for Q_1. Since the full signal is developed across the control, it can be adjusted to obtain the desired AC amplitude. Notice that this circuit uses fewer parts than the previous one. For this reason, it is usually found in the less expensive units.

Tone Controls

When an amplifier responds with equal voltage gain to all of the frequencies within its range, it is said to have a **flat frequency response.** Although this flat response is usually a necessary requirement for any good amplifier, some means of altering the response is usually incorporated to suit the different tastes of individual listeners. The gain, and therefore the response, can be boosted or cut at either end of the audio frequency range. The frequencies at the lower end of the range are called **bass frequencies,** and those at the high end are called **treble frequencies.** Bass and treble controls are generally referred to as **tone controls.** These various controls are provided on good sound systems because no two people hear the same sounds exactly the same. Some people are more sensitive to the low frequencies, while others are more sensitive to the higher end of the audio frequency band.

Figure 26-12 shows a bass and treble control circuit that can provide both boost and cut. A flat response occurs when each control is adjusted to the center of its range. The values of capacitors C_1 and C_2 are equal. The values of C_3 and C_4 are also equal. Each network forms a voltage divider whose output, when the control is centered, is one-half the value of the applied input voltage.

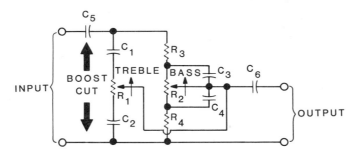

Figure 26-12 Tone control circuit.

Capacitors C_5 and C_6 are large values whose reactances are very low at all audio frequencies. These two capacitors provide DC isolation between the tone circuit and the circuits connected to its input and output. Capacitors C_1 and C_2 offer a low reactance at high (treble) frequencies and a high reactance at low frequencies. When treble control R_1 is fully clockwise, as indicated by the arrow, the balance of the circuit is upset. The treble frequencies then pass on to the output without being attenuated. When the treble control is fully counterclockwise, treble frequencies passing through C_1 and control R_1 are attenuated by R_1. Low frequencies remain unaffected because they pass through the bass circuit.

ELECTRONIC CIRCUITS

The bass control works similarly. Capacitors C_3 and C_4 bypass the high frequencies around bass control R_2, but offer a high impedance to the low frequencies and force them to pass through the control. When R_2 is fully clockwise as indicated by the arrow, all the bass frequencies pass on to the output without being attenuated. Attenuation is gradually increased as the control is turned counterclockwise. Using this combination of controls you can set the system for the sounds that you prefer.

The circuit you have just studied can be easily incorporated into a single-stage amplifier, as shown in Figure 26-13. The gain offered by the transistor compensates for the inherent signal loss in the resistor-capacitor circuit. C_1, C_2, and C_3 are large value capacitors with low reactances at audio frequencies. The capacitors also provide DC isolation. The input signal is applied to point A. The output signal from the collector of Q_1 is also fed back to point B. Because of Q_1, the signal at point B is 180 degrees out-of-phase with the input signal at point A. This circuit operates similar to the basic RC circuit previously described, using degenerative (negative) feedback for control purposes.

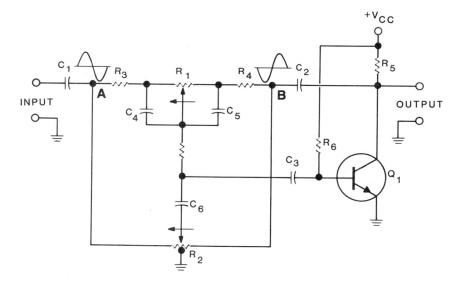

Figure 26-13 Audio amplifier with tone control.

AMPLIFIER APPLICATIONS

Video Amplifiers

You have seen how audio amplifiers are created by different coupling and biasing techniques to provide a flat frequency response for the 20 Hz-20 kHz audio range. Video amplifiers are created in the same fashion, but the range of video frequencies is much wider, and covers a large portion of very high frequencies.

Video amplifiers are commonly used in television and radar systems to amplify video, or picture information. Video signals range from 10 Hz to 5 MHz, a range much wider and higher than audio signals. A video amplifier, therefore, must be able to provide a flat frequency response from almost DC to 5 MHz. The range of video and audio amplifiers are compared in Figure 26-14.

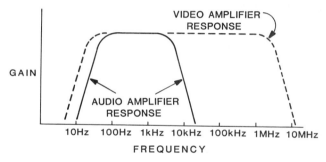

Figure 26-14 Audio amplifier response versus video amplifier response.

Only direct-coupled and RC coupled amplifiers are capable of providing this wide bandwidth. The RC amplifier has the advantage of economy, while the direct-coupled amplifier has almost perfect low frequency response. Transformer coupling is not suitable because of the wide variations in frequencies. For example, an audio transformer would be satisfactory for low frequencies, but would attenuate the higher video frequencies. Likewise, an RF transformer would not provide adequate coupling for the low frequencies.

The RC-coupled amplifier has a flat frequency response in the middle-frequency range. Also, its frequency and phase characteristics throughout the middle ranges are suitable for use in video amplifiers. Therefore, when using an RC-coupled amplifier as a video amplifier, the middle-frequency range requires no further improvement. However, the high and low frequency responses for an RC-coupled amplifier are far from adequate. Therefore, if an RC-coupled amplifier is going to be used as a video amplifier, corrective measures must be taken to compensate for these inadequate frequency responses. Before we discuss these corrective measures, let's discuss precisely what happens within an amplifier at high frequencies.

ELECTRONIC CIRCUITS

Factors Affecting Frequency Response

There are a number of factors which affect a transistor's response at higher frequencies. However, the factor that probably limits high frequency gain the most is the shunt capacitance of the circuit.

If you'll recall, a transistor has two PN junctions. A depletion region, which is created in the middle of each junction, is very small for a forward-biased junction, and very large for a reverse-biased junction. If we think of these depletion regions as open space, we can also think of the charged portions just outside of the depletion region as plates. The effect is the same as a small capacitor across the junction. This is **junction capacitance,** and it exists across each pair of terminals in a transistor. As shown in Figure 26-15A, these capacitances are C_{BE}, C_{CE}, and C_{CB}. These junction capacitances are determined by factors, such as the physical size of the junction and the spacing between the transistor's leads. Some effort is made by transistor manufacturers to keep these values as low as possible. However, they can not be completely eliminated.

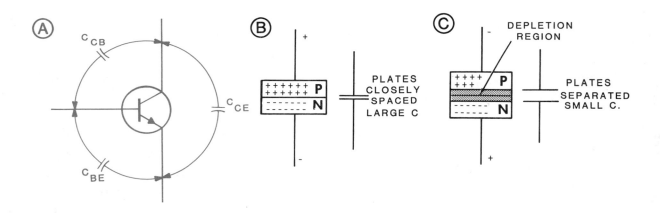

Figure 26-15   Transistor junction capacitances.

AMPLIFIER APPLICATIONS

As we hinted earlier, the amount of junction capacitance is also affected by the biasing of the PN junction. When a PN junction is forward-biased, as shown in Figure 26-15B, many charge carriers are present at the junction. This reduces the spacing between the effective capacitor plates (transistor junctions) and increases capacitance. When the junction is reverse-biased, as shown in Figure 26-15C, the carriers are forced apart, creating a wide depletion region. This has the same effect as increasing the spacing between the plates, resulting in less capacitance. Therefore, a transistor's forward-biased, base-emitter junction will have a somewhat greater capacitance than its reverse-biased collector-base junction.

The junction capacitance of the collector junction also creates another type of capacitance, known as the **Miller effect.** An example of the Miller effect is shown in Figure 26-16. As shown, the collector-to-base capacitance creates a negative feedback path between the collector and base. The resistance of the collector junction (R_{CB}) is also in this path.

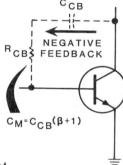

Figure 26-16 The Miller effect.

At low frequencies, this capacitance has very little effect, since X_C is rather large. However, when frequency increases, X_C decreases and the feedback path reactance is reduced. Since there is a phase inversion between the base and collector in the common-emitter configuration, feedback voltage is opposite in polarity to the base's input voltage. Therefore, this type of feedback has a degenerative effect on circuit gain at high frequencies.

ELECTRONIC CIRCUITS

Both junction capacitance and Miller capacitance adversely affect the gain of an RC-coupled amplifier at high frequencies. This is shown in Figure 26-17. As you can see, the total amount of capacitance at the output of stage 1 is the sum of collector-emitter capacitance, and any stray wiring capacitance, labeled C_S (remember, when capacitors are in parallel, total capacitance is the sum of the branches). The effect that this capacitance has on the circuit is actually quite complicated, but it essentially acts as a shunt capacitance connected to ground.

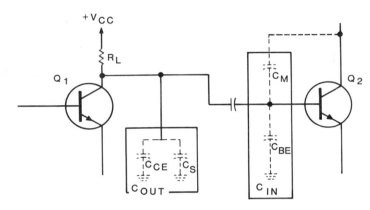

Figure 26-17   Shunt capacitance between RC coupled stages.

In addition to C_{OUT}, the input to the second stage also has a certain amount of capacitance. This is the junction capacitance of the emitter junction (C_{BE}) and Miller capacitance. Since both C_{IN} and C_{OUT} are in parallel, the total amount of shunt capacitance between the stages is the sum of C_{IN} and C_{OUT}.

Typical values of total shunt capacitance for this type circuit can run as high or higher than 180 picoforads. This may not seem like such a high value of capacitance, until you consider the fact that the voltage gain is the result of the voltage developed across the load or collector resistor. This capacitance shunts some of this output voltage to ground.

We'll use the circuit in Figure 26-18A as an example. An 8.8 kilohm load resistor R_L connected in the amplifier circuit, and a total circuit capacitance of 180 picoforads is shown across the circuit's output. The output voltage, in this case representing circuit gain, is plotted against frequency in the response curve in Figure 26-18B. The solid line represents circuit response, while the required video response curve is shown by the dashed line.

AMPLIFIER APPLICATIONS

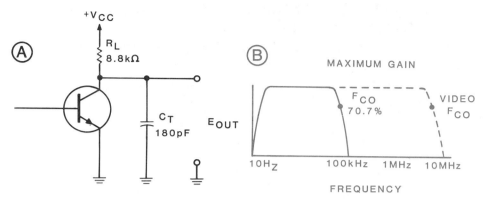

Figure 26-18 The effect of shunt capacitance.

As the response curve shows, the circuit responds equally to frequencies between 10 Hz and 90 kHz, as evidenced by the flat part of the curve. However, above 90 kHz the output voltage begins to decrease rapidly until the cutoff frequency (indicated as F_{CO}) is reached. At this point, the output voltage has decreased to 70.7% of maximum.

The total shunt capacitance causes this reduction. At those frequencies providing good response, X_{CT} is extremely high, and will have little effect on output voltage. For example, if this circuit is amplifying a frequency of 1 kHz, the reactance of C_T is approximately 900 kilohms. Therefore, the capacitance presents a relatively large impedance to ground and E_{OUT} remains high. The effect of C_T is not important until its reactance is low enough to be comparable with the resistance of R_L. When this happens, X_{CT} will shunt a large portion of the output signal to ground.

As an example, when the frequency of the amplified signal increases to around 100 kHz, X_{CT} decreases to approximately 8.8 kilohms. This equals the resistance of R_L, and the combined impedance of R_L and X_{CT} reduces Z_T to 70.7%. Consequently, this decrease in output impedance lowers E_{OUT} by approximately the same amount. This corresponds to the high frequency cutoff point, F_{CO}, which can also be defined as that frequency when $X_{CT} = R_L$. Therefore, this amplifier is useful for frequencies up to 100 kHz. This is far short of the required 5 MHz bandwidth needed for video amplifiers.

Because the high frequency cutoff occurs at the point where $X_{CT} = R_L$, the cutoff frequency can be extended if R_L is reduced. Of course, any decrease in R_L also results in a reduction in the voltage gain of the amplifier, but frequently this sacrifice is necessary in video amplifiers.

Figure 26-19 shows the amplifier just discussed with the value of R_L decreased to 2.2 kilohms. As the response curve shows, the output voltage is significantly reduced over that of the previous circuit. At .4 MHz, X_{CT} is equal to 2.2 kilohms, as indicated in the response curve.

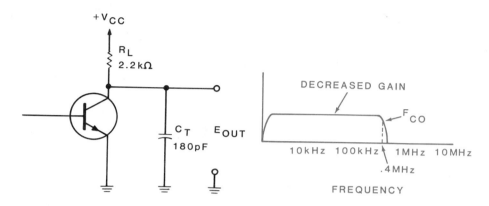

Figure 26-19   Decreasing R_L extends frequency response.

Since decreasing the value of load resistance seems such a simple way to extend frequency response, you might think that by lowering it even further we could obtain the required high frequency response. But unfortunately, since decreasing load resistance decreases amplifier gain, a point of diminishing return is quickly reached and the gain is so low that the amplifier does not provide a useful function. For example, when R_L is decreased to 1 kilohm, the amplifier's response is extended to only 844 kHz. This response is not wide enough for video frequencies, and other steps must be taken to improve frequency response.

Frequency Compensation

By far, the most popular method of increasing video amplifier high frequency response is through the use of **peaking coils**. The peaking coil is nothing more than a small inductor placed in the circuit to take advantage of the circuit's shunt capacitance. There are three different techniques for connecting these coils in the circuit: shunt peaking, series peaking, and a combination of the two.

Figure 26-20 shows the shunt peaking method. Notice that the 360 microhenry inductor, L_{SH}, is connected in series with the load resistor. Close examination of the circuit, however, shows that L_{SH} is effectively in parallel, or shunt, with C_T. This technique is known as **shunt peaking.** This small value of inductance tends to nullify the effect of the shunt capacitance. Since the inductor and the shunt capacitance are in parallel, this creates a tank circuit, and inductive properties cancel capacitive properties in a tank.

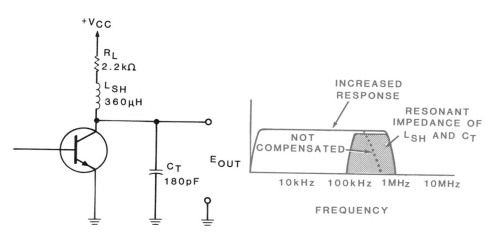

Figure 26-20   Shunt peaking.

Again, the amplifier response is shown at the right. In the low and middle frequency range, the peaking coil has little effect on the amplifier's response. However, at the higher frequencies, L_{SH} resonates with capacitance C_T to increase output impedance and boost the amplifier's gain at these higher frequencies. The dotted line in the response curve shows the response for the uncompensated amplifier. The shaded portion of the curve indicates the response produced by the resonant impedance of the peaking coil and the total capacitance. As you can see, this extends the amplifier's high frequency response.

ELECTRONIC CIRCUITS

A second way to increase high frequency response is to insert a small coil in series with the interstage coupling capacitor. Figure 26-21 shows this type of frequency compensation, called **series peaking.** Since peaking coil L_{SE} is connected between the two stages, it effectively isolates the output and input capacitances of the two stages. This separation permits the use of a larger value for R_L than was possible with shunt peaking. Actually, R_L can be increased by approximately 50 percent. Thus, higher stage gain is possible using series peaking.

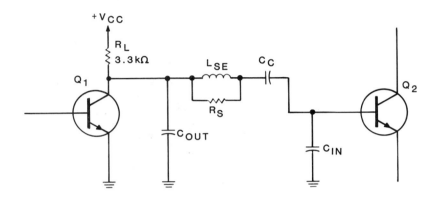

Figure 26-21 Series peaking.

Notice that the resistor, R_S, is in parallel with the series peaking coil. This is called a **swamping resistor.** If you'll recall from Chapter 16, when a resistor is added to a resonant circuit, the bandwidth flattens out and increases. This prevents the resonant circuit from reacting very sharply to the resonant frequency. Such resonance in the circuit shown in Figure 26-21, can cause the circuit to overcompensate for a narrow range of frequencies, producing a ringing effect. The swamping resistor effectively reduces this sharp reaction by flattening the bandwidth.

Shunt and series peaking are often combined to obtain the advantages of both. This type of peaking, known as **combination** or **series-shunt** peaking, is shown in Figure 26-22. In this circuit, the peaking coils assume separate roles. The shunt coil L_{SH} neutralizes the output capacitance of Q_1, while the series coil L_{SE} counteracts the input capacitance of the Q_2 stage. This combination extends the bandwidth of the amplifier up to approximately 5 MHz, which is much greater than is derived using shunt or series peaking alone. Once again, notice the swamping resistor, R_S, connected across the series coil.

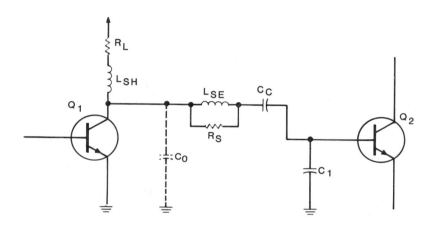

Figure 26-22 Amplifier using shunt and series peaking.

There are a number of other ways to extend the high frequency response of video amplifiers. Among these are the use of common-collector and common-base networks, rather than the common-emitters used in our discussions. For example, the common-base amplifier has a constant current gain that extends up to 100 MHz; unfortunately, this current gain is less than one. Likewise, the emitter-follower is an excellent wideband amplifier, but its voltage gain is also less than one. Therefore, it is a common practice to combine CB and CC amplifiers in the design of a total video amplifier system.

Typical Video Amplifiers

The most common use of the video amplifier is in the television receiver. Figure 26-23 shows a simplified version of a video amplifier that is used in a black and white television receiver. This amplifier employs two stages of amplification, and provides a voltage gain of 40 with a bandwidth of 3.5 MHz. Although this bandwidth is somewhat less than the 5 MHz bandwidth previously discussed, it is wide enough to amplify the video frequencies of a black and white video signal.

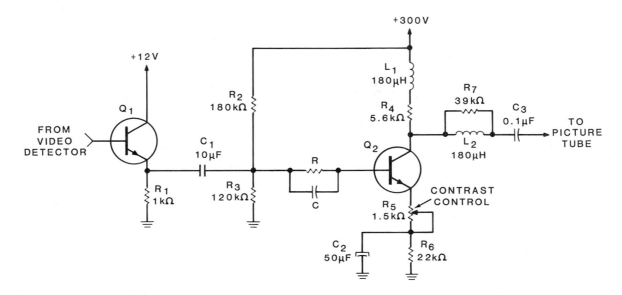

Figure 26-23   A representative video amplifier.

The first stage of the amplifier shows Q_1 connected as an emitter-follower. The input to Q_1 is from the video detector, which recovers the video (picture signal) from a much higher frequency signal called the intermediate frequency (I_F). The emitter-follower presents a high input impedance for the video detector and a low output impedance to the Q_2 stage. This low-impedance driving source improves the high frequency response of the Q_2 stage. The video signal is developed across emitter resistor R_1 and is coupled to the base of Q_2 through coupling capacitor C_1. C_1 is large enough to minimize the loss of low frequency signals. The RC network near the base of Q_2 serves as a high-pass filter to correct for the circuit's natural roll-off at high-frequencies.

The collector supply voltage may seem unusually high, but it is necessary to produce the required output voltage to drive the television's picture tube. However, such a high supply voltage permits the use of the large emitter resistor, R_6. This 22 kilohm resistor determines, to a large extent, the DC collector current of the stage. Notice that R_6 is bypassed by a large capacitor (C_2), and therefore does not affect the amplifier's performance at video frequencies.

Resistor R_5 is used to control the gain of the stage. Remember, when a resistor has no bypass capacitor, its resistance is naturally degenerative and reduces circuit gain. Gain will be maximum when the wiper arm is at the top of R_5. To reduce gain, the arm is moved toward the bottom and the circuit becomes naturally degenerative. As shown in the illustration, the gain control in the video amplifier is called the contrast control. By varying the video amplifier's gain, you can control the difference between the black and white areas of the picture on the television screen.

Connected to the collector of Q_2, we find load resistor R_4 and shunt peaking coil L_1. The output of the second stage is comprised of series peaking coil L_2 and its swamping resistor R_7. Capacitor C_3 couples the video signal to the picture tube.

It would be impossible to show all of the schemes that are used in video amplifier circuits. This example, however, is a commonly used method of extending the frequency response. Remember, video amplifiers must offset the effects of high frequency roll-off that is caused by the circuit's internal capacitance, while maintaining a relatively flat frequency response for a wide band of frequencies. This can be accomplished by sacrificing mid-frequency gain or by using peaking circuits to increase the high frequency gain.

Differential Amplifiers

A differential amplifier is a special type of amplifier which uses the outputs of two separate amplifiers, and uses them in various methods to create either two outputs, or a single output.

Single-Input, Single-Output Operation

A simple differential amplifier is shown in Figure 26-24A. This amplifier has two input terminals and the two output terminals. An input can be applied to either or both of the transistor's bases. The output can be taken from either collector with respect to ground. The output can also be taken between the two collectors.

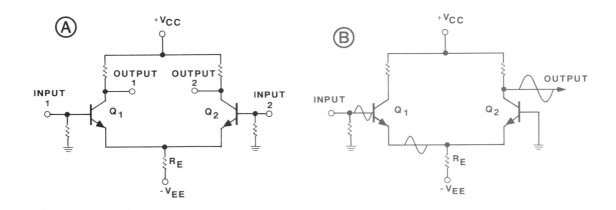

Figure 26-24 Basic differential amplifier circuit.

The simplest connection is the **single-input, single-output** circuit arrangement shown in Figure 26-24B. A lone input signal is applied to the base of Q_1, while the output is taken from the collector of Q_2.

The output of Q_1 is taken at the emitter, making it function as a common-collector amplifier. The output at the emitter is in-phase with the input, and since the common-collector has a voltage gain of essentially 1, the output has the same amplitude as the input.

AMPLIFIER APPLICATIONS

The output of Q_1 is developed by R_E, and is direct-coupled to the emitter of Q_2. With the input on the emitter, and the output on the collector, Q_2 is a common-base amplifier. The output of Q_2 is in-phase with its input, and the amplitude is increased, due to the high voltage gain of the common-base.

Notice that the final output signal is in-phase with the original input signal. This is because both the common-collector and common-base amplifiers have in-phase outputs. The overall voltage gain is approximately the same as that of the common-base amplifier stage (Q_2). Because direct coupling is used, the amplifier works for both AC and DC signals. While this arrangement does provide amplification, it does not take full advantage of the differential amplifier's characteristics.

Single-Input, Differential-Output Operation

Figure 26-25A shows another single-input circuit. This circuit operates in the same manner as the previous amplifier, with the addition of a second output at the collector of Q_1. This allows Q_1 to act as a common-emitter amplifier, producing an output voltage at the collector which is 180 degrees out-of-phase with the input signal. We therefore have two distinct outputs available.

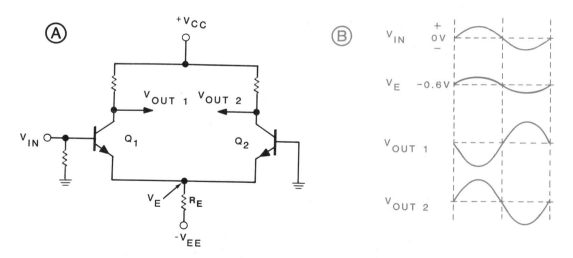

Figure 26-25 Single-input, differential-output circuit.

Figure 26-25B shows the various signals available. The top waveform is the input signal. The second waveform shows the output of Q_1 at the emitter, in-phase with the original input. This is not an output of the differential amplifier, but the input to Q_2.

The third waveform shows the output at the collector of Q_1. With this output, Q_1 acts as a common-emitter, providing a voltage gain and out-of-phase output. Last, the output of the common-base Q_2.

In order for this amplifier to function in its desired manner, the two outputs (from common-emitter Q_1 and common-base Q_2) should be equal in amplitude. In other words, the voltage levels of both outputs should be equal. To accomplish this, the collector load resistors, as well as the gain of each individual amplifier, are of equal value. If this is accomplished, the quiescent and peak voltages of the two outputs will be equal. However, the two outputs will be 180 degrees out-of-phase, since a common-emitter and a common-base are involved.

The uniqueness of this configuration occurs when an AC voltmeter is connected *across the individual outputs* to derive a final output. An example of this is shown in Figure 26-26. We'll assume our common-emitter and common-base amplifiers have an input signal of 2 volts peak-to-peak, and a gain of 4. When the two initial outputs are compared, we can see that their quiescent values are equal, as are their peak voltages. The only difference between these two outputs is that they are 180 degrees out-of phase.

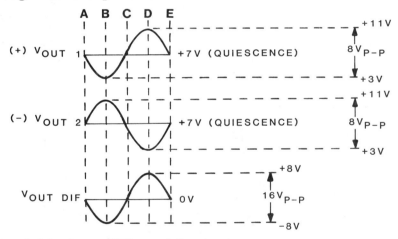

Figure 26-26 Achieving a differential output.

The bottom waveform in Figure 26-26 shows the final output when the potential difference between these two signals is taken. This would be the final output for the amplifier. At point A, both signals are at their quiescent values. The potential difference, therefore, is zero, since both quiescent values are the same.

As the two outputs move from point A to point B, $V_{OUT\ 1}$ begins its negative alternation, while $V_{OUT\ 2}$ begins its positive alternation. The potential difference between these signals begins to increase, and when the two outputs reach point B, their potential difference has increased to 8 volts. Because we have labeled $V_{OUT\ 2}$ as (-), this 8 volts is negative. Another way of determining the final output would be to say that the voltage at $V_{OUT\ 1}$, *with respect to* $V_{OUT\ 2}$, is −8 volts.

As the waveforms continue from point B to point C, they decrease from their peaks towards their quiescent values. Their potential difference therefore, decreases from −8 volts back to 0.

When the two outputs switch to their opposite alternations, the magnitude of their potential difference acts the same as the previous alternation. Therefore, the final output will, as before, change from 0 to 8 volts. However, since we are comparing $V_{OUT\ 2}$ to $V_{OUT\ 1}$, we are measuring the potential difference of +11 volts *with respect to* +3 volts. Our final output changes from 0 to +8 volts. And again, as the two outputs reach quiescence, the final output drops back to 0 volts.

Our final output is a volt peak-to-peak signal, referenced on 0 volts. An output that is derived by taking the potential difference between two signals is called a **differential output.** Our original 2 volt signal was increased to 16 volts—doubling the gain of either single amplifier. This is the primary advantage in the differential amplifier. The phase relationship between the original input and the final output is determined by which of the first outputs is labeled (−). In Figure 26-26, we used $V_{OUT\ 2}$, as (−), and the resulting output was out-of-phase with the original input. If we had used $V_{OUT\ 1}$ as (−), the output would be switched, and in-phase with the original input.

Differential-Input, Differential-Output Operation

The circuit shown in Figure 26-27A takes full advantage of the differential amplifier's characteristics. In our previous circuits, Q_2 received its input from the emitter of Q_1, and acted as a common-base amplifier. In this configuration, Q_2 receives its own input, and performs as a common-emitter amplifier. This is the most common use of a differential amplifier.

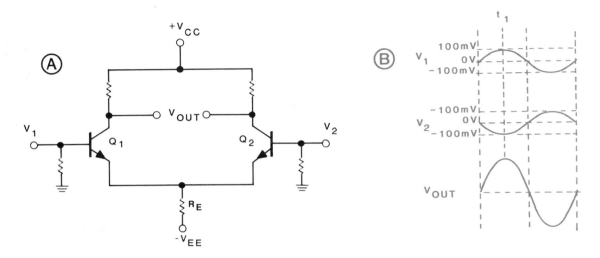

Figure 26-27 Differential-input, differential-output circuit.

Figure 26-27B shows the two original input signals, and the final output. Notice that the two inputs are equal in amplitude, and have a phase relationship of 180 degrees. When these two inputs are applied to the two common-emitters, their individual outputs will also be equal in amplitude, and 180 degrees out-of-phase. Our final output is then derived in the same manner as the single-input, differential-output amplifier we previously described: the potential difference between the two output terminals, effectively doubling the gain of either single amplifier.

Another way we can describe this operation is that it amplifies the *difference* between the two input signals. Thus, V_{OUT} is equal to $V_1 - V_2$ times the gain of the amplifier (represented by the letter "A"). That is:

$$V_{OUT} = A(V_1 - V_2).$$

AMPLIFIER APPLICATIONS

Since the two input signals are 180 degrees out-of-phase, the difference between the two is twice the amplitude of either signal. For example, at time t_1, V_1 is at +100 millivolts and V_2 is at −100 millivolts. The difference between the two signals is 200 millivolts. It is this difference that the circuit amplifies.

Common-Mode Input Operation

One of the chief advantages of the differential amplifier is its ability to reject common-mode signals. A **common-mode** signal is one which appears exactly the same at both input terminals.

The differential amplifier's output is derived by taking the potential difference between the outputs of two individual amplifiers. These amplifiers are designed to have the exact same gain, and the exact same quiescent (no-signal) value. Since the quiescent voltages determine the voltage levels of the output peaks, the highest gain occurs when the quiescent voltages are equal.

If the two input signals are equal, they not only have equal voltage levels, but are in-phase as well. These in-phase inputs are then applied to two common-emitter amplifiers. Both common-emitters will invert their particular input, so both outputs will also be in-phase.

The resulting outputs of these common-emitters are two in-phase signals with equal voltage levels. When we attempt to measure a potential difference between these signals, there is never a point where the two signals are different. Therefore, the final output is a constant 0 volts.

This illustrates how common signals are rejected by a differential amplifier. Earlier we stated that the differential amplifier amplifies the difference between the two input signals, calculated using the following formula:

$$V_{OUT} = A(V_1 - V_2).$$

Obviously, when the two inputs are equal, $V_1 - V_2$ equals 0. Thus, the differential amplifier provides a 0 volt output.

Common-Mode Rejection Ratio

We have seen that the differential amplifier performs two functions. First, it responds to and amplifies differential signals (two out-of-phase signals with equal voltage levels). Second, it rejects common-mode signals, providing a 0 volt output.

One of the most important characteristics of the differential amplifier is the ability to perform both of these functions simultaneously. That is, the differential amplifier can amplify differential inputs, while at the same time reject common inputs. This is useful in getting rid of unwanted hum and noise signals.

Figure 26-28 illustrates an extreme example of a signal which has picked up a 60 cycle hum. The signal we wish to amplify is 720 Hz. The 60 Hz is an unwanted signal and should be rejected.

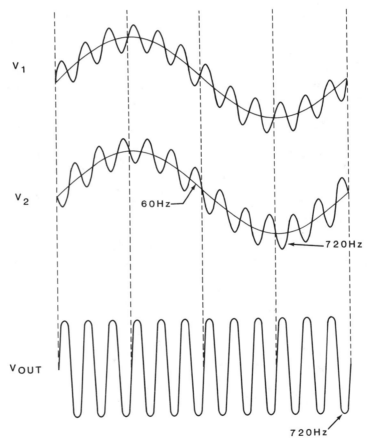

Figure 26-28 The 60 Hz common-mode signal is rejected.

Notice that the 60 Hz signal has the same phase at V_1 and V_2. Consequently, this is a common-mode signal. However, notice that the 720 Hz signal at V_1 is 180 degrees out-of-phase with the 720 Hz signal at V_2. This is a differential signal. As we have seen, the amplifier will tend to reject the common-mode signal, while amplifying the differential signal. Only the desired 720 Hz signal is produced as an output.

Actually, some tiny amount of the 60 Hz signal will appear in the output. That is, the amplifier will not entirely reject the unwanted signal. A good amplifier may amplify the desired signal by a factor of 100. At the same time, it may attenuate the unwanted signal by a factor of 100. Thus, at the output, the wanted signal may be 10,000 times as high as the unwanted signal. For example, the 720 Hz signal at the output may be 10 volts peak-to-peak, while the 60 Hz signal at the output may be only 1 millivolt peak-to-peak. This translates into a signal-to-noise ratio of 100000:1.

The ability of the differential amplifier to amplify the differential signal while rejecting the common-mode signal is expressed as a ratio. This ratio is called the common-mode rejection ratio (CMRR). It is a ratio of difference gain to common-mode gain. The difference gain (A_D) is found by comparing a change in the output voltage to the change in the differential input voltage.

When 100 millivolt p-p differential inputs causes an overall output of 10 volts p-p, the difference gain (A_D) is:

$$A_D = \frac{10V}{100\ mV} = 100$$

A common-mode input of 100 millivolts p-p causes an output of only 1 millivolt p-p. In this case, the common-mode gain (A_{CM}) is:

$$A_{CM} = \frac{1mV}{100\ mV} = 0.01$$

The common-mode rejection ratio (CMRR) is the ratio of the difference gain to the common-mode gain. Thus, the CMRR for our circuit is:

$$CMRR = \frac{A_D}{A_{CM}} = \frac{100}{0.01} = 10,000$$

ELECTRONIC CIRCUITS

Chapter Self-Test

1. What is the audio frequency range?

2. Which amplifier class of operation is preferred for an audio amplifier? Why?

3. What is the damping factor?

4. What are the input requirements for the push-pull amplifiers? What type of circuits are used to provide this input requirement?

5. An AC signal has a cycle of 360 degrees. For what portion of this input signal will a single transistor in a push-pull arrangement conduct?

6. When an NPN and a PNP transistor are combined to form an amplifier, what type of amplifier is created?

7. What is the frequency range of a standard video amplifier?

8. Describe junction capacitance, and the effect of biasing on junction capacitance.

9. What is the Miller effect?

10. When does shunt capacitance begin to affect frequency gain, and what can be done to the collector load resistor to extend this?

11. What is the difference between shunt and series peaking?

12. Which of the peaking coil configurations provide the *largest* extension of frequency response?

13. In the single-input, single-output mode of a differential amplifier, what develops the input for Q_2?

14. How is a differential output for a differential amplifier obtained?

15. When two inputs for a differential amplifier are in-phase, and have equal voltage levels, what will the final output be? Why?

16. Define the term common-mode rejection ratio, and give the formula for its calculation.

AMPLIFIER APPLICATIONS

Summary

Audio amplifiers are used to provide amplification of AC signals in the 20 Hz-20 kHz audio frequency range. In order to provide minimum distortion, class A amplifiers are most commonly used.

Voltage amplifiers are usually in the first stages of audio amplification. In this configuration, it is referred to as a pre-amplifier, or preamp. The output of a voltage audio amplifier is then coupled to a power amplifier, which provides high current flow for a low-impedance load, usually a loudspeaker.

If a single stage does not provide enough amplification, stages may be coupled together. In this manner, the stages are essentially in parallel, with separate current paths for each transistor. When stages are connected in this fashion, they are said to be cascaded. Transformer coupling is the preferred method of coupling audio amplifiers, because they allow for maximum transfer of power, stage isolation (impedance-matching), and can increase or decrease the voltage.

The audio power amplifier is usually the last amplifier before the signal is coupled to the loudspeaker. Efficient power amplifiers have a high damping factor, which is the ratio of load impedance to output impedance.

When a single active device (transistor) is used as a power amplifier, it is called a single-ended device. These are not preferred, however, because they tend to magnetize the core of the output transformer.

Push-pull amplifiers are preferred because core magnetization does not take place. Two transistors are used, alternately conducting and cutting off. An output transformer feels the current from the alternating transistors, and develops the output voltage.

The input signals—equal in amplitude and frequency, but 180 degrees out-of-phase, are required for a push-pull amplifier to properly function. Signals with this relationship are called complementary signals. Circuits known as phase splitters provide these complementary signals. Phase splitters can be comprised of a transformer, two cascaded common-emitter amplifiers, or a single amplifier, with one output from the collector, and the other from the emitter.

Push-pull operation can also be accomplished with a single amplifier using both NPN and PNP transistors. This is known as a complementary amplifier. The biasing requirements for a complementary amplifier can be accomplished by a single voltage source, or by two voltage sources of opposite polarities. A complementary amplifier must have constant and equal base voltage. This is accomplished by the use of diodes between the two base terminals, and these diodes also prevent thermal runaway.

The output volume of an audio amplifier can be adjusted by adjusting the overall gain of the amplifier. Tone controls determine which frequencies are attenuated before they reach the amplifier.

Video amplifiers provide amplification of AC signal in the 10 Hz-5 MHz video frequency range. Because of the high impedance of coils at high frequencies, video amplifiers can only use direct or RC coupling.

At extremely high frequencies, the junction within a transistor have an effective capacitance across them. This is known as junction capacitance. In addition, a negative feedback path exists between the base and collector of a transistor, known as the Miller effect. Junction capacitance and the Miller effect combine in video amplifiers to create shunt capacitance, which severely deplete the output signal. In order to counteract this effect, inductors are placed in the circuit. This is known as peaking. When the inductor is in parallel with the shunt capacitance, it is series peaking. When the inductor (usually interconnected with a swamping resistor) is placed in series with the shunt capacitance, this is shunt peaking. Both series and shunt peaking have advantages, and are usually used at the same time.

A differential amplifier employs two separate amplifiers in the same circuit. It can either accept one or two input signals, and the output can be taken from either individual amplifier, or the potential difference between the two amplifier outputs. When the potential difference is taken, the output is called the differential output.

The differential amplifier is designed in such a way that the quiescent voltages and the gain for each individual amplifier are the same. The transistors usually have the same temperature characteristics, and the collector load resistors usually have equal values.

If the inputs to a differential amplifier are out-of-phase by 180 degrees, with equal voltage levels, the overall gain will be the difference between the two input signals. This gain will also be double the gain for either individual amplifier used in the differential amplifier circuit.

In the single-input, single-output mode, one input is applied to a common-collector amplifier, and a single output is taken from a common-base amplifier. The overall gain of the circuit is equal to the gain of the common-base.

In the single-input, differential-output mode, one input is applied to a common-emitter amplifier, and its varying emitter output is applied to a common-base amplifier. A potential difference is taken between the outputs of the common-emitter and the common-base. Since these outputs are out-of-phase, the overall gain is twice the gain of either individual amplifier.

AMPLIFIER APPLICATIONS

In the differential-input, differential-output mode, two out-of-phase inputs with equal peak-to-peak voltages are applied to two common-emitter amplifiers. When the inputs share this relationship, they are said to be differential inputs. A differential output is then obtained.

Common-mode inputs have equal peak-to-peak voltages, and are in-phase. When common-mode inputs are applied to a differential amplifier, no differential output is possible because there is no potential difference between the outputs of the two common-emitter amplifiers. In this manner, the differential amplifier rejects common-mode inputs.

When differential inputs also share a common frequency, they can be applied to a differential amplifier. The resulting output will be an amplified version of the differential frequency, with the common frequency rejected. However, the common-mode frequency cannot be completely rejected. The common-mode rejection ratio expresses the ratio between differential gain and common gain, and uses the following formula:

$$\text{CMRR} = \frac{A_D}{A_{CM}}$$

ELECTRONIC CIRCUITS

CHAPTER 27

Operational Amplifiers

ELECTRONIC CIRCUITS

Contents

Introduction 719

Chapter Objectives 720

Operational Amplifier Characteristics 721

Closed-Loop Operation 726

Applications 738

Chapter Self-Test 745

Summary 747

OPERATIONAL AMPLIFIERS

Introduction

The development of the operational amplifier has had a dramatic effect on the electronic market. Years ago, operational amplifiers were so expensive that few technicians or engineers ever became involved with them. Their use was confined to special applications, such as analog computers and expensive instrumentation equipment. However, operational amplifiers are now available in an inexpensive integrated circuit form. Using operational amplifiers, it is now possible to reduce a circuit's physical size, number of components, and power requirements, while at the same time, increasing its stability at a higher gain. Moreover, all of this is accomplished at a lower price. The modern technician can ill afford to overlook this device with its many and varied applications. In fact, anyone involved with electronics should know how the "op amp" can be used, its characteristics, and its limitations.

In this chapter, you will study the advantages that op amps have over discrete components. We will examine the advantages of the op amp over discrete circuits. In addition, you will gain the knowledge required to evaluate and use these compact circuits.

ELECTRONIC CIRCUITS

Chapter Objectives

When you have completed this chapter, you will be able to:

1. Define the following terms: offset null, data sheet, slew rate, closed-loop mode, open-loop gain, breakover frequency, unity gain, unity gain frequency, and gain-bandwidth product.

2. Identify the schematic symbol for an operational amplifier.

3. List the general characteristics, and their typical values, for an operational amplifier.

4. Name the three amplifier stages in an operational amplifier.

5. Describe the operation of an inverting amplifier, and determine its gain and input impedance.

6. Describe the operation of an noninverting amplifier, and determine its gain and input impedance.

7. Describe the relationship between gain and bandwidth for both the open-loop and the closed-loop modes.

8. Describe the characteristics and purpose of the voltage follower.

9. Identify the schematic diagram and explain the operation of the summing amplifier and the scaling adder.

10. Identify the schematic diagram and explain the operation of the low-pass and high-pass active filters.

11. Identify the schematic diagram and explain the operation of the difference amplifier.

12. List at least seven different applications of operational amplifiers.

Operational Amplifier Characteristics

An **operational amplifier,** or **"op amp,"** is a special type of high gain DC amplifier. It consists of several amplifier stages cascaded together. Each amplifier stage gives the overall circuit some desirable characteristic. It is these characteristics which separate the op amp from the ordinary amplifier.

Before an amplifier can be classified as an operational amplifier, it must have certain characteristics. Three of the most important are:

1. Very high gain.
2. Very high input impedance.
3. Very low output impedance.

Other characteristics will be discussed later, but these three are essential at this point in your study of amplifiers.

Op amps come in both discrete and integrated circuit form. While the principles discussed in this section apply to both types, we will concentrate on the IC form. In particular, we will be concerned with the low-cost, general-purpose op amps which are most widely used. The IC op amp has two advantages over the discrete op amp from the designer's view point: it costs less, and it is smaller. In almost every other respect, the discrete form of the op amp is superior. The low cost of the IC op amp is responsible for its widespread popularity.

ELECTRONIC CIRCUITS

Basic Op Amp

Figure 27-1 is a block diagram of a typical operational amplifier. Because the op amp is an integrated circuit, we can use it without knowing exactly what goes on inside it. However, we can better understand the characteristics of the op amp if we have some idea of what is inside the IC.

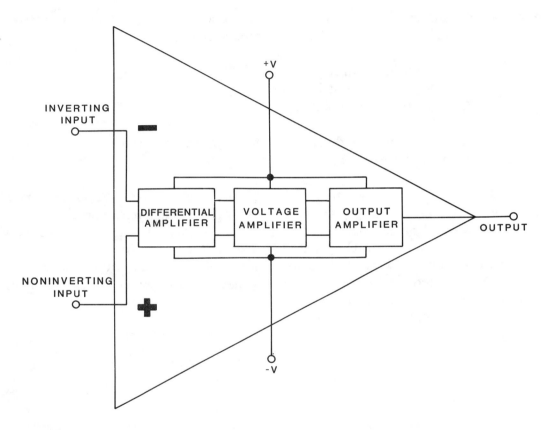

Figure 27-1 Basic operational amplifier.

The op amp is normally composed of three stages. Each stage is an amplifier which provides some unique characteristic. The input stage is a differential amplifier. This gives the op amp the advantages of high common-mode rejection, differential inputs, and a frequency response that extends down to DC. In addition, special techniques are used to give the input stage a very high input impedance (usually in the megohm range).

OPERATIONAL AMPLIFIERS

The second stage is a high gain voltage amplifier. A typical op amp may have a voltage gain of 200,000 or more when used in DC applications. This stage provides most of that gain.

The final stage is an output amplifier. This stage is often a complementary emitter-follower. It gives the op amp a low output impedance. It allows the op amp to deliver several milliamperes of current to a load. This correlates to the characteristics of the common-collector amplifier.

The op amp is powered by both negative and positive power supplies. This allows the output voltage to swing negative or positive with respect to ground. Most op amps require supply voltages that vary from 5 volts to approximately 20 volts.

The complete schematic diagram of a typical op amp is shown in Figure 27-2A. The three amplifiers are identified at the bottom. Typical integrated circuit design uses many transistors, which are easy to manufacture. Resistors and capacitors are held to a minimum, due to the difficulty in manufacturing. This makes it a little hard to identify the types of circuits being used. The input stage is a special type of differential amplifier. The output stage is a push-pull emitter-follower.

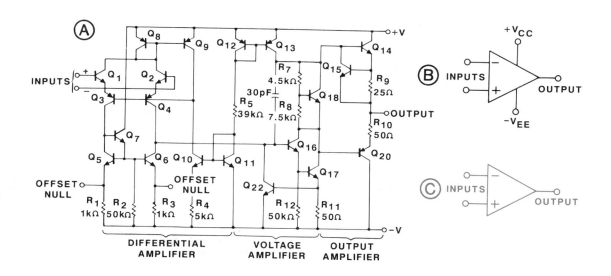

Figure 27-2 Schematic diagram and schematic symbol for the op amp.

Notice that no coupling capacitors are used. This allows the circuit to amplify DC as well as AC signals. This op amp has two terminals which were not shown on the block diagram. These are labeled **offset null.** They allow an external potentiometer to be connected to the circuit. The potentiometer is set so that the output of the op amp is exactly 0 volts when the two input terminals are at 0 volts. This is called *balancing the op amp*. The potentiometer can be used to compensate for slight circuit imbalances in the internal stages. The two input terminals are labeled "+" and "–". The "–" input is called the **inverting input,** and the "+" input is called the **non-inverting input.**

Like all differential amplifiers, the input stage can be operated in three different modes. In the differential-input mode, the inputs are two signals which are 180 degrees out-of-phase. Another arrangement is to ground the "+" input and apply the signal to the "–" input. In this case, the output will be an inverted and amplified version of the input signal. Finally, the "–" input terminal can be grounded and the signal can be applied to the "+" input terminal. In this case, the signal is amplified but *not inverted*.

The schematic symbol for the op amp is a triangle, as shown in Figure 27-2B. Often, the power supply connections are omitted, as in Figure 27-2C. Of course, power must be supplied to the op amp whether or not the connections are actually shown.

Electrical Characteristics

All of the important characteristics of an operational amplifier are included on the data sheet for that op amp. A **data sheet** specifies the operating characteristics for a component or, in the case of an integrated circuit, complete circuits. The following is a list of the important operating characteristics of an op amp, and typical values.

Input Resistance (R_{IN})—This specification tells us the resistance between the two input terminals of the op amp. Generally, the higher the input resistance, the better the op amp will perform. An input resistance in excess of one megohm is common.

Gain (A)—The voltage gain of an op amp should be very high, in most cases, the higher the gain, the better the op amp. Gains of 200,000 or higher are common with DC inputs. This is known as the **open-loop** gain, or the gain of the amplifier without feedback. Ironically, most of this gain is sacrificed in practical applications. The op amp is normally operated with very heavy degenerative or negative feedback, which drastically reduces circuit gain.

Often the gain is specified as a certain number of volts out for each millivolt input. That is, a gain of 200,000 may be specified as 200 V/mV. This means that the gain is such that a 1 millivolt change at the input will attempt to cause a 200 volt change at the output. Of course, the output cannot approach 200 volts. The amplifier saturates to an output voltage slightly less than the supply voltage. The 200 V/mV specification is just another way of saying that the output change will be 200,000 times greater than the input change.

Output Resistance (R_{OUT})—The output resistance of the op amp should be very low to provide a high amount of current to the load. The lower the output resistance, the better the op amp. An output resistance of 150 ohms or less is a typical value.

Common-Mode Rejection Ratio (CMRR)—This is the ratio of the differential voltage gain to the common-mode voltage gain. A CMRR of 30,000 or higher is typical.

Input Offset Voltage—Ideally, the output voltage of an op amp should be 0 volts when both inputs are at 0 volts. Unfortunately, because of the high gain, the slightest circuit imbalance can cause an output voltage. The output can be forced back to zero by applying a slight offset voltage to one of the inputs. Thus, the input offset voltage is the DC voltage which must be applied between the inputs to force the DC output voltage to zero. A typical value of 1 millivolt is normal with general purpose op amps.

Slew Rate—This characteristic indicates how fast the output voltage of the op amp can change. It is given in volts per microsecond. For example, a slew rate of 1 volt per microsecond means that the output voltage can change no faster than one volt during each microsecond, regardless of how fast the input voltage changes.

In addition to the characteristics mentioned previously, manufacturers may also specify such values as total power dissipation, maximum-voltage limits, etc. Most of these are either self-explanatory or are explained in the data sheets.

Closed-Loop Operation

The normal mode of operation for the op amp is the closed-loop mode. In the closed-loop configuration, the output signal is routed back to one of the input terminals. The output signal provides degenerative feedback, which always opposes the effects of the original input signal. This greatly reduces the circuit gain, but it stabilizes the op amp. There are two basic closed-loop circuits: the inverting configuration (−input) and the noninverting configuration (+input). Because the inverting configuration is more popular, it will be described first.

Inverting Configuration

The inverting configuration is shown in Figure 27-3. In its most basic form, it consists of the op amp and two resistors. The noninverting input (+) is grounded. The input signal (E_{IN}) is applied through R_2 to the inverting input. The output signal (E_{OUT}) is also fed back through R_1 to the inverting input (−). The signal at the inverting input is determined not only by E_{IN}, but also by E_{OUT}.

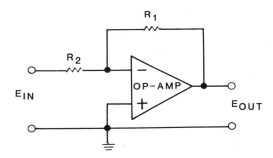

Figure 27-3 Inverting configuration.

At this point, we must differentiate between the op amp and the operational circuit. The triangle is the op amp. The operational circuit consists of the op amp and the two resistors. E_{IN} is the input to the operational circuit. However, the signal at the inverting input of the op amp is determined by both E_{IN} and E_{OUT}.

OPERATIONAL AMPLIFIERS

Feedback Operation

In a feedback arrangement of this type, events happen so quickly that they are hard to visualize. For example, assume that E_{IN} is initially at 0 volts. Since no difference of potential exists between the two input terminals, the output should also be at 0 volts. This statement assumes that the op amp is balanced.

Now, assume that E_{IN} instantly changes to +1 volt. When the voltage at the inverting input goes positive, E_{OUT} starts to swing negative. This negative-going voltage is felt back through R_1 to the inverting input where it begins to cancel the original change. Of course, the feedback signal cannot completely cancel the input signal, but it would greatly reduce the effect of the input signal. When E_{IN} changes to +1 volt at the left side of R_2, the negative feedback causes the voltage at the inverting input of the op amp (right side of R_2) to change by only a few microvolts.

Fortunately, the gain of an operational amplifier is so high, that it requires only a few microvolts to produce a large output change. The operational amplifier is normally an integrated circuit, so there is no need to discuss what happens within the amplifier itself.

To summarize, any change at the stage input is almost immediately negated by the feedback, to a change of only a few microvolts. This extremely small change is then amplified by the operational amplifier.

For any amplifier, the amount of gain is determined by the ratio of output voltage to input voltage. This is just as true for an op amp. Determining gain, however, is not that easy, because the output voltage actually affects the input voltage. A series of truths allow us to see how the gain for an operational amplifier is determined. These truths are linked together by the logical question that arises after the statement is made.

1. The amount of change at the output is determined by the amount of change at the inverting input.

    (What determines the amount of change at the inverting input?)

2. The amount of change at the inverting input is determined by the difference between the feedback voltage and the *stage* input voltage.

    (What determines the difference between the feedback voltage and the *stage* input voltage?)

3. The size of the input resistor and the feedback resistor determine the difference between the feedback voltage and the stage input voltage.

In other words, the ratio between the feedback resistor (R_1) to the input resistor (R_2) determines the amount of gain for the inverting amplifier.

Resistor Ratio

The characteristics of the inverting configuration are determined almost entirely by the values of R_1 and R_2, as shown in Figure 27-4. When an input signal is applied at E_{IN}, the feedback path prevents almost all of this voltage from being felt at the (−) input. Therefore, E_{IN} must be dropped by R_2, the input resistor. The amount of current flowing through the input resistor (I_{IN}) is equal to:

$$I_{IN} = \frac{E_{IN}}{R_2}$$

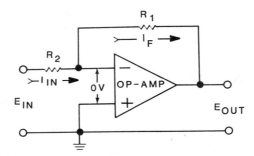

Figure 27-4 The characteristics of the circuit are determined by R_1 and R_2.

An output voltage (E_{OUT}) is developed which is opposite in phase (polarity inverted) to E_{IN}. This voltage causes a feedback current (I_F) to flow through R_1. Since the left side of R_1 is essentially at 0 volts and the right side is at E_{OUT}, this current must flow from the input to the output. The feedback current is equal to:

$$I_F = \frac{-E_{OUT}}{R_1}$$

The minus sign (−) indicates that E_{OUT} is 180 degrees out-of-phase with E_{IN}.

Due to the nature of the feedback, no current can flow through the inverting input. Therefore, any current that flows through R_2 *must* flow through R_1. Putting this into an equation, we can state that:

$$I_F = I_{IN}$$

Substituting our other equations for I_{IN} and I_F, we get:

$$\frac{E_{IN}}{R_2} = \frac{-E_{OUT}}{R_1}$$

We can rearrange this equation to form:

$$\frac{E_{OUT}}{E_{IN}} = \frac{-R_1}{R_2}$$

The gain of any amplifier (A_V) is expressed as the output over the input, or E_{OUT}/E_{IN}. Therefore, the gain of the inverting configuration is expressed as:

$$A_V = -\frac{R_1}{R_2}$$

Again, the minus sign simply indicates that the output is inverted. This equation is interpreted to mean that the gain of the inverting configuration is equal to the ratio of the feedback resistor to the input resistor. This is true because the difference between the feedback voltage and the stage input voltage is the voltage that is amplified, and the values of these voltages are determined by the input and feedback resistors, respectively. Any current introduced at E_{IN} flows through the input resistor, and through the output resistor to E_{OUT}.

ELECTRONIC CIRCUITS

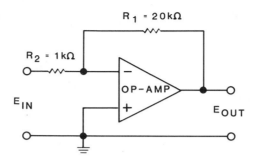

Figure 27-5 This circuit has a voltage gain of –20.

Figure 27-5 shows a typical op amp in the inverting configuration. The gain of this circuit is:

$$A_V = -\frac{R_1}{R_2} = -\frac{20k\Omega}{1k\Omega} = -20$$

Let's assume the input voltage at E_{IN} is –1 volt DC. Since the inverting input is at 0 volts, E_{IN} must be developed by R_2. Therefore, the current through R_2 (using Ohm's Law) must be:

$$I_{IN} = \frac{E_{IN}}{R_2} = \frac{-1V}{1k\Omega} = 1mA$$

Since this current cannot flow through the inverting input, it must flow through R_1. In order to force 1 mA through R_1, E_{OUT} must be:

$$E_{OUT} = (R_1)(I_{IN}) = (20k\Omega)(1mA) = 20V$$

Notice that the –1 volt input was amplified by a gain of 20, and inverted from negative to positive, for an output of +20 volts. This illustrates that the gain for this amplifier is –20. If the feedback resistor is 20 times larger than the input resistor, the gain of the inverting amplifier is 20.

While our example used a steady DC voltage, this works equally well with AC signals. If we applied a 1 volt peak-to-peak AC signal to the input, the output would be 20 volts peak-to-peak, and 180 degrees out-of-phase with the input signal.

Input Resistance

The input resistance of the op amp is very high. However, the input resistance of the inverting amplifier is not determined by the op amp. Rather, it is determined by the value of R_2.

Remember, any voltage at the stage input is almost completely negated by the feedback voltage. Therefore, any current that this voltage introduces flows through R_2, forcing R_2 to develop almost the entire input voltage.

Noninverting Configuration

The noninverting configuration is not as popular as the inverting configuration, but it is still used rather frequently. In this configuration (shown in Figure 27-6), the input (E_{IN}) is applied to the noninverting input, making E_{OUT} in-phase with E_{IN}. however, the two resistors are in the same location as the inverting amplifier, and the feedback is still applied to the inverting input.

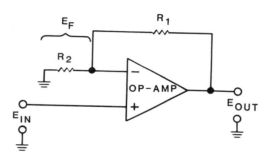

Figure 27-6 Noninverting configuration.

Feedback

In addition to E_{OUT} being applied to the load, it is also dropped by R_1 and R_2. This produces the feedback voltage (E_F), which is also in-phase with E_{IN}. At first, this may not seem like degenerative feedback. However, since the op amp responds to the difference between the two input signals, the feedback voltage always opposes the input voltage.

The feedback action is difficult to follow because it happens so quickly. When E_{IN} swings more positive, E_{OUT} tends to swing much more positive. However, as E_{OUT} increases, so does E_F. The increase in E_F tends to partially offset the increase in E_{IN}. Because the gain of the op amp is so tremendously high, only a slight difference between E_{IN} and E_F is necessary to produce E_{OUT}. Thus, E_{IN} is almost exactly equal to E_F. There is a slight difference between E_{IN} and E_F. It is this difference that is amplified. However, for most practical purposes $E_{IN} = E_F$.

ELECTRONIC CIRCUITS

Stage Gain

The gain of the noninverting amplifier is:

$$A_V = \frac{R_1}{R_2} + 1$$

Where R_1 is the feedback resistor, and R_2 is the input resistor.

Notice the similarity between this gain formula and the gain formula for the inverting amplifier. This is due to the similarity in the feedback path for the two amplifiers. The "+ 1" accounts for the slight difference between E_{IN} and E_F. Although we previously mentioned that these values are essentially equal, we also stated that there is a slight difference, and it is this difference that is amplified.

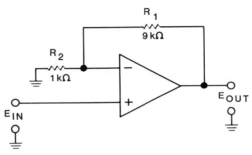

Figure 27-7 Noninverting amplifier.

Figure 27-7 shows an example of a noninverting amplifier circuit. Using the gain formula:

$$A_V = \frac{R_1}{R_2} + 1 = \frac{9k\Omega}{1k\Omega} + 1 = 9 + 1 = 10$$

The circuit produces a voltage gain of 10. The input impedance of this stage is high, while the output impedance is very low. E_{OUT} will be in phase with E_{IN}, but will be ten times higher in amplitude.

Input Impedance

The input impedance of the inverting configuration discussed earlier was equal to the input resistor's value. This is a result of the action of the feedback voltage on the inverting input.

OPERATIONAL AMPLIFIERS

The noninverting configuration does not have this characteristic. E_{IN} is applied directly to the "+" input. E_{IN} will cause a certain current to flow between the "+" and "−" input terminals of the op amp. At first, it might seem that the current is limited only by the input resistance (R_{IN}) of the op amp.

As shown before, this input resistance (R_{IN}) is very high, and holds the current to a low value. However, remember that E_F is almost equal to E_{IN}. Thus, the only voltage causing input current is the small difference between E_{IN} and E_F. Therefore, the input current is much lower than can be accounted for by E_{IN} and R_{IN} alone. The input impedance of the stage is higher than the input resistance of the op amp. Therefore, the noninverting circuit has an extremely high input impedance.

Bandwidth Limitations

The gain for both the inverting and the noninverting configurations varies with frequency. The gain figure given on op amp specification sheets is generally for DC signals. The gain at higher frequencies will be much lower.

Open-Loop Gain Versus Frequency

Figure 27-8 shows the response curve of a typical operational amplifier *without feedback*. This is known as open-loop gain. Notice that at DC and very low frequencies, the gain of the amplifier is about 100,000. However, the gain drops off very rapidly as the frequency increases. The gain is fairly flat to about 10 Hz. At 10 Hz, the gain has fallen to 70.7% of its maximum value. The point at which the gain falls to 70.7% of maximum is called the breakover point, if you'll recall from our chapters on AC filters. The frequency at this point is called the break or breakover frequency. For this op amp, the break frequency is approximately 10 Hz.

ELECTRONIC CIRCUITS

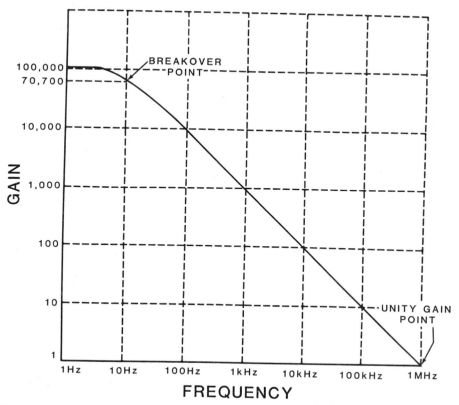

Figure 27-8 Open-loop response.

Above the break frequency, the gain drops off at a constant rate as the frequency increases. Notice that a tenfold increase in frequency causes a tenfold decrease in gain. This linear roll-off continues with the gain falling to 1 at a frequency of 1 MHz. Thus, 1 MHz is called the **unity gain frequency,** and is generally considered the maximum frequency at which the op amp is useful. We should also point out that another way of stating a gain of 1 for a stage is **unity gain.**

Bandwidth is normally measured at the point where the response falls to 70.7% of maximum. The op amp whose response is shown, has an open-loop bandwidth of only 10 Hz. Of course, if there were not some way to increase this bandwidth, the op amp could only be effectively used as a DC amplifier.

Feedback Increases Bandwidth

As you have seen, the op amp is rarely used without feedback. The heavy degenerative feedback stabilizes the circuit and makes it independent of the op amp's characteristics. Just as important, the degenerative feedback also increases the bandwidth of the circuit. Degenerative feedback is used to flatten the response curve at the expense of some of the op amp's gain.

Figure 27-9 shows how dramatically the bandwidth can be increased using feedback. Here a heavy degenerative feedback is used to decrease the gain from 100,000 to 100. That is, the feedback reduces the gain by a factor of 1000. In doing this, the bandwidth **increased** by a factor of 1000. The frequency response is now flat to approximately 10 kHz.

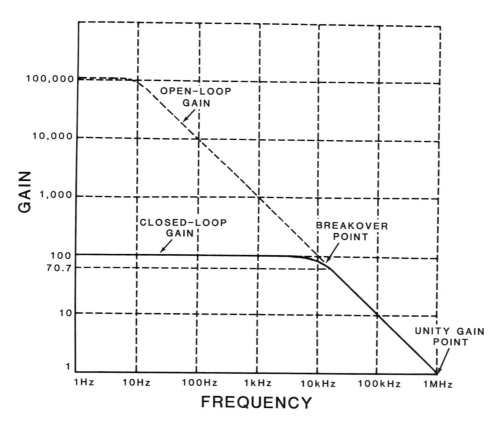

Figure 27-9 Closed-loop response (gain of 100).

By reducing the gain to 100, the breakover point occurs at a gain of 70.7 and the break frequency is approximately 10 kHz.

Gain-Bandwidth Product

For our open-loop response in Figure 27-8, op amp, the product of the gain (100.000) and the bandwidth (10 Hz), when multiplied, equals 1,000,000. Also, for our response in Figure 27-9, the product of the gain (100) and the bandwidth (10 kHz) is also equal to 1,000,000. We can also note that 1,000,000 Hz, or 1 MHz, is the unity gain frequency.

This illustrates that, for a given op amp, the **gain-bandwidth product** is constant. It also illustrates that the gain-bandwidth product is equal to the unity gain frequency.

The gain-bandwidth product is an important specification. It tells us the upper frequency that the circuit can amplify. We can use it to determine the upper frequency limit we can expect for a given gain. Some IC op amps have gain-bandwidth products of 15 MHz or higher. This is because different op amps reach unity gain at different frequencies.

Voltage Follower

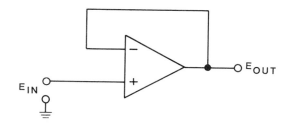

Figure 27-10 Unity gain voltage-follower.

Figure 27-10 shows a special type of noninverting amplifier. Recall that the gain formula is:

$$A_V = \frac{R_1}{R_2} + 1$$

Since R_1 doesn't exist, our fraction is equal to 0, and is not required to compute gain. Therefore, the gain for this amplifier is:

$$A_V = 1$$

With a gain of 1, the output voltage will be equal in amplitude, and in-phase, with the input. The noninverting amplifier still retains its high input impedance and low output impedance. Thus, the circuit behaves like a "super" emitter-follower. It is used for impedance-matching and isolation, similar to a typical common-collector amplifier.

Applications

The circuits we have described to this point are very basic for the operational amplifier. However, each of these circuits (the inverting amplifier, the noninverting amplifier, and the voltage follower) are widely used, despite their simplicity. We will now look at some other op amp applications. These circuits, although they are a bit more complex, are also used rather extensively.

Summing Amplifier (Adder)

An interesting circuit that illustrates the versatility of the op amp is the **summing amplifier** or **analog adder**. The schematic diagram is shown in Figure 27-11. This circuit has two inputs and one output. With the resistor values shown, E_{OUT} will be the sum of E_1 and E_2. For simplicity, we will analyze the circuit using DC input signals. However, the circuit works equally well with AC signals.

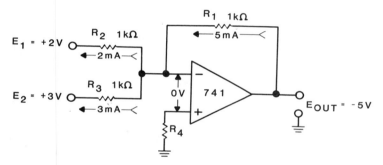

Figure 27-11 The summing amplifier or adder.

In the example shown, E_1 is a +2 volt DC signal while E_2 is a +3 volt DC signal. The point at which the three resistors connect is a virtual ground. This point is called the **summing point** or **summing node**. E_1 and E_2 are isolated from each other by this virtual ground at the summing point. E_1 causes a current to flow through R_2:

$$I_{R2} = \frac{E_{R2}}{R_2} = \frac{+2V}{1k\Omega} = 2mA$$

Likewise, E_2 causes a current to flow through R_3:

$$I_{R3} = \frac{E_{R3}}{R_3} = \frac{+3V}{1k\Omega} = 3mA$$

Therefore, 5 milliamps of current flows to the left out of the summing junction.

The current flowing out of the "−" input of the op amp is negligible because of the high input resistance. Therefore, to maintain the summing point at 0 volts, the 5 milliamps must be supplied through R_1 by E_{OUT}. To force 5 milliamps through R_1, E_{OUT} must be −5 volts. The output voltage is the inverted sum of the input voltages.

While the circuit shown in Figure 27-11 has only two inputs, this is not the maximum number of inputs possible. Figure 27-12 shows an adder with four inputs. If all of the resistors are the same value, E_{OUT} will be:

$$E_{OUT} = -(E_{R1} + E_{R2} + E_{R3} + E_{R4})$$

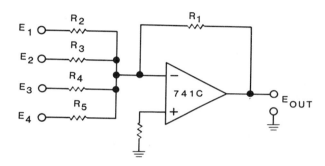

Figure 27-12 Four input summing amplifier.

The importance behind the resistor values still lies in the gain being determined by the input and feedback resistors. The amplifier in Figure 27-12 does not have a gain factor because the feedback resistance (R_1) is equal to each input resistance. As before, when the input resistor and the feedback resistor are equal, and the network is applied to the inverting input, the gain is −1.

ELECTRONIC CIRCUITS

By making the feedback resistor larger than the input resistance, a gain can be achieved. For example, in Figure 27-13, R_1 is a 10 kilohm resistor while each input resistor has a value of 1 kilohm. Thus, E_{OUT} will be:

$$E_{OUT} = -10(E_{R1} + E_{R2} + E_{R3})$$

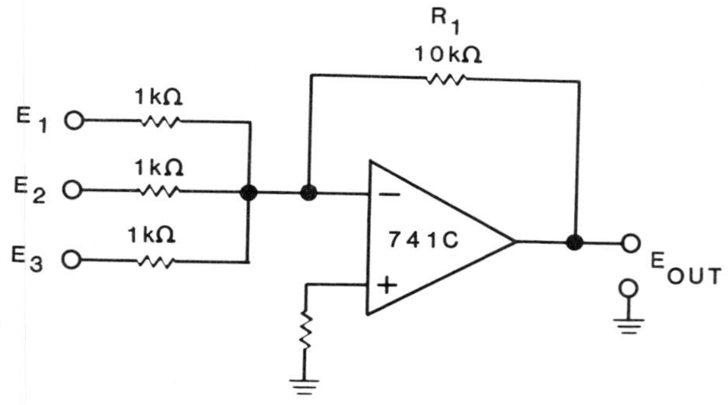

Figure 12-13   Adder with gain of –10.

In many applications, we may want one input to influence E_{OUT} more than another. In these cases, we use different value input resistors. For example, in Figure 27-14, each input resistor has a different value. The gain for each input voltage, therefore, is the ratio of its particular input resistor to the feedback resistor. These individual gains are then summed to determine E_{OUT}. The formula for calculating the gain in Figure 27-14 is:

$$E_{OUT} = -\left(\frac{R_1}{R_2}E_1 + \frac{R_1}{R_3}E_2 + \frac{R_1}{R_4}E_3 + \frac{R_1}{R_5}E_4\right)$$

This circuit is referred to as a *scaling adder,* and are common in digital-to-analog converters and in analog computers.

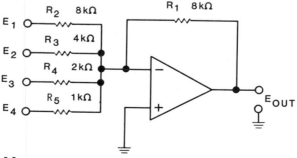

Figure 27-14   Scaling adder.

OPERATIONAL AMPLIFIERS

Active Filters

An ingenious application of the op amp is the **active filter**. Here, RC networks are used in the input or feedback circuits. This makes the op amp "frequency sensitive." That is, the op amp will produce a higher output signal at some frequencies than at others, due to the changing nature of capacitive reactance (X_C) at different frequencies.

Of course, somewhat similar effects can be achieved using only RC filters. However, the op amp gives the filter much steeper response curves. When used with one or more op amps, the response curve of an RC filter can be made as sharp as that of a good LC filter. The active filter is particularly handy at low frequencies. At low frequencies, LC filters are often impractical because of the large size of the inductor required.

Up to now, we have considered only resistors as the feedback and input networks. The gain of any stage is determined by the size of the input resistor (R_{IN}) and the feedback resistor (R_F). When capacitors are added to the input or feedback circuit, we must consider the resulting impedance as shown in Figure 27-15. Here the gain of the stage is determined by:

$$A_V = -\frac{Z_F}{Z_{IN}}$$

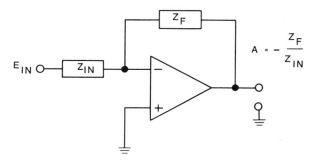

Figure 27-15 When RC networks are used in the input or feedback circuits, the gain is determined by the ratio of Z_{IN} to Z_F.

Since the impedance of an RC network varies with frequency, the gain of the stage also changes with frequency changes.

A simple high-pass filter is shown in Figure 27-16A. The input impedance (Z_{IN}) consists of an RC network. At low frequencies, the capacitive reactances of C_1 and C_2 are quite high. Therefore, Z_{IN} is high at low frequencies. Z_F is equal to R_F and is constant at all frequencies. Consequently, when Z_{IN} is high, the gain of the stage is low. Thus, E_{OUT} is low at low frequencies.

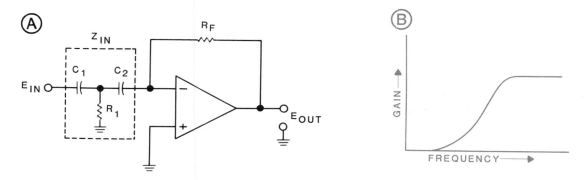

Figure 27-16 High-pass filter and its response curve.

As the frequency increases, the X_C of C_1 and C_2 decreases, which decreases Z_{IN}. As Z_{IN} decreases, the gain of the stage and E_{OUT} increase. The response curve of this circuit is shown in Figure 27-16B.

A simple **low-pass filter** is shown in Figure 27-17A. At low frequencies, X_{C1} is high. This prevents a high amount of current from flowing through C_1 to ground, and allows most of the input current to flow through the feedback component. Thus, the gain of the stage is high at low frequencies.

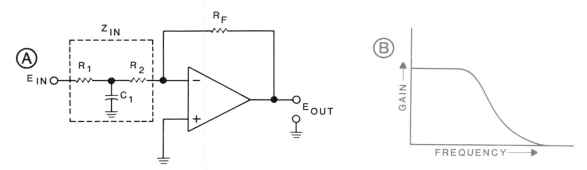

Figure 27-17 Low-pass filter and its response curve.

As the frequency increases, X_{C1} decreases. This allows more and more current to flow through the capacitor to ground, instead of through the feedback component. Consequently, the gain of the stage is low at high frequencies. The response curve is shown in Figure 27-17B.

OPERATIONAL AMPLIFIERS

Difference Amplifier

As you have seen, the input stage in the op amp is a differential amplifier. However, until now, we have not used the op amp as a difference amplifier. Let's see how the op amp behaves in this mode of operation.

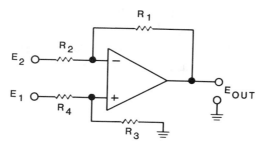

Figure 27-18 The difference amplifier.

Figure 27-18 shows the basic difference amplifier. The gain of the stage is determined by the size of the four resistors. When all four resistors are equal, E_{OUT} will be the difference between E_1 and E_2. That is:

$$E_{OUT} = E_1 - E_2$$

When used in this way, the circuit is called a **subtracter** because it literally subtracts the value of E_2 from the value of E_1.

If the ratio of the resistors is changed, the stage can provide amplification. In this case, the difference between E_1 and E_2 is amplified. However, any common-mode signal which appears at both E_1 and E_2 will still be rejected.

Normally, the ratio of R_1 to R_2 is made equal to the ratio of R_3 to R_4. That is:

$$\frac{R_1}{R_2} = \frac{R_3}{R_4}$$

In this case, E_{OUT} is determined by:

$$E_{OUT} = \frac{R_1}{R_2}(E_1 - E_2)$$

Let's assume that R_1 and R_3 are 10 kilohm resistors, and that R_2 and R_4 are 1 kilohm resistors. This would give the stage a gain of 10.

Other Applications

There are many other applications of the operational amplifier. It can be used in a variety of oscillator circuits to produce sine or square waves. It can be used as an integrator to produce linear ramps and sawtooth waveforms. It can also produce a number of other waveshapes. In analog computers, op amps are used to perform multiplication, to extract square roots, and to perform logarithmic functions. Some additional applications will be discussed later. From these examples you can see how valuable the op amp is to the electronic industry. Its use greatly reduces the size of equipment, and in many cases the number of required discrete components, which results in lower power requirements and cost.

OPERATIONAL AMPLIFIERS

Chapter Self-Test

1. Name the three stages within a typical operational amplifier.

2. What are the characteristic input and output impedances for an op amp?

3. Name the two inputs to an op amp.

4. An op amp has a gain of 1000. What is the other method of stating this gain value?

5. What type of feedback is used when an op amp is in the closed-loop operation?

6. A closed-loop op amp is in the inverting configuration. The feedback resistor is 10 kilohms, and the input resistor is 100 ohms. What is the gain of this stage?

7. A 2 millivolt DC voltage is applied to E_{IN} in the circuit described in Question #6. Answer the following questions:

    a. How much voltage is developed by the input resistor?
    b. How much current flows through the input resistor?
    c. How much current flows through the feedback resistor?
    d. What is the voltage at the output?

8. What is the formula for calculating the gain of the noninverting amplifier?

9. Name the differences in circuit configuration and circuit operation between the inverting and noninverting amplifiers.

10. An operational amplifier is operating with the output shorted to the "−" input and the input signal applied to the "+" input of the op amp. What is this circuit known as? What is its primary function? What is its gain?

11. The gain for any op amp in the open-loop configuration is very high. Unfortunately, the range of frequencies (bandwidth) is very small; usually in the very low AC frequency range. How can the bandwidth be improved, and what effect will this improvement have on the gain?

12. How does the term unity gain frequency apply to an op amp's frequency response characteristics?

13. Define gain-bandwidth product.

ELECTRONIC CIRCUITS

14. What type of amplifier has multiple inputs to the inverting amplifier?

15. When all of the resistors in the circuit described in Question #14 are equal, what is the gain?

16. When the input resistors have different values in the circuit described in Question #15, what type of amplifier is created? How is its gain determined?

17. What does an op amp become when RC networks are used as the input and feedback networks? How is the gain determined for this type of amplifier?

18. When an op amp is connected as a difference amplifier, how is E_{OUT} determined?

19. What is another name for the difference amplifier?

20. List 5 different applications for the op amp, other than those detailed in the chapter.

OPERATIONAL AMPLIFIERS

Summary

The op amp is a multiple-stage amplifier that has a very high DC gain, very high input impedance, and very low output impedance.

Usually some of the gain is sacrificed by using heavy negative feedback to stabilize the circuit. This is called closed-loop operation and increases bandwidth (frequency response).

In an op amp when the inverting input is more positive, the output is a negative voltage. When the noninverting input is more positive, the output is a positive voltage.

The most common use of the op amp is as an amplifier. Both inverting and noninverting amplifiers are common in modern equipment.

In its simplest form, the inverting amplifier consists of the op amp and two resistors. The characteristics of the stage are determined mainly by the resistors.

In the inverting configuration R_{IN} connects the input signal to the "−" input of the op amp. The other resistor (R_F) connects the output voltage to the same point. The stage automatically adjusts its gain so that any change at the inverting input is only a few microvolts.

The gain of the inverting amplifier is determined by the ratio of R_F to R_{IN}. That is:

$$A_V = -\frac{R_F}{R_{IN}}$$

In the noninverting amplifier, the input is applied to the "+" input of the op amp. A voltage divider consisting of two resistors is connected between the output of the op amp and ground. The point between the two resistors is connected to the "−" input of the op amp. The noninverting input has the capability of providing the higher gain.

The gain of the noninverting amplifier is determined by the ratio of the two resistors in the voltage divider. The formula is:

$$A_V = \frac{R_F}{R_{IN}} + 1$$

An important characteristic of the op amp is its unity gain frequency. This is the frequency at which the gain drops to unity or 1, and is generally considered the highest frequency at which the op amp is useful. It is also referred to as the op amp's gain bandwidth product.

ELECTRONIC CIRCUITS

For any feedback arrangement, the bandwidth of the amplifier can be determined by dividing the unity-gain frequency by the gain of the stage. Stated another way, multiplying the gain times the bandwidth yields the unity-gain frequency. For that reason, this frequency is also called the gain-bandwidth product. General purpose op amps have a gain-bandwidth product of 1 MHz.

The op amp finds numerous applications in electronics. One popular circuit is the voltage follower. This circuit has the characteristics of a nearly ideal emitter-follower (common-collector configuration). It has an extremely high input impedance (usually in megohms), an extremely low output impedance, and a gain of 1.

Other applications include the summing amplifier (adder), the active filter, and the difference amplifier (subtracter). The summing amplifier can be used to add several different input signals together. By selecting proper resistor values, the inputs can be multiplied by a constant or each can be weighted as needed. When each is weighted, the amplifier is referred to as a scalar amplifier. The active filter allows simple RC networks to produce very sharp response curves. The difference amplifier can be used to subtract one input from another or to amplify differential input signals.

CHAPTER 28

Power Supplies

ELECTRONIC CIRCUITS

Contents

Introduction 751

Chapter Objectives 752

Rectifier Circuits 753

Power Supply Filters 763

Voltage Regulation 769

Series Voltage Regulation 772

Chapter Self-Test 780

Summary 782

POWER SUPPLIES

Introduction

It comes as no surprise that nearly all electronic circuits require a voltage and current source. This power requirement is usually for a direct current source. The most well known direct current source is the battery, but with the exception of small transistorized radios and other light duty circuits, the battery is an impractical choice. However, we all have access to 115 volt, 60 cycle power from the wall outlets of our homes and businesses. Thus, what is needed is a circuit to convert this 115 volt alternating current to the required direct current. The circuit that performs this function is called a **power supply.**

A power supply provides a steady DC voltage under varying load conditions. In other words, any circuit connected to a power supply may have a change in resistance or current flow, but the power supply's voltage will remain steady. In addition, this voltage remains steady although the input to the power supply is an AC signal. We can therefore refine our definition to say that **a power supply provides a steady DC voltage under varying input and load conditions.**

The three basic circuits within a power supply are the **rectifier,** the **filter** and the **voltage regulator.** The purpose and operation of each of these circuits are explained in this chapter. There are a number of ways to perform each of these functions, and the most popular are explored in this chapter.

ELECTRONIC CIRCUITS

Chapter Objectives

When you have completed this chapter, you will be able to:

1. Define the following terms: rectification, ripple, pulsating DC.
2. List the three main circuits within a power supply.
3. Identify the primary purpose of a rectifier.
4. Describe the operation and characteristics of the half-wave rectifier.
5. Describe the operation and characteristics of the full-wave rectifier.
6. Describe the operation and characteristics of the bridge rectifier.
7. Identify the purpose of an input transformer.
8. Identify the primary purpose of a filter.
9. Describe the operation of a basic capacitor filter.
10. Explain the advantages of RC and LC filtering over a single capacitor.
11. Identify the primary purpose of a regulator.
12. Describe the operation of a basic zener diode regulator.
13. Describe the operation of an emitter-follower regulator.
14. Describe the operation of the feedback regulator.
15. Identify the functions of an operational amplifier within a feedback regulator.
16. Describe how a regulator can be protected from short-circuit overload.

POWER SUPPLIES

Rectifier Circuits

Most types of electronic equipment get their power from the 115 VAC, 60 Hz power line. Unfortunately, most power requirements for this same equipment is a DC voltage source. In order to provide this DC voltage source, we must take our AC sine wave, and remove one alternation. This is the process of rectification.

The heart of the power supply is the rectifier circuit. It converts the AC sine wave to a pulsating DC voltage, which is the term used to describe a waveform with a positive or negative alternation, with the other alternation removed. The rectifier is the first step in producing the smooth DC voltage required by electronic circuits. In this section, we will consider three different types of rectifier circuits. They can be classified as half-wave rectifiers, full-wave rectifiers, and bridge rectifiers. Because it is the simplest, we will consider the half-wave rectifier first.

Half-Wave Rectifiers

The most basic form of the half-wave rectifier is shown in Figure 28-1A. The load which requires the direct current is represented by the resistor. A diode is placed in series with the load so that the current flows in one direction, but not in the other.

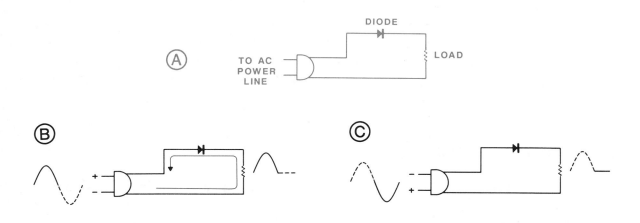

Figure 28-1 Basic half-wave rectifier circuit.

ELECTRONIC CIRCUITS

Figure 28-1B illustrates the operation of the circuit during the positive half-cycle of the sine wave from the power line. The anode of the diode is positive. Consequently, the diode conducts, allowing current to flow through the load as shown. Since the diode acts as a closed switch (forward-biased) during this time, the positive half-cycle is developed across the load.

Figure 28-1C shows the next half-cycle of the input sine wave. Here, the anode is negative and the diode cannot conduct (reverse-biased), acting like an open switch. Consequently, no current flows through the load and no voltage is developed across the load. As you can see, the AC sine wave is changed to a **pulsating DC voltage.** The output of any rectifier has only one polarity. In this circuit, the output voltage is positive.

If we were to discuss the output of this rectifier assuming a 115 VAC input, we would require a long and detailed explanation of average, effective, and peak AC values. Therefore, for the purposes of discussion we will assume a 40 volt peak-to-peak sine wave as the input signal for all rectifiers.

Refer again to Figure 28-1. Assuming the input is 40 volts peak-to-peak, the first alternation (20 volts peak) would be felt across the load, while the second alternation (also 20 volts peak) would not be felt at all. Therefore, a half-wave rectifier transfers one-half of the input sine wave to the load.

For a rectifier, this is good. If any more than half of the sine wave were transferred, we would have current flowing in both directions. The maximum amount of a sine wave that can be rectified to provide a single direction of current flow is one-half. This is also assuming that the 0.6 voltage drop of the forward-biased diode is so small that it can be ignored.

Rectifier With Transformer

Unfortunately, a rectifier is only capable of producing a pulsating DC signal in direct relation to the input sine wave. This becomes a disadvantage if circuits require a number of different voltages, and these voltages are not proportional to the input sine wave.

This problem can be overcome by connecting a transformer between the input sine wave and the rectifier. If higher voltages are required, a step-up transformer can be used. When lower voltages are required, as with most transistorized equipment, a step-down transformer is used.

For example, let's assume a circuit requires 10 volts DC to operate. Our half-wave rectifier, directly connected to 40 volts peak-to-peak, is only capable of producing a 20 volt pulsating DC signal. However, if we place a 2:1 step-down transformer between our input signal and the rectifier, as shown in Figure 28-2, our original 20 volt peak is now stepped-down to a 10 volt peak, which is developed by the load.

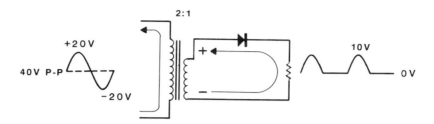

Figure 28-2 Half-wave rectifier with transformer.

In addition to allowing different voltages to be developed, a transformer provides another important function. If a transformer were not used, the rectifier, when connected to an AC power line, would connect ground *and* the chassis of the equipment to one side of AC. Since a two-prong plug can be connected two ways, we could never know which side of AC is connected to the chassis. This presents a dangerous shock hazard.

A transformer prevents this from happening. AC power is connected to the primary winding, and chassis ground is connected to the secondary. The primary and secondary windings are sufficiently isolated from each other, thereby ensuring no potential shock hazard.

Output Polarity

A rectifier can produce either a negative or a positive output voltage. The output polarity depends on what point is connected to ground and which way the diode is connected.

ELECTRONIC CIRCUITS

In Figure 28-3A, current can flow only in the direction indicated. This produces a positive voltage at the top of the load with respect to ground. Thus, this rectifier produces a positive output voltage.

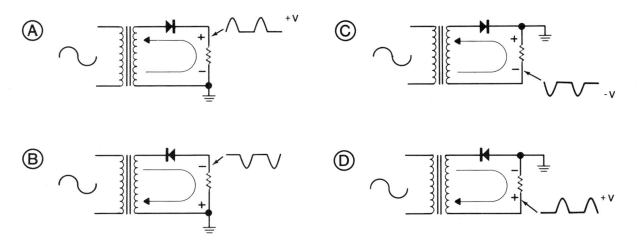

Figure 28-3 You can change the output polarity by turning the diode around or by moving ground.

If a negative voltage is required, there are two possible ways of accomplishing this. First, the diode could be reversed, forcing current through the load to flow in the opposite direction. This is shown in Figure 28-3B.

Another way this could be accomplished is by placing ground at the opposite end. This is shown in Figure 28-3C. However, if both ground is moved *and* the diode is reversed, you return to a positive output voltage, as shown in Figure 28-3D.

Other Characteristics

There are a number of different aspects which must be examined when discussing any rectifier. The primary aspect is the output voltage which, for the half-wave rectifier, is one-half of the input voltage. This is assuming a 1:1 transformer is used.

However, the portion of the input which causes current to flow through the rectifier is another important aspect. For the half-wave, the reverse-biased diode allows current to flow for only half of the input sine wave. It is this current aspect which gives the half-wave rectifier its name.

POWER SUPPLIES

Also, the **ripple frequency** of the rectifier should also be considered. Figure 28-4 shows how ripple frequency is determined. When one cycle of the input sine wave and the output signal are compared, we can see that each cycle occurs within the same time frame. In other words, for each cycle of input to a half-wave rectifier, one cycle of pulsating DC is produced. The frequency of this pulsating DC signal is known as the ripple frequency. If the input frequency were 100 Hz, the output would be 100 **rps,** or **ripples-per-second.**

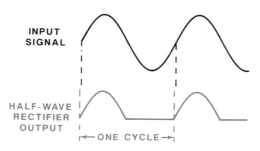

Figure 28-4   Determining ripple frequency.

We stated earlier that the output voltage of the half-wave rectifier is considered very good for a rectifier. However, current flows for only one-half on the input period, and the ripple frequency is equal to the input frequency. These aspects are not considered advantageous. You will see that other rectifiers produce current flow for the full input sine wave, which is preferred. When current flows for the full input, an output signal is developed which increases the ripple frequency. The higher the ripple frequency, the easier it is to filter the signal. Filtering, if you'll recall, is the second step in a power supply.

To summarize, the important aspects of the half-wave rectifier:

1. Output voltage one-half the input voltage (good).

2. Current flows for one-half the input signal (poor).

3. Ripple frequency is equal to the input frequency (poor).

Full-Wave Rectifier

The full-wave rectifier, as the name implies, allows current to flow for the full input signal, which allows us to increase ripple frequency.

The full-wave rectifier circuit is shown in Figure 28-5. It uses two diodes and a center-tapped transformer. When the center-tap is grounded, the voltages at the opposite ends of the secondary are 180 degrees out-of-phase with each other. Thus, when the voltage at point A swings positive with respect to ground, the voltage at point B swings negative.

Figure 28-5A shows that the voltage applied to the anode of D_1 is positive (forward-bias), while the voltage applied to the anode of D_2 is negative (reverse-bias). Obviously, only D_1 can conduct with the conditions set at this instant of time. As shown, current flows from the center-tap up through the load, and D_1 to the positive potential at the top of the secondary. When D_1 conducts, it acts like a closed switch so that the positive half-cycle is developed across the load.

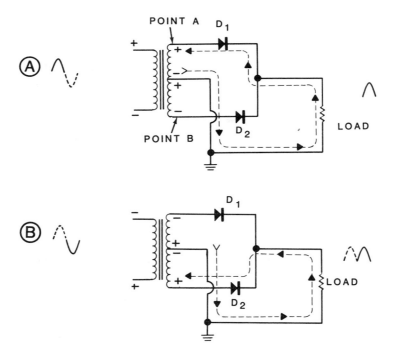

Figure 28-5 The full-wave rectifier.

POWER SUPPLIES

Figure 28-5B shows what happens during the next half-cycle, when the polarity of the voltage reverses. The anode of D_2 swings positive while the anode of D_1 swings negative. Therefore, D_1 cuts off and D_2 conducts. As shown, current flows from the center tap, through the load and D_2 to the positive voltage at the bottom of the secondary.

Notice that current flows through the load during both half-cycles. This is the prime advantage of the full-wave rectifier over the half-wave rectifier.

Figure 28-6 shows how the input and output waveforms for the full-wave rectifier, as well as how the magnitude of the output voltage is determined. For simplicity, assume that the transformer has a 1:1 turns ratio. If our 40 volt peak-to-peak signal is applied to the primary, each alternation applies 20 volts. This 20 volts is then directly felt at the secondary. However, the center-tap splits this 20 volts, causing 10 volts to be felt between the center-tap and the ends of the transformer. This 10 volts is then developed by the load. Therefore, a full-wave rectifier can only transfer a maximum of **one-fourth** the input voltage. When compared to the half-wave rectifier, the output voltage of the full-wave is considered poor.

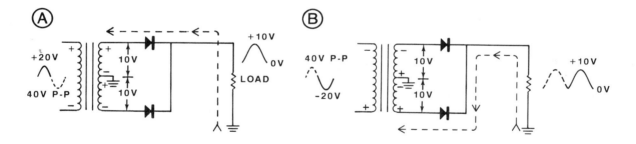

Figure 28-6   Voltages in a center-tapped transformer.

We have shown you to this point that a full-wave rectifier provides current flow for the full input signal, and that the output voltage (because of the center-tap) is one-fourth the input voltage. The last aspect of the full-wave is its ripple frequency. If our input signal has a frequency of 250 kHz, each cycle of input would be 4 microseconds long. This is illustrated in Figure 28-7.

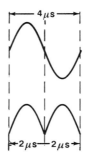

Figure 28-7   The ripple frequency for a full-wave rectifier is double the input frequency.

In that same 4 microseconds, the output of the full-wave rectifier has completed *two complete* pulsating DC signals. The time for one cycle of output, therefore, must be 2 microseconds. Since time and frequency are inversely proportional, if you'll recall from Chapter 11, less time means a higher frequency. Since time is cut in half, frequency must double. The output frequency, therefore, must be 500 kHz. The ripple frequency for a full-wave rectifier is double the input frequency.

Let's review the important aspects of the full-wave rectifier:

1. The output voltage is one-fourth the input voltage (poor).

2. Current flows for the full portion of the input signal (good).

3. The ripple frequency is double the input frequency (good).

When the half-wave and full-wave rectifiers are compared, we can see that each have good and poor aspects. An ideal rectifier would combine the good voltage output of a half-wave rectifier, and the good current and ripple frequency of a full-wave rectifier. Fortunately, such a rectifier exists.

Bridge Rectifier

An example of a bridge rectifier is shown in Figure 28-8. Notice that there is no center-tap on the secondary winding of the transformer. Assuming a 1:1 turns ratio, this allows half of the input sine wave to be felt across the load. The unique configuration of the diodes, when broken down, is two full-wave rectifiers. Current will therefore flow for the entire input sine wave. This also allows the load to develop a voltage for the entire input signal, which doubles the ripple frequency. The bridge rectifier combines all the preferred aspects of a rectifier.

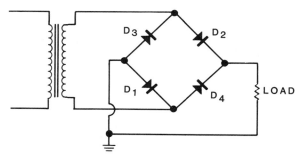

Figure 28-8   The bridge rectifier.

Figure 28-9A shows how current flows through the full-wave bridge rectifier during the positive half-cycle of the sine wave. Current flows from the bottom of the secondary through D_1, the load, and D_2 to the positive voltage at the top of the transformer's secondary. With D_1 and D_2 acting as closed switches, the entire secondary voltage is developed across the load.

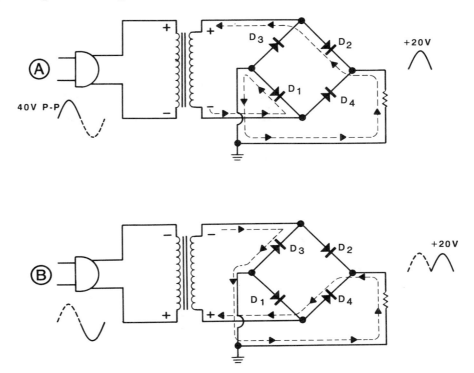

Figure 28-9 Current flow through the bridge rectifier.

On the next half-cycle, the polarities reverse as shown in Figure 28-9B. The top of the secondary is now negative and the bottom is positive. Current flows from the top of the secondary through D_3, the load, and D_4 to the bottom of the secondary. Notice that current flow through the load is always in the same direction. With D_3 and D_4 acting as closed switches, the entire secondary voltage is again developed across the load.

Notice how for both alternations of the input, the bridge rectifier develops an output voltage equal to the peak value of that alternation. Also, since the output waveform is a full-wave output, the ripple frequency is double the input frequency. Because the bridge rectifier combines all of the preferred aspects of a rectifier (high voltage, 100% current flow, and a high ripple frequency) it is the most popular type of rectifier used.

POWER SUPPLIES

Power Supply Filters

As we have seen, the output of the rectifier is a pulsating DC voltage. Such a DC voltage is unsuitable for most electronic applications. Electronic circuits usually require a very smooth, constant supply voltage. For this reason, virtually all power supplies have the rectifier circuit followed immediately by a **filter.** The purpose of the filter is to convert the pulsating DC provided by the rectifier into the smooth DC voltage required by electronic circuits.

This filter differs from the filters you learned about in Chapters 14-16, and Chapter 27. In general, a filter is something which prevents unwanted items from passing a certain point, while allowing desired items to pass. A coffee filter allows the water to pass through to the pot, while keeping the coffee grounds from passing. An electronic frequency-sensitive filter allows certain frequencies to pass, while rejecting (attenuating) undesirable frequencies. A power supply filter removes those portions of the pulsating DC signal which have drastic voltage changes over a short period of time.

The Capacitor as a Filter

In its simplest form, the power supply filter may be nothing more than a capacitor connected across the output of the rectifier. This arrangement is shown in Figure 28-10A. The addition of the capacitor modifies the operation of the circuit.

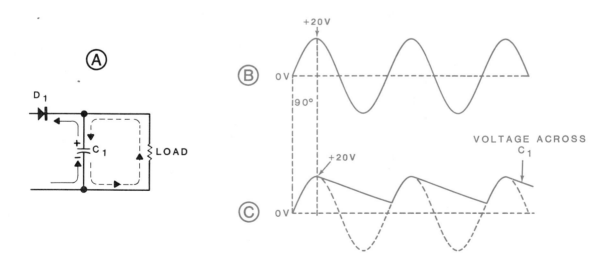

Figure 28-10 Adding a capacitor helps smooth out the ripple.

ELECTRONIC CIRCUITS

The input voltage waveform is shown in Figure 28-10B. Notice that the peak value is 20 volts. Let's examine the action starting with the point at which the sine wave initially swings positive.

When the anode of D_1 swings positive, D_1 conducts, allowing current to flow through the load. Simultaneously, the capacitor charges to the polarity shown through the small conducting resistance of the diode (R_F). Since there is very little resistance in the charge path, C_1 charges immediately to the value of the input voltage. C_1 charges for the first one-quarter cycle. After 90 degrees of the input sine wave, C_1 is charged to the peak value of 20 volts.

When the input sine wave begins to drop off, C_1 cannot charge to a higher value, and begins to discharge. The path for discharge is through the load.

The time of discharge for a capacitor is determined by the RC time constant of the discharge path. This includes the size of the capacitor and the amount of load resistance. The size of a filtering capacitor is very large, as is the amount of load resistance. This makes the RC time constant and the discharge time very high. In fact, the discharge occurs at a much slower rate than the drop-off of the input voltage.

The rectifying diode feels this discharge voltage on its cathode. Because the discharge is so much slower than the drop-off of the input, a point is quickly reached where the anode of the diode (at the input voltage) is more negative than the cathode (the discharge voltage). The diode becomes reverse-biased, and cuts off. This normally occurs almost immediately after 90 degrees.

Once the diode cuts off, the load resistance feels only that voltage of the discharging capacitor. Throughout the rest of the input signal, the capacitor is discharging. Because the discharge time is so high, the capacitor never completely discharges. The voltage felt at the load drops off very slowly, as shown in Figure 28-10C. Eventually, another positive alternation is felt at the input, and the voltage at the anode of the diode increases until it becomes forward-biased again. Then the capacitor recharges, and the entire process starts over.

Figure 28-11 compares the outputs of an unfiltered rectifier, and filtered outputs using both a small capacitor and a large capacitor. In Figure 28-11A, the unfiltered output is a complete pulsating DC signal. This, as we've stated before, is unsuitable. The shaded area represents that portion of the output that the diode is conducting.

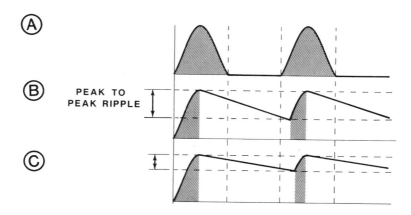

Figure 28-11   A larger capacitor decreases peak-to-peak ripple.

In Figure 28-11B, a relatively small capacitor is used as the filtering component. Notice that the filtering effect we described occurs, and the output voltage is improved. Notice also how the maximum and minimum values of the filtered output can be measured as a peak-to-peak value. This allows us to compare two different filtered outputs.

In Figure 28-11C, a larger capacitor is used as the filtering component. When a larger capacitor is used, the RC time constant, and therefore the discharge time, increases. This means that a large capacitor cannot discharge as much voltage until it is recharged again. The output voltage remains closer to a steady output than when a smaller capacitor is used. This is indicated by the smaller peak-to-peak value. Generally speaking, the larger the capacitor, the better the filtering component.

Full-Wave Supply with Capacitor Filter

As we discussed rectifiers, you'll recall that the ripple frequency of the pulsating DC signal is double the input frequency for both the full-wave and full-wave bridge rectifiers. We also stated that if the ripple frequency is higher, the pulsating DC signal is easier to filter. Let's examine how the filtering capacitor reacts to a full-wave output.

The action of the capacitor with a full-wave signal is the same as we described for the half-wave signal. The difference is the amount of time between when the capacitor begins to discharge, and when the input recharges it. Figure 28-12 compares the pulsating DC and filtered outputs for both a half-wave and full-wave signal. The same capacitor is assumed to be used for both circuits. Figure 28-12A shows the two signals for the half-wave rectifier. Again, the shaded portion represents the time the rectifying diode is conducting. The filtered output has a specific peak-to-peak value.

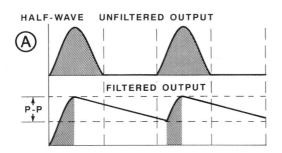 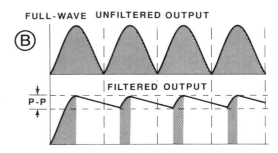

Figure 28-12   The output of the full-wave or bridge rectifier provides better filtering.

Figure 28-12B shows the unfiltered and filtered output of a full-wave (or bridge) rectifier. Because the peaks occur more frequently with the full-wave, the filtering capacitor cannot discharge much voltage until the next peak occurs, recharging it. This retains the filtered output closer to the peak voltage, and a smaller peak-to-peak ripple value is produced. When the filtering capacitor discharges and charges more often, better filtering usually occurs. In addition, using a higher value capacitor would also improve the filtered output.

RC Filters

It would seem that the best filtering can occur with an extremely large capacitor, whose RC time constant is so high that it barely discharges before it recharges again. While this is indeed true, a capacitor that large would cause a reverse-bias on the rectifying diode(s) which could easily damage them. We are therefore limited to the size of the capacitor we can use.

Figure 28-13 shows a full-wave rectifier with a resistor-capacitor (RC) filter connected across its output. An RC filter of this type does a much better job than the single capacitor filter. Let's see how this filter works.

C_1 performs exactly the same function that it did in the single capacitor filter. It is used to reduce the percent of ripple to a relatively low value. Thus, the voltage across C_1 might be a ripple voltage of 10 volts peak-to-peak. This voltage is passed on to the R_1-C_2 network, which reduces the ripple even further.

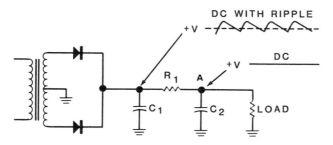

Figure 28-13   Full-wave rectifier with RC filter.

R_1, since it is in series, decreases the voltage somewhat. However, this can be easily compensated by starting with a voltage slightly higher than the required output voltage. C_2 performs a basic function of a capacitor—opposing a change in voltage. As the ripple voltage reaches point A, C_2 will provide a low impedance path for the alternating current. C_2 alternately charges and discharges, providing a path for the alternating current to ground. This maintains the voltage at point A to only the DC component of the waveform, as shown in Figure 28-13.

In extreme cases, where the ripple must be held to an absolute minimum, a second state of RC filtering can be added. In practice, the second stage is rarely required. The RC filter is extremely popular because smaller capacitors can be used with good results.

LC Filters

The next step in filters is the inductor-capacitor (LC) filter. A common type of LC filter is shown in Figure 28-14. C_1 performs the same functions as discussed earlier. It reduces the ripple to a relatively low level. L_1 and C_2 form an LC filter which reduces the ripple even further. The inductor appears as a short to DC while adding opposition to the AC component of the waveform.

L_1 is a special iron-core inductor called a **choke**. It has an extremely high inductance value and, therefore, a high value of X_L. That is, it offers a high opposition to the ripple frequency. At the same time, C_2 offers a very low opposition to the AC ripple. L_1 and C_2 form an AC voltage divider. Because the reactance of L_1 is much higher than the reactance of C_2, most of the ripple voltage is dropped across L_1. Only a slight trace of the ripple appears across C_2 and the load.

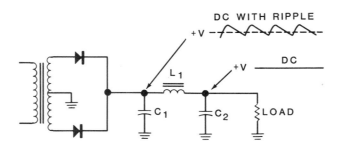

Figure 28-14   Full-wave rectifier using an LC filter.

While the L_1-C_2 network greatly attenuates the AC ripple, it has little effect on the DC. Recall that an inductor offers no reactance to DC. The only opposition to current flow is the resistance of the wire in the choke. Generally, this resistance is very low and the DC voltage drop across the coil is negligible. Thus, the LC filter overcomes the disadvantages of the RC filter.

The LC filter provides good filter action over a wide range of currents. The capacitor filters best when the load is drawing very little current. In this case, the capacitor discharges very slowly and the output voltage remains almost constant. On the other hand, the inductor filters best when the current is highest. The complementary nature of these components ensures good filtering over a range of currents.

The LC filter has two disadvantages. First, it is more expensive than the RC filter because an iron-core choke costs more than a resistor. The second disadvantage is size. The iron-core choke is bulky and heavy. Thus, the LC filter may be unsuitable for many applications.

Power Supplies

Voltage Regulation

To this point, we have taken one alternation out from our alternating signal, and have smoothed (somewhat) the remaining polarity. We are fairly close to achieving our goal of a steady, unchanging DC voltage.

We stated that this DC voltage must remain constant under varying input and load conditions. Let's determine what might have caused these changes. First, power supplies are connected to the standard AC line voltage. This *should* have a constant value of 115 VAC. However, the line voltage can vary from about 105 VAC to 125 VAC. This means that the peak AC voltage to which the rectifier responds can vary from about 148 VAC to 177 VAC. This alone causes almost a 20 percent change in the DC output voltage.

Second, while most filters are very good at their task, it is impossible to achieve a steady, unchanging value. A good filter has a ripple peak-to-peak value of approximately 0.5 volts. While this is very close to a steady DC value, it is not perfect. Both the varying line voltage and the ripple voltage are considered to be the "varying input" condition.

Varying load conditions is final, and perhaps the largest consideration for a power supply. Remember, we are discussing the very circuit which supplies the source voltage to all of the circuits we've described in previous chapters. All of our filters and amplifiers have source voltages which come from a power supply. These circuits have characteristics which alter their resistance, and also their current.

A power supply, in addition to supplying a steady DC voltage, also supplies current to these circuits. If the current requirements within the load change, current flow within the power supply changes. This changing current changes the voltage drops of the resistors in the filter, which could possibly change the output voltage. Therefore, a power supply must furnish a steady DC voltage, while at the same time furnish a varying current.

The last portion of our power supply is the regulator, sometimes referred to as an electronic voltage regulator (EVR).

The Zener Regulator

Chapter 20 introduced you to the zener diode. This is a component perfectly suited for voltage regulation. Let's briefly review the basic operation and construction of the zener diode.

A normal diode, when reverse-biased, will act as an open switch. No current can flow through the diode, and the diode offers an infinite amount of resistance to the circuit. If enough reverse-bias potential is applied, however, the diode will conduct in the reverse-bias direction due to the action of the minority carriers within the junction. The normal range for this reverse-bias voltage is 20 to 45 volts.

A zener diode is created by injecting a large portion of minority carriers. With the number of minority carriers increased, a substantially smaller voltage is required to initiate reverse current. Actually, the zener diode is designed to operate in the reverse-bias mode. The **breakdown voltage** for a zener is that voltage which allows reverse current to flow. If a zener diode is reversed-biased below this breakdown voltage, it behaves like a normal reverse-biased diode, preventing current flow and offering maximum resistance. Once this breakdown voltage is exceeded, reverse current flows, and the voltage across the zener is equal to the breakdown voltage. Any further increase in voltage increases current flow through the zener, but *the voltage drop of the zener diode remains at the breakdown voltage.* The breakdown voltage of a zener diode is determined during the manufacturing process. Zeners are commonly available with breakdown voltages that extend from approximately 2 volts to over 200 volts.

The ability of a zener to retain its voltage drop with varying current flow is what makes the zener diode so suitable for a regulator. Let's examine exactly how a zener diode operates as a regulator.

Figure 28-15A shows a very simple zener regulator circuit. A ripple voltage which varies above and below +10 volts is applied. A zener diode with a breakdown voltage of 6 volts is connected so that it is constantly reverse-biased. Since the voltage applied exceeds the breakdown voltage, a constant 6 volts is developed by the zener, leaving a constant 4 volts to be developed by R_S. As the input voltage varies above and below +10 volts, zener current will also vary. However, since the voltage drop of the zener is constant, the difference must be dropped by R_S. The voltage drop of R_S changes in relation to the change in input voltage, and the output voltage remains constant.

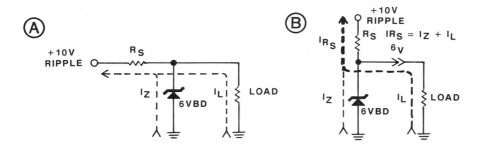

Figure 28-15 Zener diode voltage regulator.

In addition to compensating for a varying input voltage, this zener regulator must also compensate for varying load currents. We'll use a different drawing of the same circuit, shown in Figure 28-15B.

Notice how the zener diode and R_S form a voltage divider. We stated earlier that R_S must compensate for any change in the input voltage, because the voltage drop of the zener remains constant. If the input voltage had no ripple, however, the voltage drop of R_S would also remain constant. If no load were attached, the only current that would flow through R_S would be zener current (I_Z).

Based on this premise, let's attach the load. Now R_S feels an additional current from the load (I_L). Normally, any resistor whose current flow increases, also increases its voltage drop. This cannot happen however, because the zener diode has a steady voltage, which keeps the voltage drop of R_S steady, also. (Again, we're assuming no input ripple) In order to keep the voltage drop of R_S steady, current through R_S must remain the same. When the load current is attached, zener current will drop by the same amount to keep current through R_S constant.

Any change in load current, therefore, is compensated for in the same fashion. If the load current increases, for example, current through R_S attempts to increase. However, to maintain a steady voltage, I_Z decreases by the same amount, keeping current through R_S steady. The additional current required by the load is supplied, and the zener compensates by decreasing its current by the same amount.

Unfortunately, this type of regulator is limited to the abilities of the zener diode. Large changes in load current cannot be compensated for in this fashion, since the zener diode is incapable of responding to that large a change. The next section discusses more complex regulators which are able to compensate for any amount of input or load change.

Series Voltage Regulation

There are two basic types of voltage regulators: **series regulators** and **shunt regulators**. The type of regulator required is determined by the location of the primary regulating component. The zener diode regulator we discussed earlier is a shunt regulator, because the zener diode is normally connected in parallel (shunt) with the load. We will discuss more complex forms of shunt regulators later. For now, let's examine the operation of series regulators.

In a series regulator, the regulating component is placed in series with the load. The resistance of the regulating component is automatically adjusted so that the voltage dropped across the load remains constant.

Figure 28-16 shows how this works. E_{IN} represents the unregulated DC input from the rectifier-filter circuit. R_{INT} represents the internal resistance of the rectifier-filter. R_{VAR} represents the control device. Since the control's resistance changes as conditions change, it is shown as a variable resistor. In reality, it is usually a silicon transistor. E_{OUT} is the regulated voltage across the load.

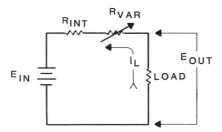

Figure 28-16 Basic series regulator.

Let's see how this voltage is held constant when the input voltage or the load current changes. Let's assume that the input voltage increases. When E_{IN} increases, E_{OUT} attempts to increase. However, if the value of R_{VAR} increases in proportion, its voltage drop also increases, which decreases E_{OUT}. Thus, E_{OUT} remains fairly constant regardless of changes in E_{IN}.

POWER SUPPLIES

The series regulator can also compensate for changes in load current. If the load current increases, R_{INT} and R_{VAR} will drop more voltage, leaving less voltage to develop across the load. That is, E_{OUT} tends to decrease. However, when the resistance of R_{VAR} is made to decrease, just as the current increases, then the voltage dropped across R_{INT} and R_{VAR} will remain constant. This allows E_{OUT} to remain constant even though the load current changed.

As mentioned, R_{VAR} is not a resistor, but a transistor connected in series with the load current. By changing the base current, the transistor can be made to conduct more or less. Stated another way, the resistance of the transistor can be changed by varying the base current (bias). Additional components are used so that the circuit is self-adjusting. This is called **dependent bias**. In this case, the bias acts on the changes and varies the transistors conduction rate. This type of bias can either aid or oppose circuit changes. This allows the resistance of the transistor to change automatically to compensate for changes in E_{IN} or load current.

The Emitter Follower Regulator

The simplest series regulator is the emitter follower type. The basic circuit consists of a zener diode, a transistor, and a resistor connected together as shown in Figure 28-17. The input to the circuit is an unregulated DC voltage. The output is a regulated DC voltage that is lower in value.

Q_1 acts as an emitter follower. The load is connected between the emitter of Q_1 and ground. The voltage on the base of Q_1 is set by the zener diode. Thus, the output voltage is equal to the zener voltage (E_Z minus the small voltage drop across the base-emitter junction (V_{BE}) of Q_1. For example, if E_Z is 10 volts and V_{BE} is 0.7 volts, then the output voltage (E_Z) will remain at approximately 9.3 volts over a wide range of load and input voltage variations. Also, V_{BE} does not change very much from its 0.7 volt value.

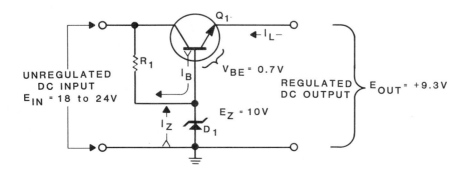

Figure 28-17 Emitter-follower regulator.

ELECTRONIC CIRCUITS

To see how the circuit works, let's assume that the unregulated input voltage changes. When E_{IN} increases, E_{OUT} attempts to increase. However, Q_1 is an NPN transistor, and its base voltage is set by the zener to +10 volts. If its emitter swings even slightly more positive, the amount of forward-bias decreases. This accomplishes two things: first, the amount of emitter and base current decreases; second, the resistance across the transistor increases. This means Q_1 conducts less, and acts as a higher value resistor between E_{IN} and E_{OUT}. Consequently, most of the increase in E_{IN} is dropped across Q_1. Only a very slight increase in E_{OUT} occurs.

As you can see, any attempt by the output voltage (at the emitter of Q_1) to change starts this circuit into regulation. This output (emitter) voltage performs the same function when load current changes. Let's assume that the load current increases. This draws more current flow through Q_1. Since we haven't yet changed the bias on Q_1, its voltage drop will increase, which tries to decrease the regulated output voltage. If the emitter goes even slightly negative, forward-bias increases. This allows more current to flow through Q_1 to the load, while at the same time decreasing the resistance of Q_1. With less resistance, Q_1's voltage drop decreases, maintaining a constant output voltage.

The main disadvantage in the emitter-follower regulator is that it cannot handle very high currents. The regulating component Q_1 would conduct very hard, and get very hot. This type of regulator works well when current requirements are low.

The Feedback Regulator

One of the most popular types of voltage regulating circuits is the **feedback regulator.** This circuit monitors the output voltage, and generates a control signal when the output voltage changes. This signal controls the conduction of the regulating component (a transistor) through which the load current passes. The bias on the transistor is adjusted so that the original change in the output voltage is canceled.

Block Diagram

The block diagram of a feedback regulator is shown in Figure 28-18. It consists of five basic circuits. An unregulated DC voltage is applied on the left. A somewhat lower, but regulated DC voltage appears across the output terminals on the right.

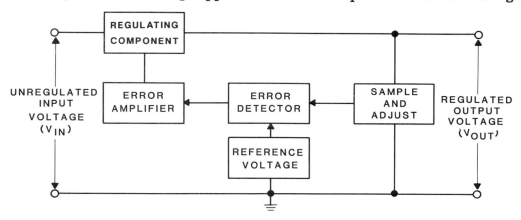

Figure 28-18 Block diagram of a feedback regulator.

A **sampling circuit** is connected across the output terminals. Generally, this is nothing more than a voltage divider. A voltage off this divider is applied to an **error detector**. Any change in the output voltage results in a proportional change to the voltage applied to the error detector. The other input to the error detector is a **DC reference voltage**. Normally, a zener diode is used to produce the reference voltage and to hold it constant.

The error detector compares the sampled voltage with the reference voltage. If there is a difference, the error detector applies a proportional voltage to the **error amplifier**. As the name implies, the error amplifier amplifies the error voltage. The output of the error amplifier is connected to the regulating component, which conducts more or less, depending upon the initial change in output voltage.

ELECTRONIC CIRCUITS

Basic Circuit

A simple feedback voltage regulator is shown in Figure 28-19. The sampling voltage divider is comprised of R_1, R_2, and R_3. The base of Q_1 is tapped directly into R_2, to receive any change in the output voltage. Q_1 performs as both the error detector and the error amplifier. It compares any change to the reference voltage, provided by zener diode D_1. It then amplifies any difference, and applies this signal to the base of Q_2, the regulating component. Since this voltage is applied to the base, any amplified error voltage alters the amount of forward-bias for Q_2, and adjusts the current flow.

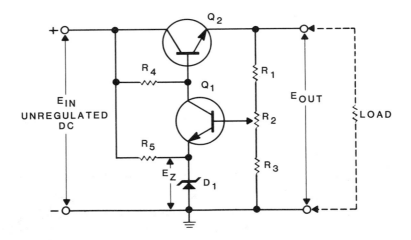

Figure 28-19 Feedback voltage regulator.

To see how the circuit works, let's assume that E_{OUT} attempts to increase. This can occur because of an increasing input voltage, or if load current decreases. For our purposes, we'll say the load current decreases.

When E_{OUT} increases, E_{R2} increases in proportion. Q_1 detects this change, and compares it to the reference voltage. Any change in the base voltage of Q_1 causes a change in forward-bias. An increase in base voltage increases forward-bias, causing the transistor to conduct harder. As Q_1 forces more current through R_4, the voltage at the collector of Q_1 and the base of Q_2 decreases. This decreases the forward-bias on Q_2.

When the forward-bias of the regulating component decreases, it adjusts to allow the proper amount of load current to flow, and the increased resistance increases its voltage drop. The output voltage therefore, drops. This all happens so fast that as soon as the output attempts to increase, the regulator compensates for the change in current, and adjusts to maintain a constant output voltage.

An important feature of this regulator is the ease with which E_{OUT} is adjusted. Variable resistor R_2 allows you to adjust E_{OUT} over a wide range. For example, to increase the value of the output voltage, you simply move the arm of R_2 down. This increases the voltage to the base of Q_1. Since the emitter voltage is constant, the forward bias is decreased and Q_1 conducts less. This causes the collector voltage of Q_1 and the base voltage of Q_2 to increase. This increases the forward-bias on Q_2, decreasing its voltage drop, and increasing the output voltage.

This also causes more current to flow through the load, but that is expected. When you increase the source voltage to a circuit (which is precisely what we're doing), more current flows in the circuit. The voltage adjust is a handy feature. It provides an easy way of compensating for component tolerances. It also allows us to periodically readjust the output voltage to compensate for component aging. This simple supply can be used to drive loads that require a variety of voltages.

The Op Amp in a Feedback Regulator

The feedback regulator is suitable for many applications. It is also fairly sensitive. The circuit can be made more sensitive by increasing the gain of the error amplifier. An easy way to do this is to substitute an op amp for Q_1. An operational amplifier makes an ideal error detector and amplifier. Its high gain allows it to detect the most minute change in the output voltage. Also, the differential inputs give the op amp an automatic error sensing capability.

Figure 28-20 shows this regulator circuit. Notice that the noninverting input of the op amp is connected to the reference voltage. The inverting input is connected to the sampling network, labeled R_A, R_B and R_C.

ELECTRONIC CIRCUITS

Notice that the output of the op amp is fed back to the inverting input through Q_1, R_A, and R_B. This allows the op amp to function in the noninverting configuration. The noninverting configuration, if you'll recall from the previous chapter, amplifies any difference between the input voltage and the feedback voltage. For this circuit, the input voltage is E_Z, and the feedback voltage is felt at the inverting input. From here, the circuit behaves as our previous circuit.

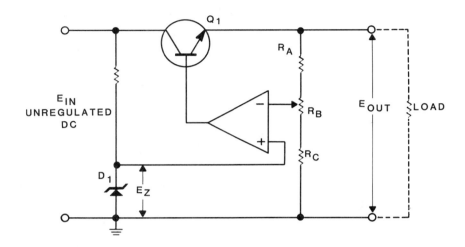

Figure 28-20 Feedback regulator with operational amplifier.

Short Circuit Protection

One disadvantage of a series regulator is that the regulating component is in series with the load. The load resistance may consist of dozens of circuits all connected in parallel. If a short develops in any one of these circuits, an extremely large current can flow through the regulating component. This transistor can easily be destroyed by the heat caused by the large overload current. One way to prevent this is to use a circuit which instantly senses overloads and automatically limits the current to a safe value.

POWER SUPPLIES

A series regulator with a current limiting circuit is shown in Figure 28-21. Notice that the circuit is identical to the one shown earlier in Figure 28-19, except that Q_3 and R_6 have been added. These two components form the automatic current limiting circuit.

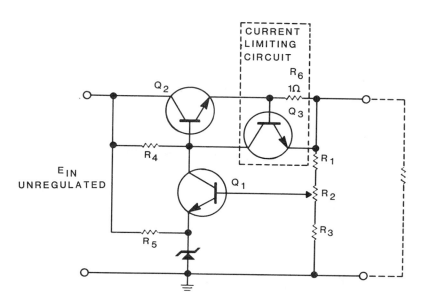

Figure 28-21 Regulator with current limiting.

Q_3 conducts when the voltage drop of R_6 is high enough to forward-bias the emitter junction. Recall that a typical transistor requires a forward-bias of approximately 0.6 volts. R_6, with an ohmic value of 1 ohm, will develop 0.6 volts when the current is equal to:

$$I_{R6} = \frac{E_{R6}}{R_6} = \frac{0.6V}{1\Omega} = 0.6 \text{ A or } 600 \text{ mA}$$

As long as the current through R_6 (the load current) is below 600 milliamps, R_6 does not develop enough voltage to forward-bias Q_3. However, when current attempts to rise above 600 milliamps, the voltage drop of R_6 rises to the 0.6 volts required to forward-bias Q_3. Notice that the path for Q_3's collector current is through R_4 to V_{CC}. When E_{R6} reaches 0.6 volts, Q_3 conducts through R_4, and R_4's voltage drop increases. This decreases the voltage on the base of Q_2, and Q_2 conducts less. Q_3 ensures that the current through the power supply cannot rise above 600 milliamps.

By using a larger resistance for R_6, you can limit maximum current to a lower value. For example, a 10 ohm resistor develops a voltage drop of 0.6 volts with 60 milliamps of current flow.

ELECTRONIC CIRCUITS

Chapter Self-Test

1. What is the purpose of a power supply?

2. What are the three main parts of a power supply?

3. What is the primary purpose of a rectifier?

4. For the half-wave, full-wave, and bridge rectifiers, answer the following questions:

    a. What is the output voltage? Is this good or poor voltage?
    b. For what portion of the input does current flow through the rectifier? Is this good or poor?
    c. What is the output frequency in comparison to the input frequency? Is this good or poor?

5. Name the two functions a transformer provides in a rectifier.

6. How can the output polarity of a rectifier be changed?

7. Which of the three rectifiers is used most often? Why?

8. What is the primary purpose of a filter?

9. When a filter is connected to the output of a rectifier, what effect will it have on the conduction of the rectifying diode(s)? Why?

10. Why does a full-wave signal allow for better filtering?

11. Why does a larger capacitor make a better filtering component?

12. Describe the function of the second capacitor in a basic RC network.

13. What advantages does an LC filtering network have over an RC filtering network?

14. What is the primary purpose of a regulator?

15. Name the two ways the input voltage can vary in a regulator.

16. Why is a zener diode a good regulator?

17. How does the basic zener diode regulator (originally shown in Figure 28-15A) compensate for a decrease in the input voltage?

18. What is the primary disadvantage of the emitter-follower regulator?

19. Name the five basic circuits in a feedback regulator.

20. Using Figure 28-19, describe how the feedback regulator maintains a steady output voltage when the input voltage decreases.

21. What function(s) is(are) performed by an operational amplifier when used in a feedback regulator?

22. Describe *in general* the method of overload protection.

ELECTRONIC CIRCUITS

Summary

The primary purpose of a power supply is to provide a steady, unchanging DC voltage under varying input and load conditions. The three main parts of a power supply are the rectifier, the filter, and the regulator.

The purpose of a rectifier is to completely remove one alternation from the AC input, converting it to a pulsating DC signal suitable for filtering. The three main types of rectifiers are the half-wave rectifier, the full-wave rectifier, and the bridge rectifier.

Rectification is accomplished through the use of a diode (or network of diodes) which allows current to flow in only one direction. In this way, that alternation of the input AC signal which would cause current to flow in the *opposite* direction is removed.

The half-wave rectifier, consisting of a single diode, provides an output voltage one-half that of the input voltage, which is considered to be a good aspect in a rectifier. However, current only flows through the half-wave for 50% of the input signal, which is considered poor. In addition, the output frequency (called the ripple frequency) is equal to that of the input frequency, which is also considered poor.

A transformer placed between the AC input and the rectifier allows the rectifier to develop voltages which are higher or lower than the AC signal. In addition, it also isolates equipment ground from the AC signal, thus removing a potential safety hazard.

The full-wave rectifier is comprised of two diodes placed with the same terminal facing the transformer. The output voltage is only one-fourth that of the input, due to a center-tapped secondary winding. However, the full-wave rectifier allows current to flow for 100% of the input signal, and has a ripple frequency which is double that of the input frequency.

The bridge rectifier combines the good aspects of both the half-wave and the full-wave rectifiers. It consists of a network of four diodes, placed in such a way that current can only flow in one direction through the load. The output voltage of a bridge rectifier is one-half that of the input signal, and the ripple frequency is double the input frequency. Current flows through the load for 100% of the input signal.

POWER SUPPLIES

A pulsating DC output is ill-suited for electronic circuits. A filter is connected to the output of a rectifier, whose purpose is to smooth out the pulsating DC signal. The primary component involved in filtering is the capacitor. As the pulsating DC signal reaches its peak, the capacitor charges to that voltage. When the pulsating DC signal starts to decrease, however, the capacitor starts to discharge. The high resistance of the load, combined with a high capacitance value, make the discharge time of the capacitor very slow. This allows the voltage felt by the load to decrease at a rate much slower than the pulsating DC signal. The voltage felt by the load is very close to the peak voltage, but the difference between fully charged and discharge creates a small peak-to-peak voltage, rising above and below the eventual DC voltage.

A larger capacitor provides better filtering, but is limited to such a value which will not damage the rectifying diode(s). RC and LC network provide much better filtering than a single capacitor. In addition, a full-wave signal also allows for better filtering because it recharges the capacitor more times per second. This allows for a smaller peak-to-peak ripple.

The output of a filter is fed to a regulator, whose purpose is to maintain a steady DC output, regardless of varying input or load conditions. The input varies simply because some ripple still exists, and the 115 volt AC signal normally applied to a rectifier can vary anywhere from 105 to 125 volts AC. In addition, any circuit connected to the regulator whose characteristic operation includes a constantly changing resistance causes a changing amount of current to come from the power supply.

The simplest of regulators is the zener diode, which maintains a steady breakdown voltage output while its reverse current varies. A series resistor must be placed with the zener to develop that voltage not developed by the zener.

A more complex regulator is an emitter-follower, which uses a transistor as the regulating component. The base voltage of the transistor is supplied by a zener diode, and any attempted change in the output voltage changes the amount of forward-bias, thus compensating for changing input and load conditions.

The most popular type of regulator is the feedback regulator. This is comprised of a sampling network, a reference voltage, an error detector, an error amplifier, and a regulating component. The sampling network continually monitors the output voltage, and if any change in the output voltage occurs, the sampling network sends a voltage proportional to that change to the error detector. The error detector compares the incoming voltage to the voltage reference source, and if there is a difference, amplifies this difference, and sends it to the regulating component. The regulating component then adjusts its internal resistance to provide the proper amounts of current flow and output voltage.

An operational amplifier can be placed in the feedback regulator to function as both the error detector and the error amplifier. The op amp is in the noninverting configuration, and amplifies any difference between the incoming signal from the sampling network (which functions as R_F) and the voltage reference source (which functions as R_I).

Many power supplies offer short-circuit protection, which ensures that only a limited amount of current flow is available from the power supply. Usually a transistor with a resistor across its emitter junction is included in the regulator. When the voltage drop across this additional resistor reaches a potential high enough to forward-bias the additional transistor, the transistor decreases the forward-bias of the regulating component.

CHAPTER 29

Oscillators

ELECTRONIC CIRCUITS

Contents

Introduction 787

Chapter Objectives 788

Oscillator Fundamentals 789

Transformer Oscillators 793

LC Oscillators 798

Crystal Oscillators 804

RC Oscillators 812

Nonsinusoidal Oscillators 818

Chapter Self-Test 822

Summary 824

OSCILLATORS

Introduction

Amplifiers are extremely important in the world of electronics. Oscillators are perhaps equally important, for no matter what field of electronics you may be involved in, you will encounter oscillators.

Oscillators are used in computers, communication systems, television systems, industrial control and manufacturing processes, and even in electronic watches as the basic time-keeping device. Probably one of the most common uses of the high-frequency oscillator is in the television tuner or "channel selector." Here, the oscillator helps select the channel to be viewed.

The term *oscillator* naturally implies an oscillating or revolving motion. One example of mechanical oscillation is the pendulum in a typical grandfather clock. The pendulum "oscillates" back and forth, ticking away the minutes. In this sense, the pendulum is the basic timing mechanism of the clock. Electronic oscillators operate in a similar manner, generating a continuously repetitive output signal that can be used to synchronize operations.

In this section, you will study common electronic oscillators; how they work, and how they are identified. Each oscillator has its own distinct characteristics of identification and operation. It is important to remember these characteristics.

ELECTRONIC CIRCUITS

Chapter Objectives

When you have completed this chapter, you will be able to:

1. Define the following terms: oscillator, tickler coil, tuned-base, tuned-collector, series-fed, shunt-fed, piezoelectricity, lead-lag network, relaxation, and blocking.

2. List the general requirements for an oscillator.

3. Describe the general operation of both the tuned-base and tuned-collector Armstrong oscillators.

4. Describe the general operation of the series-fed and the shunt-fed Hartley oscillators.

5. Describe the general operation of a Colpitts oscillator.

6. Name the identifying feature for both the Hartley and the Colpitts oscillators.

7. Describe how a crystal can be used to enhance the operation of a Hartley and Colpitts oscillator.

8. Define the difference between a crystal wafer by itself, and a crystal wafer mounted in its metal casing.

9. Describe the general operation of the Butler and Pierce oscillators.

10. Describe the general operation of the phase-shift and Wien-bridge oscillators.

11. Describe the general operation for both the blocking and sawtooth blocking oscillators.

OSCILLATORS

Oscillator Fundamentals

Frequently, electronic circuits require AC signals that can range from a few hertz to many millions of hertz. Oscillators are usually used to generate these frequencies. Because these frequencies have such a large range, hundreds of different oscillators exist. However, all oscillators operate on the same basic principles. If you understand these principles, you will be able to analyze the operation of most common oscillators.

What is an Oscillator?

An **oscillator** is a circuit that generates a repetitive AC signal with a frequency range of a few hertz to several gigahertz. Up to this point, the only circuit we've discussed that generates an AC signal is the AC generator, which supplies a 60 hertz signal at 115 volts AC. The primary purpose of this signal is to supply power; the low frequency cannot be used very often. For this reason, oscillators are used to supply other frequencies.

So why not use an AC generator for each different frequency? Keep in mind that a generator is, for all practical purposes, a motor. This makes it relatively large and expensive. Also, it would be satisfactory only for low frequency applications. Imagine a motor turning at a rate of 20 million revolutions per second. This is the rate required for a generator to produce a 20 MHz signal. A generator is unsuitable for high frequency applications.

The electronic generator or "oscillator" is an alternative to the mechanical generator. It has no moving parts and is capable of producing AC signals ranging from a few hertz to many millions of hertz. Such an oscillator is shown in Figure 29-1. It operates from a DC power supply, and generates an AC signal. The oscillator's output can be a sine wave, rectangular wave, or a sawtooth, depending on the type of oscillator. The major requirement of a good oscillator is that the output is uniform. In other words, the output *must not* change in frequency or amplitude. Each cycle of an oscillator must be identical to the other cycles. In practical applications there are slight variations from cycle to cycle. This is because there are no ideal (perfect) components. However, it is possible to create nearly perfect oscillators.

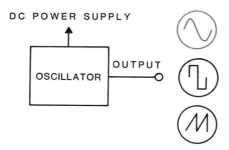

Figure 29-1 The electronic oscillator.

The Basic Oscillator

Back in Chapter 16, we described the flywheel effect of a parallel resonant circuit, or tank. A tank is comprised of an inductor and a capacitor in parallel. When current flows in a tank circuit, it has a tendency to oscillate. Circulating current flows inside the tank, producing a back and forth oscillating motion. However, resistance of the tank circuit dissipates energy, and oscillations are damped. This is shown in Figure 29-2A.

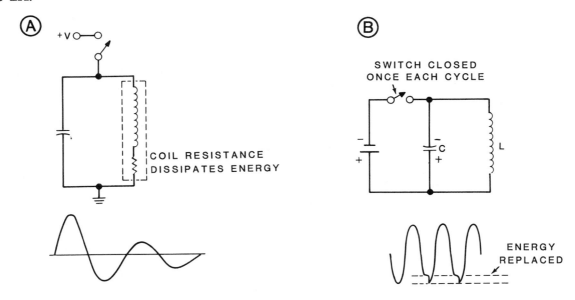

Figure 29-2   Overcoming a damped signal.

For the tank circuit to continue oscillating, any energy that is lost must be replaced. A crude method of replacing energy is to close the power switch once each cycle, as shown in Figure 29-2B. It is important for the switch to close so the energy is replaced at *exactly* the instant that reinforces the charge on the capacitor. This ensures that the replacement energy has a reinforcing effect, and also causes the replacement energy to be in-phase with the tank waveform. This is an example of **positive feedback,** which is the most important prerequisite for oscillation.

It may seem that an oscillator can be created by simply taking a portion of an amplifier's output and feeding it back to its input. In some cases this is true, but not always. For the circuit to oscillate, the feedback signal must be **in-phase** with the input signal.

OSCILLATORS

Figure 29-3 illustrates this point. A common-emitter amplifier is shown in Figure 29-3A. Recall that the output of a common-emitter is 180 degrees out-of-phase with its input signal. If the output signal were fed back directly to the input signal, it would negate the input. This is negative feedback, and the amplifier would not operate properly.

Figure 29-3 Feedback in the common-emitter amplifier.

However, if you place a 180 degree phase-shift network between the output and input, the feedback will be of the same phase as the input and the circuit will oscillate. This is shown in Figure 29-3B. Notice that the feedback signal is in-phase with the input signal, and adds to the input. The result is positive feedback, and the amplifier can now operate properly.

Once the amplifier begins to operate, the input signal can be removed and the circuit will continue to oscillate. This circuit meets almost all of the requirements of an oscillator, except a continuous, steady frequency of oscillation. This circuit could oscillate at any number of frequencies. Even minor noise pulses could change this oscillator's frequency. An oscillator should have a constant output, so a means of stabilizing the frequency is necessary.

This is where the tank we previously discussed becomes useful. Because the tank resonates at only a specific frequency, we can place it in our amplifier and select one frequency of oscillation. In addition, the reactance of the inductor and capacitor produce the 180 degree phase shift required for positive feedback. In Figure 29-4A, a tank circuit is placed in the positive feedback loop. The tank resonates at its natural frequency and the amplifier stage replaces the energy lost in the tank circuit's resistance.

Figure 29-4 LC and RC networks set frequency.

A similar result is obtained if the regenerative feedback loop contains enough RC networks to produce the desired 180 degrees phase shift. This is shown in Figure 29-4B. Here, RC time constants determine the oscillator's frequency and the amplifier replaces energy lost across the RC networks.

Up to this point, oscillators have been described as amplifiers with input signals applied to start the circuit. In actual practice, oscillators must start on their own. In other words, they must be self-starting. When a circuit is first turned on, energy levels do not instantly reach maximum, but gradually approach it. This produces many noise pulses that can be phase shifted and fed back to the input, as shown in Figure 29-5. The amplifier steps up these pulses, which are again supplied to the input. This action continues until both alternations begin to appear, and oscillation begins. Therefore, an oscillator must be naturally self-excited, meaning it starts on its own. The oscillator requires a few cycles before its amplitude and frequency are stabilized.

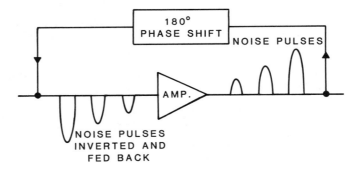

Figure 29-5 Oscillators are usually self-starting.

Let's review the most important requirements for an oscillator:

1. An amplifier is necessary to replace circuit losses.

2. Frequency determining components are necessary to set the frequency of oscillation.

3. Positive feedback supplies a regenerative feedback to sustain oscillation.

4. The oscillator must be self-starting.

Transformer Oscillators

The simplest oscillator that applies the principle of positive feedback, a tuned LC circuit, and self-starting is the **Armstrong** or **tickler coil** oscillator shown in Figure 29-6. This oscillator requires a phase shift for positive feedback. One of the easiest methods to obtain the 180 degree phase shift is to use a transformer. The 180 degree phase shift between primary and secondary makes the transformer ideal for oscillator circuits.

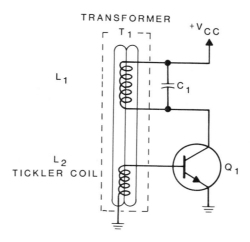

Figure 29-6 The basic transformer oscillator.

The term **tickler coil** comes from the basic operation of this oscillator. The inductor is the tank circuit, L_1, uses its magnetic field to "tickle" secondary winding L_2, and thereby controlling the conduction of the amplifier Q_1.

The biasing components for the amplifier are not shown for the sake of simplicity. We can assume, however, that the transistor is properly biased for operation. Transformer T_1 is comprised of primary winding L_1 and secondary winding L_2. Although they are shown underneath each other, they are connected across each other in actual circuitry. Capacitor C_1 and L_1 make up the tank, which determines the frequency.

ELECTRONIC CIRCUITS

We will describe the sequence of events in this oscillator in a step-by-step format. We'll use Figure 29-7 for reference.

1. Power is applied. The biasing components allow Q_1 to conduct, and collector current flows towards $+V_{CC}$. (Figure 29-7A)

2. Collector current flows through L_1 and C_1. C_1 charges in the polarity shown in Figure 29-7A, and L_1 has a negative voltage induced.

3. The negative voltage at L_1 is inverted by secondary winding L_2, and the increasing positive voltage is felt at the base of Q_1.

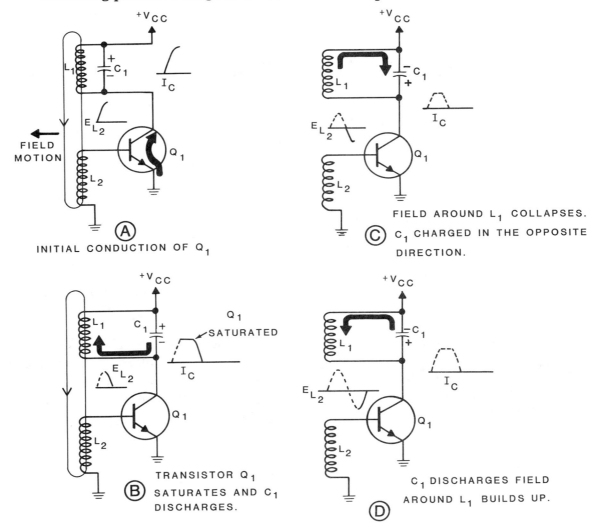

Figure 29-7 Operation of the Armstrong oscillator.

4. The increasing base voltage of Q_1 increases forward-bias and collector current. Q_1 conducts at a faster rate, continuing to charge C_1, and L_1 has a higher negative voltage induced.

5. Steps 3 and 4 are repeated until Q_1 saturates. At this point, C_1 is charged to $+V_{CC}$, and the magnetic field around L_1 is fully expanded. Collector current flows entirely through L_1, since C_1 is fully charged.

6. Since there is no more *change* in current, the expanded magnetic field around L_1 can no longer induce a voltage into L_2. This removes base voltage, and Q_1 cuts off. (Figure 29-7B)

7. C_1 discharges through L_1, sending current through L_1 in the same direction as collector current. The magnetic field around L_1 remains expanded, and still no voltage is induced into L_2.

8. When C_1 completes its discharge, Q_1 is still cut off, and an expanded magnetic field exists in L_1. The magnetic field collapses, charging C_1 in the opposite direction, and sending an opposite voltage to the base of Q_1 through L_2. This sends the transistor further into cutoff.

9. When the magnetic field completely collapses, C_1 is charged to a voltage *under* V_{CC}, due to energy loss. C_1 discharges through L_1.

10. The expanding magnetic field around L_1 induces a voltage into L_2 which is sufficient to forward-bias Q_1, and the cycle is repeated.

In summary, the transistor conducts only to replenish the lost energy within the tank. The transistor is turned on and off by the tank, through the secondary winding of the transformer. Figure 29-8 shows a similar circuit. When the energy within the tank is depleted, the switch (the transistor) closes, sending current through the capacitor, and recharging it to V_{CC}. The switch opens and closes at a rate equal to the resonant frequency.

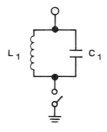

Figure 29-8 The equivalent circuit of an Armstrong oscillator.

ELECTRONIC CIRCUITS

When the frequency determining components are tied to the collector of the amplifier, the oscillator is called a **tuned-collector.** A schematic of the tuned-collector oscillator we just discussed is shown in Figure 29-9A, with the biasing and output components added. The output is taken across L_1 by an additional secondary. Figure 29-9B shows the output waveform. The sine wave is labeled with the steps we discussed to show you what occurs during each portion of the output waveform. Remember, this additional secondary has an induced voltage when the magnetic field around L_1 is *changing*. When the magnetic field around L_1 remains completely expanded, (Steps 6 and 7), the output voltage decreases.

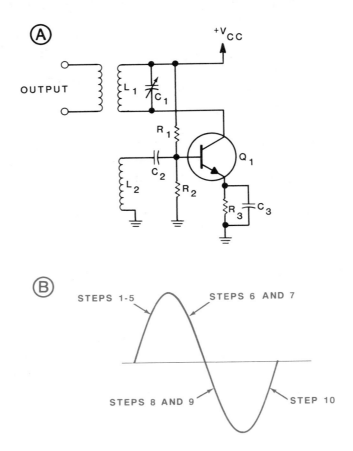

Figure 29-9 A tuned-collector oscillator and its output.

OSCILLATORS

The Tuned-Base Oscillator

Another variation of the Armstrong oscillator is shown in Figure 29-10. Here, the frequency determining components are tied to the base of the amplifier, which makes this a **tuned-base oscillator.** Component functions are similar to those in the tuned-collector oscillator. Transformer winding L_1 and capacitor C_1 form the tuned circuit that determines the oscillator's frequency. C_1 is variable to permit frequency adjustment. Tickler coil L_2 is in the collector circuit and is inductively coupled to L_1 to provide positive feedback. Again, the output is taken by adding another winding to the transformer's secondary.

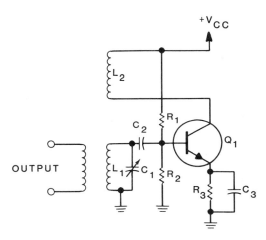

Figure 29-10 A tuned-based oscillator.

The initial bias of R_1 and R_2 turns on transistor Q_1. The transistor conducts and current flows through R_3, Q_1, and L_2, and returns through the $+V_{CC}$ supply. The changing current through L_2 induces a voltage into L_1 that is shifted 180 degrees, charging capacitor C_1. This voltage is coupled to the base of Q_1 by capacitor C_2, further forward-biasing Q_1. The resulting regenerative action continues until Q_1 saturates. With Q_1 saturated, a steady current exists through L_2 and its field is no longer changing. Therefore, no voltage is induced into L_1. Capacitor C_1 now begins to discharge, and tank circuit oscillations begin. C_1 discharges and drives transistor Q_1 into cutoff. As the tank completes one cycle of oscillation, the base of Q_1 is again forward-biased. Q_1 conducts, and the cycle continues.

LC Oscillators

Most oscillators work on the positive feedback principle, which means that feedback is necessary to sustain oscillation. Oscillators are generally classified according to their frequency determining components. These three general classifications are **LC oscillators, RC oscillators,** and **crystal oscillators.**

LC oscillators use a tuned circuit consisting of either a parallel-connected or series-connected capacitor and inductor to set the frequency. The Armstrong oscillator just discussed is an LC oscillator, since the transformer primary and shunt capacitor form a parallel resonant circuit. In this section, the basic concern is with LC oscillators that produce sine wave outputs.

The Series-Fed Hartley

One of the undesirable features of the Armstrong oscillator is that the tickler coil has a tendency to resonate with the distributed capacitance in the circuit. This results in oscillator frequency variations. If the tickler coil is made a part of the tuned circuit, the unstable effect of the tickler coil can be overcome. In the **Hartley oscillator,** shown in Figure 29-11, just such an arrangement is used. The inductor is tapped to form two coils, L_{1A} and L_{1B}. Tuning capacitor C_1 is connected across inductor L_1, making the entire coil part of a tuned circuit.

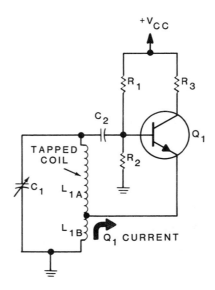

Figure 29-11 A series-fed Hartley oscillator.

Resistors R_1 and R_2 forward-bias the emitter junction of Q_1 when the circuit is initially turned on. Transistor Q_1 conducts and collector current travels through the lower section of coil $L_1(L_{1B})$, through Q_1, and through load resistor R_3. The current through L_{1B} induces current into L_{1A} because of the mutual inductance of the two coils. The result is a positive potential at the top of L_{1A}, that is coupled to the base of Q_1 by capacitor C_2. This increases the forward-bias on Q_1, causing it to quickly saturate.

Once Q_1 saturates, the current through L_{1B} is no longer changing, and no voltage is induced into L_{1A}. This removes forward-bias from Q_1, and conduction rapidly decreases. The field around L_{1B} collapses, and again induces current into L_{1A}. The polarity of this induced current is such that the top of L_{1A} is negative. The negative potential is felt on the base of Q_1, reverse-biasing Q_1 and quickly driving it into cutoff. During this time, tank capacitor C_1 charges to a negative potential. When Q_1 is completely cut off, capacitor C_1 begins to discharge, and tank action begins.

During the cycle of tank oscillation, when the upper plate of C_1 begins to accumulate a positive charge, Q_1 is again forward-biased and conducts through L_{1B} of the tank coil. Conduction through the lower section of the tank coil replaces energy lost in the tank, providing the positive feedback necessary for oscillation. The amount of feedback can be controlled by varying the position of the coil's tap. Varying the tap varies the inductive ratio of the circuit.

Since emitter current flows through a portion of the tank coil, the oscillator is said to be **series-fed.** This series-fed arrangement and the tapped coil are the identifying features of a series-fed Hartley oscillator. The disadvantage of the series-fed Hartley is that DC current flows through a portion of the tank, increasing power losses in the circuit. This causes the oscillator to become unstable.

The Shunt-Fed Hartley

Figure 29-12 is a schematic of another type of Hartley oscillator known as the **shunt-fed Hartley.** The tapped coil, L_1, immediately identifies this as a Hartley oscillator. Unlike the series-fed Hartley, no DC current passes through the tank coil, hence the name "shunt-fed." This keeps circuit Q high, resulting in better frequency stability than can be achieved in the series-fed Hartley.

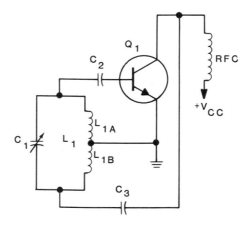

Figure 29-12 A shunt-fed Hartley oscillator.

The bias circuit, which is similar to the series-fed Hartley, has been deleted for the sake of simplicity. The parallel network of L_1 and C_1 set the frequency at which the circuit oscillates. Capacitor C_1 is variable so that the oscillators frequency can be adjusted. Capacitor C_2 is a coupling capacitor between the resonant tank and the base of Q_1. The radio frequency choke (RFC) acts as a collector load and effectively blocks high frequency AC oscillations from the power supply. Collector AC variations are coupled to the tank through capacitor C_3.

The Colpitts Oscillator

The Colpitts oscillator is similar to the shunt-fed Hartley except two capacitors are used instead of a tapped coil. Essentially, the Colpitts is shunt-fed, so DC collector current does not flow through the tank circuit. Since the Colpitts is more stable than the Hartley, it is used in many signal generators.

Figure 29-13 shows the schematic for a Colpitts oscillator. Again, the bias networks have been eliminated for the sake of simplicity. However, they are similar to the transistor biasing networks discussed earlier. The tapped capacitor arrangement identifies this oscillator as a Colpitts oscillator. As with all LC oscillators, frequency is determined by inductor L_1 and the series combination of capacitors C_1 and C_2. Capacitor C_3 couples AC collector voltage to the tank, while blocking DC.

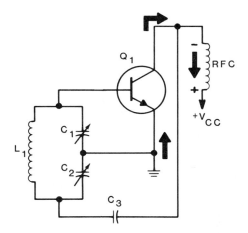

Figure 29-13 A simplified Colpitts oscillator.

Since the oscillator uses a common-emitter, the collector is 180 degrees out-of-phase with the base. The arrangement of capacitors C_1 and C_2 in a voltage-divider network produces the desired 180 degree phase shift across capacitor C_1, resulting in positive feedback. Although the combination of C_1 and C_2 determines oscillator frequency, C_2 has the most pronounced effect on frequency.

The Colpitts is shock-excited into oscillation much like the other oscillators. Initial forward-bias is furnished by the bias network and Q_1 begins to conduct. The DC collector current path is from emitter to collector, through the RFC, returning to the power supply. This initial surge of current causes a negative voltage drop across the RFC, since the change is rapid. Capacitor C_3 couples this negative voltage to the lower plate of capacitor C_2.

ELECTRONIC CIRCUITS

Let's see what happens within the tank when this negative voltage is felt. Figure 29-14A shows the tank circuit with a negative potential at the bottom plate of C_2. This sends current through both capacitors, and C_1 and C_2 charge. The net result is the top plate of C_1 becoming positive, which provides regenerative feedback. Q_1 becomes more forward-biased and quickly saturates.

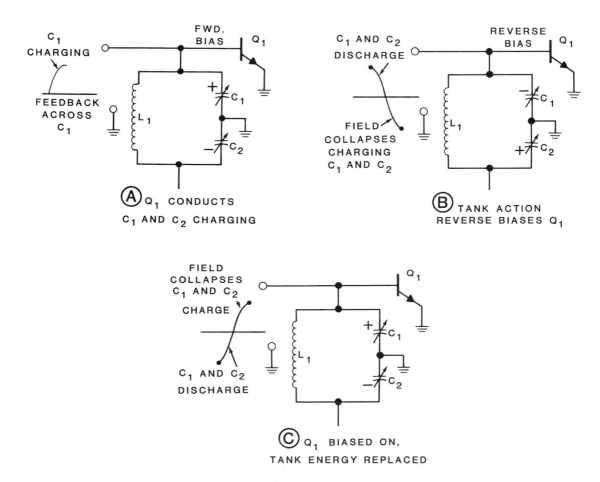

Figure 29-14 Analysis of the Colpitts LC circuit.

With Q_1 saturated, there is no voltage drop across RFC, because current is no longer changing. The flywheel effect of the tank then takes over (Figure 29-14B) as C_1 and C_2 act as one capacitor, discharging through L_1 and building up the magnetic field.

When the capacitors are completely discharged, the field collapses and charges the top plate of C_1 negative, reverse-biasing Q_1. Q_1 is driven into cutoff. When feedback capacitor C_1 is fully charged, it discharges through L_1.

Again, a field is built up around L_1 that subsequently collapses and charges C_1 in the opposite direction (Figure 29-18C). Transistor Q_1 is now forward-biased and conducts. Thus, energy lost in the tank is replaced. A similar action occurs each cycle as positive feedback replenishes lost energy.

The series combination of C_1 and C_2 determines the oscillator's frequency. However, C_1, the feedback capacitor, controls the amount of feedback. This is the result of the series voltage-divider arrangement of C_1 and C_2. If the capacitance of C_1 is decreased, the amount of feedback is increased.

Usually the tapped capacitors are variable, and are ganged (mechanically connected) together. This arrangement permits adjustment of the oscillator's frequency. The ganged capacitor arrangement has a disadvantage in that, as the frequency is varied the feedback changes. At one end of the frequency range, there is too much feedback and the output waveform is distorted. At the other end, feedback is small and cannot sustain oscillation. These factors, combined with the distributed capacitance of the circuit, limit the adjustable frequency range of the Colpitts oscillator.

Crystal Oscillators

The LC oscillators just discussed are commonly used. However, in applications where extreme oscillator stability is required, the LC oscillator is unsatisfactory. Temperature changes, component aging, and load fluctuations cause the oscillator to drift, which makes the oscillator unstable. When a high degree of stability is required, **crystal oscillators** are generally used.

Stability is discussed in percent of drift. A typical drift for a crystal oscillator is 0.001%. Compare this with the stability of the LC oscillators just studied, where 1% frequency drift is common. If an electronic wristwatch had a timing oscillator that drifts 1%, the watch could either gain or lose 14 minutes each day and still be within the oscillator's tolerance. However, if the watch oscillator is stable to within 0.001%, the maximum time lost or gained each day is .09 seconds or 32 seconds each year. To achieve this accuracy, electronic watches use crystal oscillators as the basic timing device.

Crystal Characteristics

Back in Chapter 2, we discussed the method of producing a voltage by pressure. This is known as **piezoelectricity.** When mechanical pressure (stress) is applied to a crystal, a potential difference develops. In Figure 29-15A, a normal crystal has its charges evenly distributed and is therefore neutral. If force is applied to the sides of the crystal, as shown in Figure 29-15B, the crystal is compressed and opposite charges accumulate on the sides. Thus, a potential difference is developed.

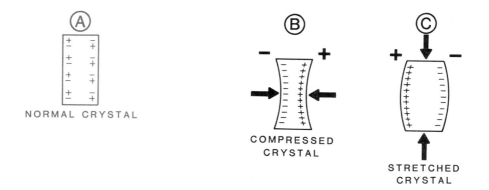

Figure 29-15 Mechanical stress applied to a crystal.

If pressure is applied to the top and bottom, as in Figure 29-15C, the crystal is stretched and again opposite polarities appear across the crystal. Thus, if the crystal is alternately compressed and stretched, an AC voltage can be generated. Therefore, a crystal can convert mechanical energy into electrical energy.

Just the opposite effect occurs if AC voltage is applied to a crystal. The electrical energy from the voltage source is converted to mechanical energy in the crystal. Figure 29-16A illustrates this point. The AC input signal causes the crystal to stretch and compress, which creates mechanical vibrations that correspond to the frequency of the AC signal.

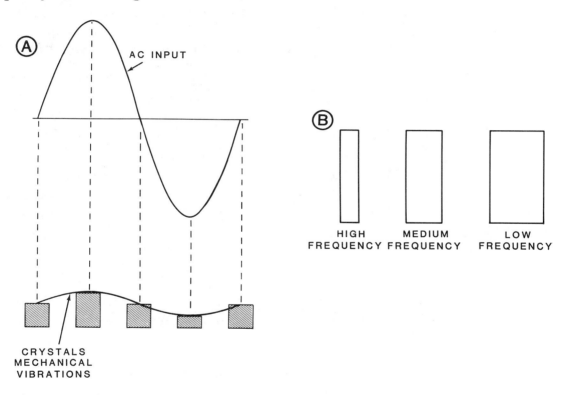

Figure 29-16 AC signal applied to a crystal, and frequency characteristics.

Because of their structure, crystals have a natural frequency of vibration. If the frequency of the applied AC signal matches this natural frequency, the crystal will stretch and compress by a large amount. However, if the frequency of the exciting voltage is slightly different than the crystal's natural frequency, little vibration is produced. In addition, the natural vibration of the crystal is extremely constant. These characteristics make the crystal ideal for oscillator circuits.

ELECTRONIC CIRCUITS

The natural frequency of a crystal is usually determined by its thickness. A thinner crystal produces a higher frequency. Conversely, a thicker crystal produces a lower frequency. This is shown in Figure 29-16B. Of course, there are practical limits on just how thin a crystal can be cut, without it becoming extremely fragile. This places an upper limit on the crystal's natural frequency of approximately 50 MHz.

Equivalent Crystal Circuits

The crystal by itself acts like a series-resonant circuit, as shown in Figure 29-17A. In the series equivalent circuit, inductance (L), represents the crystal's mass that effectively causes vibration; C represents crystal stiffness, which is the equivalent of capacitance; R is the electrical equivalent of internal resistance caused by friction.

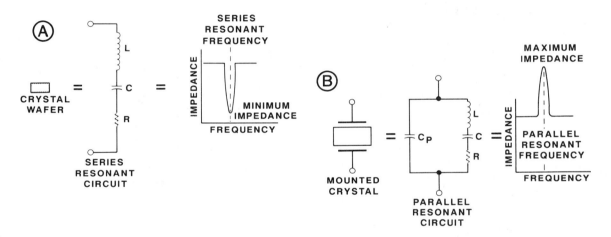

Figure 29-17 The equivalent circuits of an unmounted and a mounted crystal.

Therefore, at the crystal's natural mechanical resonant frequency, the electrical circuit is series-resonant, and offers minimum impedance to current flow. When the circuit's characteristics are plotted on an impedance-frequency curve, it shows very sharp transitions and minimum impedance at the series-resonant frequency. The sharp transitions indicate the highly-selective frequency characteristic of the crystal.

In most applications, however, the crystal is mounted between two metal plates, and a spring applies mechanical pressure to the plates. The metal plates secure the crystal and also provide the electrical contact. The crystal is then placed in a metal casing or holder.

OSCILLATORS

When the crystal is mounted between metal plates, the equivalent circuit is modified as shown in Figure 29-17B. The metal mounting plates now appear as a capacitor (C_P) in parallel with the series-resonant circuit of the crystal. The value of C_P is relatively high and, at lower frequencies, does not appreciably effect the series-resonant crystal.

However, at frequencies above the crystal's series-resonant frequency, the inductive reactance of the crystal is greater than the crystal's capacitive reactance, and the crystal appears **inductive.** At these higher frequencies, a point is reached where the inductive reactance of the crystal equals the capacitive reactance of the mounting plates ($X_L = X_{CP}$). In this case the equivalent circuit is parallel-resonant and impedance is maximum. The electrical equivalent of the crystal at this frequency is a parallel-tuned LC circuit. Therefore, a crystal has two resonant frequencies.

At the natural mechanical frequency of the crystal, the crystal is series-resonant and impedance is minimum. At a slightly higher frequency, the crystal and the capacitance of its mounting plates form a parallel-resonant circuit and impedance is maximum. The overall crystal response curve is illustrated in Figure 29-18.

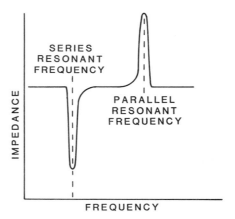

Figure 29-18 The response curve of a crystal.

As mentioned before, a crystal is highly frequency selective as indicated by the sharp skirts on the response curve. This is natural, since a crystal has an extremely high Q, sometimes approaching a Q of 50,000. When Q's of this value are compared with the Q of an LC circuit, usually 100, it is clear why crystal oscillators are more stable than normal LC oscillators. You will also find that with crystal oscillators you can use either the series-resonant or parallel-resonant characteristic of the crystal.

ELECTRONIC CIRCUITS

The schematic symbol for a crystal is derived from the way it is mounted and represents the crystal slab held between two plates, as shown in Figure 29-17B. The word crystal is often abbreviated "XTAL" or "Y" on schematics.

Crystal-Enhanced Oscillators

We stated earlier that typical LC oscillators have a high amount of drift. A crystal can be used within these LC oscillators to enhance their operation. In this manner, the amount of drift can be greatly reduced.

The Hartley is a typical LC oscillator and, although fairly stable, frequency drifts of 1% are common. If the Hartley oscillator is to be operated at a specific frequency and a high-degree of stability is required, a crystal can be placed in the circuit. Figure 29-19A is such a circuit. However, if the frequency of this oscillator is to be changed, even by a fraction of a percent, the crystal must be replaced.

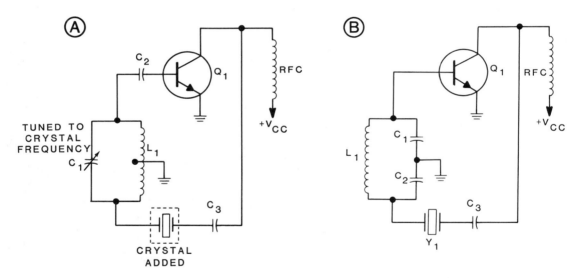

Figure 29-19 Crystal-enhanced Hartley and Colpitts oscillators.

Notice that in Hartley crystal oscillators, the crystal is connected in series with the feedback path. Therefore, the crystal operates at its series-resonant frequency. Also, the LC tank network must be tuned to the series-resonant frequency of the crystal.

OSCILLATORS

When the oscillator is operating at the crystal frequency, the crystal's equivalent series-resonant circuit offers minimum opposition to current and feedback is maximum. If the oscillator drifts away from the crystal frequency, the impedance of the crystal increases drastically, reducing feedback. This forces the oscillator to return to the natural frequency of the crystal. Therefore, when the crystal is series-connected, it controls feedback. Note the tapped inductors which identify this circuit as a Hartley oscillator.

The Colpitts oscillator can be crystal controlled in the same manner as the Hartley. Again, crystal Y_1 is connected in series with the feedback path, as shown in Figure 29-19B.

Since the crystal is series-connected, it controls feedback and the LC tank circuit is tuned to the crystal's frequency. Otherwise, its operation is identical to the basic Colpitts oscillator studied earlier. Notice that the tapped capacitors identify this circuit as a Colpitts oscillator.

The Butler Oscillator

The Butler crystal oscillator combines a tuned LC circuit with the frequency selectivity of a crystal. As Figure 29-20 indicates, the Butler oscillator employs two transistors. Transistor Q_2 operates as a common-base amplifier with a tuned collector circuit, while Q_1 functions as an emitter-follower. Crystal Y_1 is connected between the emitters of the two transistors and operates in its series-resonant mode to control feedback. Bias components have again been omitted for simplicity.

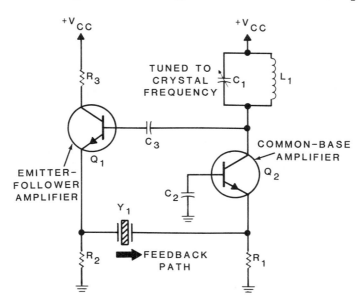

Figure 29-20 A simplified Butler crystal oscillator.

When the circuit is first energized, transistor Q_2 is forward-biased by the initial bias circuit (not shown). Q_2 conducts, developing a negative-going voltage at the bottom of the collector's tank circuit. Capacitor C_3 couples this negative potential to the base of Q_1 cutting Q_1 off.

Transistor Q_2 quickly saturates. At this point, transistor Q_1 begins to conduct. The resulting positive voltage drop across the emitter resistor R_2 is coupled to the emitter of Q_2 by the crystal and is developed across resistor R_1. This positive potential on the emitter of Q_2 reverse biases Q_2 and it begins to cut off. Consequently, the collector voltage of Q_2 starts to go more positive. This change is coupled to Q_1 through C_3, increasing the forward bias on Q_1. Q_1 conducts harder and Q_2 is driven into cutoff. The tank action of the tuned LC circuit takes over and reverse biases Q_1, cutting it off. As Q_1 ceases conduction, Q_2 is forward-biased and the cycle repeats.

Since positive feedback is through crystal Y_1, the oscillator is operating at the crystal's series-resonant frequency, the crystal presents a low impedance path between the emitters and feedback is maximum. If frequency drifts, however, the crystal decreases feedback and forces the oscillator back on frequency.

The tuned LC circuit is important because, if the tank circuit is not tuned to the crystal frequency, the oscillator will not work. The combined effect of the tuned circuit and the crystal results in good oscillator performance.

The Pierce Oscillator

In the Hartley, Colpitts, and Butler oscillators, the crystal operated only as a feedback component. If the frequency of these oscillators begins to drift, the crystal severely reduces feedback, and the oscillator returns to its optimum frequency.

However, because of the resonant characteristics of the crystal, it can operate as the frequency determining device in an oscillator, as well as provide feedback. One example of this type of oscillator is the Pierce oscillator.

OSCILLATORS

The Pierce oscillator is similar to the basic Colpitts, except the tank inductor is replaced with a crystal, which operates at its parallel-resonant frequency. Figure 29-21 shows crystal Y_1 replacing the tank coil. Remember, the crystal's parallel-resonant frequency is slightly higher than its series resonant frequency and appears as an inductor.

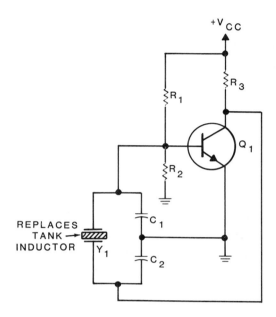

Figure 29-21 The Pierce crystal oscillator.

The voltage divider arrangement of capacitors C_1 and C_2 provides the 180 degree phase shift between the collector and emitter of Q_1, resulting in positive feedback. The ratio of these two capacitors also determines the feedback ratio and therefore, the crystal's excitation voltage. Since the crystal's response is extremely sharp, it will vibrate only over a narrow range of frequencies, producing a stable output. Therefore, the Pierce oscillator is more stable at higher frequencies than either the basic Colpitts or Hartley oscillators.

The crystal operates in its parallel-resonant mode, and controls the tuned circuit's impedance. At resonance, tank impedance is maximum and a large feedback voltage is developed across capacitor C_1. If frequency drifts above or below resonance, crystal impedance decreases rapidly which decreases feedback. By controlling the tank's impedance, the crystal effectively determines circuit feedback and the stability of the oscillator.

RC Oscillators

Up to this point, LC and crystal oscillators that are commonly used in RF applications have been discussed. However, in the low and audio frequency ranges. these oscillators are usually not practical. For example, inductors for the low frequency range, around 60 Hz, will be large and expensive. And, the practical low limit on crystals is usually around 50 kHz. So an inexpensive approach to oscillator design for these low frequency ranges is to use RC oscillators.

The RC oscillator uses resistance-capacitance networks to determine the oscillator's frequency. This makes the oscillator inexpensive, easy to construct, and relatively stable. There are basically two types of RC oscillators that produce sine wave outputs. They are the phase-shift oscillator and the Wien bridge oscillator.

The Phase-Shift Oscillator

The basic phase-shift oscillator, as the name implies, is a conventional amplifier and a phase shifting RC feedback network. It is typically used in fixed frequency applications. As in the conventional LC oscillator, the collector output signal must be shifted 180 degrees to produce the required regenerative feedback. The phase-shift oscillator accomplishes this with a series of RC networks connected in the collector-to-base feedback loop.

Remember that in a purely capacitive circuit, current leads voltage by 90 degrees. However, in an RC network the phase difference between current and voltage falls between 0 and 90 degrees, because resistance affects the phase relationship. Thus the phase difference in an RC circuit is a function of the capacitive-reactance (X_C) and resistance of the network. By carefully selecting the resistance and capacitive values, the amount of phase shift across an RC network can be controlled.

We also stated earlier that it is easy to cascade identical networks. So in order to obtain 180 degree phase shift, filters whose phase shift are multiples of 180 degrees are required. In Figure 29-22A, three identical RC networks are connected in series. An input signal is applied whose frequency produces a 60 degree phase shift through one network. When this network is cascaded, the input signal produces a total phase shift of 180 degrees.

OSCILLATORS

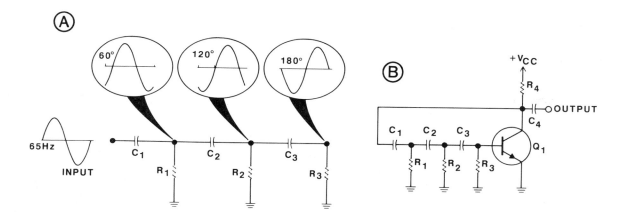

Figure 29-22 A simple RC phase shift oscillator.

If we place this network between the collector and base of a common-emitter amplifier, the result is a phase-shift oscillator. Figure 29-22B shows such a circuit. The phase-shift network is connected between the collector and base of Q_1, and provides the 180 degree phase shift that makes the circuit regenerative. Since there is a considerable power loss across the RC networks, the transistor's gain must be high enough to compensate for these losses. Usually a voltage gain of between 30 and 50 is required to overcome circuit losses.

The operation of this oscillator is very simple. In fact Q_1 simply switches between saturation and cutoff. Initial conduction causes a decrease in collector voltage. Collector voltage is shifted 180 degrees by the RC networks, placing a positive potential on the base of Q_1, further biasing it into saturation. When Q_1 saturates, the forward bias of Q_1 decreases until Q_1 cuts off. This action is repeated continually and as a result, the collector voltage varies in a sinusoidal manner, producing a slightly distorted sine wave output. Since each phase-shift network must produce a 60 degree phase shift, the circuit will naturally oscillate at the frequency at which this phase shift occurs.

ELECTRONIC CIRCUITS

The phase-shift oscillator functions best at fixed frequencies, since any variation of resistance or capacitance upsets the phase shift. However, it is possible to change the frequency over a small range by varying the resistance or capacitance of the RC networks. Stability can be improved by increasing the number of RC networks, thereby reducing the phase shift across each network. In other words, four 45 degree phase shift networks provides better stability than three 60 degree phase shift networks.

The Wien-Bridge Oscillator

Like the phase-shift oscillator, the Wien-bridge uses RC networks. However, in the Wien-bridge oscillator, the RC networks are part of a bridge circuit that produces both regenerative and degenerative feedback. The result is an excellent sine wave oscillator that can be used to generate frequencies ranging from 5 Hz to 1 MHz.

In the phase-shift oscillator just discussed, the RC networks produce the desired 180 degree phase shift for regenerative feedback. In the Wien-bridge oscillator, the RC networks select the frequency at which maximum feedback occurs, but do not shift the phase of the feedback voltage.

The Wien-bridge oscillator uses a basic circuit known as the lead-lag network. A lead-lag network is shown in Figure 29-23. It is a band-pass filter comprised of a series RC network ($R_1 \backslash C_1$) and a parallel RC network (R_2, C_2). It is called a lead-lag network because the output phase angle leads for some frequencies and lags for others. However, at the resonant frequency, the phase shift is exactly equal to zero degrees. This important characteristic allows the lead-lag network to determine the oscillating frequency of the bridge oscillator.

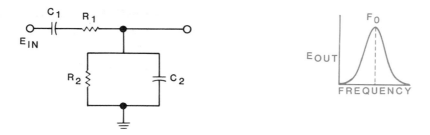

Figure 29-23   A lead-lag network for the Wien-bridge.

OSCILLATORS

At low frequencies, series capacitor C_1 has such high impedance that it acts as an open, and prevents an output. At very high frequencies, the parallel capacitor C_2 shunts the output to ground and again, there is no output. However at the resonant frequency, the output voltage is maximum. This is illustrated by the voltage-output-versus-frequency curve at the output of the circuit. Output is maximum at F_O, therefore, the RC network is frequency selective. On both sides of F_O the output decreases significantly.

At low frequencies, the phase angle is positive and the circuit acts like a lead network. At high frequencies, the output phase angle is negative and the circuit acts like a lag network. At the circuit's resonant frequency, the phase shifts of the series and parallel circuits cancel. Since the phase shifts are equal, but of opposite polarity, the resultant output is in-phase with the input.

Apply this lead-lag network to a Wien-bridge oscillator. Figure 29-24 illustrates a Wien-bridge oscillator using an operational amplifier as the active device. The lead-lag network, comprised of R_1/C_1 and R_2/C_2, makes up one side of the bridge. A voltage divider, R_3 and R_4 is the remaining leg of the bridge. The inverting and noninverting inputs of the op amp make it ideal for use in the Wien-bridge oscillator, since both regenerative and degenerative feedback are required. The op amp's high gain is also very useful in offsetting circuit losses.

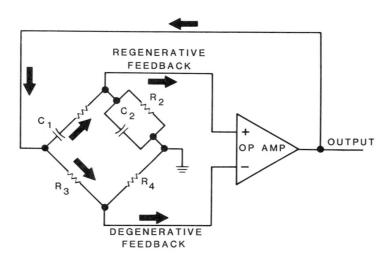

Figure 29-24 An IC Wien-bridge oscillator.

The op amp's output is fed back to the bridge input. Regenerative feedback is developed across the lead-lag network and applied to the noninverting input. Therefore, regenerative feedback is in-phase with the output signal. Degenerative feedback is developed across resistors R_3 and R_4 and is applied to the inverting input. For the circuit to sustain oscillation, regenerative feedback must be greater than degenerative feedback.

Since degenerative feedback is developed by the resistors alone, it remains constant regardless of the change in frequency. However, regenerative feedback *does* depend on the frequency response of the lead-lag network, which is frequency sensitive.

Component values are selected so that, at the desired oscillator frequency, regenerative feedback is larger than degenerative feedback and oscillation occurs. When the oscillator's frequency attempts to increase, the reactance of capacitor C_2 decreases and shunts more voltage to ground, reducing the regenerative feedback. A decrease in oscillator frequency increases the reactance of C_1 and less voltage is developed across the $R_2 \backslash C_2$ network, which reduces regenerative feedback. Only over a narrow range of frequencies, set by the lead-lag network, is regenerative feedback large enough to sustain oscillation. Thus, the oscillator is forced to stay on frequency by the lead-lag network.

The oscillator's frequency may be varied by changing either the resistance or capacitance in the lead-lag network. Usually, resistors R_1 and R_2 are potentiometers, permitting frequency variations.

The IC Wien-bridge oscillator is simple to construct and relatively inexpensive. Before integrated circuits were widely used in electronic designs, Wien-bridge oscillators were assembled using discrete components. Figure 29-25 is such a circuit. The circuit is a two-stage common-emitter amplifier. The Wien-bridge is connected across the base of transistor Q_1. Regenerative feedback is applied to the base of Q_1, while degenerative feedback is applied to its emitter. Q_1 is the oscillator transistor. Capacitor C_4 couples the oscillator's output to the base of Q_2, where the signal is amplified and phase-shifted the required 180 degrees. Capacitor C_3 provides feedback to the bridge network. Otherwise, operation is identical to the IC oscillator shown in Figure 29-24.

OSCILLATORS

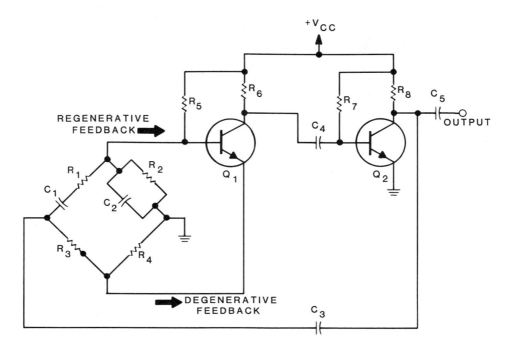

Figure 29-25 A discrete Wien-bridge oscillator.

Nonsinusoidal Oscillators

You have studied various types of oscillators that generate sinusoidal waveforms. Another broad class of oscillators is **nonsinusoidal oscillators.** As the name implies, the output from these oscillators is not a sine wave.

No specific wave shape is characteristic of all nonsinusoidal oscillators. They are usually a collection of many circuits, each with its own characteristic wave shape. The nonsinusoidal output may be square, sawtooth, rectangular, triangular, or even a combination of two such waveshapes. However, one common characteristic is that they are usually some form of **relaxation oscillator.** This terms applies to a portion of the oscillation cycle where energy is rapidly stored in one of the reactive circuit components, and gradually released during the "**relaxation**" part of the cycle.

The Blocking Oscillator

The blocking oscillator, shown in Figure 29-26, is a perfect example of the relaxation principle, since it keeps itself cut off during most of the cycle. The connection of transformer T_1 may remind you of the Armstrong oscillator studied earlier. However, in this circuit, neither of the transformer's windings are tuned by a capacitor, as in the Armstrong oscillator. This is the distinguishing feature that makes the two oscillators different.

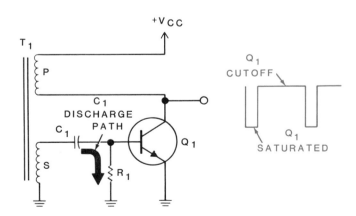

Figure 29-26 A basic blocking oscillator.

When the circuit is initially energized, transistor Q_1 is forward-biased by the bias network (not shown), and collector current is through the primary of T_1. This current induces a positive voltage into the transformer's secondary. This current is capacitively coupled to the base of Q_1. This increases forward-bias, driving Q_1 into saturation.

With Q_1 saturated, no voltage is induced into the secondary of T_1, because the field around the primary is no longer changing. The field around the secondary collapses, developing a negative potential that C_1 couples to the base of Q_1. Transistor Q_1 is reverse-biased, and quickly driven into cutoff.

At this point, C_1 is charged to a negative potential. The only discharge path is through the high resistance of R_1. Therefore, C_1 discharges very slowly, keeping Q_1 at cutoff. When the charge on C_1 is significantly reduced, Q_1 is again forward-biased, and the oscillator's action is repeated.

This oscillator is called a blocking oscillator because the transistor is easily driven into the **blocking** mode. This blocking condition is determined by the slow discharge of capacitor C_1, which holds the transistor at cutoff. Capacitor C_1 charges rapidly through the low resistance of the emitter-base junction of Q_1 when it is conducting. However, with Q_1 cut-off the only discharge path is through the relatively high resistance of R_1. Therefore, the RC time constant of R_1 and C_1 determines how long the capacitor discharges, which in turn determines the amount of time Q_1 is blocked, or cut-off. This sets the frequency of oscillation. A long time constant results in a low frequency oscillation. Subsequently, a short time constant will produce high frequency oscillation. This frequency is usually made adjustable by using a potentiometer as R_1.

ELECTRONIC CIRCUITS

A Sawtooth Blocking Oscillator

The output from the previous circuit was taken across the transistor and resembles a rectangular waveshape. However, the blocking oscillator is frequently used to generate sawtooth wave shapes. Such a circuit is shown in Figure 29-27.

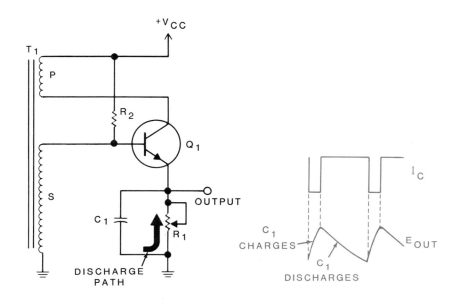

Figure 29-27 Sawtooth blocking oscillator.

Here, the output is taken across the RC network in the emitter circuit. In this oscillator, the RC network performs a dual function. It determines the frequency of oscillation, and it produces the sawtooth output.

Basically, the circuit operates much like the oscillator shown in Figure 29-26. Q_1 is forward-biased by resistor R_2, and conducts through the primary of T_1. Voltage induced into the transformer's secondary increases forward-bias, and Q_1 saturates, as shown by the low point of the collector's output waveform. As Q_1 conducts, capacitor C_1 charges rapidly and, as shown by the output wave shape, its charge rate is almost linear.

OSCILLATORS

Once Q_1 saturates, no voltage is induced into the transformer's secondary winding and the base of Q_1 is no longer forward-biased. The positive potential on the top plate of C_1 now reverse-biases the emitter junction, and Q_1 quickly cuts off. Capacitor C_1 discharges through R_1, producing the trailing portion of the output sawtooth. When C_1 is completely discharged, Q_1 is again forward-biased, and the action is repeated to provide the next cycle of the output waveform.

Notice that the capacitor C_1 and potentiometer R_1 determine the frequency of oscillation. R_1 is made variable for frequency adjustment. Again, if R_1 is set to a high resistance, a long RC discharge time constant produces low frequency oscillation. Likewise, if R_1 is set to a low resistance, providing a short RC time constant, the oscillator's frequency increases.

ELECTRONIC CIRCUITS

Chapter Self-Test

1. What is the primary function of an oscillator?

2. What type of feedback is used in most oscillators?

3. List the general requirements for an oscillator.

4. What is the difference between a tuned-base and a tuned-collector Armstrong oscillator?

5. What function does the amplifier serve in the Armstrong oscillator?

6. What is the identifying feature of a Hartley oscillator?

7. What is the difference between a series-fed and a shunt-fed Hartley oscillator?

8. Is a Colpitts oscillator series-fed or shunt-fed?

9. What is the identifying feature of a Colpitts oscillator?

10. Define piezoelectricity.

11. Which will have a higher natural frequency; a thinner crystal wafer, or a thicker crystal wafer?

12. In the crystal-enhanced Hartley and Colpitts oscillators, what function does the crystal perform?

13. What functions does the crystal perform in the Pierce oscillator?

14. A phase-shift oscillator uses two RC networks as the feedback. What is the phase shift for each of these RC networks? Why?

15. What type of circuit serves as both the feedback circuit and the frequency determining circuit in the Wien-bridge oscillator?

OSCILLATORS

16. In order for the Wien-bridge oscillator to properly function, what must be the relationship between the regenerative and the degenerative feedback?

17. What is the difference between the Armstrong oscillator and the blocking oscillator?

18. Why do the capacitor discharge paths in both the blocking and sawtooth blocking oscillators have such high resistance?

Summary

An oscillator is an electronic circuit that generates a repetitive AC signal. Oscillator output must be uniform (not varying in frequency or amplitude.) The basic oscillator requires an amplifier to replace circuit losses, frequency determining components, and positive feedback to sustain oscillation. In addition, they must also be self-starting.

Feedback oscillators are classified by the frequency determining components. The three classifications are LC, RC, and crystal. The parallel LC circuit oscillates when shock-excited by a DC source, but internal resistance quickly dampens the oscillations. LC oscillators use the resonant frequency characteristic of the LC "tank" circuit. You can find the frequency of an LC oscillator by using the resonant frequency formula for the tank circuit.

Commonly used LC oscillators are the Armstrong, the series- and shunt-fed Hartley, and the Colpitts. The Armstrong oscillator is identified by the "tickler" coil and tuned LC tank circuit. Hartley oscillators are easily identified by the tapped coil in the LC tank. In the series-fed Hartley, transistor current flows through a portion of the tapped inductor. The shunt-fed Hartley is so named because feedback is AC coupled through a capacitor. Therefore, DC current is not present in the tank circuit of the shunt-fed Hartley.

The Colpitts oscillator is easily identified by the tapped voltage-divider arrangement of two capacitors in the LC circuit. These capacitors develop regenerative feedback necessary for oscillation.

Crystal oscillators are used where a high degree of stability is required. Crystals have a high Q and, therefore, good selectivity.

A crystal may be operated in the series-resonant or parallel-resonant mode. In both modes, the crystal controls oscillator feedback. A crystal operating in the series-resonant mode has minimum impedance at the series-resonant frequency and therefore, permits maximum feedback current. Subsequently, a crystal operating in the parallel-resonant mode has maximum impedance at resonance and develops maximum feedback voltage.

Most crystal oscillators are modifications of basic LC oscillators. The Pierce oscillator is a modified Colpitts. The modification is to replace the inductor with a crystal. The Butler oscillator combines an LC tuned collector circuit with the frequency selectivity of the crystal. This allows the oscillator to be used as an overtone (frequency multiplier) circuit.

OSCILLATORS

RC oscillators use resistance-capacitance networks to determine the oscillator's frequency. The two basic types of sine wave RC oscillators are the phase-shift and Wien bridge.

The phase shift oscillator produces the required 180 degrees phase shift for regenerative feedback, through a series of RC networks, connected between the collector and base of the transistor amplifier. Each RC network contributes to the total phase shift and the oscillator operates only at the frequency where the total phase shift is 180 degrees.

The Wien-bridge oscillator uses a lead-lag network which is frequency sensitive. The lead-lag network controls regenerative feedback and makes up one leg of the bridge.

Resistors make up the other leg of the bridge and controls degenerative feedback. Oscillations are produced when regenerative feedback exceeds degenerative feedback. The lead-lag network controls the frequency at which this condition is present and, therefore, controls the oscillator's frequency. Regenerative feedback is frequency sensitive. Degenerative feedback is resistor controlled and thus not frequency sensitive.

The blocking oscillator is a perfect example of the relaxation oscillator. A capacitor stores energy during a small portion of the oscillation cycle and then gradually releases this energy during the relaxation portion of the cycle. This type of oscillator produces a nonsinusoidal output such as the sawtooth or rectangular wave shape. Like the Armstrong oscillator it uses a tickler coil, but unlike the Armstrong oscillator it does not contain a tuned tank.

ELECTRONIC CIRCUITS

CHAPTER 30

Waveshaping Circuits

ELECTRONIC CIRCUITS

Contents

Introduction 829

Chapter Objectives 830

Waveshaping. 831

Rectangular Wave Generators 845

Ramp Generators 857

Chapter Self-Test 863

Summary. 865

WAVESHAPING CIRCUITS

Introduction

In our final chapter, we will introduce you to the many different types of circuits which alter the shape of a waveform. For example, many computers and other sophisticated equipment require timing pulses of various frequencies to ensure proper operation. You have seen that an oscillator will produce an accurate sine wave, but a sine wave cannot provide the instant transitions required for digital operation. Therefore, the sine wave must be altered or changed to a square wave. This is accomplished through a waveshaping circuit.

In addition, you have seen how many components and circuits require controlling pulses to operate. In order to produce pulses that have a specific length and frequency, waveshaping circuits are used.

This chapter will show you many different types of waveshaping circuits, from the simple to the complex. You will see simple RC networks producing complex waveforms, and circuits similar to rectifiers completely changing the shape of an input signal. **Differentiators, integrators, clippers, clampers,** and **slicers** are examples of these circuits.

You will then learn about **multivibrators,** which use square waves of various shapes and frequencies for both input and output signals. As digital electronics becomes more and more popular, these circuits are used more often. We will show you the **astable,** the **monostable,** and the **bistable** multivibrator.

The final portion of this chapter will show you the various circuits used to create different triangular waveforms. We have discussed sawtooth and triangular waves previously, and these are the circuits used to create these waveforms.

Chapter Objectives

When you have completed this chapter, you will be able to:

1. Define the following terms: periodic waveform, aperiodic waveform, pulse width, duty cycle, rise time, fall time, overshoot, ringing, undershoot, steering diode, gate, set, reset, and trapezoid waveform.

2. Describe the operation of differentiator and integrator circuits.

3. Given the input waveform, draw the output waveform of biased and unbiased, series and shunt diode clippers.

4. Describe the general operation of a slicer circuit.

5. Describe the general operation of a clamper.

6. Describe the operation of astable, monostable, and bistable multivibrators.

7. Explain the operation of a simple Schmitt trigger circuit.

8. Show how an operational amplifier can be made to produce a linear ramp.

9. Explain the operation of a transistor sawtooth generator.

WAVESHAPING CIRCUITS

Waveshaping

Frequently in electronics, it is necessary to change the shape of a waveform. A sine wave may be changed to a square wave, a rectangular wave may be changed to a pulse waveform, etc. Generally, the waveshaping is done intentionally. However, it sometimes happens accidentally due to poor design, component changes, and other factors. In this chapter, you will study several different types of circuits which can change the shape of a waveform.

Terminology

Before beginning our study of these circuits, let's define some of the terms we will be using. The first of these is **periodic waveforms.** A periodic waveform is any waveform which has the exact same shape from one cycle to the next. Figure 30-1A shows a sine wave, a square wave, and a sawtooth wave. Each of these are periodic waveforms.

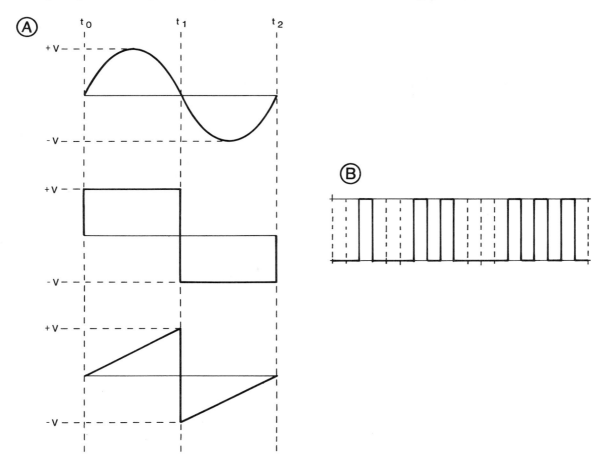

Figure 30-1 Periodic and aperiodic waveforms.

ELECTRONIC CIRCUITS

Just as there are periodic waveforms, there are also **aperiodic** or **nonperiodic waveforms.** These, as you may guess, are signals which do *not* retain the same shape from one cycle to the next. A good example of an aperiodic waveform is a data signal, shown in Figure 30-1B. A data signal is made up of an altered square wave, which switches from its positive to its negative voltage level at different times. A specific series of the positive and negative voltages represent a piece of data. Since data constantly changes, the switching time for this signal changes. Thus, a data signal does not retain the same shape from one cycle to the next.

Additional terminology stems from the use of pulse circuits; one of the circuits we'll be discussing in this chapter. The first of these terms is a **pulse width.** We already know that the period of a waveform is the amount of time required to make one cycle. We also know that the frequency of a signal is inversely proportional to its period. A pulse width, however, is different. In Figure 30-2A, the period of the entire cycle is 100 microseconds. However, the square wave only stays in the positive portion for 250 microseconds. If this positive pulse is the pulse that is being used, then this signal would have a pulse width of 250 microseconds.

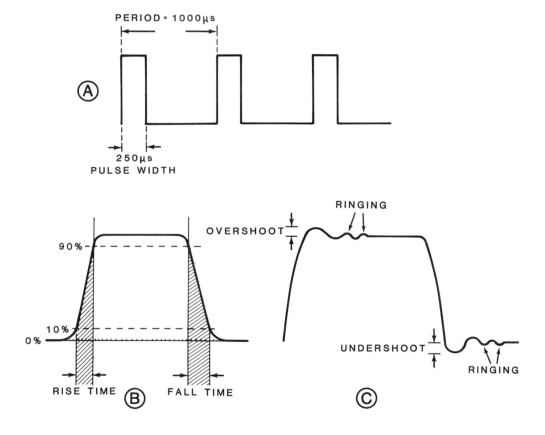

Figure 30-2 Pulse characteristics.

WAVESHAPING CIRCUITS

The square waves we've discussed to this point all had positive and negative portions that were equal. Pulse circuits, on the other hand, may have a very small positive or negative portion. Thus, the pulse width becomes a very important characteristic.

Duty cycle is the ratio of pulse width to period. It can be thought of as the percent of each cycle that the pulse exists. Duty cycle may be computed using the formula:

$$\text{duty cycle} = \frac{\text{pulse width}}{\text{period}}$$

In Figure 30-2A, the duty cycle is:

$$\text{duty cycle} = \frac{250\mu s}{1000\mu s} = 0.25$$

However, duty cycle is normally expressed as a percent. Thus, the duty cycle is 25%. Duty cycle is sometimes defined as the transistor's on-time compared to its off-time. In this case, it is the percentage of each cycle during which the transistor is active (conducting). In this course, we will refer to duty cycle as the time of the positive pulse compared to time of one complete cycle (waveform's period).

Figure 30-2B demonstrates another set of very important set of terms. An ideal square wave would change instantly from the positive to the negative portion. Let's say this square wave is the output of an overdriven amplifier. Obviously, there is a certain amount of time required for the amplifier to change from the positive to the negative portions, and vice-versa. The time required to change from the positive to the negative portion is known as **fall time,** and the time required to change from the negative to the positive portion is known as **rise time.** The rise time is defined as the time required for the pulse to rise from 10% to 90% of its maximum amplitude. By the same token, the fall time is the time required for the pulse to fall from 90% to 10% of its maximum amplitude. These transition portions are also known as **leading edge** and **trailing edge.**

Overshoot, undershoot, and **ringing** frequently accompany high frequency pulses. These conditions are illustrated in Figure 30-2C. Notice that the leading edge initially overshoots its normal maximum value. The overshoot is often followed by damped oscillations known as ringing. Finally, upon returning to its normal minimum value, the trailing edge undershoots this value, and again, some ringing is evident. These conditions are normally unwanted, but occur because of imperfect circuit conditions.

ELECTRONIC CIRCUITS

RC Waveshaping

Now that we've discussed some of the basic terms involved in waveshaping circuits, let's discuss our first type of waveshaping. A simple resistor-capacitor (RC) network can change the shape of complex waveforms so drastically that the output barely resembles the input. The amount of distortion is determined by the RC time constant. The nature of the distortion is determined by the component across which the output is taken. If the output is taken across the resistor, the circuit is called a **differentiator**. When the output is taken across the capacitor, the circuit is called an **integrator**.

Differentiator

A differentiator is shown in Figure 30-3A. This circuit is the exact same circuit as those used as RC voltage dividers and phase shifters back in Chapter 14. When a sine wave is applied to the input, the output will also be a sine wave, but lower in amplitude, and with a phase shift. Figures 30-3B and 30-3C show the input and output waveforms.

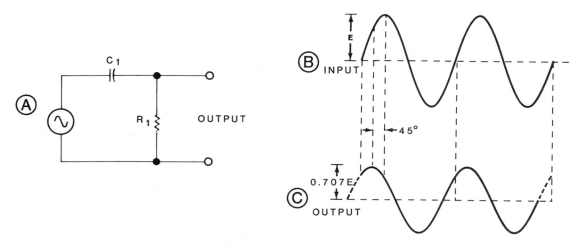

Figure 30-3 The differentiator does not distort a sine wave.

834

However, when a square wave is applied to this exact same circuit, a completely different output is obtained. This is shown in Figure 30-4. By following the charge and discharge of C_1, we can understand why the square wave is distorted.

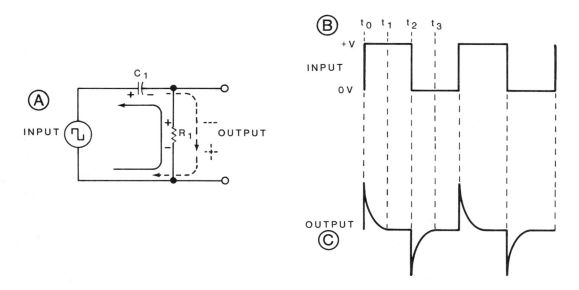

Figure 30-4 The differentiator changes a square wave to spikes.

The differentiator will produce the results shown in Figure 30-4C only if the RC time constant is short when compared to the period of the input square wave. A short RC time constant is when the RC time constant is approximately equal to 1/10 of the time of one cycle. Let's assume that this is true for the circuit in Figure 30-4A.

Notice how the input square wave for this circuit (shown in Figure 30-4B) has a negative portion of 0 volts. The significance of this becomes apparent when we consider circuit operation. When the input signal is initially applied, the positive voltage is immediately felt at the input. The capacitor immediately begins to charge to this positive voltage, as shown by the solid arrow.

The charge of C_1 is controlled by the RC time constant. As C_1 charges, it forces current through R_1, developing a positive voltage at the output. However, the current through R_1 quickly decreases as the capacitor becomes charged. In fact, when C_1 is fully charged, the current through R_1 ceases altogether. Thus, the output voltage quickly decreases, falling back to 0 volts when C_1 is completely charged. The output voltage remains at 0 volts until the input changes to its negative portion.

ELECTRONIC CIRCUITS

We should point out that the voltage drops of the capacitor and the resistor must always equal the applied voltage. When power is first applied, maximum current flows because the capacitor has not charged. This means that the resistor drops the applied voltage, and the capacitor develops no voltage. As the capacitor charges, it develops more and more voltage. At the same time, current flow through the resistor decreases, which decreases the resistor's voltage drop. By the time the capacitor has fully charged, it develops the applied voltage. Current flow ceases, and the resistor develops no voltage.

When the input changes to the negative portion, the input is essentially 0 volts. C_1 immediately begins its discharge through R_1 as shown by the dotted arrow. This develops a negative voltage across R_1. Thus, the output suddenly goes sharply negative as shown in Figure 30-4C. As C_1 discharges, the current through R_1 quickly decreases. The output voltage returns to 0 volts when the capacitor is fully discharged. As you can see, the RC circuit converts the square wave to positive and negative spikes when the RC time constant is very short.

The importance of the RC time constant is illustrated in Figure 30-5. Figure 30-5A shows a 1000 Hz square wave that is applied to a differentiator. Figure 30-5B shows the sharp output spikes when the RC time constant is very short. When the time constant is made equal to one-half the period of the input, as shown in Figure 30-5C, less distortion results. The reason for this is that the capacitor never becomes fully charged. Even so, the output is still clearly distorted.

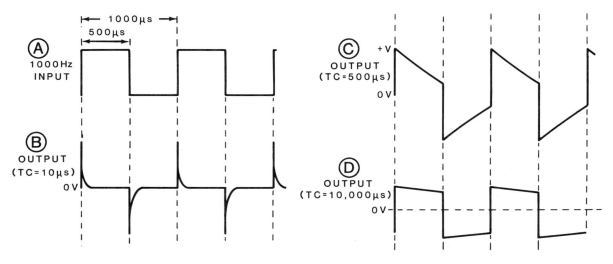

Figure 30-5 Effects of RC time constant.

When the time constant is much longer than the period of the input as shown in Figure 30-5D, the output is only slightly distorted. This illustrates that a differentiator circuit must have a very long time constant if it is to pass complex waveforms without distorting them.

Integrator

An integrator circuit is shown in Figure 30-6A. Its appearance is similar to that of the differentiator, except that the output is taken across the capacitor. Like the differentiator, the integrator cannot distort a pure sine wave. However, it will distort a complex waveform.

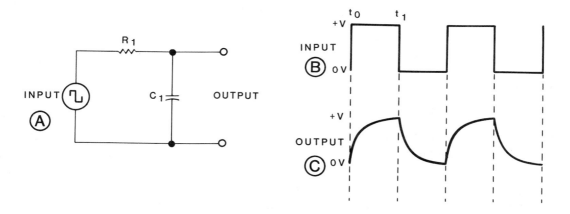

Figure 30-6 The integrator and its waveforms.

We'll apply the same square wave to the integrator as we did to the differentiator. The positive portion is some positive voltage, while the negative portion is equal to 0 volts. We'll also assume that the RC time constant is about one-tenth the period of the square wave. When the input square wave is first applied, the capacitor begins to charge. Initially the voltage across C_1, which is the output voltage, is 0 volts. As C_1 charges, the output voltage rises. C_1 is completely charged before the input changes to 0 volts.

When the input changes to 0 volts, the capacitor discharges. The output voltage drops to 0 volts at the rate of discharge. Again, the capacitor is completely discharged before the input voltage changes to the positive portion. As you can see, the integrator distorts the input but in a different way than the differentiator.

ELECTRONIC CIRCUITS

Diode Clipping Circuits

RC circuits change the shape of the waveform by charging and discharging a capacitor. Another circuit that can change the shape of a waveform is the **clipping circuit,** or **clipper.**

A clipping circuit is used to cut off an unwanted portion of a waveform. The unwanted portion may be a high amplitude noise pulse, an overshoot produced by a reactive component, or a natural part of the waveform that we wish to eliminate. The clipper can also be used to prevent a voltage from exceeding certain limits. When used in this way, the circuit is often called a **limiter.**

A diode makes an ideal clipper since it passes current in one direction, but not in the other. The diode can be used to clip off any voltage above or below a certain reference level. The half-wave rectifier, which was discussed in Chapter 28, is a good example of diode clipping.

The Series Clipper

Figure 30-7A shows a basic diode clipper along with its input and output waveforms. Notice how the output is similar to the half-wave rectifier we discussed in Chapter 28, with R_1 taking place of the load. The positive input alternation forward-biases the diode, and current flows through the resistor, developing the output. The forward-biased diode develops only 0.7 volts, which can be ignored. Therefore the full alternation is developed at the output. The negative input alternation reverse-biases the diode, preventing current flow in the circuit. Therefore, no output voltage is developed.

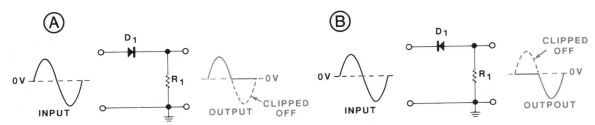

Figure 30-7   Series clippers.

In Figure 30-7B, the diode is turned around. The circuit behaves exactly the same, but the diode will be biased on the opposite alternation. The positive alternation causes reverse-bias, and no output is developed. The negative alternation forward-biases the diode, and the resistor develops the full negative alternation.

WAVESHAPING CIRCUITS

These circuits are called **series clippers** when they are not used as half-wave rectifiers. They are called series clippers because the component which causes the clipping is in series with the input and output signals. These circuits are also sometimes referred to as **diode detectors.** It detects (passes) one polarity pulse and blocks the opposite polarity pulse. By reversing the diode in the circuit it can detect (pass) the other half cycle.

Biased Series Clippers

In the circuits discussed, the clipping level was at 0 volts. Depending on how the diode was installed in the current path, everything above or below 0 volts was clipped off. This is because the voltage on the opposite end of the diode (as opposed to the input voltage) was ground, or 0 volts. In the biased clippers, the clipping level is changed by biasing the side of the diode opposite the input at a voltage other than 0 volts. In the examples shown in Figure 30-8, the bias voltage is represented by a battery.

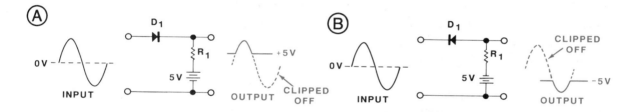

Figure 30-8 Biased series clippers.

In Figure 30-8A, the battery voltage is +5 volts. In the unbiased clipper shown in Figure 30-7A, the *entire* positive alternation forward-biased the diode because the entire alternation was above 0 volts. With the cathode of the diode at +5 volts in Figure 30-8A, only that portion of the positive alternation *above* +5 volts will forward-bias the diode, and allow current flow. Obviously then, the diode cannot conduct until the input signal swings to +5.7 volts (which allows the 0.7 volts required for forward-bias).

ELECTRONIC CIRCUITS

Since the battery is also at the output, the output remains at +5 volts, except for that portion of the positive half-cycle which swings above this level. During the most positive part of the waveform, D_1 conducts and passes that portion of the input signal to the output. As you can see, all of the negative half-cycle is clipped off.

By reversing the diode and the polarity of the bias voltage, the same but opposite effect is achieved. Because of the biasing of the anode of the diode, only a small portion of the negative half-cycle is developed at the output, as shown in Figure 30-8B. The anode of D_1 is held at −5 volts. Thus, the diode cannot conduct until the input signal drops below −5.7 volts. This occurs only during the most negative portion of the waveform. The diode conducts, allowing current to flow through the resistor, which develops that part of the negative input signal.

Shunt Clipper

The shunt clipper behaves exactly the same as the series clipper, but the output is taken across the diode. When the diode is reverse-biased, it acts as an open and the input signal is passed to the output. However, when the diode conducts, the output voltage drops to the 0.7 volts dropped across the diode. In other words, a shunt clipper limits when the forward-biased diode cannot develop a significant output voltage, and develops an output across the reverse-biased diode.

A positive shunt clipper is shown in Figure 30-9A. When the sine wave swings positive, the diode conducts, and 0.7 volts is developed across the output. On the negative half-cycle, the diode cuts off. Thus, the negative half-cycle is simply coupled through R_1 to the output. This circuit clips off most of the positive half-cycle. If we wish to clip the negative half-cycle instead, we simply reverse the diode as shown in Figure 30-9B.

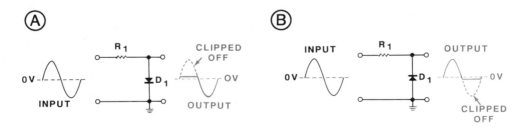

Figure 30-9 Shunt clipper.

WAVESHAPING CIRCUITS

Biased Shunt Clipper

The clipping level can be adjusted by introducing a bias voltage. For example in Figure 30-10A, the cathode of the diode is set to +5 volts. Obviously then, the diode cannot conduct until the input signal exceeds +5.7 volts. Until the input signal reaches this voltage level, the diode is reverse-biased, and the input signal is coupled through R_1 to the output. When the input signal exceeds +5 volts, the diode conducts, and +0.7 volts is developed at the output. If we wish to clip the most negative portion of the input signal, we simply switch the diode, as shown in Figure 30-10B.

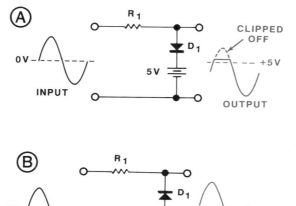

Figure 30-10 Biased shunt clippers.

For simplicity, the waveforms have been shown as sine waves. However, all of the clippers shown will work for any type of waveform. Since there are no reactive components, they are relatively insensitive to frequency.

Slicer Circuits

The limits of the clipper circuit is that it can only clip one alternation at a time. The slicer circuit is used to limit both extremes of the input waveform. A slicer circuit uses two diodes, each in a different parallel branch. The diodes are connected oppositely; that is, the anode of one diode faces the input, while the other diode has the cathode facing the input. Such a circuit can be either unbiased (with the opposite ends of the diodes at ground) or biased with a battery.

ELECTRONIC CIRCUITS

In Figure 30-11A, a biased slicer circuit is shown. The batteries connected between ground and the diodes indicate that this circuit is biased. If the diodes were tied directly to ground, the circuit would be unbiased. The cathode of D_1 is held at +5 volts, while the anode of D_2 is held at −5 volts (note the short end of the battery). When the input swings above +5.7 volts, D_1 conducts, clipping the output. When the input swings below −5.7 volts, the anode of D_2 conducts, and clips the output to this level. When the input is below +5.7 volts, but above −5.7 volts, neither diode conducts and the input signal is coupled to the output. This type of circuit could be used in equipment and integrated circuits to ensure that voltages don't exceed safe limits.

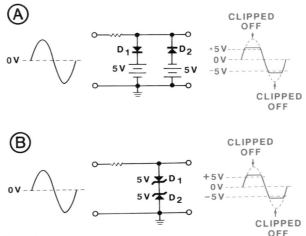

Figure 30-11 Slicer circuits.

Figure 30-11B shows a different method of biasing a slicer circuit. This circuit is considered more practical because zener diodes are used, eliminating the need for batteries. The zener diodes will conduct if forward-biased, or if reverse-biased beyond 5 volts, either positive or negative. Let's assume the input is at +3 volts. D_1 is forward-biased, and conducts. D_2 is reverse-biased, but not beyond +5 volts. Therefore, +3 volts is coupled to the output. When the output attempts to go beyond +5 volts, D_2 enters the breakdown region, and conducts. The +5 volts breakdown voltage is coupled to the output.

Let's see what happens during the negative alternation. As long as the negative alternation is below −5 volts, D_1 is simply reverse-biased, and the input is coupled to the output. The fact that D_2 is forward-biased doesn't matter, because D_1 prevents current flow. However, once the input exceeds −5 volts, D_1 enters the breakdown region. Now both D_1 and D_2 conduct, and the output feels the −5 volt breakdown voltage of D_1.

WAVESHAPING CIRCUITS

To summarize, any time the input is between +5 and −5 volts, one or the other diode is reverse-biased, and the input is coupled to the output. When the input attempts to exceed these voltage levels, one or the other diode enters the breakdown region, and + or −5 volts is developed at the output.

Clampers

A clamping circuit is used to change the DC reference voltage of a waveform. It clamps the top or bottom of a waveform to a DC voltage. Unlike the clipper, the clamping circuit does not change the shape (distort) of the waveform, it simply inserts a DC reference voltage. For this reason, the clamper is sometimes called a DC restorer. Unlike the clipper circuit, the clamper passes all of the input waveform, but offset to a new reference.

A simple diode clamping circuit is shown in Figure 30-12A. In this example, a square wave (Figure 30-12B) is used as the input signal. The purpose of this circuit is to clamp the top of the square wave to 0 volts, without changing the shape of the waveform.

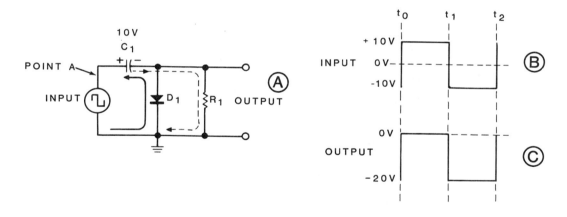

Figure 30-12 Clamping the top of a waveform to ground.

The key to the clamping circuit is the unequal charge and discharge paths of the capacitor. During the positive portion of the input (t_0 to t_1), the diode is forward-biased. This provides a low-resistance charge path for C_1. C_1 quickly charges to the applied voltage, which is the voltage level of the positive portion of the input. In this example, C_1 charges to +10 volts. At the same time, the output voltage is 0 volts, because of the forward-biased diode.

Before we discuss the negative portion of the input, keep in mind that C_1 is charged to 10 volts. Now the input switches to the negative portion, which is −10 volts for this example. With the negative plate of C_1 at −10 volts, R_1, and the output, feels the −10 volts of the input signal *and* the −10 volts of the charged capacitor. The output, therefore, drops to −20 volts. Normally, as the capacitor discharges, its voltage starts to drop. However, with the clamper, the ohmic value of R_1 is very high, which prevents C_1 from discharging hardly any voltage. This allows the output to remain at −20 volts. When the input signal swings positive again, any voltage C_1 lost is regained when it recharges.

As you can see, the shape of the waveform is still a square wave, but the 0 volt reference has been shifted to the top of the waveform. That is, the top of the waveform has been clamped to 0 volts. This is called **positive peak clamping** because the positive peak has been clamped.

WAVESHAPING CIRCUITS

Rectangular Wave Generators

Rectangular waveforms play important roles in electronics. Because of their sharp leading and trailing edges, the rectangular wave is ideal for timing purposes. Rectangular waveforms are easily generated and then changed to other shapes. For this reason, inside most electronic equipment, most sawtooth and triangle waveforms begin as rectangular waveforms.

In this section we will look at several circuits which produce rectangular waveshapes. Some, like the astable multivibrator, are free-running and produce an output without being triggered by an input signal. This is similar to an oscillator. Others, like the one-shot multivibrator, produce an output only when triggered by an input. Still others, like the Schmitt trigger, change the characteristics of the input signal to produce a rectangular output, regardless of the shape of input waveform.

Astable Multivibrator

The **astable multivibrator** produces a rectangular waveform without requiring an input signal. For this reason, it is often called a **free-running multivibrator.** It is a type of RC oscillator which uses two transistor stages. A heavy regenerative feedback causes the transistor to alternate between cutoff and saturation. Consequently, the output is a square or rectangular waveform rather than a sine wave. The frequency of oscillation is determined by two RC time constants. An RC network controls the conduction time of each transistor.

ELECTRONIC CIRCUITS

The basic circuit is shown in Figure 30-13. It consists of two transistors, with the output of one connected to the input of the other. R_2 and R_3 bias the transistors into saturation. Capacitor C_1 couples the collector of Q_1 to the base of Q_2. In the same way, C_2 couples the collector of Q_2 to the base of Q_1.

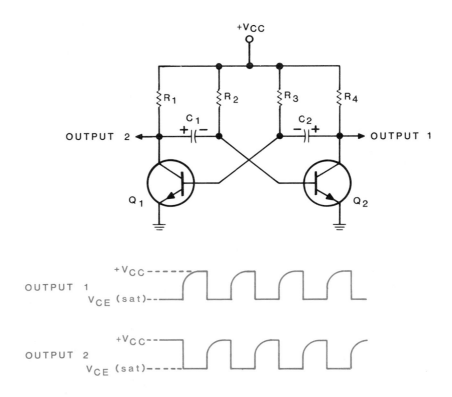

Figure 30-13 The astable multivibrator and its waveforms.

In normal operation, one transistor is cut off while the other is conducting. After a brief interval, the circuit changes states: the transistor which was conducting cuts off, while the transistor which was cut off starts conducting. The circuit oscillates back and forth between these two states. Since the transistors always operate either saturated or cutoff, the transistors will develop either 0 volts (saturated), or V_{CC} (cutoff). This makes the output a rectangular wave which can be taken from the collector of either transistor. The frequency of oscillation is primarily determined by the values of R_2, R_3, C_1, and C_2.

When power is initially applied to the circuit, normal component tolerance (mismatches) will cause one of the transistors to conduct harder than the other. It doesn't matter which transistor conducts first, so we'll assume that Q_1 initially conducts harder. As Q_1 conducts, its collector voltage decreases. This decrease in voltage is coupled through C_1 to Q_2. This negative change (less positive) causes Q_2 to conduct less, causing its collector voltage to increase. This increase is tied to the base of Q_1, which conducts even harder. While Q_1 conducts, C_2 charges to V_{CC}. This continues until Q_1 goes to saturation, and the 0 volt output of Q_1 is coupled to Q_2, which cuts it off.

When the collector of Q_1 reaches 0 volts, C_1 begins to charge. As soon as C_1 charges to 0.7 volts, this voltage is felt at the base of Q_2, and Q_2 begins to conduct. The collector voltage of Q_2 begins a sudden decrease, and eventually cuts Q_1 off. C_2, which was charged to V_{CC}, now begins to discharge, while C_1 charges to V_{CC}. This continues until C_2 finishes its discharge, and begins to charge in the opposite direction. As soon as the charge on C_2 reaches 0.7 volts, Q_1 is forward-biased, and begins to conduct again.

To summarize, the transistors alternately conduct and cut off, and switch when the capacitors complete their discharge. Once a transistor is turned on, it quickly goes to saturation because of the heavy regenerative action involved. The output alternates between V_{CC} and 0 volts. The time the transistors alternate is therefore determined by the RC time constant of the two capacitors.

The discharge path of the capacitors is through the base resistors (R_2 and R_3 in Figure 30-13), the power supply, and through the collector junction where they are tied. In other words, C_1 discharges when Q_1 conducts and C_2 discharges when Q_2 conducts.

The output waveforms are also shown in Figure 30-13. The outputs switch between V_{CC} and 0 volts, although the transistors will drop a small voltage. The positive pulse is produced at output 1 when Q_2 is cut off. The curved leading edge of the pulse is caused by the charging action of C_2. The positive pulse at output 2 is produced when Q_1 is cut off. Since Q_1 and Q_2 are cut off at different times, the two outputs are out-of-phase.

The RC components in the base circuits determine the frequency of operation. R_2 and R_3 (the base resistors) are generally selected in order to ensure saturation of Q_1 and Q_2. Capacitors C_1 and C_2 are then chosen to produce the desired operating frequency.

When the two transistor amplifier sections are balanced (identical), the positive portion of the outputs will be equal to the negative portion. Unequal values of capacitors can be used to produce a wider or more narrow positive pulse. When unequal capacitors are used, the outputs are considered to be unbalanced.

Occasionally, the base resistors will be made variable, allowing you to change the size of either pulse. With variable resistors, you can rebalance the circuit, upset the balance of the circuit, or change the frequency of operation. The value of the collector load resistors can be made variable to increase or decrease the amplitude (vertical height) of the waveform.

The astable (free-running) multivibrator is often referred to as being quasi-stable. This means that it has no stable state. After a pre-determined time, it switches back and forth between the conducting transistors.

Monostable Multivibrator

The astable multivibrator is so named because it has no stable state. By the same token, the monostable multivibrator gets its name from the fact that it has one stable state. The circuit is also called a one-shot multivibrator or a single-shot multivibrator, because it usually produces one output pulse for each input pulse. The multivibrator can also be used to stretch pulses, or as a frequency divider.

Figure 30-14A shows the schematic diagram of the monostable multivibrator. Notice how this circuit has the same essential look as the astable, with some noticeable differences. The output for this circuit is available from the collector of Q_2. For this reason, the stable state for this circuit is when Q_2 conducts, and Q_1 is cut off. Consequently, the unstable state for this circuit is when Q_2 is cut off, and Q_1 conducts. The circuit rests in its stable state when it is not being triggered. The unstable state is initiated when the circuit receives an input trigger pulse. The circuit stays in its unstable state for a period of time, determined by the RC time constant of C_1 and R_2. When the capacitor discharges, the circuit returns to its stable state and awaits the next input trigger.

WAVESHAPING CIRCUITS

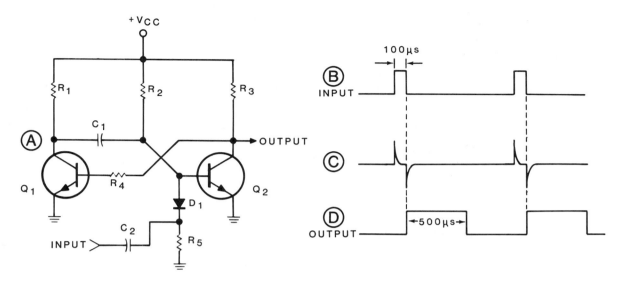

Figure 30-14 The monostable multivibrator and its waveforms.

When the circuit is in its stable state, the biasing network around Q_2 forces it to quickly saturate. This takes the output to 0 volts, which holds Q_1 at cutoff. With Q_1 cutoff, its collector remains at V_{CC}, which allows C_1 to charge to V_{CC}, also.

The circuit remains in its stable state until it receives an input triggering pulse. A typical trigger pulse is shown in Figure 30-14B. Notice that the input is applied through C_2 and R_5. This is a differentiator circuit like the ones we discussed earlier in this chapter. The differentiator circuit changes the trigger pulse to positive and negative spikes, as shown in Figure 30-14C.

These spikes are then applied to diode D_1. Because the cathode of D_1 faces the input, only the negative spike will have any effect on the circuit. This makes the circuit appear as if it acts upon the trailing edge of the input pulse. (Note the dotted line, aligning the negative spike, and the trailing edge of the trigger pulse.)

The negative spike quickly cuts off Q_2. The rising collector voltage from Q_2 is felt at the base of Q_1, as well as the output. This voltage permits Q_1 to conduct, forcing its collector voltage down. The decreasing collector voltage of Q_1 allows C_1 to discharge through R_2. The discharging capacitor holds Q_2 cut off, which is the unstable state for the circuit.

Q_2 remains cut off until C_1 completes its discharge, and charges in the opposite direction. When C_1 charges to 0.7 volts, Q_2 is again forward-biased, the output voltage drops to 0 volts, and the circuit returns to its stable state. It will remain in this state until another trigger pulse is applied.

Because the monostable multivibrator produces one output pulse for each negative going input pulse, the output frequency is the same as the input frequency (again, note the dotted line in Figure 30-14B). The circuit remains in its unstable state for as long as it takes the capacitor to discharge, and slightly charge. The output pulse width is formed when the circuit changes from its stable state to its unstable state, and back again. Therefore, the output pulse width is determined by the RC time constant of the capacitor.

This circuit provided a positive pulse because the output of Q_2 was taken as the output voltage. However, if a negative pulse were required, the output voltage could be taken from the collector of Q_1, which acted complementary to Q_2.

The diode in this circuit is used as a negative pulse detector and is sometimes referred to as a **steering diode,** because it steers (couples) the negative pulses to the base of Q_2. If we required this circuit to act on the leading edge of the input trigger, we could have simply reversed the diode, and allow the positive spike to change the circuit to its unstable state. For this to occur, however, we would also need a PNP transistor, and a negative source voltage.

The monostable multivibrator is used to produce a pulse of some specific duration. For this reason, it is sometimes called a **pulse stretcher.** Notice how the output pulses in Figure 30-14D are much longer than the input trigger pulses in Figure 30-14B.

The one-shot circuit can also be used to delay a pulse. Figure 30-15A shows how this can be done. Suppose, for example, that we wish to delay a pulse by 1000 microseconds. A simple way to do this is to use the pulse to trigger a one-shot. The original pulse is shown in Figure 30-15A. Component values are chosen so that the one-shot produces a pulse that is 1000 microseconds wide, as shown in Figure 6-25B. The 1000 microsecond pulse can then be converted to negative and positive spikes by a separate differentiator circuit. The result is a negative pulse which occurs 1000 microseconds later than the original pulse, as shown in Figure 30-15C.

WAVESHAPING CIRCUITS

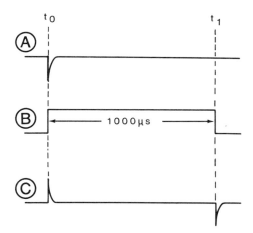

Figure 30-15 Delaying a pulse.

Since this pulse has the same characteristics as the original pulse, but occurs 1000 microseconds later, we have, in effect, delayed the pulse by 1000 microseconds. The diode only passes negative pulses, and a negative pulse will not effect the circuit's operation until after it returns to its stable state (Q_1 off and Q_2 on).

The monostable multivibrator can also be used as a frequency divider. We know that an input pulse causes the monostable to change to its unstable state. Once the circuit is in its unstable state, no input pulses can affect it. In fact, nothing can affect this circuit until it returns to its stable state. Suppose an input signal of a very high frequency were applied to the monostable. The first pulse causes the circuit to change to its unstable state. 5 more input pulses are applied, yet the circuit remains in its unstable state. Before a 6th pulse is felt, the circuit returns to its stable state. The 6th pulse then causes the circuit to return to its unstable state.

In this manner, there would be an output pulse for every 6 pulses in. This is a frequency division by a factor of 6. By adjusting the RC time of our circuit, we can either change the duration of the waveform's pulse widths (make the positive pulse either longer or shorter), or change the division ratio.

As you can see, by using simple RC components and proper biasing, you can vary the operation and application of most circuits to provide a wide range of practical circuits. This is why it is necessary to understand the component relationships and not just memorize facts about circuits. When you have mastered the basic relationships, you will be able to apply them to almost any circuit that you encounter when using schematic diagrams.

Bistable Multivibrator

The third type of multivibrator is the bistable circuit. As the name implies, this circuit has two stable states. Like the astable and the monostable, the bistable employs two transistors, and has two available outputs. Because of its operation, the bistable is also known as a flip-flop.

The flip-flop has an endless variety of applications in computers and other sophisticated equipment. Often times, a signal requires a temporary start and stop, allowing it to pass at one point, and preventing it from passing at another point. A signal which controls this is known as a gate. A flip-flop is perfect for this type of application.

By modifying the flip-flop, it can be made to divide an input frequency by two. In fact, a number of flip-flops connected together can divide a frequency by powers of two. Also, a flip-flop can be used as a memory element. It can "remember" its last input. This is the principle behind one type of computer memory.

The basic flip-flop circuit is shown in Figure 30-16A. Notice that the collector of each transistor is coupled to the base of the other. Notice that the circuit also has two inputs known as set and reset. These are applied directly to the base of each transistor. When a positive set pulse is applied, Q_1 conducts. The collector voltage of Q_1 (which is decreasing) is felt at the base of Q_2. This causes Q_2 to conduct less, and the increasing collector voltage of Q_2 causes Q_1 to conduct even harder.

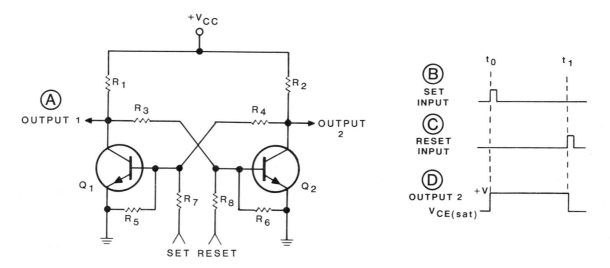

Figure 30-16   The bistable multivibrator and its waveforms.

WAVESHAPING CIRCUITS

The result is a quick saturation and cutoff, similar to the astable and monostable. Once a set pulse has been applied, no other set pulse will affect circuit operation. However, if a reset pulse is applied, Q_2 saturates, and Q_1 cuts off. Notice that output 2 in Figure 30-16D is the voltage drop of Q_2. When the set pulse is applied, Q_2 is cutoff, and output 2 rises to V_{CC}. When the reset pulse is applied. Q_2 conducts, and the output drops to 0 volts. Again, this is assuming the voltage drop of the saturated transistor is so small that it can be ignored.

Because of component tolerances and imbalances, there is no telling which output will go high when power is first applied. In sophisticated equipment, this type of guesswork cannot occur. For this reason, the set and reset inputs can be used when power is first applied in a piece of equipment. Just after power-on, the equipment will send a set or reset pulse (depending on the output desired) to the flip-flop placing it in a predetermined condition. Once there, its operation can be predicted. Also, there may be times when the equipment would like to return the flip-flop to this state. Since the set and reset inputs are located directly at the base of the transistor, it will automatically cause the flip-flop to to be placed in this state, regardless of what the flip-flop is doing at the time. When the set and reset inputs are used, this is known as **setting** or **resetting** the flip-flop. Also, because of this, the bistable multivibrator is also known as an **RS** (set-reset) **flip-flop.**

In order for the flip-flop to function as a frequency divider, it must be somewhat modified. This is shown in Figure 30-17A. Notice that in this circuit, there is only one input. To see how the circuit works, let's assume that initially Q_1 is cut off and Q_2 is conducting. With Q_1 cut off, the collector of Q_1 and the anode of D_1 will be at or near $+V_{CC}$. The voltage at the cathode of D_1 is determined by the input signal.

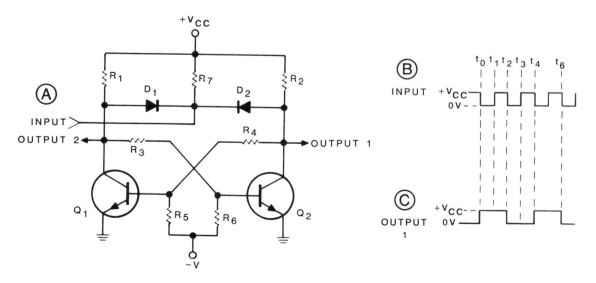

Figure 30-17   This flip-flop can divide the input frequency by two.

The input signal is shown in Figure 30-17B. Prior to t_0, the input is at $+V_{CC}$. At time t_0, the input voltage falls to 0 volts. D_1 conducts, clamping the collector of Q_1 to a low voltage. This low voltage is felt on the base of Q_2, cutting Q_2 off.

When Q_2 cuts off, its collector voltage rises to $+V_{CC}$. This increases the voltage on the base of Q_1, driving it into saturation. At t_1, the input signal returns to $+V_{CC}$. However, this has no effect on the circuit since Q_1 is held at saturation by the high collector voltage from Q_2. Q_2 is held cut off by the low collector voltage of Q_1. The output taken from the collector of Q_2 is shown in Figure 30-17C.

At time t_2, the input again drops to 0 volts. This time D_1 cannot conduct because its anode is at too low a voltage. However, D_2 can conduct because its anode is at $+V_{CC}$. D_2 conducts, placing the collector of Q_2 near 0 volts. This decrease in voltage is felt on the base of Q_1, cutting Q_1 off. Thus, the collector voltage of Q_1 rises, driving Q_2 into saturation.

With Q_2 saturated, its collector voltage will stay very low even when the input signal returns to $+V_{CC}$ at time t_3. Notice that the circuit has now returned to its original state with Q_1 cut off and Q_2 conducting. Thus, additional input pulses will cause the cycle to repeat over and over again.

D_1 and D_2 are steering diodes which are used to steer the negative-going edge of the input signal to the base of the transistor that is conducting. The flip-flop changes states only on the negative-going edge of the input pulse. Therefore, two complete cycles of the input signal are required to produce one complete cycle at the output. That is, the output frequency is one-half the input frequency. This makes the bistable a divide-by-2 circuit. Notice that the output signal has a cycle twice as long as the input. Since time and frequency are inversely proportional, a cycle twice as long means a frequency half as large.

When the stages of a bistable device are alternately triggered with the same polarity pulse, the output frequency will be one-half the input frequency. When the same side is triggered with alternating positive and negative pulses, its output frequency will be the same as its input frequency.

Discrete versions of the flip-flop are seldom used in modern equipment. However, IC versions of the flip-flop are extremely popular. The flip-flop has been elaborated and refined until it is perhaps the most important circuit used in the world of digital electronics. It is used for frequency division, storing date, counting, etc.

WAVESHAPING CIRCUITS

Schmitt Trigger

The Schmitt trigger is an important circuit used for pulse shaping purposes. It can be compared to a flip-flop because it is normally a bistable device. A common application of the Schmitt circuit is to convert a sine wave or trigger spikes to rectangular waveforms.

The circuit for doing this is shown in Figure 30-18A. As you can see, the circuit is comprised of two common-emitter amplifiers. Normally, when a sine wave is applied to a common-emitter amplifier, it produces an amplified output which is 180 degrees out-of-phase. If this output is applied to a second common-emitter, the output becomes even more amplified, and is placed back in-phase with the original input.

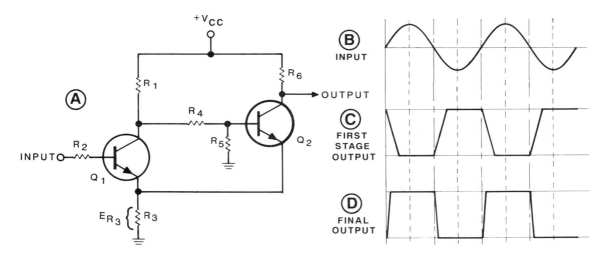

Figure 30-18 The Schmitt trigger and its waveforms.

What makes this circuit perform as it does is the biasing networks. The resistors used to determine the biasing voltages for both amplifiers are chosen so that the transistors are easily saturated and cut off.

The operation of this circuit is very simple. The original input is shown in Figure 30-18B. This sine wave is applied to the first common emitter amplifier. The output of this first stage is an overdriven waveform, 180 degrees out-of-phase with the sine wave. The voltage levels vary from V_{CC} to 0 volts, as shown in Figure 30-18C. However, you should notice that this is not a perfect square wave, as the leading and trailing edges are not very sharp.

For this reason, the output of the first common-emitter is placed at the input of the second common-emitter. Again, the second stage will easily saturate and cutoff. And because the input changes so drastically from maximum reverse-bias to maximum forward-bias, the output of the second stage is much closer to an actual square wave. This is shown in Figure 30-18D.

Because the stages are set so that they are easily saturated and cutoff, trigger pulses will cause a similar output. Figure 30-19A shows an input trigger pulse. The circuit operates in the same fashion, but the output has uneven positive and negative portions, because the input pulses have such small positive portions. If a more symmetrical output is desired, the input pulses would have to have longer positive portions.

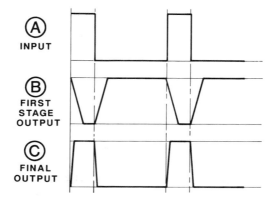

Figure 30-19 A Schmitt trigger with input pulses.

WAVESHAPING CIRCUITS

Ramp Generators

In electronics a ramp is that part of a waveform that changes linearly with time. Figure 30-20 shows three different types of ramp waveforms. Figure 30-20A is the sawtooth waveform we've discussed in various portions of this book. This type of waveform is used to sweep the electron beam across the screen of an oscilloscope.

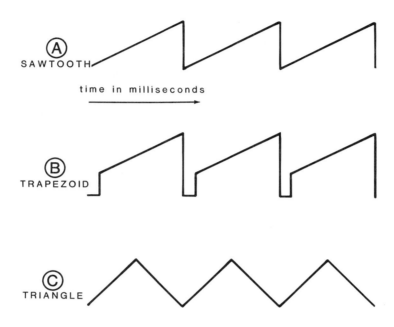

Figure 30-20   Ramp waveforms.

The trapezoid waveform shown in Figure 30-20B also has a ramp portion. This type of waveform is used in radar indicators and TV receivers to sweep the electron beam across the screen. The step (square wave portion) is used to saturate the large coils used in electromagnetic deflection systems. Then the ramp is used as the linear sweep.

Figure 30-20C is the triangular waveform, which we've also discussed previously. The triangle waveform contains both a positive-going and a negative-going ramp, and is used in digital voltmeters and other types of analog-to digital converters.

Circuits which produce waveforms of this type are often called integrators, derived from a mathematical operation called integration. For example, this waveform could be obtained using the RC integrator circuit in conjunction with a square wave input. When the square wave goes positive, the capacitor charges to a positive potential. When the square wave goes negative, the capacitor discharges.

Forming The Ramp

A ramp is formed by charging or discharging a capacitor at a linear rate. Figure 30-21A shows a simple RC circuit which can convert a square wave to a crude triangle wave. Figure 30-21B shows the square wave input. When the input jumps positive at time t_0, C_1 begins charging through R_1. The time required for C_1 to charge is determined by the R_1/C_1 time constant. If the time constant is relatively short, the output voltage taken across the capacitor will appear as shown in Figure 30-21C. The reason for this is that the capacitor does not charge linearly, it charges exponentially.

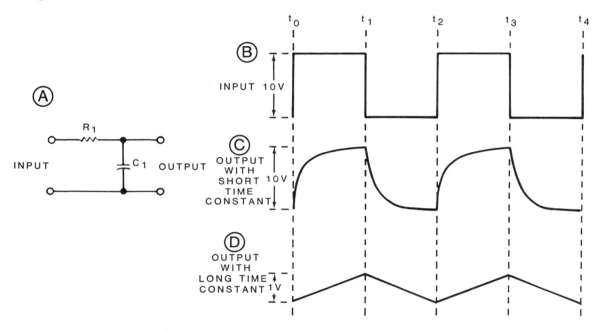

Figure 30-21 Simple RC ramp generator and its waveforms.

However, the same circuit can be made to produce a fairly linear ramp if the RC time constant is greatly increased. Assume that the value of R_1 and C_1 are increased so that C_1 charges to only 1 volt between times t_0 and t_1. In this case, the output voltage will be much lower in amplitude. However, the waveform produced will be fairly linear. Figure 30-21D shows that the output will be a low voltage triangle waveform with linear ramps. Although the capacitor still charges at an exponential rate, the very first portion of this will be fairly linear.

Operational Amplifier Integrator

Another way to produce a ramp is to charge a capacitor with a constant current. If the current flowing into and out of the capacitor can be held constant, the voltage across the capacitor will increase linearly. The trick is to hold the current constant. In most circuits, the current decreases as the capacitor charges. This is what causes the familiar time constant curve to show an exponential wave shape. One way to hold the current constant is to use an operational amplifier, as shown in Figure 30-22A.

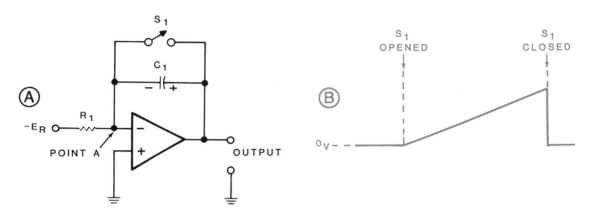

Figure 30-22 Basic integrator circuit and its output waveform.

The input to the stage is a DC reference voltage (E_R). For simplicity, S_1 is shown as a mechanical switch. In reality, a mechanical switch is much too slow. The actual switch is usually some type of transistor. C_1 is connected in the feedback path between the output and the inverting input of the op amp. The operation of this circuit is easy to understand if you remember the two principles discussed in Chapter 27. First, the feedback voltage causes the voltage at the inverting input to be *extremely* small. Second, practically no current flows into or out of the inverting input of the op amp.

Assume that S_1 is closed, that E_R is −1 volt, and that R_1 is a 1 megohm resistor. With S_1 closed, C_1 is shorted so that it completely discharges through the switch. With −1 volt applied across a 1 megohm resistor, there is 1 microamp of current flow. This current flows through the closed switch. Since the feedback resistance is 0 ohms due to the closed switch, the output remains at 0 volts.

Now, assume that S_1 is momentarily opened. Now the feedback resistance is determined by C_1. With a steady 1 microamp of current, C_1 charges very slowly. As it charges, its resistance increases. This creates an effect of an increasing feedback resistance, which increases the gain of the op amp. Therefore, the output increases. The charge on C_1 increases linearly, because there is a steady 1 microamp of current through C_1. This causes the output voltage to also increase linearly.

When a negative-going sawtooth is required, the polarity of the reference voltage (E_R) is reversed. When C_R is changed to +1 volt, the current will still be constant at 1 microampere, but the direction of current flow is now reversed. This charges C_1 to a negative voltage with respect to ground. Thus, the ramp will start at 0 volts and be negative-going.

For specific values of R_1 and C_1, the slope of the ramp is determined by the magnitude of E_R. When E_R is increased, a larger current flows through R_1 and C_1. This causes C_1 to charge more rapidly, producing a steeper ramp.

A somewhat different circuit can also be used to produce a triangle waveform. The circuit and its associated waveforms are shown in Figure 30-23. The switch is removed, and the input is changed to a rectangular waveform. The integrator converts the rectangular input waveform to a triangle waveform. The leading and trailing edges of the input waveform controls the switching action of the operational amplifier. The frequency of the input pulse is set so that the capacitor does not charge to the peak of the waveform.

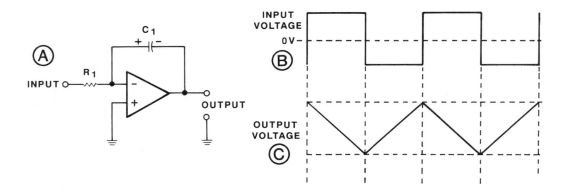

Figure 30-23 Triangular wave generator and its waveforms.

When the input waveform swings positive, current flows from the output terminal through C_1 and R_1 to the positive input voltage. This charges C_1 to the polarity shown. The magnitude of the current is determined by the input voltage (amplitude) and the value of R_1. Because this current is constant, a linear negative-going ramp is formed. When the input swings negative, the direction of the current reverses and the positive-going ramp is formed.

Sawtooth Generator

Another version of a ramp generator is the sawtooth generator shown in Figure 30-24A. This circuit is driven by a rectangular waveform, and the output is a negative-going linear ramp.

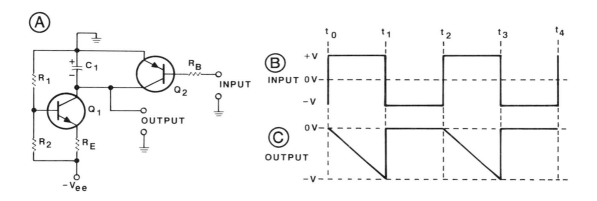

Figure 30-24 Sawtooth generator and its waveforms.

C_1 is the capacitor which develops the ramp, and Q_1, R_1, R_2, and R_E form a constant current source. Because there is no input signal to amplifier Q_1, it will remain at quiescence, and provide a constant current to C_1. Q_2 and R_B form an electronic switch that periodically shorts the capacitor, allowing it to discharge. Q_2 is a PNP transistor. The rectangular input is shown in Figure 30-24B. Notice that the input goes above and below 0 volts.

At time t_0, the input swings positive enough to cut off the PNP transistor. Thus, during the period from t_0 to t_1, the current from Q_1 flows into the capacitor. The capacitor charges to the polarity shown. As long as the current is constant, the voltage across C_1 builds up in a linear manner. Thus, an output ramp is developed at the output, as shown in Figure 30-24C.

Of course, the voltage across the capacitor cannot continue to increase indefinitely. Notice that NPN transistor Q_1 is biased using a negative voltage source. With the negative plate of the capacitor tied to the collector, there will eventually be some point where the voltage on the negative plate of C_1 will equal the base voltage of Q_1. At this point, Q_1 will cut off, and C_1 will cease to charge, leveling off the output voltage. To prevent this from happening, the period of the input waveform must be short enough to prevent the capacitor from charging to the base voltage.

At time t_1, the input waveform swings negative, forcing Q_2 to conduct. This provides a discharge path for C_1, which will immediately discharge through Q_2, dropping the output voltage to nearly zero volts. For the time of t_1 to time t_2, the saturated Q_2 provides a path for current for Q_1, and current flows away from the capacitor. The output remains near zero volts for this period of time.

At time t_2, the input voltage again swings positive, which allows C_1 to begin its linear charge once more. The frequency of the sawtooth is determined by the input frequency. However, the slope of the ramp is determined by the current and the value of C_1. A steep ramp is caused by relatively large values of current or by a small valued capacitor.

WAVESHAPING CIRCUITS

Chapter Self-Test

1. What is a periodic waveform?

2. If a positive pulse lasts for only 200 microseconds out of a 1000 microsecond cycle, what is the duty cycle of the pulse?

3. What is the difference between rise time and fall time? By what other names are rise time and fall time known as?

4. What is required to make a simple RC network into a differentiator circuit? (input waveform, output component)

5. Why is a high RC time constant ill-suited for a differentiator circuit, and perfectly suited for an integrator circuit?

6. Why does a differentiator circuit produce positive and negative output pulses, when only a square wave is applied?

7. What type of circuit (studied in a previous chapter) is an unbiased diode clipper similar to?

8. What is the difference between series and shunt clipping?

9. How does a biased clipper differ from an unbiased clipper?

10. What is the advantage in using zener diodes for a slicer circuit, as opposed to two biased diodes?

11. If the breakdown voltage for the zener diodes in a slicer circuit is 3 volts, at what voltage will the output be clipped, ignoring diode voltage drops?

12. Does a clamper circuit alter the output waveshape?

13. What is the purpose of a clamper?

14. For a clamper circuit, which is higher resistance: the *charge* path for the capacitor, or the *discharge* path of the capacitor? Why?

ELECTRONIC CIRCUITS

15. What determines the frequency of an astable multivibrator?

16. By what other name(s) is the astable known as?

17. What other name(s) is a monostable multivibrator known as?

18. Name two functions of the monostable multivibrator.

19. What determines the output frequency of the monostable multivibrator? What determines the pulse width?

20. What type of circuit is used as the input for a monostable multivibrator?

21. What is a steering diode?

22. What is the best-known term for a bistable multivibrator?

23. Name two functions of a bistable multivibrator.

24. What is the purpose of set and reset in a bistable?

25. What type of amplifiers are used in a Schmitt trigger? How do they differ from normal amplifiers?

26. What type of input(s) can a Schmitt trigger accept?

27. What are the three types of triangular waveforms?

28. What is the primary component used in creating a ramp waveform?

29. In order for an op amp ramp generator to produce a triangular waveform, what type of input is required?

30. Why can't the input period for a sawtooth generator be very long?

WAVESHAPING CIRCUITS

Summary

There are many different circuits which can alter the shape of a waveform. A simple RC circuit can distort a square wave. A differentiator usually has a short time constant. It will convert a square wave to negative and positive spikes. An integrator normally has a long time constant. It will convert a square wave to a triangular shaped waveform. In the differentiator circuit, the output is taken across the resistor. In the integrator circuit, the output is taken across the capacitor.

Diode clippers can clip off that portion of a waveform which extends above or below 0 volts. A series clipper limits the waveform when the diode is cut off. The shunt clipper limits the waveform when the diode conducts. By biasing the diode, the clipping level can be set to any point on the waveform. Zener diodes can also be used as clippers.

Clampers reproduce the input waveform in the output, but offset (clamped) to a new reference. Clamper circuits are also referred to as DC restorers. The clamper circuit uses a capacitor, and its time constant is controlled by charging through the conducting resistance of the diode (short RC time) and discharging through the load resistor (long RC time).

There are three basic types of multivibrators. The astable multivibrator free runs at a frequency determined primarily by the RC components in the base circuits. It is sometimes referred to as a square wave generator or rectangular wave oscillator.

The monostable, or one-shot multivibrator, usually produces one pulse out for each pulse in. The output pulse width is set by the RC components in one of the transistor's base circuits. The monostable multivibrator is a driven device and can be used as a pulse stretcher or a frequency divider.

The bistable multivibrator has two inputs. It is set to one state by one input. It remains in this state until reset by the other input. It is normally used as a digital device. In digital devices, it is used for storage and binary mathematical operations. It is widely used as a divide-by-two circuit.

The Schmitt trigger is an important pulse shaping circuit. It can convert sine waves and other varying inputs into pulse (rectangular) waveforms. It is most widely used as a waveshape restorer or waveform converter to change spikes and sine waves to rectangular waveforms.

ELECTRONIC CIRCUITS

Ramp waveforms are used in television receivers, oscilloscopes, and digital voltmeters. Ramps are generally formed by charging a capacitor at a linear rate. This can be done by charging a capacitor with a constant voltage so long as the capacitor is not allowed to charge to more than ten percent of the applied voltage. However, a more practical approach is to charge a capacitor from a constant current source. Ramp generators using discrete transistors and operational amplifiers are common. These linear ramp generator circuits are modified integrator circuits. When driven by a square wave they can produce triangular waveforms.

POWERS OF TEN
RIGHT TRIANGLES AND
    TRIGONOMETRY
TRIGONOMETRIC TABLE
ANSWERS TO ODD NUMBERED
    QUESTIONS
GLOSSARY

SECTION V

APPENDICES

Appendix A
Powers of Ten

In electronics, very small and very large numbers are used constantly. For example, one coulomb is equal to an electrical charge of 6,250,000,000,000,000,000 electrons. in addition, it is not uncommon to have components whose values are 0.000000000005 or less.

Men and women in the wide variety of electronics industries must work with numbers like these every day. If you want to understand electronics, you must not only understand these numbers, but how to use them in complex calculations. For example, current can be measured in coulombs per second. If you wanted to find out how many electrons flow when 5 coulombs travels in 1 minute, the following calculations would be necessary:

First, multiply one coulomb by 5:

$$\begin{array}{r} 6{,}250{,}000{,}000{,}000{,}000{,}000 \\ \times \phantom{6{,}250{,}000{,}000{,}000{,}000{,}00} 5 \\ \hline 31{,}250{,}000{,}000{,}000{,}000{,}000 \end{array}$$

Then, divide that number by 60:

$$\frac{31{,}250{,}000{,}000{,}000{,}000{,}000}{60} = 520{,}000{,}000{,}000{,}000{,}000 \text{ electrons}$$

While this specific calculation probably won't be necessary, calculations similar to this are required constantly. As you can see, there are a *lot* of zeros, which makes the opportunity for error very large. Therefore, a method of writing these numbers was created which eliminated the need for all those zeros, and made both expressing and calculating these numbers much easier. This method is based on **powers of ten.**

Exponents

When the number 10 is raised to a specific power, it simply means that 10 is multiplied by itself a specific number of times. For example, if 10 were raised to the third power, it would equal:

$$10 \times 10 \times 10 = 1000$$

APPENDIX A

In order to write a number that is raised to a power, **exponential notation** is used. In our previous example, 10 raised to the third power is written in exponential notation as follows:

$$10^3$$

Where the number "10" is the **base**, and the number "3" is the **exponent**. Thus:

$$10^3 = 10 \times 10 \times 10 = 1000 \text{ or } 10^3 = 1000$$

Positive powers of 10 are very easy, because we can simply count the number of zeros after the "1" to determine to what power 10 was raised. In our example, we raised 10 to the third power, and three zeros were placed after the "1". Some common positive powers of ten are shown in the following list. Notice how the positive sign is not used. When there is no sign before the exponent, it is assumed to be positive.

$10^0 = 1$ (*NOTE:* $X^0 = 1$, so $5^0 = 1$, $10^0 = 1$, $25^0 = 1$, etc.)
$10^1 = 10$ (*NOTE:* $X^1 = X$, so $5^1 = 5$, $10^1 = 10$, $25^1 = 25$, etc.)
$10^2 = 100$
$10^3 = 1000$
$10^4 = 10,000$
$10^5 = 100,000$
$10^6 = 1,000,000$

Notice how the number in the exponent is equal to the number of zeros after the "1".

Just as there are positive powers of ten (shown in the previous list), there are *negative* powers of ten. Positive powers of ten are used for very large numbers, and negative powers of ten express very small numbers. The method for determining the equivalent value for a negative power is almost as easy as for the positive powers of ten, but slightly different. For example, 10 raised to the "–4" power is equal to:

$$10^{-4} = 0.0001$$

In order to make any number smaller, the decimal point is moved to the *left*. In the previous example, we raised 10 to the –4 power, and moved the decimal point 4 spaces to left. This is assuming the decimal starts after the "1". This rule holds true for any negative power of ten—the decimal place is moved to the left the same number of spaces as the exponent.

$$10^{-4} = 00000.0001 = 0.0001$$

POWERS OF TEN

Common negative powers of ten are shown in the following list:

$10^{-1} = 0.1$
$10^{-2} = 0.01$
$10^{-3} = 0.001$
$10^{-4} = 0.0001$
$10^{-5} = 0.00001$
$10^{-6} = 0.000001$

Now let's place both lists in descending order to see what happens to the decimal point using different powers of ten:

$10^{6} = 1,000,000.$
$10^{5} = 100,000.$
$10^{4} = 10,000.$
$10^{3} = 1000.$
$10^{2} = 100.$
$10^{1} = 10.$
$10^{0} = 1.$
$10^{-1} = 0.1$
$10^{-2} = 0.01$
$10^{-3} = 0.001$
$10^{-4} = 0.0001$
$10^{-5} = 0.00001$
$10^{-6} = 0.000001$

As you can see, a power of ten merely moves the decimal place to the left or the right. For positive powers of ten, though, we recommend you think of it as the number of zeros, because it's a little easier.

Scientific and Engineering Notation

In the world of electronics, calculations would be incredibly easy if all the numbers involved were powers of ten. Unfortunately, this doesn't happen quite that often. But we can use these powers of ten to express *any number* as a power of ten. Two specific methods that you should become familiar with are called scientific notation and engineering notation.

Let's write the number 29,000 using scientific notation. First, we already know the following:

$29,000 = 2.9 \times 10,000$

APPENDIX A

Second, since 10,000 is equal to 10^4, we can say:

$$29{,}000 = 2.9 \times 10^4$$

So the number 29,000 is expressed in scientific notation as "2.9×10^4". Scientific notation simply takes any large or small number, and expresses that number using a power of ten.

Let's see how this is used for a very small number. Let's convert 0.0000000037 to scientific notation.

$$0.0000000037 = 3.7 \times 0.000000001$$

We found the number 0.000000001 by counting the number of places the decimal point had to move from its original location, to the place between the "3" and the "7". The decimal moved a total of nine places, so 3.7 is multiplied by 0.000000001 (the decimal point is moved from after the "1", nine places to the left.)

We also know that 0.000000001 is equal to 10^{-9}, which gives us our scientific notation:

$$0.0000000037 = 3.7 \times 10^{-9}$$

Keep in mind that in powers of ten notation, the number that is being multiplied by the power of ten is called the *mantissa*. In scientific notation, the mantissa is *always* some number between 1 and 10. For example, 37×10^{-10} isn't valid scientific notation, while 3.7×10^{-9} is.

Engineering notation is similar to scientific notation, with two exceptions. First, the mantissa is not limited to the range between 1 and 10. Any number between 0.1 and 1000 can be used as the mantissa in engineering notation. Second, the exponent must be in multiples of 3. That is, 10^3, 10^{-9}, 10^6, etc. For instance, the number 290,000 is written 2.9×10^5 in scientific notation. In engineering notation, this number could be written either as 290×10^3 or 0.29×10^6. While both examples are acceptable, 290×10^3 would probably be easier to use because it eliminates the fraction. There are times, however, when 0.29×10^6 could be preferred.

So which method should you use? As the names suggest, each method has specific uses. You could use either one, but because you are studying electronics, it is best to concentrate on engineering notation. The metric prefixes that are used throughout electronics represent powers of ten that are multiples of three.

POWERS OF TEN

The following is a list of examples of numbers converted to engineering notation.

$$68,000 = 68 \times 1000 \qquad = 68 \times 10^3$$
$$5100 = 5.1 \times 1000 \qquad = 5.1 \times 10^3$$
$$28,000,000 = 28 \times 1,000,000 \qquad = 28 \times 10^6$$
$$3,300,000 = 3.3 \times 1,000,000 \qquad = 3.3 \times 10^6$$
$$970,000 = 0.97 \times 1,000,000 \qquad = 0.97 \times 10^6$$
$$\text{or} = 970 \times 1000 \qquad = 970 \times 10^3$$

$$0.0000000035 = 3.5 \times 0.000000001 \qquad = 3.5 \times 10^{-9}$$
$$0.054 = 54 \times 0.001 \qquad = 54 \times 10^{-3}$$
$$0.000063 = 63 \times 0.000001 \qquad = 63 \times 10^{-6}$$
$$0.00000000002 = 0.02 \times 0.000000001 \qquad = 0.02 \times 10^{-9}$$
$$\text{or} = 20 \times 0.000000000001 \qquad = 20 \times 10^{-12}$$

Exercise #1

In order to give you a little practice converting numbers to engineering notation, the following exercise is provided. We recommend you write the answers on a separate piece of paper. To help make sure you are on the right track, the answers to the odd-numbered questions are located at the end of Appendix D.

Convert the following numbers to engineering notation:

1. 3400
2. 125,000,000
3. 52,000,000
4. 8,200,000,000,000
5. 94,000
6. 761,000
7. 39,000
8. 0.00000000003546
9. 0.00053
10. 0.0000179
11. 0.0000000002
12. 0.00000044
13. 0.00000186
14. 0.00000000000000943

APPENDIX A

Metric Prefixes

Powers of ten notation allows us to write very large and very small numbers with relative ease. Engineering notation is a variation of powers of ten notation, where the exponent of ten is always a multiple of 3. The common multiples of 3 can also be expressed with **metric prefixes.** Metric prefixes are Greek and English letters used to designate a power of ten that is a multiple of 3. The table of the most common metric prefixes used in electronics is below:

Metric Prefix	Power of Ten
Mega (M)	10^6, or 1,000,000
kilo (k)	10^3, or 1000
milli (m)	10^{-3}, or 0.001
micro (μ)	10^{-6}, or 0.000001
nano (n)	10^{-9}, or 0.000000001
pico (p)	10^{-12}, or 0.000000000001

The letter designation for micro is *not* a small "u", but the Greek letter **mu** (μ). These prefixes are simply used to replace the powers of ten for which they represent. For example:

$$39 \times 10^6 = 39M$$

$$2.5 \times 10^{-12} = 2.5p$$

$$10 \times 10^3 = 10k$$

If you have a whole number that you wish to express with a metric prefix, you must first change that number to engineering notation, ensuring that the exponent of 10 is one that can be replaced with a metric prefix. For example:

$$0.000024 = 2.4 \times 0.00001 = 2.4 \times 10^{-5} \quad \textit{No prefix possible!}$$

However,

$$0.000024 = 24 \times 0.000001 = 24 \times 10^{-6} = 24\mu$$

POWERS OF TEN

Exercise #2

Convert the following numbers to a metric prefix. Again, we recommend you use a separate piece of paper. The answers to the odd-numbered questions are located in Appendix D.

1. 68×10^{-3}
2. 4.45×10^{-5}
3. 722×10^{3}
4. 0.414×10^{-10}
5. 59×10^{7}
6. 0.0000000033
7. $7,780,000$
8. 10×10^{4}
9. 12×10^{-8}
10. 275×10^{6}

Calculations Involving Powers of Ten Notation

Now that you understand powers of ten notation and metric prefixes, it is time to learn how to make calculations involving these numbers. The rules that you must follow are very simple, but you must make sure that they are never forgotten, otherwise your answers may be extremely inaccurate. Remember we are dealing with very large and very small numbers—any small mistake becomes a large mistake with numbers of these proportions.

Addition and Subtraction

When adding or subtracting two (or more) numbers in powers of ten notation, the only requirement is that the exponent of 10 is the same for all the numbers. The mantissas can then be combined, and the answer is expressed in the same power of ten as the other numbers. For example:

$$(32 \times 10^2) + (19 \times 10^2) = 51 \times 10^2$$

or

$$(32 \times 10^2) - (19 \times 10^2) = 13 \times 10^2$$

APPENDIX A

When working with electronics values, you will be combining metric prefixes more often than numbers in powers of ten notation. In this case, the rule is the same: ensure all of the numbers have the same metric prefix, then the answer is also expressed with the same metric prefix. Another example demonstrates this:

$$5k + 12k + 100k = 117k$$

or

$$25.4p - 12.2p = 13.2p$$

But what if the numbers you must combine do not have the same exponent or metric prefix? Quite simply, alter the numbers so that their metric prefixes or exponents are the same, then combine. For example, let's add the following two numbers:

$$(0.47 \times 10^{-4}) + (8.2 \times 10^{-5}) = ?$$

Since the exponents are not the same, we cannot add these numbers together as they stand. But if we change one of the values so that its exponent is the same as the other, we can add the values together. It doesn't matter which number is changed, so long as the outcome is the same. It might depend, however, on which exponent you would like your answer to be expressed. We'll demonstrate changing either value:

$$(0.47 \times 10^{-4}) + (8.2 \times 10^{-5}) = (4.7 \times 10^{-5}) + (8.2 \times 10^{-5}) = 12.9 \times 10^{-5}$$

or

$$(0.47 \times 10^{-4}) + (8.2 \times 10^{-5}) = (0.47 \times 10^{-4}) + (0.82 \times 10^{-4}) = 1.29 \times 10^{-4}$$

Incidentally, we should point out that either method provides the same answer, as indicated below:

$$12.9 \times 10^{-5} = 12.9 \times 0.00001 = 0.000129$$
$$1.29 \times 10^{-4} = 1.29 \times 0.0001 = 0.000129$$

Equal values!

Multiplication

Although adding/subtracting two numbers in powers of ten notation is easy, multiplying them is even easier. When multiplying two numbers in powers of ten notation, the mantissas are multiplied together, and the exponents are added together. The formula for this calculation is as follows:

$$(A \times 10^X) \times (B \times 10^Y) = (A \times B) \times 10^{X+Y}$$

POWERS OF TEN

Let's prove how this formula works by multiplying two numbers, using both the written out and powers of ten notation methods. To keep things simple, we'll use rather simple numbers:

$$\begin{array}{cc} 3000 & 3 \times 10^3 \\ \times 200 & \times 2 \times 10^2 \\ \hline 600000 & 6 \times 10^5 \end{array} \quad (3 \times 2 = 6;\ 3 + 2 = 5)$$

It's that simple. Even when both positive and negative powers of ten are added together, the rule is the same. You should be careful, though, when adding together positive and negative numbers. A simple rule to follow for this is to take the difference between the two numbers, and use the sign (positive or negative) of the larger number. Let's see how this works:

$$(5.2 \times 10^4) \times (3.6 \times 10^{-7}) = ?$$

$$(5.2 \times 3.6 = 18.72);\ [4 + (-7)] = -3$$

$$= 18.72 \times 10^{-3}$$

Division

You know that division is the exact opposite of multiplication. It logically follows then, that the rule for dividing numbers in powers of ten notation is the exact opposite of the rule for multiplication.

In other words, since the sum of the exponents is used when multiplying, the *difference* between the exponents is used when dividing. The formula which expresses this rule is as follows:

$$\frac{A \times 10^X}{B \times 10^Y} = \left(\frac{A}{B}\right) \times 10^{X-Y}$$

Let's look at a simple example:

$$\frac{32 \times 10^5}{4 \times 10^3} = \left(\frac{32}{4}\right) \times 10^{5-3} = 8 \times 10^2$$

APPENDIX A

Unfortunately, subtracting two numbers when you're dealing with positive *and* negative numbers can be a bit confusing. Let's try another example to see what we mean:

$$\frac{39.6 \times 10^{-6}}{3.3 \times 10^{-9}} = \left(\frac{39.6}{3.3}\right) \times 10^{(-6) - (-9)} = 12 \times 10^{??}$$

That is where our problem lies. Somehow, we need a way to subtract two numbers with relative ease *regardless* of their signs. Basic algebra states that:

$$A - B = A + (-B)$$

This shows us that when we subtract B from A, it's the same as *adding* (−B) to A. Therefore, let's return to our exponents in the previous example:

$$(-6) - (-9) = (-6) + (+9) \qquad \text{\textit{The sign of the second exponent (at the denominator) is reversed!}}$$

From there, we know that to add two numbers with different signs, we take the difference between the two, and the sign of the larger number:

$$(-6) - (-9) = (-6) + (+9) = +3$$

So in our division problem before, the answer is 12×10^3.

A typical problem in electronics involves using resistance and current with voltage. Resistance, if you'll recall from Chapter 4, is a relatively high value, usually expressed in kilohms or megohms. Current, on the other hand, is a small value, normally expressed in milliamps or microamps. These values are combined with voltage, which rarely has a metric prefix. Let's divide 48 volts by 12 kilohms, and see how this problem is worked out:

$$\frac{48}{12k} = \frac{48 \times 10^0}{12 \times 10^3}$$

48 is multiplied by 10^0, or 1. This is done simply to give 48 an exponent, although it is not normally expressed with one. The rest of the problem is worked out using our "reverse-and-add" formula:

$$\frac{48 \times 10^0}{12 \times 10^3} = \left(\frac{48}{12}\right) \times 10^{0 - (3)} = 4 \times 10^{0 + (-3)} = 4 \times 10^{-3} = 4m$$

Notice that we left out the volts, ohms, and amps symbols. This was done to keep the formula as simple as possible. You should always keep track of each electrical value.

POWERS OF TEN

Combining Multiplication and Division

Occasionally, you are required to find the answer to more complex problems which include both multiplication *and* division. A formula in Chapter 14 is a perfect example of this type of calculation. The formula states that:

$$X_C = \frac{159m}{FC}$$

Let's assume that f = 318k, and C = 10p. When these values are inserted into the formula, our problem becomes:

$$X_C = \frac{159m}{(318k)(5p)} = \frac{159 \times 10^{-3}}{(318 \times 10^3)(5 \times 10^{-12})}$$

Now, what do you do first? The simplest method of calculating this problem is to do *all* possible multiplication first, then divide. (When two numbers are placed next to each other with no multiplication sign, it is assumed that the two numbers are multiplied). Let's see how this works out:

$$X_C = \frac{159 \times 10^{-3}}{(318 \times 10^3)(5 \times 10^{-12})} = \frac{159 \times 10^{-3}}{(318 \times 5) \times 10^{3 + (-12)}}$$

$$= \frac{159 \times 10^{-3}}{1590 \times 10^{-9}} = \left(\frac{159}{1590}\right) \times 10^{(-3) - (-9)} = 0.1 \times 10^{(-3) + (+9)}$$

$$= 0.1 \times 10^6 = 0.1M \text{ or } 100k$$

Problems like these are the whole reason power of ten notation and metric prefixes exist. This type of problem is a typical example of the calculations you must make in electronics. Once you've mastered them, understanding electronics becomes much easier.

APPENDIX A

Exercise #3

The following problems are examples of calculations involving power of ten notation and metric prefixes. Take your time in completing these, to be sure you have a good grasp of the ideas presented in this Appendix. As always, we recommend you perform these on a separate piece of paper, and the answers to the odd-numbered problems are located at the end of Appendix D.

1. $(3.5 \times 10^{-4}) + (0.89 \times 10^{-3})$
2. $(4.8 \times 10^{5}) + (0.2 \times 10^{3})$
3. $(96 \times 10^{-6}) - (0.079 \times 10^{-3})$
4. $0.21M - 195k$
5. $(4 \times 10^{-6}) \times (16 \times 10^{-3})$
6. $22k \times 250m$
7. $(8 \times 10^{3}) \times (4.5 \times 10^{-2})$
8. $\dfrac{3.6 \times 10^{2}}{9 \times 10^{6}}$
9. $\dfrac{2.4 \times 10^{-12}}{0.12 \times 10^{-3}}$
10. $\dfrac{125 \times 10^{3}}{2.5 \times 10^{-9}}$
11. $\dfrac{0.72 \times 10^{-2}}{8 \times 10^{7}}$
12. $\dfrac{56k}{0.7n}$
13. $\dfrac{(7 \times 10^{-3})(0.4 \times 10^{8})}{2.8 \times 10^{3}}$
14. $\dfrac{(0.16 \times 10^{5})(3 \times 10^{-4})}{(60 \times 10^{7})(0.002p)}$

Appendix B
Right Triangles and Trigonometry

As we enter into alternating current and voltage, our methods for determining circuit totals for current, voltage, impedance, and power are more complicated than simple addition. Because an AC voltage constantly varies, these other values cannot be simply added together. In addition, many AC components offer a phase shift between the particular voltage drops, as well as circuit voltage and current. These components also offer an impedance to an AC circuit that would not exist if these components are used in a DC circuit.

Fortunately, the manner is which an AC signal varies can be predicted. Chapter 11 mentions that an AC voltage varies in a sinusoidal manner, or based on the sine function. This is a function of trigonometry, a branch of mathematics that works with right triangles. For this reason, we can use trigonometric functions to solve for many other values in an AC circuit.

This Appendix is designed to introduce you to the theories behind a right triangle, and the fundamentals of trigonometry. Our purpose is not to overwhelm you with hundreds of mathematical terms and formulas, but merely to show you the mathematics involved in solving for AC circuit values.

In addition to trigonometry, we will also show you the relationships between the sides of a right triangle, known as Pythagorean's Theorem.

Right Triangles

A triangle is a geometric figure with three sides and three angles. Figure B-1 shows a triangle. The angles are labeled a, b, and c, while the sides opposite the angles are labeled A, B, and C. The sum of the three angles in a triangle must equal 180 degrees: when angles a, b, and c are added together, their sum is 180 degrees.

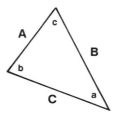

Figure B-1

APPENDIX B

A **right triangle** is a triangle in which one of the angles is a right angle, which equals 90 degrees. Figure B-2 shows one such triangle. The angles and sides are labeled similarly. Angles a, b, and c must still equal 180 degrees.

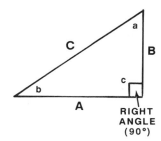

Figure B-2

Because angle c is a right angle, there are certain conditions within this triangle that we can already assume. First, sides A and B are perpendicular from each other, because they are the sides which form the right angle. Second, since angle c equals 90 degrees, the sum of angles a and b must also equal 90 degrees. This ensures us that angles a and b are **acute angles,** or angles that are less than 90 degrees. In addition, the right angle is always noted by the small square within the angle.

The final important term surrounding right triangles concerns the side opposite the right angle. This side is *always* the longest side of any right triangle, and is known as the **hypotenuse.** The hypotenuse of a right triangle is an extremely important measurement; it is used extensively in both the Pythagorean Theorem, and in fundamental trigonometry. In Figure B-2, side C is the hypotenuse.

Pythagorean's Theorem

Because all right triangles have the same essential shape, we can predict the relative length of the sides. An ancient Greek mathematician named Pythagoras determined the relationship between the sides in a right triangle. This is known as the **Pythagorean Theorem,** which has formed the basis for a large portion of geometry.

RIGHT TRIANGLES AND TRIGONOMETRY

The Pythagorean Theorem states that in a right triangle, the length of the hypotenuse squared is equal to the sum of the squares of the two other sides. For the right triangle in Figure B-3, the theorem states:

$$C^2 = A^2 + B^2$$

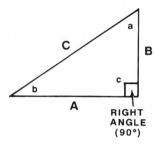

Figure B-3

Looking at the formula, you can see that this relationship exists between the *squares* of the sides. If we wanted to find the length of the hypotenuse, we must change the formula:

$$C = \sqrt{A^2 + B^2}$$

For example, if the two shorter sides in a right triangle are equal to 3 and 4 centimeters respectively, the hypotenuse is equal to:

$$C = \sqrt{A^2 + B^2} = \sqrt{(3)^2 + (4)^2} \quad \sqrt{9 + 16} = \sqrt{25} = 5 \text{cm}.$$

This formula is especially useful when working with series RL or RC circuits. Inductors and capacitors introduce a 90 degree phase shift into the circuit, and therefore change the time at which voltage drops occur. Because the voltage drops occur at different times, they cannot be added together algebraically to find the voltage applied. However, because the 90 degree phase shift can be compared to the 90 degree angle in a right triangle, Pythagorean's Theorem can be used to find the voltage applied. The formula is as follows:

$$E_A = \sqrt{E_R^2 + E_C^2}$$

Where E_A is the voltage applied to the circuit, E_R is the voltage drop of the resistor(s), and E_C is the voltage drop of the capacitor(s).

Because the voltage drops occur at different times, the power consumed by the capacitor and the resistor also occur at different times. So Pythagorean's Theorem is also used to find total power:

$$P_T = \sqrt{P_R^2 + P_C^2}$$

Where P_T is the total power consumed in an electrical circuit, P_R is the power consumed by the resistor(s), and P_C is the power consumed by the capacitor(s).

Finally, the capacitive reactance offered in a series RC circuit changes with the AC input signal, and therefore occurs at a different time than the steady opposition offered by the resistance. For this reason, the total amount of opposition (total impedance) is also calculated using Pythagorean's Theorem:

$$Z_T = \sqrt{R^2 + X_C^2}$$

Where Z_T is total impedance, R is the total resistance in the circuit, and X_C is the total amount of capacitive reactance in the circuit.

As you can see by the previous formulas, Pythagorean's Theorem is very useful for finding circuit totals in an AC circuit. This is because the 90 degree phase shift introduced by inductors and capacitors is similar to the 90 degree right angle in a right triangle.

Trigonometry

As you can see, Pythagorean's Theorem can be used to find a variety of circuit totals in an AC circuit. **Trigonometry** provides not only a different method for finding these values, but can also find the phase angle for an AC circuit. As you have (or will) learn in Chapters 14 and 15, the phase angle for an RC or RL circuit will be some angle between 0 and 90 degrees. Trigonometry is used when evaluating a circuit using vector diagrams.

Generally speaking, trigonometry is the branch of mathematics which studies the relationships between the angles and the sides of a right triangle. This becomes helpful when working with vector diagrams, which form a right triangle.

RIGHT TRIANGLES AND TRIGONOMETRY

A basic right triangle is shown in Figure B-4. The right angle is indicated by the small square, and the hypotenuse is the side directly opposite the right angle. The angle to the left of the right angle is labeled with the Greek letter **theta** (θ). This is done because trigonometry deals with angles as well as sides, and the angles must be labeled. Notice also that the other two sides in the triangle are labeled in their relation to the angle θ. The side *opposite* the angle θ is called opposite, or O. The side which, in addition to the hypotenuse, form the angle θ, is labeled adjacent, or A. These are the labels we will use throughout this appendix.

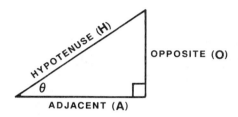

Figure B-4

We stated earlier that because all right triangles share a certain similarity in shape, Pythagorean's Theorem can express the relationship between the three sides. Trigonometry can be used for the exact same reason. Regardless of the shape of the right triangle, certain relationships must always exist between the sides and the angles. In order to express these relationships, we must give them a certain reference, which is the angle θ. There are a total of six relationships, but we'll concentrate on the three which are most popular: These relationships are expressed as functions, called **sine, cosine,** and **tangent**. These functions are listed below:

$$\text{sine } \theta = \frac{\text{opposite}}{\text{hypotenuse}} = \frac{O}{H}$$

$$\text{cosine } \theta = \frac{\text{adjacent}}{\text{hypotenuse}} = \frac{A}{H}$$

$$\text{tangent } \theta = \frac{\text{opposite}}{\text{adjacent}} = \frac{O}{A}$$

APPENDIX B

These functions express the ratios between the three sides of a right triangle when one of the other angles is equal to θ. The angle θ, because it must be an acute angle, must be somewhere between 0 and 90 degrees. Therefore, each angle, from 0 degrees to 90 degrees, defines the ratios that exist between the sides of a right triangle, regardless of the actual length of the sides. Appendix C is a table of trigonometric functions that give the values for sine, cosine, and tangent for each angle from 0 to 90 degrees. As an example, refer to Appendix C and find the value for sine when angle θ is 30 degrees:

$$\text{sine } \theta \, (30) = 0.5$$

This means that the ratio of the side *opposite* the angle θ to the hypotenuse is 0.5. This process can also be reversed. For example, if the hypotenuse is 5 centimeters long, and the side opposite the angle θ is 2 centimeters long, then we can determine the size of angle by the following method:

$$\frac{O}{H} = 0.4$$

When 0.4 is found within the sine table (actually, Appendix C shows the closest value to 0.4 to be 0.407), the angle is 24 degrees. This means that when the ratio of the opposite side to the hypotenuse is 0.407, the angle θ *must* be 24 degrees.

This is how trigonometry works. But how can it be used in electronics? Let's refer to Figure B-5, which is an impedance vector diagram for a typical series RC circuit. Notice that the values for resistance and capacitive reactance are known. The total impedance for the circuit can be found using Pythagorean's Theorem:

$$Z_T = \sqrt{R^2 + X_C^2} = \sqrt{5^2 + 12^2} = \sqrt{25 + 144} = \sqrt{169} = 13$$

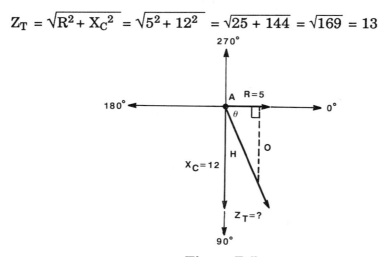

Figure B-5

RIGHT TRIANGLES AND TRIGONOMETRY

We now know the length of all three sides of the "triangle", and we can now use trigonometry to determine the phase angle of the circuit. In Figure B-5, the phase angle for the circuit is designated by theta (θ). If we take a moment to ignore the fact that this is a vector diagram for an AC circuit, we can see it simply as a right triangle. The hypotenuse is Z_T (13), the side opposite θ is equal to 12, and the side adjacent to θ is 5.

Since the length of all three sides are known, we can use any of the three trigonometric functions to determine the phase angle. The method is this: choose which function to use, and calculate the ratio between those particular sides; then look up the resultant value in Appendix C, and see which angle corresponds to that value. We'll work out all three, starting with the sine function:

$$\text{sine } \theta = \frac{O}{H} = \frac{12}{13} = 0.923$$

Looking at Appendix C, we can see that the angle whose sine is approximately equal to 0.923 is 67 degrees (0.921). Now let's see if we end with the same angle with the cosine function:

$$\text{cosine } \theta = \frac{A}{H} = \frac{5}{13} = 0.384$$

Again, Appendix C shows 67 degrees also has a cosine approximately equal to 0.384 (0.391).

The tangent function is the last function to work out. Actually, since tangent is the ratio of the opposite and adjacent sides, we didn't have to find Z_T for the phase angle, because the lengths of the opposite and adjacent sides were already known. Normally, though, total values are required when analyzing a circuit, so if we hadn't worked out Z_T when we did, we would have to find it eventually. Anyway, let's determine whether the tangent function arrives at the same answer:

$$\text{tangent } \theta = \frac{O}{A} = \frac{12}{5} = 2.4$$

Once again, Appendix C verifies that 67 degrees has a tangent approximately equal to 2.4 (2.356).

APPENDIX B

Through this method, the phase angle for any AC circuit can be found. This is the only application of trigonometry in this textbook, but it is not the only way trigonometry can be used when analyzing an electrical circuit. For example, trigonometry can be used to find the instantaneous value of an AC voltage when the armature in an AC generator is rotated through to an angle θ. Also, the cosine function can also find the power factor of an AC circuit, which is further described in Chapter 14. These are just two of the many applications that trigonometry has in electronic analysis; but we'll keep our needs down to only phase angles.

Appendix C
Table of Trigonometric Functions

Degrees	Sine	Cosine	Tangent	Degrees	Sine	Cosine	Tangent
0	0.000	1.000	0.000				
1	0.017	1.000	0.017	46	0.719	0.695	1.036
2	0.035	0.999	0.035	47	0.731	0.682	1.072
3	0.052	0.999	0.052	48	0.743	0.669	1.111
4	0.070	0.998	0.070	49	0.755	0.656	1.150
5	0.087	0.996	0.087	50	0.766	0.643	1.192
6	0.105	0.995	0.105	51	0.777	0.629	1.235
7	0.122	0.993	0.123	52	0.788	0.616	1.280
8	0.139	0.990	0.141	53	0.799	0.602	1.327
9	0.156	0.988	0.158	54	0.809	0.588	1.376
10	0.174	0.985	0.176	55	0.819	0.574	1.428
11	0.191	0.982	0.194	56	0.829	0.559	1.483
12	0.208	0.978	0.213	57	0.839	0.545	1.540
13	0.225	0.974	0.231	58	0.848	0.530	1.600
14	0.242	0.970	0.249	59	0.857	0.515	1.664
15	0.259	0.966	0.268	60	0.866	0.500	1.732
16	0.276	0.961	0.287	61	0.875	0.485	1.804
17	0.292	0.956	0.306	62	0.883	0.469	1.881
18	0.309	0.951	0.325	63	0.891	0.454	1.963
19	0.326	0.946	0.344	64	0.899	0.438	2.050
20	0.342	0.940	0.364	65	0.906	0.423	2.145
21	0.358	0.934	0.384	66	0.914	0.407	2.246
22	0.375	0.927	0.404	67	0.921	0.391	2.356
23	0.391	0.921	0.424	68	0.927	0.375	2.475
24	0.407	0.914	0.445	69	0.934	0.358	2.605
25	0.423	0.906	0.466	70	0.940	0.342	2.748
26	0.438	0.899	0.488	71	0.946	0.326	2.904
27	0.454	0.891	0.510	72	0.951	0.309	3.078
28	0.469	0.883	0.532	73	0.956	0.292	3.271
29	0.485	0.875	0.554	74	0.961	0.276	3.487
30	0.500	0.866	0.577	75	0.966	0.259	3.732
31	0.515	0.857	0.601	76	0.970	0.242	4.011
32	0.530	0.848	0.625	77	0.974	0.225	4.332
33	0.545	0.839	0.649	78	0.978	0.208	4.705
34	0.559	0.829	0.675	79	0.982	0.191	5.145
35	0.574	0.819	0.700	80	0.985	0.174	5.671
36	0.588	0.809	0.727	81	0.988	0.156	6.314
37	0.602	0.799	0.754	82	0.990	0.139	7.115
38	0.616	0.788	0.781	83	0.993	0.122	8.144
39	0.629	0.777	0.810	84	0.995	0.105	9.514
40	0.643	0.766	0.839	85	0.996	0.087	11.43
41	0.656	0.755	0.869	86	0.998	0.070	14.30
42	0.669	0.743	0.900	87	0.999	0.052	19.08
43	0.682	0.731	0.933	88	0.999	0.035	28.61
44	0.695	0.719	0.966	89	1.000	0.017	57.29
45	0.707	0.707	1.000	90	1.000	0.000	

Appendix D
Answers to Odd-Numbered Questions

To assist you in your chapter reviews, we have included the answers to the odd-numbered questions from the Chapter Self-Tests. For those questions requiring calculations, the complete calculation has been provided. For others, a brief descriptive answer is included where necessary. Within these calculations of descriptive paragraphs, the actual answer is highlighted to avoid confusion.

Section I: DC Electronics

Chapter 1

1. The proton.
3. Electrostatic force.
5. The distance between two bodies.
7. A positive ion.
9. Positive.

Chapter 2

1. Valence electrons.
3. The negative terminal.
5. Electromotive force.
7. Yes, from B to A.
9. Piezoelectricity.

Chapter 3

1. A simple electrical circuit consists of a voltage source connected to a load through the use of conductors.
3. The negative terminal.
5. 2.1V + 2.1V + 2.1V + 2.1V = **8.4 volts**. (Because it is a series-aiding configuration, the voltages are added together).
7. The sum of the voltage drops must equal the sum of the voltage rises.
9. Chassis ground.

Chapter 4

1. A material's opposition to current flow.
3. Conductors.

APPENDIX D

5. Length.

7. If a material has a positive temperature coefficient, its resistance will change in direct proportion to a change in temperature. If temperature increases, resistance increases — if temperature decreases, resistance decreases.

9. Resistors have a tolerance rating because they are extremely difficult to manufacture with specific, unchanging ohmic values. The tolerance rating defines the specific range within which the ohmic value for the resistor may be, and still function properly.

11. $R_T = R_1 + R_2 + R_3$, etc.

13. In order to compute total resistance for this circuit, the parallel portion must be combined to an equivalent series resistance (R_A), and then added to the actual series resistance.

$$R_A = \frac{\text{Value}}{\text{Number}} = \frac{30\Omega}{2\Omega} = 15\Omega$$
$$R_T = R_A + R_S = 15\Omega + 30\Omega = 45\Omega.$$

Chapter 5

1. False.
3. False.
5. False.
7. True.
9. True.

Chapter 6

1. Voltage, current, and resistance.

3. $I = \frac{E}{R} = \frac{5V}{10\Omega} = .5\,A.$

5. $E = IR.$

7. Megohms.

9. $R = \frac{E}{I} = \frac{12V}{0.75A} = 16\Omega.$

Chapter 7

1. Power is defined as work over a period of time—work is simply force over a distance, without regard to time.

3. The watt.

5. $P = IE = (3.2V)(0.75A) = 2.4$ watts.

7. $P = I^2 R$

9. $P = \frac{E^2}{R} = \frac{(5)^2 V}{300\Omega} = \frac{25V}{300\Omega}$
   $= 0.0833\,W\,(83.3\,mW).$

Chapter 8

1. Current.
3. Voltage.
5. When there is no potential difference between the output terminals.

ANSWERS TO ODD-NUMBERED QUESTIONS

7. The wheatstone bridge.

9. Around a closed loop, the algebraic sum of all the voltages equal zero.

Chapter 9

1. Flux lines.

3. a. Flux lines have direction or polarity.
   b. Lines of force always form complete loops.
   c. Flux lines cannot cross each other.
   d. Flux lines tend to form the smallest possible loops.

5. a. It brings the flux lines closer together.
   b. It concentrates the majority of the flux lines in the center of the loop.
   c. It creates north and south poles.

7. Residual magnetism.

9. A conductor, a magnetic field, and relative motion between the two, allowing the lines of flux to cut across the conductor.

Chapter 10

1. The action of inducing an EMF into a conductor when there is a change of current in the conductor.

3. The henry.

5. $L_T = L_1 + L_2 + L_3$, etc.

7. As a capacitor charges, current in the circuit slowly decreases, until the capacitor is fully charged. At that point, current ceases to flow.

9. 63.2%.

Section II: AC Electronics

Chapter 11

1. The strength of the magnetic field, the speed of conductor movement, and the length of the conductor.

3. One cycle or period.

5. One cycle (or period) of AC.

7. Maximum instantaneous voltages, or peak values.

9. $I_{RMS} = I_{PEAK} \times 0.707 = 3A_{PEAK} \times 0.707 = $ **2.121** A_{RMS}.

11. $F = \frac{1}{T}$ or $T = \frac{1}{F}$

13. The triangular waveform.

APPENDIX D

Chapter 12

1. You do not have to observe polarity when measuring AC current (or voltage).

3. The graticule.

5. The vertical axis.

7. In order to compute the frequency of this waveform, you must multiply the number of squares that represent one cycle by the setting of the TIME/CM knob.

   4 squares × 2ms/square = 8 ms.

   Once the period of the waveform is known, use the time/frequency equation:

   $$F = \frac{1}{T} = \frac{1}{0.008 \text{ s}} = 125 \text{ Hertz}.$$

9. Lissajous patterns.

Chapter 13

1. Current and voltage are in-phase in an AC resistive circuit.

3. AC current is not a negative value because you cannot have a *negative* amount of current flow. The area in which the graph goes below the reference line indicates that current is flowing in the *opposite* direction.

5. $R_T = 10\Omega + 20\Omega = 30\Omega$.

   $$I_T = \frac{E_A}{R_T} = \frac{90 V_{RMS}}{30\Omega} = 3 A_{RMS}.$$

7. As long as current flows, and there is force, there is power being dissipated. This means that power can never be a negative value.

Chapter 14

1. 90 degrees.

3. Capacitive reactance is the steady opposition to current flow offered by a capacitor in an AC circuit.

5. If frequency increases, capacitive reactance decreases.

7. No. Capacitive reactance and resistance are not in-phase in a capacitive circuit, and cannot therefore be added together by normal means.

9. In series RC vector diagrams, current is used as a reference, while voltage is used as the reference in parallel RC vectors.

11. When a signal is attenuated, its amplitude is decreased below a usable value.

13. When a low frequency signal is applied to an RC high-pass filter, capacitive reactance is very high. This prevents a large amount of current from flowing in the circuit. The output signal (taken across the resistor) is very low.

    As frequency increases, capacitive reactance decreases, and the output voltage increases.

15. A lagging output phase shift network.

Chapter 15

1. Constantly.

3. 180 degrees.

5. Inductive reactance is the steady opposition to current flow offered by an inductor in an AC circuit.

7. $X_L = 2\pi fL$.

9. Reactive power in an inductive circuit is that power which is consumed by the inductor as its magnetic lines of flux expand, and returned back to the circuit as the lines of flux collapse.

11. Vector diagrams for parallel RL circuits use voltage as a reference, while current is the reference in a series RL circuit.

13. A leading-output phase shift network.

Chapter 16

1. $Z_T = \sqrt{R^2 + X_X^2}$, with X_X being the difference between X_L and X_C.

3. Resonance is that point in any RLC circuit where X_L is equal to X_C.

5. Bandwidth is the band of frequencies surrounding the resonant frequency that provide a minimum of 70.7 percent of the maximum output voltage or current.

7. If Q decreases, the bandwidth increases.

9. If a resistor is placed in series with the reactive components in a series RLC circuit, Q will decrease, which increases the bandwidth.

11. When a resonant frequency is applied to a tank circuit, the capacitor charges and discharges, sending current through the inductor. The inductor's magnetic field expands and contracts, sending current back to the capacitor. The entire cycle then repeats. This is known as the flywheel effect.

13. The response curve of a parallel resonant circuit has a larger bandwidth when a swamping resistor is used.

APPENDIX D

Chapter 17

1. The primary and secondary windings.

3. The voltage ratio between the primary and secondary windings is directly proportional to the number of windings in the primary and secondary, ignoring losses.

5. The coefficient of coupling is a number between zero and one that expresses the percentage of primary flux lines that cut the secondary winding. When all the primary flux lines cut the secondary, the coefficient of coupling is one.

7. The four most common applications of a transformer are phase-shifting, phase-splitting, isolation, and the autotransformer.

9. Polarity inversion in a transformer is not a consideration when the AC signal will be rectified to create a DC voltage.

11. A center-tap is a grounded point on the secondary winding which splits the secondary signal into two signals. when the center-tap is at the exact center of the secondary, the amplitude of the two output signals is half the amplitude of the normal secondary signal.

Section III: Semiconductors

Chapter 18

1. Small size, low operating voltages, ruggedness, and the elimination of filaments.

3. Germanium and silicon.

5. The nucleus of a semiconductor atom "sees" eight electrons because the electrons from the adjacent atoms share their four valence electrons with each other. This is known as covalent bonding.

7. At low temperatures, the covalent bonds within semiconductor materials are very tight, preventing the creation of many free electrons. At high temperatures, the valence electrons become more agitated, and a large number of free electrons can be created.

9. False. Electrons move opposite from holes because as electrons move in one direction they leave holes behind. This creates the illusion of holes flowing in the opposite direction.

11. A pure semiconductor is doped by injecting an impurity material, which alters the crystal-lattice structure.

13. Electrons.

ANSWERS TO ODD-NUMBERED QUESTIONS

Chapter 19

1. Mobile charges.

3. The depletion region.

5. When the diode is forward-biased.

7. To forward-bias a diode, a potential difference must exist on either side of the diode which is larger than the barrier voltage. This potential difference must also have the more negative potential at the N-type material, and the more positive potential at the P-type material.

    A forward-biased diode acts like a closed switch, supporting current flow with very little resistance.

9. Leakage-current or reverse-current.

11. Since the positive lead of the source voltage is connected to the P section of the diode, and the potential exceeds the barrier voltage for a germanium diode, the diode is forward-biased. The amount of current flow is equal to the difference between the source voltage and the bias voltage, divided by the external resistance.

    $$I_F = \frac{E_A - E_F}{R} = \frac{5V - 0.3V}{1\,k\Omega} = \frac{4.7V}{1k\Omega}$$
    $$= 4.7\text{ mA}.$$

13. The silicon diode has a higher barrier voltage, and a lower reverse current.

15. The arrow.

Chapter 20

1. True.

3. Through the use of zener breakdown voltage tolerance values.

5. As temperature decreases, the power dissipated by a zener diode increases.

7. The maximum reverse current for a zener diode is calculated by dividing the power rating of the zener by the *maximum* zener voltage, as opposed to the breakdown voltage rating.

9. A temperature-compensated zener diode is formed by connecting a zener diode in series with an ordinary PN junction diode, ensuring that a) the temperature characteristics for the two devices are exactly opposite, and b) the two devices are connected back-to-back: either anode-to-anode or cathode-to-cathode.

APPENDIX D

11. A zener diode operates exactly like an ordinary PN junction diode when forward-biased because the majority carriers behave the same when either device is forward-biased.

13. A regulator circuit compensates for changes in voltage.

Chapter 21

1. A PNP transistor.

3. On the emitter.

5. For a bipolar transistor to be properly biased, the emitter junction must be forward-biased, while the collector junction is reverse-biased.

7. All of the current within a transistor starts at the emitter.

9. PNP bias polarities are opposite of NPN bias polarities because opposite polarity voltages are required to forward-bias the different transistors.

11. Common-base, common-emitter, and common-collector.

13. In a common-base amplifier, the input is applied to the emitter, and the output is taken across the collector. The primary function of the common-base is to provide voltage gain, and the amount of current gain is less than one — approximately .95-.99.

15. The common-collector amplifier has the input applied to the base, and the output taken across the emitter. Its primary function is to provide impedance matching between two circuits, which is accomplished due to the high amount of current gain in the common-collector.

17. Manufacturers will specify each transistor's alpha and beta values, in addition to collector breakdown voltage, emitter breakdown voltage, and the maximum collector dissipation rating.

Chapter 22

1. The substrate.

3. Reverse-biasing.

5. The size of the depletion region within the channel.

7. If VGS increases, the amount of reverse-bias between the gate and the source increases, which will make the depletion region within the channel **decrease.** This will decrease the amount of I_D.

9. N-type material.

11. Although the depletion-mode IGFET and the JFET both operate in the depletion-mode, the IGFET has the capability of allowing either polarity of voltage to be applied to the gate, whereas the JFET can only operate with one polarity.

13. Yes.

15. The drain.

Chapter 23

1. Four.

3. Holding current.

5. An SCR can be used to control an AC signal, but two SCRs must be placed in parallel, and connected in opposite directions.

7. To lower the triac's breakdown voltage, the gate voltage must be made more positive (or negative, depending upon the TRIAC) with respect to MT1.

9. TRIACs are used instead of SCRs because a single TRIAC can support current flow in either direction.

11. The intrinsic standoff ratio, designated with the Greek letter eta (η).

13. The minimum value of V_E is known as the valley voltage, designated as V_V.

15. The cathode, anode, and the gate.

Chapter 24

1. Electromagnetic radiation.

3. Photons.

5. Electrical energy.

7. Photovoltaic cells must be connected in series to create a higher output voltage. When connected in parallel, they have a higher current capability.

9. The emitter and the collector.

11. Yes.

13. As the amount of forward-bias increases, the amount of current flow through the LED increases, which increases the amount of light produced.

15. A seven-segment LED.

Section IV: Electronic Circuits

Chapter 25

1. Distortion.

3. Amplifiers are used in a wide variety of equipment, including radios, televisions, stereo amplifiers, telephones, and satellite communications.

5. Base-biasing.

APPENDIX D

7. Base-biasing.

9. Emitter feedback biasing.

11. The common-base amplifier has:

    a. The input at the emitter and the output at the collector.
    b. Low current gain.
    c. High voltage gain.
    d. Low input resistance
    e. High output resistance.
    f. No phase shift between input and output signals.
    g. A primary function of a voltage amplifier.

13. The alpha value for a common-base amplifier also describes the amount of current gain.

15. The common-emitter has:

    a. The input at the base and the output at the collector.
    b. High current gain.
    c. High voltage gain.
    d. High input resistance.
    e. Low output resistance.
    f. 180 degree phase shift between input and output.
    g. A primary function of a power amplifier.

17. A transistor's beta value also determines the current gain for the common-emitter amplifier.

19. Current gain for the common-collector is equal to the transistor's beta plus 1.

21. Clipping is when an output reaches a minimum or maximum voltage before the peak of the alternation.

23. The classes of operation differ by the percentage of input time that the amplifier conducts. A class A amplifier conducts for a full 100% of the input waveform's period; a class AB amplifier conducts for more than 50%, but less than 100%; a class B amplifier conducts for 50%; and the class C amplifier conducts for less than 50% of the input waveform's period.

25. Transformer coupling also provides impedance-matching.

27. DC coupling is most efficient when the input signal is a DC voltage, or a very low AC frequency (10 Hz or lower).

Chapter 26

1. 20 Hz to 20 kHz.

3. The ratio of load impedance to output impedance of a power amplifier.

5. 180 degrees.

7. 10 Hz to 5 MHz.

9. The Miller effect is the negative feedback path across a transistor's collector junction at high frequencies.

11. The difference between shunt and series peaking is the location of the inductor. In shunt peaking, the inductor is in parallel with the output; in series peaking, the inductor is in series with the collector load resistor.

13. In the single-input, single-output mode for a differential amplifier, Q_2 receives its input from the voltage drop of R_E.

15. When the two inputs for a differential amplifier have equal voltage levels and are in-phase, the differential output will be 0 volts. If the inputs are exactly the same, the outputs for each amplifier will also be exactly the same, and no difference of potential will exist between the two outputs.

Chapter 27

1. Stage 1 is a differential amplifier, stage 2 is a voltage amplifier, and stage 3 is an output amplifier, often a complementary emitter-follower.

3. The inverting input and the non-inverting input.

5. Negative.

7. a. 2 millivolts.
   b. 20 microamps.
   c. 20 microamps.
   d. 200 millivolts.

9. The primary difference between the configurations of the inverting and noninverting amplifiers is that the input signal is applied to the noninverting input, and the feedback path for both amplifiers is the same.

   The input applied to the noninverting input, however, causes the gain of the noninverting amplifier to be a) in-phase with the input, and b) the ratio of the feedback resistor to the input resistor, plus 1.

11. The bandwidth can be improved by converting to the closed-loop operating mode, but using feedback will decrease gain.

13. The gain-bandwidth product is that value which remains constant throughout an operational amplifier. When bandwidth and gain changes (by changing the ratio of the resistors), their product will always remain the same. Also, the product is that frequency that provides unity gain.

15. One.

17. When RC networks are used as input and feedback networks, an operational amplifier operates as an active filter. In this configuration, the gain is determined by the ratio of the feedback impedance to the input impedance.

19. Subtracter.

Chapter 28

1. A power supply provides a steady, unchanging DC voltage under varying input and load conditions.

3. A rectifier converts an AC signal to a pulsating DC signal.

5. An input transformer can a) step-up or step-down the input voltage if required, and b) provides isolation between the input signal and the equipment.

7. The bridge rectifier is used most often because it provides the largest pulsating DC signal, and allows current to flow through itself for 100% of the input signal period.

9. The diode will only conduct for a small portion of the alternation which would normally forward-bias it, because the capacitor holds the voltage to the diode to the point where it becomes reverse-biased.

11. Because a larger capacitor has a larger RC time constant, and therefore takes longer to discharge, which means its output voltage will stay closer to the desired DC voltage.

13. The LC filter provides better filtering action over a larger range of current values, and less voltage is dropped across the inductor, reducing the amount of power consumption.

15. The input voltage to a regulator can vary due to a varying AC signal amplitude, or the inherent ripple of the input voltage.

17. When the input voltage decreases, the amount of current flow also decreases. R_S develops less voltage, and the voltage drop of the zener diode remains the same.

ANSWERS TO ODD-NUMBERED QUESTIONS

19. The sampling circuit, the error detector, the DC reference voltage, the error amplifier, and the regulating component.

21. The operational amplifier acts as both the error detector and the error amplifier.

Chapter 29

1. An oscillator provides a repetitive AC signal using only a DC voltage source.

3. An oscillator requires an amplifier, a frequency determining network, positive feedback, and it must be self-starting.

5. The purpose of the amplifier in the Armstrong oscillator is to conduct when the energy in the tank is depleted, and returns the tank to full power.

7. A series-fed Hartley has the same current path through both the feedback components and the emitter of the amplifier, while the shunt-fed Hartley have different current paths for the feedback path and the amplifier.

9. Split capacitance in the tank.

11. The thinner crystal.

13. The crystal acts as both the frequency determining device and the feedback path.

15. The lead-lag network.

17. In the Armstrong oscillator, the transformer's windings are tuned by a capacitor, while the blocking oscillator has no tuning capacitors.

Chapter 30

1. A waveform that retains the same shape from one cycle to the next.

3. In a rectangular waveform, the rise time is the portion where the signal changes from its most negative voltage to its most positive voltage, while the fall time is the direct opposite—changing from the most positive to the most negative. These are also known as the leading edge (rise time) and the trailing edge (fall time).

5. In order for a differentiator to obtain a sharp spike, a very low RC time constant is required. For an integrator, a low RC time constant is required to obtain the wave-like output across the capacitor.

7. A clipper circuit is similar to a half-wave rectifier.

9. A biased clipper clips a portion of an AC signal at a voltage other than 0 volts, while an unbiased clipper clips the AC signal at 0 volts.

11. 3 volts.

13. The purpose of a clamper is to change the reference voltage of a waveform, without changing the frequency or amplitude of the waveform.

15. The output frequency the astable multivibrator is determined by the RC time constants within the multivibrator.

17. A monostable multivibrator is also known as a one-shot or single-shot multivibrator.

19. The output frequency is determined by the input frequency, while the pulse width is determined by the RC time constants within the multivibrator.

21. A steering diode is that diode which becomes forward-biased to allow a certain polarity of the input signal to change the output of the monostable multivibrator.

23. A bistable has a great number of functions, including gates, frequency divider, and memory elements.

25. A Schmitt Trigger employs two common-emitter amplifiers, which differ from normal common-emitters in that they are biased to easily saturate and cutoff.

27. Sawtooth, triangular, and trapezoid waveforms.

29. An input whose current maintains the same value.

Appendix A

Exercise #1

1. $3400 = 3.4 \times 1000 = 3.4 \times 10^3$

3. $52{,}000{,}000 = 52 \times 1{,}000{,}000 = 52 \times 10^6$

5. $94{,}000 = 94 \times 1000 = 94 \times 10^3$

7. $39{,}000 = 39 \times 1000 = 39 \times 10^3$

ANSWERS TO ODD-NUMBERED QUESTIONS

9. $0.00053 = 0.53 \times 0.001$
   $= 0.53 \times 10^{-3}$

   or

   $0.00053 = 530 \times 0.000001$
   $= 530 \times 10^{-6}$

11. $0.0000000002 = 0.2 \times 0.000000001$
    $= 0.2 \times 10^{-9}$

    or

    $0.0000000002 =$
    $200 \times 0.000000000001$
    $= 200 \times 10^{-12}$

13. $0.00000186 = 1.86 \times 0.000001$
    $= 1.86 \times 10^{-6}$

Exercise #2

1. $68 \times 10^{-3} = 68\text{m}$

3. $722 \times 10^{3} = 722\text{k}$ or 0.722M

5. $59 \times 10^{7} = 590\text{M}$

7. $7{,}780{,}000 = 7.78 \times 10^{6} = 7.78\text{M}$

9. $12 \times 10^{-8} = 0.12\mu$ or 120n

APPENDIX D

Exercise #3

1. $(3.5 \times 10^{-4}) + (0.89 \times 10^{-3}) =$

   a) $(3.5 \times 10^{-4}) + (8.9 \times 10^{-4}) = 12.4 \times 10^{-4}$

   b) $(0.35 \times 10^{-3}) + (0.89 \times 10^{-3}) = 1.24 \times 10^{-3}$

3. $(96 \times 10^{-6}) - (0.079 \times 10^{-3}) =$

   a) $(0.096 \times 10^{-3}) - (0.079 \times 10^{-3}) = 0.017 \times 10^{-3}$

   b) $(96 \times 10^{-6}) - (79 \times 10^{-6}) = 17 \times 10^{-6}$

5. $(4 \times 10^{-6}) \times (16 \times 10^{-3}) = (4 \times 16) \times 10^{(-6)+(-3)} = 64 \times 10^{-9}$

7. $(8 \times 10^{3}) \times (4.5 \times 10^{-2}) = (8 \times 4.5) \times 10^{3+(-2)} = 36 \times 10^{1}$

9. $\dfrac{2.4 \times 10^{-12}}{0.12 \times 10^{-3}} = \left(\dfrac{2.4}{0.12}\right) \times 10^{(-12)-(-3)} = 20 \times 10^{(-12)+3} = 20 \times 10^{-9}$

11. $\dfrac{0.72 \times 10^{-2}}{8 \times 10^{7}} = \left(\dfrac{0.72}{8}\right) \times 10^{(-2)-7} = 0.09 \times 10^{(-2)+(-7)} = 0.09 \times 10^{-9}$

13. $\dfrac{(7 \times 10^{-3})(0.4 \times 10^{8})}{2.8 \times 10^{3}} = \dfrac{(7 \times 0.4) \times 10^{(-3)+8}}{2.8 \times 10^{3}} = \dfrac{2.8 \times 10^{5}}{2.8 \times 10^{3}} = \left(\dfrac{2.8}{2.8}\right) \times 10^{5-3} = 1 \times 10^{2}$

Appendix E
Glossary

The following is an alphabetized listing of the key electronic terms that were introduced in this textbook. The purpose of these definitions is to provide a reminder or a study aid—they are not intended to be the sole source of information. For further review of a particular item in this glossary, consult the index, and refer to that page.

7-segment display A pattern of light-emitting diodes (LEDs) which, using various combinations, can display the numbers 0 through 9 and certain letters.

AC Alternating current.

AC alpha The ratio of collector and emitter current which is found when forward-bias is changed by a pre-determined amount, and the *change* in emitter and collector currents are measured. These measurements are then inserted into the following formula:

$$\alpha_{AC} = \frac{\Delta I_C}{\Delta I_E}$$

Where delta (Δ) is used to indicate change.

AC beta The ratio of collector to base current which is found when forward-bias is changed by a pre-determined amount, and the *change* in collector and base currents are measured. These measurements are then inserted into the following formula:

$$\beta_{AC} = \frac{\Delta I_C}{\Delta I_B}$$

AC generator A mechanical device consisting of a wire loop (known as an armature) which rotates within a magnetic field. As the armature rotates, its cuts the flux lines at varying rates, producing an AC signal.

acceptor atom An atom that is created by the insertion of trivalent material into a semiconductor material. Acceptor atoms readily accept electrons from other atoms, due to a hole in the valence shell.

active filter A hi-pass or low-pass filter circuit involving an operational amplifier.

air-core An inductor or transformer whose coils are not wrapped around any type of core material.

alpha (α) The Greek letter used to represent the ratio of collector current to emitter current for a bipolar transistor.

alternating current Current flow in both directions in the same circuit, resulting from the application of an AC signal to the circuit.

alternation A portion of an AC sine wave whose voltages are either all positive (positive alternation) or all negative (negative alternation).

alternator Another name for an AC generator.

ammeter A piece of test equipment which is used to measure the amount of current flow through a circuit.

ampere The unit of measurement for current, usually referred to as an amp, and represented by the letter "A." One ampere of current is equal to one coulomb of electrons passing a given point in one second.

amplifier Any circuit which increases in magnitude one particular aspect of that signal as an output. The most common amplifiers are voltage amplifiers, current amplifiers, and power amplifiers.

amplitude The level of a signal in terms of voltage or current.

analog adder An operational amplifier circuit commonly known as a summing amplifier.

anode The P-type material in a PN junction diode, a zener diode, a light-emitting diode, and an SCR. The anode is normally represented by an arrow in most schematic symbols.

aperiodic waveforms Waveforms or signals which do not retain the same shape from one cycle to the next.

apparent power The power dissipated in a circuit containing both resistive and reactive components. Apparent power is measured in volt-amps, and can be calculated two different ways: by multiplying the effective applied voltage by the effective circuit current, or by taking the true power and the reactive power, and finding the vector sum.

armature The wire loop in an AC generator which rotates within the magnetic field and had an alternating voltage induced within it.

Armstrong oscillator An oscillator which uses a transformer and the tickler coil principle.

artificial magnet Any magnet which is created by artificial means.

astable multivibrator A type of rectangular-wave generator which, through the alternate charging and discharging of two capacitors, creates a rectangular wave without the need of an input signal.

atom The smallest particle to which an element can be broken down, and still retain the characteristics of that element.

attenuation The process by which the voltage level of a signal is reduced, not allowing it to perform a useful function for the load.

GLOSSARY

audio amplifiers Amplifier circuits specifically designed to operate in the audio frequency range.

audio frequency Any frequency which is normally audible. The audio frequency range extends roughly from 20 Hz - 20 kHz.

autotransformer A special type of transformer which uses only a single coil for both the primary and secondary windings.

average power The power dissipated in a purely resistive AC circuit, calculated by using the effective values for voltage and current.

average value The value for a single alternation of an AC signal which is equal to approximately 63.6 percent of the peak value. The average value for an entire sine wave (both the positive and negative alternations) is zero.

balanced atom An atom that contains an equal number of protons and neutrons.

balanced bridge A bridge circuit which has no potential difference or current flow between its output terminals.

band-pass filter A frequency-specific RLC filter which allows only those frequencies within the resonant band to pass to the output, while attenuating all other frequencies.

band-reject filter Another term for a band-stop filter.

band-stop filter A frequency-specific RLC filter which attenuates those frequencies within the resonant band, while passing all other frequencies to the output.

bandwidth That band of frequencies surrounding the resonant frequency of a circuit whose values range from approximately 70.7% of the peak value to the peak value. Bandwidth is usually designated by the abbreviation BW.

barrier voltage The potential difference necessary to cause forward-bias current across a PN junction. For germanium junctions, the barrier voltage ranges from 0.2-0.3 volts, while the range for silicon junctions is 0.5-0.7 volts.

base 1. The region between the emitter and the collector in both a bipolar transistor and a phototransistor. 2. The N-type material used to create the two base leads (B_1 and B_2) in a unijunction transistor. 3. When working with exponents, the number which is raised to a power. In scientific notation, the number "10" is the base.

base current That current which flows through the base terminal to the source voltage in a bipolar transistor, represented by I_B.

base-biasing A method of biasing a bipolar transistor circuit by connecting a large resistor (R_B) between the base and the source voltage. That voltage which is not developed by R_B is felt at the base, and is sufficient to forward-bias the emitter junction.

bass frequencies Those frequencies at the lower end of the audio frequency range.

battery A device which creates a DC voltage through electrochemistry. The three most common types of batteries are the dry-cell, the lead-acid, and the NICAD battery.

beta (β) The Greek letter used to represent the ratio of collector current to base current in a bipolar transistor.

bi-directional triode thyristor A control device which can provide switching action for current flow in two directions. The common name for this device is the TRIAC.

biased clamper A clamper circuit which clamps a particular voltage level of the input signal to a voltage other than 0 volts.

biased clipper A clipper circuit which removes a portion of an input signal at a voltage level other than 0 volts.

biasing Placing voltage levels on all sides of a single or multiple PN junction device in order to allow or prevent current flow. Forward-biasing normally allows current flow, while reverse-biasing normally prevents current flow.

biasing network A resistive network designed to provide the proper biasing voltages for a bipolar transistor.

bipolar transistor A solid-state device which, through proper biasing, allows current to flow through both P-type and N-type materials.

bistable multivibrator A rectangular-wave generator with two stable states. The bistable will remain in a particular state (sometimes selected by a set or reset input) until an input signal provides the appropriate trigger. A bistable multivibrator will change states once for each input cycle, and therefore requires two input cycles to generate one complete output cycle. The bistable is commonly referred to as a flip-flop.

block diagram A diagram which separates a circuit into smaller circuits, each having an individual function. This is used primarily for explaining overall circuit operation.

blocking oscillator A nonsinusoidal oscillator which provides an output pulse, due to the blocking action of the transistor.

GLOSSARY

Bohr model A theoretical model of the atom which illustrates that the atom is comprised of a nucleus containing protons and neutrons, with electrons outside the nucleus revolving in orbital shells.

branch A single current path in a parallel circuit.

breakdown voltage The potential difference across a reverse-biased PN junction which is high enough to allow the minority carriers to support current flow.

breakover frequency The frequency for an operational amplifier where the gain of the amplifier is approximately 70.7% of the maximum.

bridge circuit A resistive network comprised of a pair of resistors in two parallel branches, where an output is available between the resistors within each branch. A bridge circuit is said to be balanced when no potential difference or current flow is available across the output terminals. If this condition is not met, the bridge is unbalanced, and current flows.

bridge rectifier A network of four diodes in a bridge configuration, used to remove one alternation from an AC signal. The bridge rectifier provides an output voltage one-half that of the input voltage, and allows current to flow for the full input cycle. The output ripple frequency of the bridge rectifier is double the input frequency.

brushes Highly conductive material (usually carbon) which provides a stationary contact for the slip rings in an AC generator. Because the brushes are stationary, they are the primary contacts to which the external load can be connected.

buffer amplifier Another term for a common-collector amplifier, used to describe the ability to act as a buffer between a high-impedance input, and a low-impedance load.

Butler oscillator A crystal-enhanced LC oscillator where the frequency is selected by both the tank circuit and the crystal. The crystal is part of the feedback network, and provides maximum positive feedback when the oscillating frequency is equal to the crystal's natural frequency.

bypass capacitor A capacitor connected to the common terminal in a bipolar transistor amplifier which holds the voltage at that terminal steady to allow for maximum gain.

CMRR Common-Mode Rejection Ratio.

capacitance The ability of a device to store a voltage in an electrostatic field. Capacitance is measured in farads.

capacitive reactance The steady opposition offered by a constantly charging and discharging capacitor in an AC circuit. Capacitive reactance is represented by the symbol X_C, and can be calculated using the following formula:

$$X_C = \frac{1}{2\pi fc} \text{ or } \frac{159m}{fC}$$

Where "f" is the frequency of the AC voltage, and "C" is capacitance.

capacitor A device uniquely designed to have a specific amount of capacitance.

cascaded amplifiers A number of separate amplifiers, each with a specific function, connected in series to provide a single output.

cathode The N-type material in a PN junction diode, a zener diode, a light-emitting diode, and an SCR. The cathode is normally represented by a bar in most schematic symbols.

CEMF Counter-ElectroMotive Force.

cathode ray tube The portion of an oscilloscope where the beam of electrons is controlled and displayed, normally abbreviated CRT.

center-tap A connection at the center of a winding for either a transformer or an inductor.

channel The conductive path, made of either N-type or P-type material alone, between the source and the drain in the field effect transistor.

characteristic curve A graph showing the relationship of changing values for a device or a circuit. The most common characteristic curves are voltage-current curves for semiconductor components, which allow the viewer to analyze how the device will conduct current for a given bias voltage.

charging The act of a capacitor building up a stored voltage within an electrostatic field.

chassis ground The reference point for a piece of equipment when the metal frame for the equipment is tied to earth ground. The metal frame provides both a common reference, and a common return for the equipment.

choke An inductor which is made by using multi-layered windings around an iron core.

circuit ground A point in a circuit that is used as a common reference, although it is not tied directly to earth ground.

clamper A circuit which changes the reference voltage for a signal without changing the amplitude or frequency of the signal. An unbiased clamper will change the reference voltage to 0 volts, while a biased clamper changes the reference voltage to a value other than 0 volts.

class A amplifier An amplifier which amplifies an input signal with no clipping. In other words, the amplifier never becomes saturated or cutoff.

GLOSSARY

class AB amplifier An amplifier which clips the output signal for less than 50% of the input period.

class B amplifier An amplifier which clips the output for exactly 50% of the input period.

class C amplifier An amplifier which clips the output signal for more than 50% of the input period.

class of operation The characteristic of an amplifier which describes what percentage of the input period causes clipping in the output signal.

clipper A circuit which removes a portion of an AC signal by changing the bias on a junction diode. Both series and shunt clipping circuits exist, and a clipper can either be unbiased (removing the signal above or below 0 volts) or biased (removing a portion of the input above or below some voltage other than 0 volts).

clipping A portion of the output signal from an amplifier which reaches and remains at either 0 volts or the source voltage. Clipping occurs when the amplifying device (normally a bipolar transistor) reaches saturation or cutoff.

closed-loop operation An operational amplifier which has its output fed back to either the inverting or noninverting input.

coefficient of coupling A number between 0 and 1 which describes the percentage of flux linkage between the two windings in a transformer. If all the primary flux lines cut the secondary, the coefficient of coupling is 1. For a smaller amount of flux linkage, the coefficient of coupling is less than one.

coil A term often interchangeable with inductor.

collector The terminal in a bipolar transistor which receives 95-99% of current flow from the emitter.

collector current That current which flows from the collector terminal of a bipolar transistor to the source voltage, normally labeled as I_C.

collector feedback A method for biasing a bipolar transistor where the base resistor is tied directly to both the base and the collector terminals. In this fashion, the amount of base current is determined by the collector voltage.

collector junction The junction in a bipolar transistor formed by the base and collector terminals. For a bipolar transistor to properly conduct, the collector junction must be reverse-biased.

color code The international method of identifying the ohmic value and tolerance range for a resistor by using bands of color in a specific pattern.

APPENDIX E

Colpitts oscillator A type of LC oscillator which uses two capacitors in the tank circuit.

combination peaking A method of frequency compensation for a video amplifier which uses inductors both in series and in parallel to compensate for any junction capacitance between any two amplifier stages.

common return A function of ground which allows those circuits connected to ground to have a complete path for current to or from the source voltage without doubling the number of wires in the equipment.

common-base A bipolar transistor amplifier whose input is on the emitter, and output is taken across the collector. The primary function of the common-base is to provide voltage amplification.

common-collector A bipolar transistor amplifier whose input is on the base, and output is taken across the emitter. The primary function of the common-collector is to act as an impedance-matching device, between a high-impedance source and a low-impedance load.

common-drain A FET amplifier whose input is on the gate, and output is taken across the source. The common-drain amplifier performs essentially the same function as a bipolar common-collector amplifier.

common-emitter A bipolar transistor amplifier whose input is on the base, and output is taken across the collector. Because the common-emitter amplifies both voltage and current, it is primarily used as a power amplifier.

common-gate A FET amplifier whose input is on the source, and output is taken across the drain. The common-gate amplifier performs essentially the same function as a bipolar common-base amplifier.

common-mode input An input to a differential amplifier where both signals have the same amplitude, and are in-phase.

common-mode rejection ratio The ratio of difference gain to common-mode gain when a differential amplifier is used to remove an unwanted frequency (e.g., a carrier frequency) from an input signal. It is represented by the acronym CMRR, and can be calculated using the following formula:

$$\text{CMRR} = \frac{A_D}{A_{CM}}$$

Where A_D is the difference gain, and A_{CM} is the common-mode gain.

common-source A FET amplifier whose input is on the gate, and output is taken across the drain. The common-source amplifier performs essentially the same function as a bipolar common-emitter amplifier.

complementary amplifier A type of push-pull amplifier which uses both NPN and PNP transistors, eliminating the need for input or output transformers, and phase splitting.

complementary angles Two angles whose sum is equal to 90 degrees.

complementary signals Two signals who share the same peak-to-peak voltage, but are out-of-phase by 180 degrees.

compound A substance which is created by combining two or more elements. Examples of compounds are water, salt, and steel.

conductance The property of a substance that describes how easily it allows current to flow. Conductance is the direct opposite of resistance, and is measured in mhos. The symbol for conductance is G.

conductor 1. Any substance which readily allows current flow because of a low number of valence electrons. Typical conductors are copper, aluminum, gold, and silver. 2. A path, normally a wire or strip of conductive material, which current can flow from one device to another.

constant temperature coefficient That property of a resistor whose ohmic value does not change with a change in temperature.

contrast The difference between the dark and light areas on a television screen.

copper loss One type of transformer power loss, caused by the resistance of the copper wire used for the windings. Since the total length of the wire can be very long, its resistance can be high, which will dissipate power in the form of heat.

core losses The category of transformer power loss caused by the core of the transformer. Typical core losses are eddy currents and hysteresis.

coulomb The unit of measurement for electrical charge. One coulomb is equal to a charge (either an excess or deficiency) of 6.26×10^{18} electrons.

Coulomb's Law Coulomb's Law of Electrical Charges states that like charges repel, and unlike charges attract.

counter EMF The voltage induced by the magnetic field that surrounds a traveling current flow which *opposes* that current flow. Counter EMF is normally abbreviated as CEMF.

coupling The method of connecting one amplifier to another. The most common forms of coupling are direct, RC, transformer, and impedance.

covalent bonding The structure of semiconductor material where the valence shells of surrounding atoms share their valence electrons to provide greater stability.

crystal lattice The structure of a material which is connected through covalent bonding.

crystal oscillator An oscillator which uses a crystal for both the frequency determining device and the feedback path. The Pierce oscillator is a good example of a crystal oscillator.

current amplifier An amplifier which produces a higher current at the output for a smaller input current, yet does not amplify voltage. A common-collector amplifier is a typical current amplifier.

current flow The movement of electrons from a more negative point to a more positive point across a resistive path. The symbol for current is I.

cutoff The state of a bipolar transistor when it is not properly biased, and cannot conduct current. A cutoff transistor acts as an open switch, preventing current flow because of an infinite resistance, and developing the applied voltage.

cutoff frequency The frequency for hi-pass and low-pass filters that produces an output amplitude approximately 70.7% of the input amplitude, abbreviated as F_{CO}.

cycle One complete AC sine wave, equal to 360 degrees.

DC alpha The ratio of quiescent collector current to quiescent emitter current in a bipolar transistor, abbreviated α_{DC}.

DC beta The ratio of quiescent collector current to quiescent base current in a bipolar transistor, abbreviated β_{DC}.

DC reference voltage The voltage in a voltage regulator which provides the reference for the error detector. The error detector compares any input from the sampling circuit with this DC reference voltage, and provides an output if there is a difference.

damped sine wave A sine wave whose amplitude gradually decreases due to energy loss.

damping factor The ratio of output impedance to load impedance for an audio power amplifier, designated as D.

data sheet The list of operating specifications produced by the manufacturer for a specific component.

degenerative feedback Another term used for negative feedback.

delta (Δ) The Greek letter used to represent a change in a value.

GLOSSARY

dependent bias The type of bias used in a voltage regulator. Dependent bias is achieved when the bias on the regulating component reacts to the change in the output voltage, adjusting its conduction accordingly.

depletion region The region in the middle of a PN junction which is completely void of majority or minority carriers.

depletion-mode IGFET A specific type of insulated-gate field effect transistor which is normally conducting when the gate voltage is 0 volts.

dielectric The insulating (nonconducting) material between the two plates of a capacitor. Typical dielectric materials include ceramic, mica, waxed paper, and plastic.

dielectric constant The rating for different dielectric materials which describes how well the material supports an electrostatic field. Air is given a dielectric constant of 1, and the greater the ability to store the field, the higher the constant.

difference amplifier An operational amplifier circuit which amplifies the difference between two separate input signals.

differential amplifier A special amplifier which is comprised of two identical basic amplifiers. The differential amplifier can accept either one or two inputs, and provide an output that is either the result of one stage, or the difference between the outputs of both stages.

differential-input operation The mode of operation for a differential amplifier where two inputs signals are applied, which are equal in amplitude and 180 degrees out-of-phase.

differential-output operation The mode of operation for a differential amplifier which provides a single output from the difference between the outputs of both stages within the amplifier.

differentiator An RC circuit which creates positive and negative spikes from an input square wave.

diode A semiconductor component which is created from a single PN junction.

direct coupling A method of amplifier coupling which directly connects the output of one amplifier to the input of another. Direct coupling is preferred for DC or very low AC signals.

direct current A steady, unchanging current flow in one direction, resulting from a steady, unchanging potential across a path.

directed drift Electrons within a material moving in the same general direction as the result of an applied potential difference.

discharging A capacitor releasing its stored voltage as excess electrons leave the negative plate, and electrons arrive to reduce the deficiency on the positive plate.

dissipation The method that power is released within a circuit, usually in the form of heat.

distortion Any irregularities or unwanted variations in an AC signal.

donor atom An atom from a pentavalent material which readily gives its extra valence electron to the crystal lattice of a semiconductor.

doping The process by which intrinsic semiconductor material is injected with pentavalent or trivalent material to create P-type or N-type material.

double-battery biasing The first method of biasing a bipolar transistor, with two separate batteries to provide the two required biasing voltages.

drain The terminal of a FET which is connected to the channel, and is opposite the source. In most applications, the drain is the point at which current leaves the FET. In this manner, the drain is similar to the collector in a bipolar transistor.

drain current The current which flows from the drain terminal of a FET to the source voltage.

dry cell A primary cell, usually small and mobile, which is created through the use of electrolyte paste. A dry cell can provide a DC voltage through electrochemistry.

duty cycle The percentage of each output cycle that a pulse exists. It is also the ratio of pulse width to period, calculated using the following formula:

$$\text{Duty cycle} = \frac{\text{pulse width}}{\text{period}}$$

earth ground A connection from a piece of equipment to a metal pipe that is driven into the ground.

eddy currents Current flow through the core of a transformer caused by the varying magnetic field cutting across the core.

effective value The value of an AC sine wave that will dissipate the same amount of heat as a corresponding value of DC voltage. The effective value for a sine wave can be calculated using the following formula:

$$\text{eff. value} = \text{peak value} \times 0.707$$

electrochemistry The method of producing electromotive force through a chemical reaction. The battery is the most common device to employ electrochemistry.

electrolyte A solution or paste through which a chemical reaction takes place to produce electromotive force.

electromagnet A temporary magnet that is created by a wire wound around a core. As current flows through the wire, the magnetic field cuts the core, and magnetizes the core.

electromagnetic induction The action that causes electrons to move through a conductor when the conductor passes through a magnetic field.

electromotive force The force which causes electrons to flow across a path by introducing a potential difference on either side of the path. electromotive force is abbreviated EMF.

electron The smallest of the three particles within an atom. The electron orbits around the nucleus, and can move from one atom to another. The ability for the electron to move is the primary basis for electronics.

electrostatic field The field within a capacitor which stores a potential difference.

electrostatics The branch of physics that deals with electrical charges at rest, or static electricity.

element Pure substances which cannot be broken down into other substances by chemical means. At present, there are 105 elements, comprised of 92 natural elements, and 13 man-made elements.

emitter 1. The terminal in a bipolar transistor which is the primary origin of current flow within the device. 2. The terminal in a unijunction transistor connected to the other side of the single PN junction across from the base.

emitter breakdown voltage The manufacturer's specification for a bipolar transistor, which states the amount of reverse-bias voltage necessary to cause a sharp increase in reverse current through the emitter junction.

emitter current The current that flows to or from the emitter terminal for both the bipolar and the unijunction transistors, designated as I_E.

emitter feedback A biasing method which allows a changing emitter current to determine the amount of base voltage in a bipolar transistor amplifier.

emitter junction The PN junction shared by the emitter and the base in a bipolar transistor. For a bipolar transistor to properly conduct, the emitter junction must be forward-biased.

emitter-follower Another name for the common-collector amplifier, derived from the method by which the output voltage (at the emitter) follows the input voltage (at the base).

APPENDIX E

emitter-follower regulator A basic voltage regulator which uses an emitter-follower (common-collector) amplifier as the regulating component. Any change in the output voltage is immediately felt by the emitter, which changes the forward-bias for the amplifier.

enhancement-mode IGFET A specific type of insulated-gate field effect transistor in which no current flows when the gate voltage is 0 volts.

error amplifier The portion of a feedback voltage regulator which amplifies any error detected by the error detector, and applies it to the regulating component.

error detector The portion of a feedback regulator which compares the input from the sampling circuit, and compares it to the DC reference voltage.

eta (η) The Greek letter used to represent the intrinsic standoff ratio of a unijunction transistor.

excitation current The current which flows through the primary winding of a transformer when the secondary has no load connected.

external induction loss A type of transformer loss which is caused by the magnetic field cutting some outside conductor other than the secondary winding.

FET Field effect transistor.

fall time The portion of a rectangular wave when the signal changes from its most positive voltage to its most negative voltage. The fall time is also known as the trailing edge.

farad The unit of measurement for capacitance. One farad is equal to a one coulomb of charge being felt when an EMF of one volt is applied.

Faraday's Law Faraday's Law of electromagnetic induction states that the amount of voltage induced into a conductor is directly proportional to the rate at which the conductor cuts the magnetic field. The four factors which affect this rate are the speed of the conductor, the strength of the magnetic field, the angle the conductor cuts the field, and the length of the conductor.

feedback biasing A method of biasing a bipolar transistor in which a portion of the output voltage (or current) causes a change at the input to the amplifier.

feedback regulator A common type of voltage regulator which employs a sampling circuit, an error detector, an error amplifier, and a regulating component.

field effect transistor A solid-state component which controls current by a changing depletion region. Current flows through the channel, and a potential applied between the source and the gate determines the size of the depletion region. A separate voltage between the source and the drain allows current to flow through the channel.

field magnet A strong magnet used in an AC generator.

filter 1. Any circuit which selects (passes) a specific frequency or group of frequencies to the output, while preventing (attenuating) all other frequencies. 2. A portion of a power supply which removes the majority of a pulsating DC signal, leaving only a small ripple voltage, which is applied to a voltage regulator.

fixed resistor A component with a specific, unchanging resistance.

flip-flop A common term for the bistable multivibrator.

flux The complete magnetic field of a magnet.

flux density The number of flux lines per unit of area.

flux line A single line of flux.

flux linkage The percentage of primary flux lines which cut the secondary winding in a transformer.

flywheel effect The alternating transfer of energy between an inductor and a capacitor in a tank circuit.

forward breakover voltage The voltage required on the gate of an SCR which will trigger the SCR and allow current to flow.

forward current transfer ratio Another term for the current gains of the common-emitter and the common-base bipolar amplifiers.

forward-bias The condition that occurs in a PN junction when the voltage on the N-type material is more negative than the voltage on the P-type material, and the potential is greater than the barrier voltage. When a PN junction is forward-biased, current will flow from the N-type material to the P-type material.

free-running multivibrator Another term for an astable multivibrator.

frequency The number of cycles that occur for a given signal in one second.

full-wave rectifier A type of rectifier that uses two parallel junction diodes connected on either end of a center-tapped transformer. The full-wave rectifier, because of the center-tap, produces an output voltage only one-fourth that of the input voltage. Yet, current flows in the full-wave for 100% of the input period, and the ripple frequency at the output is double that of the input frequency.

APPENDIX E

gain The ratio of output amplitude to input amplitude (voltage, current, or power) for an amplifier. Voltage gain is represented by A_V, current gain by A_C, and power gain by A_P.

gain-bandwidth product The product of the gain of an operational amplifier and the bandwidth of the operational amplifier, which specifies that frequency which provides unity gain. The gain-bandwidth product for a given operational amplifier is constant.

galvanometer A special type of ammeter which is very sensitive, and can measure current flow in either direction.

gate 1. The control terminal for a field effect transistor. 2. The terminal on an SCR and a TRIAC which receives the trigger pulse.

gate-to-source cutoff voltage The potential across the gate-source junction of a FET which causes the depletion region to expand to its largest size, preventing current flow through the channel. This value is normally represented by $V_{GS(off)}$.

germanium A semiconductor element which is often used to create solid-sate components.

giga The metric prefix which is equal to 1,000,000,000, or 10^9. Giga is represented by "G".

graticule The graph on the face of an oscilloscope screen which divides the screen into one centimeter squares for ease in measuring voltages and periods.

ground 1. The voltage reference point on a circuit, usually 0 volts. 2. The connection of a circuit to a common reference point, or to earth ground.

half-power point A point on a frequency response curve for a filter where the power in the circuit is one-half of the maximum. This is also the same point at which the output voltage is approximately 70.7% of the maximum.

half-wave rectifier A type of rectifier which changed an AC voltage to a pulsating DC voltage through the use of a single junction diode. The half-wave rectifier produces an output voltage that is one-half the input voltage, and a ripple frequency which is equal to the input frequency. However, current flows for the half-wave for only 50% of the input period.

Hartley oscillator A type of LC oscillator which uses a center-tapped inductor in the tank circuit to provide positive feedback. The Hartley oscillator may be series-fed or shunt-fed.

henry The unit of measurement for inductance, represented by "h" or "H".

hertz The unit of measurement for frequency, abbreviated by "Hz."

hi-pass filter An RC, RL, or operational amplifier circuit which passes high frequencies and attenuates low frequencies. The frequency is determined by the values of the components in the circuit, and the cutoff frequency is that frequency which produces an output approximately 70.7% of the source voltage.

holding current The minimum amount of current that must flow through a conducting SCR to maintain conduction.

hole The deficiency of an electron within the valence structure of a semiconductor material that behaves as a positively charged particle.

hole flow The apparent movement of holes in a direction opposite the movement of electrons within a semiconductor material. It is important to note that these holes do not actually move, but due to the movement of the electrons, the holes give the *appearance* of movement.

horizontal amplifier The circuit between the sweep generator and the horizontal deflection plates within an oscilloscope. The horizontal amplifier amplifies the sawtooth wave created by the sweep generator, and applies it to the horizontal deflection plates to control the rate at which the trace moves across the CRT.

horizontal deflection plates The plates that control the horizontal movement of the trace across the oscilloscope screen through the use of electrostatics.

hysteresis loss A type of transformer power loss that occurs within the core. As the magnetic field changes direction due to the AC signal applied, it must overcome friction and inertia. This causes a certain amount of power consumption.

IGFET Insulated Gate Field Effect Transistor. A special type of field effect transistor with a gate that is electrically insulated from the channel. This allows voltages of either polarity to be applied to the gate without fear of component damage.

impedance The opposition to current flow offered by the combination of resistive and reactive components. The symbol for impedance is Z, and is measured in ohms.

impedance coupling A method of amplifier coupling which employs an inductor operating as the load resistor, and a coupling capacitor between the stages. Because of the LC combination, this type of coupling is used for a narrow range of frequencies.

impedance-matching The ability of a component or circuit to provide isolation between a high-impedance circuit and a low-impedance circuit, while still maintaining an electrical connection. A transformer and the common-collector circuit are commonly used for this application.

impurities Non-semiconductor materials with an odd number of valence electrons that are injected into germanium or silicon to create N-type and P-type semiconductors.

in-phase Two signals of the same frequency which reach their maximum and minimum instantaneous values at the same time.

induced voltage A potential difference that is created in a conductor when a magnetic field cuts across the conductor.

inductance 1. The ability of a circuit or component to oppose any change in current. As current attempts to change, the varying magnetic field induces CEMF, which temporarily opposes the change. 2. The ability to store a potential difference in an electromagnetic field. The symbol for inductance is L, and is measured in henrys.

induction The action of opposing a change in current by inducing CEMF with a varying magnetic field.

inductive reactance The steady opposition to AC current flow offered by an inductor. The symbol for inductive reactance is X_L, and is measured in ohms. Inductive reactance can be calculated using the following formula:

$$X_L = 2\pi f L$$

Where f is frequency, and L is inductance.

inductor A component designed to have a specific amount of inductance through the use of coils of wire wrapped around a core.

input offset voltage The small DC voltage applied between the inputs of an operational amplifier to keep the output of the amplifier at 0 volts with no input signal applied.

input resistance The property of a circuit which defines the amount of opposition "seen" or "felt" by the input signal at the point of input.

instantaneous value The magnitude of a voltage or current sine wave at any particular instant during the signal.

insulator Any material with a large number of valence electrons, which greatly opposes any current flow.

integrator An RC circuit which produces a wave-like output with a square wave input.

interstage transformer A transformer used to connect two amplifier stages together.

intrinsic materials Pure semiconductor materials with no impurities added.

intrinsic standoff ratio The ratio between the voltage at the B_1 terminal and the source voltage of a unijunction transistor. It is represented by the Greek letter eta (η), and can be calculated using the following formula:

$$\eta = \frac{V_{B1}}{V_{BB}}$$

Where V_{B1} is the voltage at the B_1 terminal, and V_{BB} is the source voltage.

inverting configuration An operational amplifier circuit where the output is fed back to the inverting input.

inverting input One of the two inputs to an operational amplifier which will cause a 180 degree phase shift output when an input is applied.

ion An unbalanced atom, or an atom with an unequal number of protons and electrons.

iron-core An inductor or transformer whose windings are wound around a ferromagnetic material, normally soft iron.

isolation The separation of two circuits with different impedance values, yet still share an electrical connection.

JFET Junction Field Effect Transistor.

joule A unit of energy. One joule will cause one ampere of current to move across a resistance of one ohm when one volt is applied.

junction capacitance The essential capacitance created across a PN junction.

kilo The metric prefix which stands for 1,000, or 10^3. Kilo is represented by "k."

Kirchhoff's Current Law The law which states that the current entering any point in a circuit must equal the current leaving that same point.

Kirchhoff's Voltage Law 1. Around a closed loop, the sum of the voltage rises must equal the sum of the voltage drops. 2. Around a closed loop, the algebraic sum of all the voltages is zero.

knee of the curve The curve or rounded portion of a voltage-current characteristic curve.

LC circuit A circuit containing both capacitors and inductors.

LC oscillator An oscillator which uses a tank circuit as the frequency determining device.

LCD Liquid Crystal Display.

LED Light-Emitting Diode.

lambda (λ) The Greek letter used to represent one wavelength of light.

laminated core A type of transformer core which uses thin sheets of core material placed adjacent to each other. The result is an equal magnetic property, with a severe decrease in eddy current.

lead-acid cell A type of secondary cell which uses a liquid solution of electrolyte to produce EMF.

lead-lag network A basic RC circuit which produces a leading output for some frequencies, and a lagging output for other frequencies. It is the basic circuit used in the Wien-bridge oscillator.

leading edge The portion of a rectangular wave in which the signal changes from its most negative voltage to its most positive voltage. Also known as rise time.

leakage current the small amount of reverse current that flows through a reverse-biased PN junction.

left-hand rule for conductors A rule for determining the direction of the flux lines surrounding a current-carrying conductor. If you point your left thumb in the direction of current flow through the conductor, your fingers indicate the direction of the flux line surrounding the conductor.

left-hand rule for generators A rule for determining the polarity of EMF induced in a conductor from a magnetic field. The rule is applied with the thumb pointing in the direction of conductor movement, and the index finger pointing in the direction of the flux lines. From there, the middle finger points to the negative end of the conductor.

light-emitting diode A special PN junction diode made from a semi-transparent material. Photons are emitted through the recombination of electrons and holes when current is forced through the junction.

limiter Another term for a diode clipper circuit.

liquid crystal displays An alpha-numeric display that uses the properties of nematic liquid crystals. These crystals reflect light when a potential is applied.

lissajous pattern Patterns that appear on an oscilloscope screen when two out-of-phase signals are applied to both the vertical and the horizontal inputs of the scope.

load A circuit or component that is connected either to a voltage source of a signal source in order to perform some useful function.

low-pass filter An RL, RC, or operational amplifier circuit which passes low frequencies, while attenuating high frequencies. The cutoff frequency is determined by components values, and is that frequency which produces an output approximately 70.7% of the maximum.

MOSFET Metal-Oxide Semiconductor Field Effect Transistor. Another term for an insulated-gate field effect transistor.

magnetic field The region of magnetic influence that extends outward from a magnet, made up of magnetic lines of flux.

magnetic induction The ability to induce a magnetic field into a previously unmagnetized object.

magnetoelectricity The process of producing an EMF by passing a conductor through a magnetic field.

magnitude The size of a value without regard to its polarity. For example, +10 volts and −10 volts have the same *magnitude*, although their polarity is different.

majority carriers The principal carrier in a doped semiconductor. Electrons are the majority carriers for N-type material, and holes are the majority carriers for P-type material.

maximum collector dissipation A manufacturer's rating for a bipolar transistor which defines the maximum amount of power dissipated by the transistor.

mega The metric prefix used to designate 1,000,000, or 10^6, represented by "M."

metric prefixes Symbols used to represent powers of ten, used for ease in writing and calculations with very large and very small numbers.

mho The unit of measurement for conductance, represented by an inverted omega (℧).

micro The metric prefix used to designate 0.0000001, or 10^{-6}, represented by the Greek letter mu (μ).

Miller effect The effect created by the junction capacitance of the collector junction of a bipolar transistor, which creates a negative feedback path from the collector to the base.

milli The metric prefix used to represent 0.001, or 10^{-3}, represented by "m."

minority carriers The lesser carrier in doped semiconductor material. Electrons are the minority carriers in P-type material, and holes are the minority carriers in N-type material.

molecule The smallest particle of a substance that retains the characteristics of that substance.

APPENDIX E

monostable multivibrator A rectangular-wave generator with a single steady state. Also known as the one-shot or the single-shot multivibrator, this circuit is forced away from its stable state by an input signal, and returns to its steady state upon the complete discharge of the internal capacitor.

mu (μ) The Greek letter used to represent the metric prefix micro.

mutual inductance The property that exists between two current-carrying conductors when their respective lines of flux cut each other.

N-channel FET A field effect transistor with a channel made of N-type material.

N-type material Semiconductor material that has been doped to create extra valence electrons.

NICAD battery A secondary cell comprised of a nickel-cadmium combination.

NPN transistor A bipolar transistor with a P-type base, and an N-type collector and emitter.

nano The metric prefix used to represent 0.000000001, or 10^{-9}, represented by "n."

natural magnet Materials which, in their natural state, are surrounded by a magnetic field.

negative alternation That portion of an AC sine wave whose voltages are all negative, or below the reference line.

negative feedback A process by which the output of a circuit if fed back to the input 180 degrees out-of-phase with the input, thus decreasing amplification.

negative ion An unbalanced atom with a greater number of electrons than protons.

negative peak The negative maximum instantaneous value of an AC sine wave.

negative resistance A resistance which exhibits characteristics opposite to normal resistance; when voltage increases, current will decrease.

negative temperature coefficient The property of a resistor whose resistance decreases with an increase in temperature.

neutron One of the three primary particles of an atom, which resides in the nucleus, and has no electrical charge.

noninverting configuration An operational amplifier circuit where the output is fed back to the noninverting input.

GLOSSARY

noninverting input One of two inputs to an operational amplifier which produces an in-phase output when a signal is applied.

nonperiodic waveforms Waveforms which do not retain the same shape from one cycle to the next.

nonsinusoidal oscillator Any type of oscillator which produces a non-sinusoidal waveform.

nonsinusoidal waveform Any waveform which is not an AC signal. For example, square waves, triangular waves, sawtooth waves, rectangular waves, spikes, pulses, and sine waves referenced on a voltage other than 0 volts are all nonsinusoidal waveforms.

nucleus The center of an atom which contains protons and neutrons.

offset null Additional inputs (not normally shown) to an operational amplifier to ensure its output remains at 0 volts when there is no input signal applied.

ohm The unit of measurement for resistance, inductive and capacitive reactance, and impedance, represented by the Greek letter omega (Ω).

Ohm's Law The law which states the relationship between current, voltage, resistance, and power in an electrical circuit. Basically speaking, Ohm's Law states that current is directly proportional to voltage, and inversely proportional to resistance.

ohmmeter A piece of test equipment used to measure DC resistance of a component or circuit.

omega (Ω) The Greek letter used to represent any resistive value, including resistance, inductive and capacitive reactance, and impedance.

one-shot multivibrator Another name for a monostable multivibrator.

open A condition of a conductor, component, or circuit in which no current flows. An open offers an infinite resistance, and often develops the applied voltage as a voltage drop.

open-loop gain The amount of signal gain offered by an operational amplifier when the output is not fed back to one of the inputs.

open-loop operation An operational amplifier which operates without any type of feedback.

operational amplifier An integrated circuit which provides a very high amount of gain for DC and very low-frequency AC signals.

operator interpolation A error in reading needle-type measurement equipment when the eye of the operator is not directly in front of the needle.

optical coupler A circuit which uses an infrared LED to connect the output of one circuit to another circuit that contains a phototransistor.

optical limit switch A circuit comprised of a photodiode and a phototransistor which senses the presence or absence of an object.

optoelectronic devices Devices which use light as the primary medium for sensing or emitting.

orbital distortion The effect on the orbital paths of an atom when a potential is felt across two sides of a capacitor.

orbital shell The location of an electron's orbital path within an atom.

oscillator A circuit which generates a sinusoidal or nonsinusoidal waveform from a direct voltage source. The frequency of the waveform is determined by the component values within the circuit.

oscilloscope A piece of test equipment which measures voltage and time through the use of a controlled electron beam. The beam strikes a phosphor surface within a cathode ray tube to create the signal.

out-of-phase Two signals whose maximum and minimum instantaneous values do not occur at the same time.

output resistance The property of a circuit that defines the amount of opposition "seen" or "felt" at the output terminals of the circuit.

overdriving The act of biasing an amplifier to the point of saturation or cutoff.

overshoot The portion of a rectangular wave where the leading edge temporarily rises above the maximum positive voltage.

P-channel FET A field effect transistor whose channel is composed of P-type material.

P-type material Semiconductor material that has been doped with a trivalent material, leaving a number of extra holes within the material.

PIV Peak Inverse Voltage.

PN junction That which is formed when N-type and P-type materials are bonded together.

PNP transistor A bipolar transistor with an N-type base, and a P-type collector and emitter.

PUT Programmable Unijunction Transistor.

parallel circuit A circuit which has two or more paths for current across a single potential.

peak value The positive or negative maximum instantaneous values for an AC sine wave.

peak voltage The potential at the emitter of a unijunction transistor which is required to forward-bias the junction and cause the unijunction transistor to conduct. This is represented by V_P, and can be calculated using the following formula:

$$V_P = \eta + V_F$$

Where eta (η) is the intrinsic standoff ration, and V_F is the barrier voltage of the PN junction.

peak-inverse voltage A manufacturer's rating which specifies the maximum amount of reverse-bias potential that can be placed across a PN junction without damage.

peak-to-peak value The overall potential of an AC sine wave, measured from the positive peak to the negative peak.

peaking coils A method of compensating for transistor junction capacitance in an amplifier by placing inductors in various positions within the amplifier.

pentavalent material Material with five valence electrons, used for doping purposes to create N-type material.

period The length of one cycle of a sinusoidal or nonsinusoidal waveform.

periodic waveforms Waveforms which retain the same shape from one cycle to the next.

permanent magnet Materials which retain their magnetic fields for long periods of time.

permeability The ease at which a material accepts or passes a magnetic field.

phase angle The angle, measured in degrees, of phase difference between two signals.

phase shift The phase difference, normally expressed as an angle, between two signals.

phase shift network An RL or RC circuit which produces an output with a phase difference to the input.

phase splitter A circuit whose sole purpose is to produce two signals with equal amplitude and a 180 degree phase shift.

phase-shift oscillator An RC oscillator which uses a series of RC phase shift networks for both the frequency determined device and the feedback path.

phi (Φ) The Greek letter used to represent the phase difference between the applied voltage and the voltage drop of a reactive component within a circuit. Phi is normally complementary to theta (θ), the phase angle for the circuit.

photocell A popular term designating a photoconductive cell.

photoconductive cell A resistor whose resistance changes nonlinearly with changing intensity of light.

photodiode An optoelectronic diode which normally operates with a reverse-bias voltage, and determines current flow depending upon the intensity of light which strikes it.

photoelectricity The method of producing an EMF through the use of light.

photon A particle of light.

phototransistor An optoelectronic device which operates as a transistor, with the change in emitter junction forward-bias being caused by the changing intensity of light.

photovoltaic cell An optoelectronic device which generates a voltage when struck by light.

pico The metric prefix used to represent 0.000000000001, or 10^{-12}.

Pierce oscillator A crystal oscillator in which the crystal operates as both the frequency determining device and as the primary feedback path.

piezoelectricity The ability to produce an EMF by applying pressure to a crystal.

pinch-off The condition in a field effect transistor when the depletion region extends across the entire channel, preventing current flow.

pinch-off voltage The gate-to-source voltage which causes pinch-off to occur in a field effect transistor.

polarity The positive or negative condition of a voltage; that which determines the direction of current flow.

polarized capacitor A capacitor that has each plate placed at a particular potential by the manufacturer. Polarized capacitors can only accept current flow in one direction, from the negative plate to the positive plate.

positive alternation That portion of an AC sine wave whose voltages are all positive, or above the reference line.

positive feedback The process by which the output of a circuit is fed back to the input, making sure the feedback signal is in-phase with the input, thereby increasing amplification.

GLOSSARY

positive ion An unbalanced atom with a deficiency of electrons, or a greater number of protons than electrons.

positive peak The positive maximum instantaneous value of a sinusoidal waveform.

positive temperature coefficient The property of a resistor whose resistance changes in direct proportion to a change in temperature.

potential The possibility of doing work. In electronics, any single voltage is a potential.

potential difference Two voltages which are unequal, commonly known as EMF or voltage.

potentiometer A three-terminal resistor whose resistance can be varied by moving the middle terminal (called the wiper arm) across a resistive material.

power The rate at which work is done in an electrical circuit, or the amount of work done in a specific length of time. The symbol for power is "P", and is measured in watts.

power amplifier Any amplifier which increases the amplitude of both voltage and current for a given input signal.

power factor The ratio of apparent power to true power in a reactive circuit. Power factor is abbreviated PF, and can be calculated using the following formula:

$$PF = \frac{\text{true power}}{\text{apparent power}} \text{ or } \frac{\text{watts}}{\text{volt-amps}}$$

power supply A circuit that provides a steady, unchanging DC voltage under varying input and load conditions. A power supply is comprised of a rectifier, filter, and a voltage regulator.

preamp A voltage audio amplifier which is used to increase the voltage level of an audio signal before it is sent to an output power amplifier.

primary cell A type of battery cell that cannot be recharged.

primary current That current which flows through the primary winding of a transformer.

primary winding The initial winding of a transformer which receives the input signal, and transfers it to the one or more secondary windings through electromagnetic induction.

product/sum A method of calculating total resistance in a parallel circuit with two branches, using the formula:

$$R_T = \frac{R_1 \times R_2}{R_1 + R_2}$$

programmable UJT A unijunction transistor whose peak voltage can be controlled or predetermined.

proton One of three particles within an atom, which resides in the nucleus, and has a positive electrical charge.

pulsating DC A DC voltage which looks and operates similar to a single alternation of an AC signal.

pulse A periodic waveform which rises or drops to a voltage for a very brief portion of the cycle.

pulse delay An application of a monostable multivibrator which receives an input pulse at one time, and provides an output which delays the pulse without affecting its amplitude or pulse width.

pulse stretcher An application of a monostable multivibrator which extends the duty cycle or pulse width of an input pulse.

pulse width The length, in terms of time, that a signal remains at the voltage level of the pulse.

push-pull amplifier A type of output power amplifier which uses two transistors which alternately conduct and cutoff, providing a signal to an output transformer.

Pythagorean Theorem A theorem which expresses the relationships between the three sides of a right triangle. Pythagorean's Theorem states:

$$C^2 = A^2 + B^2$$

Where side C is the hypotenuse of the triangle, and sides A and B are the other two sides. This theorem is very important in making voltage, current, and impedance calculations in a resistive/reactive circuit.

Q Quality of merit. 1. For an inductor, quality of merit is the ratio of inductive reactance to the DC resistance of the component, found by using the formula:

$$Q = \frac{X_L}{R}$$

This number is used to represent the efficiency of an inductor—the higher the Q, the more efficient the inductor. 2. For RLC circuits, it is the magnification factor that determines how much the voltage drop of the reactive components rise above the applied voltage. In addition to the previous formula, Q can also be found for this application by using another formula:

$$Q = \frac{E_L}{E_{IN}} \text{ or } Q = \frac{E_C}{E_{IN}}$$

Where E_L and E_C are the voltage drops of the reactive components, and E_{IN} is the applied voltage.

quantum theory The theory which states that although light has many wave-like characteristics, it also behaves as if it were comprised of tiny particles called photons.

quiescence The condition of an amplifier when there is no input signal applied.

quiescent voltage An output or biasing voltage of an amplifier when no input signal is applied.

RC circuit A circuit containing resistive and capacitive components.

RC coupling A method of amplifier coupling where an RC circuit is placed between the output of one stage and the input of another. The advantage of RC coupling is that the reference voltage for a signal can be changed between stages without affecting the signal's amplitude.

RC filter An RC circuit which passes some frequencies, while attenuating others.

RC oscillator An oscillator which uses an RC circuit for both the frequency determining device and the feedback path. Common RC oscillators are the phase-shift oscillator and the Wien-bridge oscillator.

RC time constant The time constant that defines the rate of capacitive charge and discharge. After 5 time constants, the capacitor is said to be fully charged or discharged.

RF choke A special inductor used in a Colpitts oscillator which has enough DC resistance to develop a voltage, yet offers enough impedance to prevent any AC signal from reaching the voltage source.

RL circuit A circuit containing both resistive and inductive components.

RL filter An RL circuit which passes some frequencies, while attenuating others.

ramp A portion of a waveform that has a linear increase in amplitude. Sawtooth, triangular, and trapezoid waveforms all have ramp portions.

ramp generator A circuit which provides some type of ramp waveform as an output signal.

random drift Electrons within a material that does not have a potential applied. Electrons drift around the material in a random manner.

reactive component A component whose characteristics react to and change with different voltages applied to it.

reactive power Any power that appears to be consumed by a reactive component. Reactive components will dissipate some power, but will also return that power back to the circuit.

reciprocal method A method for determining the total resistance of a parallel resistive circuit that has no limitations in its use. The reciprocal method uses the following formula:

$$R_T = \frac{1}{\frac{1}{R_1} + \frac{1}{R_2} + \frac{1}{R_3}}, \text{etc.}$$

rectangular wave generator A circuit which generates a rectangular wave as an output signal.

rectification The process by which a portion of or an entire alternation is removed from an AC sine wave.

rectifier A circuit which receives an AC input and, through the process of rectification, produces a pulsating DC output.

reference A voltage level—usually ground or zero volts—which is used as a basis for measuring the potential difference at other points in the circuit.

reference line The voltage level on which a sinusoidal waveform rides, or the point precisely between the positive and negative peak values.

regenerative feedback Another term for positive feedback.

regulating component The solid-state component in a voltage regulator whose bias voltage varies to control the amount of current flow to the load, and the output voltage applied to the load.

relaxation oscillator Another term used for a nonsinusoidal oscillator.

reluctance The resistance of a material to accept a magnetic field. Reluctance is the exact opposite to permeability.

reset One of two alternate inputs to a bistable multivibrator used to place the two outputs in a predetermined condition, normally used in equipment just after power has been applied.

residual magnetism A magnetic field which remains within a temporary magnet after a permanent magnet has been removed.

resistance The property of a component or circuit which provides an opposition to current flow and dissipates power. The symbol for resistance is R, and is measured in ohms.

resistive component A component whose resistive characteristics remain constant regardless of the voltage applied.

resistivity The specific resistance of a substance which is measured through a one-foot piece of wire with a 0.001 inch diameter at 20 degrees centigrade.

GLOSSARY

resonance The condition in an RLC circuit when capacitive reactance and inductive reactance are equal.

resonant band The band of frequencies which surround the resonant frequency, and cause an output in an RLC filter which is approximately 70.7% or higher of the maximum possible output.

resonant frequency The frequency which causes an RLC circuit to reach the point of resonance. The formula for calculating the resonant frequency (designated as F_O) is:

$$F_O = \frac{1}{2\pi LC} \text{ or } \frac{.159}{LC}$$

Where L is the inductance of the circuit, and C is the capacitance of the circuit.

retentivity The property of a substance which defines how well the substance retains its own magnetic field after the source of the magnetic field has been removed.

reverse current The current which flows through a reverse-biased PN junction.

reverse-bias The condition of a PN junction where the N-type material has a more positive voltage than the P-type material. This widens the depletion region, and prevents current flow. Reverse-bias is the opposite of forward-bias.

ringing Small, damped oscillations which occur after an overshoot or undershoot of a rectangular wave as the signal attempts to reach the proper voltage level.

ripple frequency The frequency of the pulsating DC voltage or ripple voltage at the outputs of a power supply rectifier or filter, measures in ripples-per-second.

ripples-per-second The unit of measurement for ripple frequency.

ripple voltage A varying DC voltage with a small peak-to-peak voltage, which is the output of a power supply filter.

rise time The portion of a rectangular wave where the signal rises from its most negative voltage to its most positive voltage. Also known as the leading edge.

rms value Another term for effective value of an AC sine wave.

SCR Silicon Controlled Rectifier.

sampling circuit The circuit within a feedback voltage regulator which continuously samples the output voltage, and sends any change in the output voltage to the error detector.

APPENDIX E

saturation The point in a transistor where maximum forward-bias is achieved. A saturated transistor acts like a closed switch, providing maximum current with little or no resistance, and developing 0 volts.

sawtooth blocking oscillator A nonsinusoidal oscillator which employs a capacitor to gradually charge and quickly discharge to produce an output sawtooth waveform.

sawtooth generator A ramp generator which produces an output sawtooth waveform.

sawtooth wave A nonsinusoidal waveform with a gradual linear increase in voltage and a sharp decrease in voltage, or vice-versa.

scaling adder A summing amplifier where one input will produce a greater gain than another.

schematic diagram A diagram which uses schematic symbols for components to represent an electrical circuit.

schematic symbol A universal symbol for an electrical component which represents the component in a schematic diagram.

Schmitt Trigger A rectangular wave generator which receives either an AC or pulsing input, and produces a square wave output.

scientific notation A method of writing or displaying numbers in terms of powers of ten.

secondary cell A battery cell which can be recharged.

secondary current The current which flows through the secondary winding in a transformer.

secondary winding The output winding of a transformer, whose number of turns when compared to the primary winding determines the amount of amplitude change for the input signal.

self-induction The process by which a current-carrying conductor induces CEMF into itself due to the moving magnetic field of a changing current.

semiconductor A substance with four valence electrons, which has properties of both conductors and insulators.

series-aiding connection A method of connecting batteries which can increase the overall available voltage by adding together the individual battery values.

series circuit An electrical circuit which has a single path for current flow.

series clipper A clipping circuit where the diode is in series with the input and output signals.

series-opposing connection A method of connecting two batteries where the individual battery voltages cancel each other out.

series peaking A method of compensating for junction capacitance in a cascaded video amplifier by connecting an inductor in series between the stages.

series voltage regulation A method of voltage regulation where the primary regulating component is in series with the input and output voltages.

series-fed oscillator An oscillator where the feedback path and the primary current path are the same.

series-parallel circuit An electrical circuit where both series and parallel components are used. The series components feel the total current for the circuit, while the parallel components feel only that current which flows through that particular branch.

series-shunt peaking A method of compensating for junction capacitance in a video amplifier which employs both series and shunt peaking coils.

set One of two alternate inputs to a bistable multivibrator which, when used, places the bistable in a predetermined condition. The opposite of this input is the reset, where the set input will place the bistable in one condition, and the reset places the bistable in the opposite condition.

short An abnormal condition that exists in electrical circuit where little or no resistance exists between two points where there should either be high resistance, or no connection at all. A short results in a high (often damaging) amount of current flow between these two points.

shunt clipper A clipper circuit where the diode is in parallel with the input and output signals.

shunt peaking A method of compensating for junction capacitance in a video amplifier where an inductor is placed in parallel between the output of one stage and the input of another.

shunt regulator A voltage regulator where the primary regulating component (often a zener diode) is in parallel with the input and output voltages.

shunt-fed oscillator An oscillator where the feedback path and the primary current path are in parallel, and do not share the same path.

signal-to-noise ratio The ratio between the desired frequency output amplitude and the rejected frequency amplitude in a differential amplifier in the common-mode operation.

silicon A semiconductor element which is often used to create solid-state components.

silicon-controlled rectifier A four-layer PNPN or NPNP device which conducts only when forward-biased and a trigger pulse of the required amplitude is applied.

sine wave An AC waveform, referenced on 0 volts, which varies according to the trigonometric function sine. This waveform is produced by an AC generator.

single-ended power amplifier An audio power amplifier which employs a single solid-sate component, and has a single input.

single-input operation A differential amplifier with a single input applied.

single-shot multivibrator Another term for a monostable multivibrator.

sinusoidal oscillator An oscillator which produces a sinusoidal waveform.

slew rate The characteristic of an operational amplifier which defines how fast the output will change voltage levels when the input changes.

slicer circuit A clipper circuit which employs either two parallel diodes or two series zener diodes, and clips a portion of both alternations of an AC sine wave.

slip rings The rings attached to the end of the armature in an AC generator which make contact with the brushes to transfer the generated voltage to the load.

solar cell Another term for a photovoltaic cell.

solid-sate components Components which are created through the use of semiconductor materials and PN junctions.

source-follower Another name for the common-drain FET amplifier.

square wave A rectangular wave in which the waveform remains in the positive and negative voltage levels for an equal amount of time.

stage A single amplifier in a multi-amplifier configuration.

static electricity Electrical charges at rest, or without relative motion.

steering diode A diode placed at the input to a bistable multivibrator which determines which polarity of input voltage will cause the multivibrator to change states.

step-down A transformer with fewer secondary windings than primary windings, resulting in a decrease in output amplitude.

step-up A transformer with more secondary windings than primary windings, resulting in an increase in output amplitude.

substrate The material upon which semiconductor components are created.

subtracter Another term used for an operational amplifier in the difference amplifier configuration.

summing amplifier An operational amplifier with two or more inputs whose output is equal to the sum of the inputs.

summing point The point in an operational amplifier in the summing amplifier configuration where the multiple inputs connect.

swamping resistor A resistor which is added to an RLC circuit which will affect the Q and bandwidth values for the circuit, but will not affect the circuit's resonant frequency.

sweep generator A circuit within an oscilloscope which generates the sawtooth waveform which is applied to the horizontal deflection plates.

sweep time The time required for the oscilloscope trace to travel across the entire screen.

switching time The manufacturer's specification which defines the time required for a transistor (or other solid-state component) to change from saturation to cutoff.

sync circuit A circuit within an oscilloscope which compares the frequency of the input signal to the frequency of the sweep generator, and ensures that the two signals begin their cycles at the same time.

TRIAC The designation for a bi-directional triode thyristor.

tank circuit A parallel circuit with a capacitor and an inductor.

temperature-compensated zener diode A special zener diode whose voltage remains constant over a wide range of temperature.

temperature-sensing bridge A bridge circuit which uses a thermistor as one of the four resistors in the network, and whose output varies as temperature varies.

temporary magnet A material which quickly loses its magnetic field.

thermal instability The characteristic of a transistor amplifier where the internal resistance of the transistor varies with temperature, resulting in changing quiescent voltages. Thermal instability is an undesired characteristic, and can be negated through various amplifier biasing techniques.

thermal runaway A condition in a complementary amplifier where an increase in temperature causes the transistors to conduct harder, resulting in a further increase in temperature, and so on. Thermal runaway can be prevented by the insertion of two diodes between the bases of the two transistors.

APPENDIX E

thermistor A temperature-sensitive resistor, whose resistance changes with a change in temperature. Most thermistors operate with a negative temperature coefficient, with an increase in temperature causing a decrease in thermistor resistance.

thermoelectricity The process of producing an EMF through heat. A thermocouple is a device which employs thermoelectricity.

theta (θ) The Greek letter which designates the phase angle for an RC or RL circuit.

thyristor A broad range of solid-state components which are used as electronically controlled switches.

tickler-coil oscillator An oscillator which employs two separate inductors, where the magnetic field from one inductor "tickles" the other, thereby continuing oscillations.

tolerance A resistor rating which determines the range within which the ohmic value can be, yet still perform its proper function.

trailing edge The portion of a rectangular wave where the signal changes from the most positive voltage to the most negative voltage. The trailing edge is also known as fall time.

transconductance The ratio of drain current to the gate-to-source voltage for a field effect transistor. Designated as "gm", transconductance can be calculated using the following formula:

$$gm = \frac{I_D}{V_{GS}}$$

Where I_D represents drain current, and V_{GS} represents gate-to-source voltage.

transformer A device which transfers an AC voltage through electromagnetic induction between a primary and secondary windings. The ratio of the number of turns for the primary and secondary windings can also change the amplitude of the signal, making it larger or smaller.

transformer coupling A method of coupling two amplifiers together through a transformer, which provides impedance-matching capability. Two disadvantages of transformer coupling are the bulk and expense of the transformer, and the inability of the transformer to couple a DC signal.

transformer efficiency The ratio of output power to input power of a transformer. This value is expressed as a percentage, and is calculated using the following formula:

$$eff = \frac{output\ power}{input\ power}$$

transformer losses Various occurrences in normal transformer action which prevents 100% of the input power to be transferred to the output. Examples include eddy current losses, copper loss, hysteresis, and external induction loss.

transformer oscillator An oscillator which uses a transformer and the tickler-coil principle as the feedback path.

trapezoid waveform A waveform which has both rectangular wave characteristics and a linear change.

treble frequencies Audio frequencies in the higher end of the audio frequency range.

triangular wave A nonsinusoidal wave which increases and decreases at a linear rate.

triboelectricity A method of producing EMF through friction. A Van de Graff generator employs triboelectricity.

trivalent material A substance with three valance electrons, which is injected into intrinsic semiconductor material to create P-type material.

true power The power that is actually dissipated by any DC resistance in a resistive/reactive circuit.

tuned-base oscillator An oscillator whose frequency determining device is located at the base of the amplifier.

tuned-collector oscillator An oscillator whose frequency determining device is located at the collector of the amplifier.

turns ratio The ratio of the windings in the primary and secondary of a transformer. For example, if there were twice as many turns in the secondary winding as in the primary, the turns ratio would be 2:1.

UJT UniJunction Transistor.

UJT **relaxation oscillator** An oscillator which employs the inherent action of a unijunction transistor to produce a wave-like signal, and positive pulses.

unbalanced bridge A bridge circuit whose resistor ratios are unequal, resulting in a potential across the output terminals.

unbiased clamper A clamper circuit which clamps the input signal entirely above or below 0 volts.

unbiased clipper A clipper circuit which clips the input signal at 0 volts.

undershoot That which occurs in a rectangular wave during the trailing edge, when the transition goes slightly beyond the most negative portion of the signal.

unijunction transistor A three-terminal semiconductor device with a single PN junction. The three terminals are base 1, base 2, and emitter. A small current constantly flows between the two base terminals, and when the emitter voltage goes beyond the barrier voltage for the PN junction, current also flows from base 1 to the emitter.

unity gain An amplifier gain of 1, or no change between input and output amplitude.

unity gain frequency The frequency in an operational amplifier which causes unity gain within the amplifier.

VAR Volt-Amps Reactive.

valence electron An electron which orbits the nucleus in the valence shell.

valence shell The outermost shell of an atom.

valley voltage The voltage beyond the peak voltage of a conducting UJT, when the voltage across the emitter junction has decreased to its lowest point as the voltage applied to the emitter increases. Beyond this point, a further increase in the voltage applied to the emitter will cause the voltage across the emitter junction to increase.

valley current The current that flows through the PN junction of a unijunction transistor when the valley voltage is reached.

value/number A method of calculating for total resistance in a parallel circuit whose resistances are all equal. The formula is:

$$R_T = \frac{\text{value of one resistor}}{\text{total number of resistors}}$$

vector diagram A diagram used to add similar values that differ in phase.

vertical attenuator control The fine adjustment for the amount of vertical deflection for an oscilloscope.

vertical control The coarse adjustment for the amount of vertical deflection for an oscilloscope.

vertical deflection plates The plates that control the vertical movement of the trace on the oscilloscope screen through the use of electrostatics.

vertical input terminals The terminals which accept a voltage input for an oscilloscope.

video amplifiers Amplifiers specifically designed to amplify video signals.

video frequency A frequency used to transmit video signals, ranging from 5Hz-5 MHz.

volt The unit of measurement for electromotive force.

volt-amps The unit of measurement for apparent power.

GLOSSARY

volt-amps reactive The unit of measurement for reactive power.

voltage A popular term for electromotive force, or the potential difference between two points.

voltage amplifier An amplifier which increase the amplitude of a signal.

voltage divider A circuit which produces a number of smaller voltages with a single source voltage. The components simply develop their individual voltages, and the output voltages are taken from the same reference point.

voltage divider biasing A method of amplifier biasing which uses voltage divider circuit to develop to required biasing voltages for a solid-state component.

voltage drop The potential difference across a component through which current is flowing.

voltage follower An operational amplifier circuit whose feedback path has no resistance. This causes unity gain, yet allows the op amp to function as a "super" emitter-follower, providing excellent impedance-matching capability.

voltage regulator The last circuit within a power supply, whose specific purpose to provide a steady, unchanging DC voltage with an input ripple voltage, and a varying output load current flow.

voltage rise Any voltage introduced into an electrical circuit by another circuit or device that serves as a voltage source.

voltage source A circuit or component which produces an EMF, and supplies that EMF to another electrical circuit.

voltmeter A piece of test equipment used to measure voltage.

volume A term used to describe the magnitude or amplitude of an audio signal.

watt The unit of measurement for power.

Wheatstone bridge An application of the basic bridge circuit which employs a variable resistor and a galvanometer. A resistor of unknown value is placed across a pair of terminals, and the variable resistor is adjusted until the galvanometer reads 0 amps. When this point is reached, the value of the variable resistor is equal to the value of the unknown resistor.

Wien-bridge oscillator An RC oscillator which employs a lead-lag network as the frequency determining device and as the path for feedback.

work The product of a force applied to an object, and the distance the object is moved. Electrically, when electrons are forced to move from a negative to a positive terminal, work is done.

Zener breakdown voltage The potential required for the zener diode to have a relatively high amount of reverse current. This is also the voltage drop of the zener is the reverse-bias voltage increases beyond the breakdown voltage.

zener diode A semiconductor device which is doped with a high amount of minority carriers. This allows the device to efficiently operate in the reverse-bias configuration, dropping a relatively stable voltage under varying reverse current values.

zener impedance The opposition to current offered by a zener diode, calculated by the formula:

$$Z_{ZT} = \frac{\Delta V_Z}{\Delta I_Z}$$

Where Z_{ZT} is zener impedance, ΔV_Z is the change in zener voltage, and ΔI_Z is the change in zener current.

zener knee impedance The impedance of a zener diode at the point of breakdown voltage, where the knee of the characteristic curve occurs.

zener regulator A voltage regulator where the zener diode is the primary regulating component, usually in parallel with the input and output voltages.

zener voltage temperature coefficient The positive or negative temperature coefficient assigned to a zener diode. Zener diodes with a breakdown voltage of 5 volts or higher have positive temperature coefficients, while those zener diodes below 5 volts have negative temperature coefficients.

Index

7-segment display, 608

AC, *see* Alternating current
AC generator, 216, 221-228
AC signals, 217
Acceptor atom, 452
Active filters, 741-742
Adder, *see* Summing amplifier
Air-core transformer, 413
Alpha, 517, 522, 637
Alternating current, 215-216
Ammeter, 95-97, 255-257
Ampere, 28, 94
Amplification, 504, 510-512
Amplifier biasing, 623-631
Amplifier coupling, 660-667
Amplifier definition, 619
Anode, 476, 563
Aperiodic waveforms, *see* Nonperiodic waveforms
Apparent power, 308-309
Armature, 222
Armstrong oscillator, 793-796
Artificial elements, 5
Artificial magnet, 167
Astable multivibrator, 845-848
Atom, 6
Attenuation, 316
Audio amplifier, 520
Autotransformer, 431-433
Average power, 281-282, 306
Average value, 237

Balanced atom, 8
Balanced bridge, 151-153
Bandwidth, 376, 378-379, 380-382, 388-391
Band-pass filter, 392-394
Band-stop filter, 394-395
Barrier voltage, 463
Base, 503, 577
Base-biasing, 624-625
Bass frequencies, 691
Battery, 27, 48
Beta, 521-523, 643--644
Biasing network, 632
Bistable (UJT), 584
Bistable multivibrator, 852-854
Blocking oscillator, 818-819
Bohr model, 7
Breakdown voltage, 472

Breakdown voltage tolerance, 487
Breakover frequency, 733
Breakover point, 733
Bridge circuits, 151-156
Bridge rectifier, 761-762
Brushes, 222
Buffer amplifier, 653
Bulk photoconductor, *see* Photoconductive cell
Butler oscillator, 809-810

Capacitance, 195-198, 291
Capacitive filters, 763-766
Capacitive phase shift, 294-295
Capacitive reactance, 296-298
Carbon-composition resistor, 73
Cascading, 322, 679
Cathode, 476, 563
Cathode ray tube, 260
CEMF, 189-190, 331
Center-tap, 431
Channel, 535
Chassis ground, 57
Circuit ground, 57
Clamper circuits, 843-844
Class A amplifier, 658
Class AB amplifier, 658
Class B amplifier, 658
Class C amplifier, 658
Class of operation, 620
Classes of operation, 658
Clipping, 656
Closed-loop gain, 735
Closed-loop operation, 726-737
CMRR, *see* Common-mode rejection ratio
Coefficient of coupling, 407
Coil, 332, *see* Inductor
Collector, 503
Collector breakdown voltage, 526
Collector characteristic curves, 515-516, 520-521
Collector feedback, 626
Collector junction, 505
Color code, 74-75
Colpitts oscillator, 801-803
Combination peaking, 701
Common-base amplifier, 514-517, 632-640
Common-collector amplifiers, 647-653
Common-drain, 554

INDEX

Common-emitter amplifiers, 518-523, 640-647
Common-gate, 553-554
Common-mode operation, 709
Common-mode rejection ratio, 710-711, 725
Common-source, 552-553
Complementary amplifier, 686-689
Complementary angles, 321
Complementary signals, 683
Compound, 5
Conductance, 69
Conductor
    battery connections, 50
    circuit connector, 45
    type of substance, 27
Constant temperature coefficient, 71
Copper loss, 427-428
Cosine function, 310, 311, 349, 350
Coulomb, 93, 129
Coulomb's Law, 11
Covalent bond, 446
CRT, see Cathode ray tube
Crystal lattice, 446
Crystal oscillators, 804-811
Current, 23-28, 93, 129, 171-174
Current flow, 28
Cutoff, 654-657
Cycle, 228, 233-234

Damping, 386-387
Damping factor, 680
DC amplifier, 620
DC reference voltage, 774
DC voltage, 215
Degenerative feedback, see Negative feedback
Depletion region, 462
Depletion-mode, 543-547
Deposited-film resistor, 73
Dielectric, 195, 289
Difference amplifier, 743
Differential amplifiers, 704-711
Differential-input, differential-output operation, 708-709
Differentiator, 834-837
Diode clipping circuits, 838-843
Diodes, 441
Direct coupling, 664-665
Direct current (DC), 34
Directed drift, 28
Distortion, 619
Donor atom, 451
Double-battery biasing, 623
Drain, 536
Drain characteristic curves, 539-540, 546, 550
Drain current, 536

Drain-to-source voltage, 536
Dry cell, 48
Duty cycle, 833

Earth ground, 57
Eddy current loss, 425-426
Effective value, 238-239
Electrical charge, 9
Electrical circuit, 45
Electrochemistry, 33
Electrodes, 34
Electrolyte, 34
Electrolytic capacitor, 201
Electromagnet, 167
Electromagnetic induction, 177-179, 218-221
Electromotive force (EMF), 29
Electron, 6
Electron-hole pair, 448
Electrostatic field, 10, 195
Electrostatic force, 9
Electrostatic induction, 14
Electrostatics, 9
Element, 5
ELI the ICE man, 295, 338, 373
EMF, 29-37
Emitter, 503, 577
Emitter breakdown voltage, 526
Emitter feedback, 627
Emitter junction, 505
Emitter-follower, see Common-collector
Emitter-follower regulator, 773-774
Enhancement-mode, 547-551
Error amplifier, 774
Error detector, 774
Excitation current, 414
External induction loss, 428

Fall time, 836
Farad, 199, 291
Faraday's Law, 179, 219
Feedback bias, 625-628
Feedback regulator, 774-779
FET, see Field effect transistor
Field effect transistors, 535-554
Field magnet, 222
Flip-flop, see Bistable multivibrator
Fluctuating DC wave, 246
Flux, 175
Flux density, 175
Flux lines, 169
Flux linkage, 407
Flywheel effect, 385-386
Forward breakover voltage, 565
Forward current transfer ratio, see Alpha

INDEX

Forward-bias, 464-467
Free electrons, 14
Free-running multivibrator, *see* Astable multivibrator
Frequency, 240-243
Friction, 14, 35
Full-wave rectifier, 758-761

Gain, 513, 636, 724-725
Gain-bandwidth product, 736
Gate, 536, 852, 563
Gate-to-source voltage, 536
Germanium, 441
Graticule, 263
Ground, 57

Half-power points, 379-380
Half-wave rectifier, 753-758
Hartley oscillators, 798-800
Henry, 190, 333
Hertz, 241
History effect, 597
Hi-pass filter, 319-320
Holding current, 567
Hole flow, 448, 508-509
Holes, 448
Horizontal deflection plates, 261
Horizontal time base, 261
Hysteresis, 427

IF amplifier, 620
IGFET, 543-551,
Impedance, 305-306, 345-346
Impedance coupling, 663-664
Impedance-matching, 652-653
Induced voltage, 178
Inductance, 187-195, 332-334
    definition, 190
Induction, 14, 176
Inductive reactance, 339
Inductive time constant, 335-336
Inductor, 191
Input offset voltage, 725
Input resistance, 639-640, 646, 724
Instantaneous value, 233
Insulator, 27
Integrated circuits, 441
Integrator, 837
Interstage transformer, 680
Intrinsic materials, 446
Intrinsic standoff ratio, 579
Inverter, 657
Inverting configuration, 726-731
Inverting input, 724
In-phase, 266, 275
Ion, 13

Ionization, 13
Iron-core transformer, 412

JFET, 535-542
Joule, 130
Junction capacitance, 694-698

Kirchhoff's Law, 157-159
Knee of the curve, 486

LC filters, 392-395
    power supplies, 767-768
LC oscillators, 798-804
Leading edge, 833
Lead-acid battery, 49
LED, *see* Light-emitting diode
Left-hand rule for conductors, 172
Left-hand rule for generators, 179
Light, 593
Light-emitting diode, 604-608
Liquid crystal displays, 609-611
Lissajous patterns, 267-268
Load, 45

Magnetic field, 167
Magnetic induction, 176-177
Magnetism, 167-179
Magnetoelectricity, 33
Magnitude, 30
Majority carriers, 451
Matter, 5
Maximum amplitude, *see* Peak amplitude
Maximum collector dissipation, 526
Mho, 69
Miller effect, 695
Minority carriers, 455
Mobile charges, 461
Molecule, 5
Monostable multivibrator, 848-851
Mutual inductance, 333, 405-411

Natural elements, 5
Natural magnet, 167
Negative feedback, 625
Negative ion, 13
Negative peak, 235
Negative resistance, 581
Negative temperature coefficient, 71
Nematic liquids, 609
Neutral atom, 8
Neutron, 6
NICAD battery, 50
Noise immunity, 550

947

INDEX

Noninverting configuration, 731-733
Noninverting input, 724
Nonperiodic waveforms, 832
Nonsinusoidal oscillators, 818-821
Normal atom, 8
North pole, 168
NPN transistor, 504
N-channel FET, 535
N-type material, 450-451

Ohm, 67
Ohmmeter, 74, 101-102
Ohm's Law, 111-121
Ohm's Law, 276, 299, 340-341
Omega, 67
One-shot multivibrator, *see* Monostable multivibrator
Open-loop gain, 733-734
Operational amplifier integrator, 859-861
Operational amplifiers, 721-744
Optical coupler, 608
Optical limit switch, 608
Orbital distortion, 293
Orbital shells, 23
Oscillator requirements, 792
Oscillators, 789-821
Oscilloscope, 259-268
Output resistance, 639-640, 646-647, 725
Out-of-phase, 266
Overdriven amplifier, 657
Overshoot, 833

Parallel circuits, 79, 145-146, 279-280
Parallel connection (battery), 52
Parallel RC circuits, 312-314
Parallel resonance, 383-391
Parallel RL circuits, 351-352
Parallel RLC circuits, 368-370
Peak amplitude, 236
Peak value, 235-236
Peaking coils, 698-701
Peak-to-peak value, 236
Pentavalent material, 450
Period, 228, 240
Periodic waveforms, 831
Permanent magnet, 167
Permeability, 175, 334
Phase angle, 343
Phase relationship, 266-267
Phase shift, 301, 310-312
Phase-shift oscillator, 812-814
Phase-splitting, 431
Phase-splitting circuits, 683-686
Phi, 321, 344

Photocell, *see* Photoconductive cell
Photoconductive cells, 596-597
Photodiode, 600-601
Photoelectric effect, 35
Photon, 594-595
Photoresistive cell, *see* Photoconductive cell
Phototransistor, 602-604
Photovoltaic cells, 598-600
Pierce oscillator, 810-811
Piezoelectric effect, 36
Piezoelectricity, 804
Pinch-off, 538-539
Pinch-off region, 540
Pinch-off voltage, 540
PIV, 475
PNP transistor, 503
Polarity, 30
Polarized capacitor, 202
Positive feedback, 790
Positive ion, 13
Positive peak, 235
Positive temperature coefficient, 70
Potential, 29, 129
Potential difference, 29
Potentiometer, 77
Power, 129-135, 281-282, 306-310, 346-348
Power amplifiers, 620, 680-689
Power factor, 309-310
Power supplies, 753-779
Power supply filters, 763-768
Powers of ten, 93-94, 113-115
Preamplifier, 677
Primary cell, 48
Primary current, 407
Primary winding, 407
Product/sum, 81
Programmable UJT, 584-586
Proton, 6
PRV, *see* PIV
Pulsating DC, 753
Pulse, 568
Pulse stretcher, 850-851
Pulse width, 832-833
Push-pull amplifiers, 682-689
PUT, *see* Programmable UJT
Pythagorean Theorem, 304
P-channel FET, 535
P-type material, 452-453

Q, *see* Quality of merit
Quality of merit (Q), 340, 376-378, 380-382, 387-388, 389-391
Quantum theory, 594
Quiescence, 633
Quiescent state, 634

INDEX

Ramp generators, 857-862
Random drift, 28
RC circuits, 300-322
RC coupling, 660-662
RC filters, 316-320
RC filters, 766-767
RC low-pass filter, 317-319
RC oscillators, 812-817
RC phase shift networks, 320-322
RC time constants, 203-206, 292
RC voltage divider, 315-316
RC waveshaping, 834-837
Reactive components, 342
Reactive power, 307
Reciprocal method, 82
Rectangular wave generators, 845-856
Rectification, 216, 753
Rectifiers, 753-762
Regulating component, 774
Relaxation oscillator ujt 583
Reluctance, 175
Reset, 852
Residual magnetism, 177
Resistance, 67-85
Resistivity, 67
Resonance, 371-372
Resonant frequency, 371-372
Retentivity, 177
Reverse breakdown voltage, scr 567
Reverse-bias, 467-469
RF amplifier, 620
Rheostat, 77
Ringing, 833
Ripple frequency, 757
Rise time, 833
RL filters, 353-354
RL hi-pass filter, 354
RL low-pass filter, 353
RL phase shift, 343, 348-350
RL phase shift networks, 354-356
RLC circuits, 365-395
Rms value, 238

Sampling circuit, 774
Saturation, 654-657
Sawtooth blocking oscillator, 820-821
Sawtooth generator, 861-862
Sawtooth wave, 245
Scaling adder, 740
Schematic diagram, 46
Schmitt Trigger, 855-856
Scientific notation, 94
SCR, 563-570
Secondary cell, 48
Secondary current, 407
Secondary winding, 407
Self-induction, 187-190, 331
Semiconductors, 27, 441

Series aiding connection (battery), 50
Series circuits, 143-144, 277-278
Series clippers, 838-840
Series connection (resistors), 78
Series opposing connection (battery), 51
Series peaking, 700
Series RC circuits, 300-312
Series resonant circuits, 373-382
Series RL circuits, 342-350
Series RLC circuits, 365-368
Series voltage regulation, 772-779
Series-fed Hartley oscillator, 798-799
Series-parallel circuits, 83, 147
Series-parallel connection (battery), 52
Series-shunt peaking, 701
Set, 852
Short circuit protection, 778-779
Shunt capacitance, 696-698
Shunt clippers, 840841
Shunt peaking, 699
Shunt resistor, *see* Swamping resistor
Shunt-fed Hartley oscillator, 800
Signal inversion, 406
Signal-to-noise ratio, 711
Silicon, 441
Sine function, 231-233, 310, 311, 349, 350
Sine wave, 229, 230-234
Single-ended power amplifier, 681
Single-input, differential-output operation, 705-707
Single-input, single-output operation, 704-705
Single-shot multivibrator, *see* Monostable multivibrator
Sinusoidal waveform, 229
Slew rate, 725
Slicer circuits, 841-843
Slip rings, 222
Solar cell, *see* Photovoltaic cell
Solid-state components, 442
Source, 536
Source-follower, *see* Common-drain
South pole, 168
Specific resistance, 67
Square wave, 244-245
Static electricity, 9
Steering diode, 850
Stray electrons, 14
Substrate, 535
Subtracter, see Difference amplifier
Summing amplifier, 738-740
Summing node, see Summing point
Summing point, 738
Swamping resistor, 390, 700
Sweep oscillator, 261
Sweep time, 264
Sync circuit, 262

INDEX

Tangent function, 310, 312, 349, 356
Tank circuit, 386, *see* Parallel RLC circuit
Temperature compensated zener diode, 490-491
Temperature sensing bridge, 156
Temporary magnet, 167
Thermal instability, 625
Thermistor, 71
Thermocouple, 36
Thermoelectric effect, 37
Theta, 311, 344
Tickler-coil, *see* Armstrong oscillator
Tolerance, 76
Tone controls, 691-692
Trailing edge, 833
Transconductance, 540-541
Transformer, 405-433
Transformer core losses, 425-427
Transformer coupling, 666-667
Transformer current ratio, 420-423
Transformer impedance ratio, 423-424
Transformer losses, 425-428
Transformer oscillators, 793-797
Transformer power ratio, 420
Transformer voltage ratio, 418-420
Transistor amplifiers, 619-667
Transistor switching action, 657
Transistors, 441, 503-526
Trapezoid waveform, 857
Treble frequencies, 691
TRIAC, 571-576
Triangular wave, 245
Triboelectric effect, 35
Trivalent material, 450
True power, *see* Average power
Tuned-base oscillator, 797
Turns ratio, 418-420

UJT, 557-586
Unbalanced bridge, 153-154
Undershoot, 833
Unity gain, 734
Unity gain frequency, 734

Valence electrons, 24, 444
Valence shell, 24, 444
Valley current, 582
Valley voltage, 582
Value/number, 80

Van de Graff generator, 35
Variable capacitor, 200-201
Variable resistor, 76
VARS, *see* Volts-amp reactive
Vector diagram, 302-309, 313, 344-345, 348, 349, 351-352, 356, 366, 367, 369
Vertical attenuator, *see* Vertical sensitivity
Vertical deflection plates, 261
Vertical input terminals, 260
Vertical sensitivity, 264
Video amplifier, 620, 693-703
Video frequency response, 693-698
Voltage, 129
Voltage, 32, 129
Voltage amplifiers, 620, 677-680
Voltage divider biasing, 629-631
Voltage dividers, 148-150
Voltage drop, 55, 129
Voltage gain, 638-639
Voltage regulation, 769-779
Voltage rise, 54, 129
Voltage-follower, 736-737
Voltmeter, 98-100, 257-258
Volts-amp reactive, 307
Volt-amps, 308
Volume controls, 690
V-I characteristic curves
    diode, 471
    SCR, 565
    LED, 606
    UJT, 582
    TRIAC, 573

Watt, 129
Wattage rating, 76
Wavelength, 593
Waveshaping circuits, 831-862
Wheatstone bridge, 154-155
Wien-bridge oscillator, 814-817
Wire-wound resistor, 72
Work, 129
Work, 29

Zener diode, 485-495
Zener diode regulator, 770-771
Zener impedance, 492, 493
Zener test current, 487
Zener voltage regulator, 493-495

WHAT DO YOU THINK?

In order to make this and future Heath textbooks more effective, we'd like your opinion on various aspects of this book. Please take the time to complete this short survey and mail it to the address shown at the end of the survey. Your opinion is very important to the improvement of our products, and we hope you will respond.

Name (optional) _____

Address_____

City_____ State_____ Zip _____

School _____

Course Title_____

Instructor's Name _____

Course Length: # of weeks_____ Hrs/week _____

Was there lab time? (yes/no) If yes, how much? (hrs./week) _____

Your Class Rank: Freshman_____ Sophomore_____

Junior_____ Senior_____ Graduate Student_____

Overall, how do you rate this textbook to others you've used?

Superior_____ Better Than Average_____ Average _____

Less Than Average_____ Poor_____

Which chapter(s) did you like? _____

Why?

(cut along dotted line)

(Please turn over)

Which chapter(s) *didn't* you like?_____

Why?

List any chapters NOT covered in your course.

List any topics covered by your instructor which were not covered in the textbook.

List any additional books or materials your instructor used in addition to this textbook.

What did you like least about this book?

What did you like most about this book?

Cont'd on next page

Please rate the textbook in the following areas:

	Superior	Better Than Most	Average	
Writing Style	___	___	___	___
Readability	___	___	___	___
Organization	___	___	___	___
Topic Selection	___	___	___	___
Figures/Tables	___	___	___	___
Objectives	___	___	___	___
Summaries	___	___	___	___
Self-Tests	___	___	___	___
Appendices	___	___	___	___
Currentness of Material	___	___	___	___

(cut along dotted line)

Did you buy this book, or was it issued to you?   (buy/issue)

Did you buy this book new or used?   (new/used)

If used, how much did you pay? _____

Will you sell this book?   (yes/no)

If yes, how much do you expect to receive? _____

(Please turn over)

re any other courses or topics you would like to see as a Heathkit textbook?

Please include any comments or suggestions you may have to improve this book (use additional paper if necessary).

Please mail to: Heath Company
                  Benton Harbor, MI 49022
                  Attn: Education Department/Philip Wheeler

Thank you very much!

Please rate the textbook in the following areas:

	Superior	Better Than Most	Average	Poor
Writing Style	___	___	___	___
Readability	___	___	___	___
Organization	___	___	___	___
Topic Selection	___	___	___	___
Figures/Tables	___	___	___	___
Objectives	___	___	___	___
Summaries	___	___	___	___
Self-Tests	___	___	___	___
Appendices	___	___	___	___
Currentness of Material	___	___	___	___

Did you buy this book, or was it issued to you?   (buy/issue)

Did you buy this book new or used?   (new/used)

If used, how much did you pay? _____

Will you sell this book?   (yes/no)

If yes, how much do you expect to receive? _____

(cut along dotted line)

(Please turn over)

Are there any other courses or topics you would like to see as a Heathkit textbook?

Please include any comments or suggestions you may have to improve this book (use additional paper if necessary).

Please mail to: Heath Company
                  Benton Harbor, MI 49022
                  Attn: Education Department/Philip Wheeler

Thank you very much!